$\dot{q}$	volumetric flow rate [m^3/s for a liqu... standard cubic meters per second (s... (ideal gas measured at 273 K, 1 atm...
$\dot{Q}$	rate of heat transfer (J/s)
$r_{p,i}$	compression ratio (absolute pressures) in a particular stage of a compressor (dimensionless)
$r_{p,T}$	overall compression ratio (absolute pressures) in a multistage compressor (dimensionless)
R	gas constant, 0.0831 $m^3 \cdot$ bara/mol $\cdot$ K; also reflux ratio L/D; also reduction ratio (see Table 4-5); also a parameter for mixed-flow heat exchangers (see Equation 4-70)
S	net heat of solution (J/kg, see Equation 4-29); also a parameter for mixed-flow heat exchangers (see Equation 4-70); also allowable tensile stress in vessel design (Pa or bar)
t	thickness (m); also, temperature ($^\circ$C); also time (s)
T	temperature ($^\circ$C or K as indicated by context)
ΔT_{lm}	logarithmic-mean temperature difference ($^\circ$C or K)
ΔT_m	mean temperature difference ($^\circ$C or K)
$\Delta T'_m$	pseudo-mean temperature difference ($^\circ$C or K)
u	velocity (m/s)
u'	superficial velocity in a vessel, that is, the volumetric flow rate divided by the vessel cross-sectional area (m/s)
u_t	terminal velocity in particle settling (m/s)
u'_t	pseudo-terminal velocity in particle settling (m/s, see Equation 4-110)
U	overall heat transfer coefficient (J/$m^2 \cdot$ s $\cdot$ K)
U'	overall heat transfer coefficient per unit volume (J/$m^3 \cdot$ s $\cdot$ K)
V	volume (m^3); also vapor rate in tray and packed towers (mol/s)
$\dot{w}_f$	power delivered to a pumped or compressed fluid (W or kW)
$\dot{w}_i$	ideal reversible power (W or kW)
$\dot{w}_o$	overall power consumed by a driver such as an electric motor, expander, or turbine (W or kW)
$\dot{w}_s$	power transmitted by a shaft to or from equipment, also known as brake power (W or kW)
w	weight fraction
x	mole fraction in liquid phase (dimensionless); also a variable employed generally
X	polytropic parameter (dimensionless); also moisture content in solids drying calculations (kilograms of moisture per kilogram of dry solids)
y	mole fraction in vapor phase (dimensionless); also vapor yield; the kilograms of vapor produced in a multiple-effect evaporator per kilogram of vapor condensed in the heat exchanger (see Equation 4-24)
Y	polytropic parameter (dimensionless)
z	elevation (m); also gas compressibility factor to correct for nonideality (dimensionless); also mole fraction composition of feed in distillation

GREEK LETTERS

α	relative volatility (dimensionless)
λ	latent heat (J/kg)
ϵ_d	drive efficiency; the fraction of power or work supplied by electricity, steam, or other utility that is converted in a drive and transmitted by a shaft to the equipment (dimensionless)
ϵ_i	intrinsic efficiency; ratio of theoretical power to shaft power (dimensionless)
ϵ_o	overall efficiency ($\epsilon_o = \epsilon_i \epsilon_d$), theoretical work or power divided by that provided by a utility such as electricity or steam (dimensionless)
ϵ_s	stage efficiency in distillation and absorption
σ	surface tension (N/m)
ϕ	Underwood parameter (dimensionless)
ρ	density (kg/m^3 or mol/m^3 depending on context)
μ	viscosity (Pa · s)
θ	residence or processing time (s); also the size of an angle (degrees)

SUBSCRIPTS

1, 2, . . .	stream number designations, usually denoting streams entering and leaving a system; often, the subscript corresponds to a stream number on a flow sheet
$a, b, c, \ldots$	path designations, as in Figure 3-5
ave	average
B	refers to bottoms product in a distillation column
c	critical property; also denotes cold stream
C	continuous phase in a mixture
D	refers to overhead product in a distillation column, also dispersed phase in a mixture
f	film, fouling
g	gas or vapor
h	denotes hot stream
hk	"heavy key" in multicomponent distillation
i	intrinsic, ideal, inside, one of several in a summation or tabulation
l	liquid
lk	"light key" in multicomponent distillation
lm	logarithmic mean
m	mean
o	overall, outside
p	particle, constant pressure
r	reduced property
s	shaft
S	constant entropy
t	terminal
v	constant volume

A GUIDE TO CHEMICAL ENGINEERING PROCESS DESIGN AND ECONOMICS

GAEL D. ULRICH
University of New Hampshire

JOHN WILEY & SONS
New York   Chichester   Brisbane   Toronto   Singapore

To the Curfew Crew

Library of Congress Cataloging in Publication Data:

Ulrich, Gael D.
    A guide to chemical engineering process design and economics.

    Includes index.
    1. Chemical processes. I. Title.
TP155.7.U46 1984   660.2'81   83-6919
ISBN 0-471-08276-7

Printed in the United States of America

10 9 8 7 6 5 4 3 2 1

Preface

For an engineer, the transfer from college to industry is much like immigration to a new country. In neither case are intelligence and training sufficient. One must also speak the language and understand the customs. Fortunately, entering the engineering profession is not as traumatic as entering another culture, and the transition can be eased considerably with adequate preparation. This book was conceived and developed as a text for use in teaching process design to chemical engineers in their senior undergraduate year, to smooth and abet the professional transformation.

Even though firm "traditions" have become established in the chemical engineering profession, many of them are not sufficiently emphasized in the classroom. For example, the process flow sheet with its accompanying tabular material balance is a standard process design document. It has gained widespread acceptance because it is a superior method for displaying important process specifications clearly and in a format that permits the easy detection of errors. Nonetheless, except for the rare cases of teachers who recognize its value and introduce it into a course, the engineering flow sheet is seldom encountered by undergraduate students until they enter a senior process design course. Even then, the format often departs from that of its industrial counterpart, with a sacrifice of clarity and power. Thus, a valuable tool is underemployed and the academic–industrial transition is made more difficult.

A similar argument can be made for the teaching of basic economic concepts. These pervade the profession, yet rarely do they enter the classroom until the final undergraduate year. As the impetus for most of our professional activity, economics deserves an earlier introduction in the curriculum. Although this text cannot, of itself, revolutionize the early undergraduate program, it can serve as a reference for faculty members and students who wish to broaden it.

As most experienced practitioners recognize, the chemical engineering curriculum is largely design oriented. In numerous core courses, students calculate the sizes and capacities of pumps, exchangers, columns, and reactors. They execute material and energy balances and consider various process possibilities, although alternatives are often limited by the instructor. The process design course does not introduce new concepts; rather, it places those already learned in perspective. It impels students to hone their technical skills and to learn or review process economic principles. It helps them to develop judgment and confidence. If successful, it is one of the most rewarding courses in the curriculum for both teacher and student.

Because the design course employs previously learned skills, this book does not review techniques for detailed equipment design. This information already exists on the student's bookshelf in familiar texts. (During this course and throughout their careers, engineers should use those personal libraries.) So-called short-cut techniques and other abbreviated and useful methods for specifying equipment and isolating important elements of a design project are, on the other hand, presented. Miscellaneous other valuable bits of information and rules of thumb are also included.

The first four chapters of this book emphasize project definition, flow sheet development, and equipment specification. Techniques for determining capital costs plus an assembly of equipment cost charts comprise Chapter Five. Methods for estimating operating expenses such as utilities, labor, raw materials, overhead, and other costs are described in Chapter Six. Techniques for transforming these parameters into a coherent economic statement are outlined in Chapters Seven through Nine. This is the chronological sequence one usually follows in attacking a design problem. Some teachers, however, may prefer to cover Chapters Five through Nine first, since process economics is new to most students. Case studies can then be employed to sharpen the skills outlined in Chapters One through Four.

As taught by me and other professors throughout the United States, the design course is centered around one or more extensive case studies.[1] This text was designed to prepare students for these major projects and to serve as a reference during execution. Beginning with Chapter Three, the end-of-chapter exercises form a continuous thread throughout the text. Each represents, in essence, a case study. Four excellent case studies adapted from past competitions of the American Institute of Chemical Engineers (AIChE) are found at the ends of Chapters Seven and Eight. The instructor may also wish to substitute other exercises more appropriate to the resources and experience available.

Though intended for classroom use, this treatise can be employed informally by any individual who has the necessary engineering background and wishes an introduction to process design techniques. Even engineers other than chemical, in related fields, will find much of the material enlightening and beneficial.

The computers is an indispensable tool in detailed process design. Designers, however, must be aware of its limitations and those of programs they are using. For this reason, fundamental techniques amenable to hand computation are emphasized in this text. Even when computer resources are readily available, I find these short-cut methods valuable in checking results.

Regarding units of measure, perhaps the only system worse than any of the possibilities if a *combination* of two or more. It seems, for example, that the public would develop a "feel" for degrees Celsius much more readily if weather forecasters excluded the Fahrenheit equivalent from their reports. The AIChE has officially adopted the International System (SI—Système Internationale). In the process of converting data from all systems into SI units for this book, I have come to

---

[1] *Chemical Engineering Education*, Vol. 16, Winter 1982, and *Chemical Engineering Process*, pp. 76–78 (June 1980), feature the shared experiences of several distinguished teachers of process design.

appreciate the wisdom of that decision and will not undermine the movement by including English equivalents. For the ambitious and incorrigible, conversion factors are contained in Appendix A.

With growing sophistication, practicing technologists are preparing process designs and economic evaluations of increasing precision. The balance between capital expenditures and operating costs, corporate history, inflation, taxes, political climate, and numerous intangible factors affect a company's decision of whether to proceed with a project. In the process of conceiving, building, and operating a plant, hundreds of individuals and three or more companies may be involved. Thus, there is a need for increasing uniformity in the techniques used by various parties as they evaluate and discuss a project. Because it is designed primarily to prepare the neophyte or student, this book cannot delve into all the intricacies of project evaluation. Nevertheless, through the standard design techniques described here, I hope to encourage more unity of language and practice within the profession.

In acknowledgment, I thank my son Nathan who typed most of the manuscript. Other typists participated in the project. Among these, Alice Greenleaf deserves special mention for typing most of the difficult tables of Chapter Four. I appreciate Laurel, who was concurrently writing her own book and nurturing our five children. She inspired by giving no less and taking no more than she expected of me. Robert Beattie, Raymond Desrosiers, Jordan Loftus, Cynthia Jones Riley, Eugene Tucker, and Ronald Willey (friends, colleagues, former students and practitioners all) provided corrections and suggestions to improve this work. I appreciate the assistance of my friend Robert Fisher, a successful process designer, who helped define the scope and content of this text. Finally, I thank former teachers and colleagues who demonstrated to me that engineering has a practical dignity and elegance all its own.

This book is dedicated to the precept that any answer, even a hard-earned *wrong* answer, is better than *no* answer. An imperfect answer can be improved. By the same token, as a first attempt, there may well be flaws and some outright incorrect information herein. With your help, these too can be remedied and corrected.

Gael D. Ulrich

Contents

Section 1
PROCESS DESIGN

PROCESS DESIGN

Someone once said, "The scientist makes things known; the engineer makes things work." This axiom is well illustrated by the serendipitous discovery and tortuous development of Teflon [1]. In early 1938, Anthony Benning, a group leader, Roy Plunkett, a chemist, and Jack Rebok, a laboratory technician, were performing research on Freon refrigerants at the duPont Jackson Laboratory in Deepwater, New Jersey. Dr. Plunkett had been assigned to produce a new composition based on tetrafluoroethylene (TFE). According to a later description [2, p. 2], Plunkett had

> made several cylinders full of gas and stored them in dry ice. On the morning of April 6, 1938, Rebok noticed there was no pressure in one cylinder, indicating it was empty. Yet it weighed the same as when almost full.
>
> Plunkett and Rebok removed the valve and tilted the cylinder. Some white powder fell out. They decided to cut the cylinder, but first they checked with Benning, who held strong views about squandering corporate assets. They found more solid material inside [Figure S1-1]. Plunkett realized there had been spontaneous polymerization of the gas, forming a new material. Benning suggested they try to dissolve it, but none of the common solvents affected it. Other tests followed, revealing more unusual properties.

The commercial development of polytetrafluoroethylene (PTFE), so dramatically made by accident in the laboratory, was fraught with engineering obstacles. The process of producing the TFE monomer itself was poorly developed, yielding a complex mixture of toxic and potentially explosive compounds. Uniformity and

Figure S1-1   Reenactment of the discovery of fluorocarbon polymers in 1938. Dr. Roy
Plunkett (right) was scientist in charge. Jack Rebok (left) was technician at
the duPont Jackson Laboratory. (duPont Company [3], by permission.)

quality of the polymerized product were difficult to achieve. Its failure to melt and
its high decomposition temperature, although superior product advantages, created
serious production problems, requiring metallurgical fabrication techniques that
were foreign to plastics technology [3,4].

As recorded by Dr. Plunkett (personal communication, July 9, 1979),
"Manufacturing costs were terrific." However, with the onset of World War II, the
unique chemical resistance and dielectric strength of PTFE, coupled with other
superior properties, created urgent demands for it within the Manhattan Project
and the defense industry. Drs. Russell Akin and Chester Rosenbaum, two pioneers
in the work, recall a buyer who, thinking "it sold for 45 cents a pound, ordered 1000
pounds of Teflon. The price was actually 45 dollars per pound and duPont did not
have 1000 pounds [2]."

Pilot plant production began in 1943. Serious full-scale commercial production
did not occur until 1948, 10 years after the laboratory discovery. Asked about the
role of chemical engineers in this drama, Dr. Plunkett wrote (personal communication, July 9, 1979):

They were intimately and extensively involved in pilot plant develop-

*ment . . . in every phase of process design . . . developing techniques for
separating, purifying, storing and handling tetrafluoroethylene . . . developing safety procedures to prevent accidents to personnel and equipment.*

As this story shows, transforming small-scale, exploratory or research procedures into large-scale commercial processes is an important and challenging responsibility of the chemical engineer. It should be no surprise to mature engineering students that they have been learning elements of this technique all along. Defining material and energy balances, calculating pressure drops and flow rates in pipeline systems, determining pump sizes, identifying heat transfer areas, calculating tower diameters and heights, determining reactor sizes and types—all are steps in the definition of a commercial chemical process.

Considering the future, a student might ask, "Why must I take a design course? I want to be in sales . . . plant engineering . . . research . . . management." Those with more experience will answer that even salespeople, to be effective, must understand the processes that generate their products. They should know cost and capacity limitations plus the variations in product quality they can anticipate. The operating or plant engineer, though not a designer, should know where important process costs are focused, which equipment items are most vital to capacity and quality, and what problem areas justify most attention. The effective researcher, in particular, is guided consciously or otherwise to avenues which promise practical or commercial return.[1] Supervisors, especially, must have a basic understanding of the plant if they are to manage it effectively.

But design is a creative process. Why frustrate this by promoting rigid rules and techniques? As in art, literature, and music, one would be foolish, indeed, to disregard the heritage of history and language built through the mistakes and trials of others.

As pointed out by O. A. Hougen in his engaging review of chemical engineering history [5], "Filtration operations were carried out 5,000 years ago during the third Egyptian dynasty." From such operations, requiring about 1 percent science, the rest being art, more sophisticated chemical processes developed through the ages. The formalization of chemical engineering as a discipline began in the 1880s as chemists and mechanical engineers in Europe and the United States began

---

[1]One perceptive researcher [6] has noted the value of process design from another angle:

The importance of design in research and development needs further discussion. An engineer working in this area must frequently design his own apparatus, and he must usually show results within a year to justify the continuation of a research project. A considerable amount of valuable research time may be wasted due to errors in the design of apparatus. There are many unpredictable factors in a research project that cannot afford to be held back by errors in design. Probably some of the recent disenchantment with research productivity can be traced back to considerable delays caused by poor equipment design. At any rate, the one way a research engineer can increase his productivity is by becoming a skilled designer of equipment. The importance of design in research and development is generally overlooked.

A colleague added: (Desrosiers, R., personal communication, March 16, 1982), "The student inevitably leaves a design course with the impression that the skills acquired are appropriate only to multimillion dollar projects. The evidence in graduate-level research is plain to see. Students fail to plan, and, aside from budget considerations, a project takes on the "chicken-wire" and "masking-tape" appearance of this design-as-you-go approach. Without exception, on the professional level, the quality of results obtained by a researcher is in direct proportion to the effort expended *before* construction of apparatus: i.e., design.

"There ought to be another 'Teflon' in there somewhere!"

(duPont Company [3], by permission.)

collaborating in the practice of industrial chemistry. A benchmark in the birth of the profession occurred in 1888 with the organization of the first curriculum in chemical engineering at the Massachusetts Institute of Technology. The profession evolved during the early 1900s with a decline in rote techniques of industrial chemistry (where numerous individual processes were studied) in favor of unit operations where equipment common to many processes are examined in depth. Chemical engineering became more soundly based on science with the introduction of basic material and energy balances (1925–1935) followed by thermodynamics, and process control (1935–1945). Courses in process design and kinetics were introduced in the decade after World War II. In the 1960s, with considerable controversy, there was a movement toward a stronger emphasis on the underlying sciences, the so-called unified transport phenomena approach. More recently, in reaction to the alleged lack of practical skills among graduates, there has been a renewed emphasis on the basic unit operations.

According to Prof. Hougen, chemical engineering has supposedly advanced to the stage where "it is only 50% art." Process design, which contains a substantial fraction of this artistic segment, has continued as a vital advanced course in most chemical engineering curricula. In early undergraduate training, creative elements of the discipline are, by necessity, suppressed in favor of technical and scientific skills necessary for one to "speak the language." In these fundamental courses, teachers usually provide basic data such as temperatures, pressures, and flow rates. By contrast, in real design situations, most of these parameters must be specified by the designer. Merely to identify the quantities that can be specified and those that must be calculated requires judgment and experience. Usually the type of unit operation to be employed is open to question. The best equipment for a particular job may not be apparent, even to an experienced engineer, until several different types have been designed and evaluated.

Chemical process engineering, through years of experience, has developed its own traditions and practices. Teaching that "culture" is the purpose of this treatise. The chapters that follow are arranged in a sequence designed to achieve this goal.

REFERENCES

1 DuBois, H.J., *Plastics History—USA*, Cahners, Boston (1972).

2 DuPont Company, *The Wide World of Teflon*, duPont, Wilmington, Del. (May 1963).

3 DuPont Company, *J. Teflon*, **4** (March–April 1963).

4 DuPont Company, *J. Teflon*, 40th Anniversary Issue (1978).

5 Hougen, Olaf A., "Seven Decades of Chemical Engineering," *Chem. Eng. Prog.* **73,** pp. 89–104 (January 1977).

6 Silla, Harry, "The Ch.E. Design Laboratory," *Chem. Eng. Educ.*, **8,** p. 149 (Summer 1973).

TECHNICAL NONMENCLATURE

A	area (m^2)
B	bottoms product flow rate in distillation columns (mol/s)
C_f	tower packing parameter (dimensionless)
C_p	constant pressure specific heat (J/kg·K)
C_v	constant volume specific heat (J/kg·K)
d_p	particle or packing diameter (mm, cm, or m)
D	diameter or width (m); also distillate flow rate in distillation columns (mol/s)
f	Fanning friction factor (Perry 5-22: see ref. 3, Chapter One); also fraction of original feed vaporized in flashing (dimensionless)
f_s	fraction of volume in a vessel occupied by solids (dimensionless)
F	feed flow rate in distillation column (mol/s)
g	gravitational constant = 9.8 m/s^2
G	superficial gas mass flux in packed or tray towers, based on total tower cross-sectional area (kg/s·m^2)
h	specific enthalpy (J/kg); also, film coefficient in heat transfer (J/m^2·s·K)
H_a	active height in a distillation or absorption tower (m)
H_t	tray separation distance in a distillation or absorption tower (m)
$\dot{H}$	rate of enthalpy change (J/s)
k	thermal conductivity (J/m·s·K); also ratio of specific heats C_p/C_v (dimensionless)
k_d	drying-rate constant (m/°C·s)
K	particle settling index (dimensionless; see Equation 4-107)
K_i	fitting losses in fluid-flow systems; also, vapor–liquid molar equilibrium ratio y_i/x_i (dimensionless)

K_{SB} constant in Souders–Brown equation (m/s)

L length, m; also liquid flow rate in distillation and absorption (mol/s)

LMTD   logarithmic–mean temperature difference (°C or K)

$\dot{m}$ mass flow rate (kg/s)

m polytropic "constant"; also the slope of the operating line in gas absorption (dimensionless)

M molecular weight (kg/mol)

MTD   mean temperature difference (°C or K)

n, N gas polytropic "constant" for compressor–expander analysis; also employed to designate the number of stages in multistage equipment (dimensionless)

N_{Re} Reynolds number (dimensionless)

p pressure (Pa or bar; bara denotes absolute pressure and barg, gage pressure)

P power provided from a utility source such as electricity, steam, or compressed air (W or kW)

q number of stages in a multistage compressor; also thermal quality of feed in distillation (dimensionless)

$\dot{q}$ volumetric flow rate [m³/s for a liquid and some gas applications; standard cubic meters per second (std m³/s) for most gas systems (ideal gas measured at 273 K, 1 atm)]

$\dot{Q}$ rate of heat transfer (J/s)

$r_{p,i}$ compression ratio (absolute pressures) in a particular stage of a compressor (dimensionless)

$r_{p,T}$ overall compression ratio (absolute pressures) in a multistage compressor (dimensionless)

R gas constant, 0.0831 m³·bara/mol·K; also reflux ratio L/D; also reduction ratio (see Table 4-5); also a parameter for mixed-flow heat exchangers (see Equation 4-70)

S net heat of solution (J/kg, see Equation 4-29); also a parameter for mixed-flow heat exchangers (see Equation 4-70); also allowable tensile stress in vessel design (Pa or bar)

t thickness (m); also, temperature (°C); also time (s)

T temperature (°C or K as indicated by context)

ΔT_{lm} logarithmic-mean temperature difference (°C or K)

ΔT_m mean temperature difference (°C or K)

$\Delta T'_m$ pseudo-mean temperature difference (°C or K)

u velocity (m/s)

u' superficial velocity in a vessel, that is, the volumetric flow rate divided by the vessel cross-sectional area (m/s)

u_t terminal velocity in particle settling (m/s)

u_t'	pseudo-terminal velocity in particle settling (m/s, see Equation 4-110)
U	overall heat transfer coefficient ($J/m^2 \cdot s \cdot K$)
U'	overall heat transfer coefficient per unit volume ($J/m^3 \cdot s \cdot K$)
V	volume (m^3); also vapor rate in tray and packed towers (mol/s)
$\dot{w}_f$	power delivered to a pumped or compressed fluid (W or kW)
$\dot{w}_i$	ideal reversible power (W or kW)
$\dot{w}_o$	overall power consumed by a driver such as an electric motor, expander, or turbine (W or kW)
$\dot{w}_s$	power transmitted by a shaft to or from equipment, also known as brake power (W or kW)
w	weight fraction
x	mole fraction in liquid phase (dimensionless); also a variable employed generally
X	polytropic parameter (dimensionless); also moisture content in solids drying calculations (kilograms of moisture per kilogram of dry solids)
y	mole fraction in vapor phase (dimensionless); also vapor yield; the kilograms of vapor produced in a multiple-effect evaporator per kilogram of vapor condensed in the heat exchanger (see Equation 4-24)
Y	polytropic parameter (dimensionless)
z	elevation (m); also gas compressibility factor to correct for nonideality (dimensionless); also mole fraction composition of feed in distillation

GREEK LETTERS

α	relative volatility (dimensionless)
λ	latent heat (J/kg)
ϵ_d	drive efficiency; the fraction of power or work supplied by electricity, steam, or other utility that is converted in a drive and transmitted by a shaft to the equipment (dimensionless)
ϵ_i	intrinsic efficiency; ratio of theoretical power to shaft power (dimensionless)
ϵ_o	overall efficiency ($\epsilon_o = \epsilon_i \epsilon_d$), theoretical work or power divided by that provided by a utility such as electricity or steam (dimensionless)
ϵ_s	stage efficiency in distillation and absorption
σ	surface tension (N/m)
ϕ	Underwood parameter (dimensionless)
ρ	density (kg/m^3 or mol/m^3 depending on context)
μ	viscosity (Pa $\cdot$ s)
θ	residence or processing time (s); also the size of an angle (degrees)

SUBSCRIPTS

1, 2, . . .	stream number designations, usually denoting streams entering and leaving a system; often, the subscript corresponds to a stream number on a flow sheet
a, b, c, . . .	path designations, as in Figure 3-5
ave	average
B	refers to bottoms product in a distillation column
c	critical property; also denotes cold stream
C	continuous phase in a mixture
D	refers to overhead product in a distillation column, also dispersed phase in a mixture
f	film, fouling
g	gas or vapor
h	denotes hot stream
hk	"heavy key" in multicomponent distillation
i	intrinsic, ideal, inside, one of several in a summation or tabulation
l	liquid
lk	"light key" in multicomponent distillation
lm	logarithmic mean
m	mean
o	overall, outside
p	particle, constant pressure
r	reduced property
s	shaft
S	constant entropy
t	terminal
v	constant volume

Chapter One

THE NATURE AND FUNCTION OF PROCESS DESIGN

LEVELS OF DESIGN ACCURACY

A process engineer is usually involved in one of two activities: building[1] a manufacturing plant or deciding whether to do so. Similar skills are required in both cases, but the money, time, and depth involved are many times greater in the former. Calvin Cronin, editor of the journal *Chemical Engineering*, wrote: "The typical large multinational chemicals producer will derive approximately one third of its sales volume from products introduced within the past ten years" [1]. Yet it has been said that on the average, only about 1 out of 15 proposed new processes is ever actually constructed. Thus wisdom in the decision stage is vital to avoid loss of money on one hand or opportunity on the other. In a well-managed organization, engineering evaluation is an essential activity that begins with the conception of and fundamental research on a new process. In continues through development to final construction of a promising project. In fact, process development is a series of action and decision steps leading from conception to construction.

It is important to decide whether a project has promise as early as possible. Not only are research and pilot plant expenses wasted on a nonviable idea, but evaluation costs themselves balloon as the project approaches maturity. The growing expense of securing more detail and increased accuracy in an economic evaluation is illustrated in Figure 1-1. Here, the costs of making estimates are plotted against their accuracy. (The case illustrated is for a project costing between \$1M and \$5M.[2])

In practice, there are five levels of estimating sophistication. An "order-of-magnitude" estimate, the most rudimentary, requires little more than identification of products, raw materials, and utilities. (This is also known, informally, as a "rule-of-thumb" or "back-of-envelope" estimate.) Such evaluations are often made by extra-

---

[1] This includes expanding, modifying, or retrofitting an existing plant.

[2] SI units in this text extend to economic as well as scientific nomenclature. Thus, "\$1M to \$5M" denotes 1 million to 5 million (mega) dollars and "\$1K or \$5K." 1 thousand to 5 thousand (kilo) dollars. An unfortunate tradition, still alive in the profession, employs m or M to denote thousand and mm or MM to denote million. This is especially common in reference to dollars but is also encountered occasionally in capacity, flow rate, or heating duty specifications.

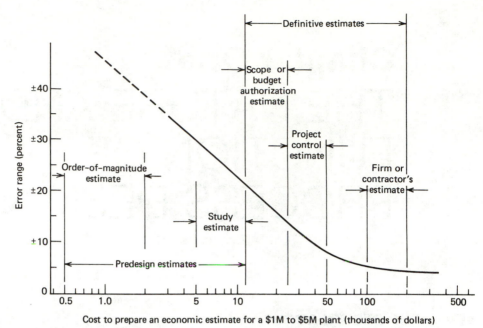
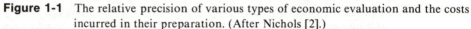

Figure 1-1   The relative precision of various types of economic evaluation and the costs incurred in their preparation. (After Nichols [2].)

polating or interpolating from data on similar existing processes. They can be done quickly, at a cost of about $1000, but with an expected error of greater than plus or minus 30 percent.

A "study" estimate, the next level of sophistication, requires a preliminary process flow sheet and an approximate definition of equipment, utilities, materials of construction, and other processing details. Accuracy improves to within plus or minus 20 percent, but more time is required, and the cost is about $5k to $12k for a $1M to $5M plant. Examination at this level normally precedes expenditures for market evaluation, pilot plant work, site selection, and detailed equipment design. If the process survives this stage, pilot plant and other activities normally begin.

The next level of economic evaluation, entitled "scope" or "budget authorization," requires a more explicit process definition, detailed process flow sheets, and prefinal equipment design; information obtained, in part, from pilot plant, marketing, and other studies. The scope or budget authorization estimate for a $1M to $5M plant would cost from $12K or $25K with an accuracy better than plus or minus 20 percent. As implied by the name, a firm decision on whether to proceed is normally supported by a budget authorization estimate.

If the decision is positive at this stage, a "project control" estimate will be prepared. More and more detail (e.g., final flow sheets, site analysis, equipment specifications, and architectural and engineering sketches) is employed to prepare this estimate, which has an accuracy of plus or minus 10 percent. It can serve as the basis for a corporate appropriation, to evaluate contractor bids, and to monitor actual construction expenses. Because of increased detail and precision, the cost of preparing a project control estimate for the plant in question falls in the range of $25K or $50K, five times greater than a study estimate and twice as large as the budget authorization estimate. (The logarithmic abscissa in Figure 1-1 tends to mask the exponentially increasing cost of improved accuracy.)

The final economic analysis, a "firm" or "contractor's" estimate is based on detailed specifications and actual equipment bids. Employed by the contractor to establish a project cost, it has the highest level of accuracy, plus or minus 5 percent. The cost of preparation, $100k to $300k, reflects additional expenses for engineering, drafting, support, and management labor. This cost, however, is offset somewhat if engineering and drafting documents from the project control estimate are integrated.[3] Because of unforeseem contingencies, inflation, and changing political and economic trends, accuracy better than plus or minus 5 percent is not expected. In fact, these factors make it impossible to guarantee that actual costs will be within such narrow limits even for the most precise estimates.

Owing to the increasing resources and decreasing creativity required for scope, project control, and contractor's estimates (frequently called "definitive" estimates), we emphasize the other, more approximate, methods known by some as "predesign techniques." These are adequate for use by researchers or managers to evaluate the viability of a proposed project, and they employ most of the intellectual skills without the tedium involved in more accurate appraisals.

In past coursework, the student has been taught to achieve accuracy befitting a detailed process design. This is appropriate and important to demonstrate the level of accuracy that can be achieved. Unfortunately, most students find it difficult to adjust from 10-digit calculator displays to the seemingly arbitrary specifications and educated guesses required to synthesize a chemical process. Each has its place. Extreme accuracy is extravagant when not justified. Its absence is devastating when key processing equipment fails to perform according to specifications. Deciding the level of accuracy appropriate for a given situation is the essence of engineering. Competence requires experience and alertness. To develop skill in this area, you should frequently ask, "What are the most important assumptions I have made? Where are major uncertainties? Is more accuracy needed or justified?"

The story is told of a student in a mass-transfer course who reported the calculated height of an absorption tower to be 107.34 feet. The teacher, with a note of irony in his voice, responded, "We certainly cannot forget that 0.34 foot."

THE "PROCESS" OF DESIGN

Contrary to many of the homework problems a chemical engineering student solves, there is no absolutely correct solution to a design problem. There is, however, usually a "better" solution. For example, in storing a cryogenic liquid, one engineer may choose to bury the storage tanks in the earth to minimize seasonal variations in temperature and to moderate capacity fluctuations in the refrigeration equipment. Another design engineer might choose to construct the tanks, with extra-heavy insulation, above ground for easier maintenance. Either design, properly executed, will function satisfactorily. In most cases, however, one alternative will prove to be economically superior to another. Sometimes, factors other than cost affect the decision. Depending on location and political climate,

---

[3]See Perry and Chilton [3] Sec 25, for a more complete specification of the information required for each type of estimate.

aesthetics, pollution, noise, intensity of lighting, traffic impact, and number of employees may be controlling factors in the design. In some projects—a nuclear power plant, for example—safety is the prime consideration. Some managers, to protect against negative public or industry reaction, voice a conservative design philosophy stating, "It doesn't matter all that much how expensive the plant is, but it had better start up properly and operate efficiently from the beginning."

Although everyone approaches design problems somewhat differently, the six major steps are similar. They are discussed below and listed in Table 1-1, in the order that they normally are employed. (As a rule, steps 1, 2, and 4 are repeated many times, with the economic analysis refined in each cycle.) You will note a parallel between the listing in Table 1-1 and the sequence of the chapters that follows.

STEP 1 Conception and Definition

Step 1 may be partially completed in advance by a supervisor, or it may evolve through a series of discussions between the engineer and others concerned with a project. An engineer must know the bases and assumptions that apply, the plant capacity, and the time allotted. Project philosophy must be defined. For example, how "tight" or precise must the result be? How much, if any, extra capacity is desired? What are the design "tradeoffs"? Should initial capital be minimized, or should it be higher to produce a more trouble-free startup? Should expensive materials of construction be employed to reduce corrosion and subsequent maintenance costs, or vice versa? Some of the answers will be obvious from past experience. Several matters of conception and definition must be reexamined for each new project, however. Quite often, a designer is asked to explore several alternatives to provide a basis for selecting the best. In many cases, these questions should be considered again later in the project when more information is available.

For the novice, it is important that uncertainties at this stage do not frustrate or impede progress. It is always possible to refine the assumptions later and quickly revise the calculations, but assumptions of some type must be made. Information contained in Chapter Two will aid in this step.

STEP 2 Flow Sheet Development

After the problem has been conceived, defined, and assigned, the mode of solution is seldom obvious. Frequently, a large number of possibilities and potential assumptions exist. Even with these uncertainties, it is generally possible to construct a process flow sheet. One should take this step early for several reasons: in generating a flow sheet, one is led to assumptions that can reduce complexity of the problem. In executing material and energy balances, the most important process variables are often exposed. It is an efficient way to become familiar with a process.

TABLE 1-1
STEPS IN THE DESIGN OF A CHEMICAL PROCESS

1. Conception and definition
2. Flow sheet development
3. Design of equipment
4. Economic analysis
5. Optimization
6. Reporting

It will identify where information is lacking. Properly executed and completed, the flow sheet will contain the data required for design of individual equipment items. Generally, even if there are later modifications, the flow sheet can be corrected and design calculations repeated with relative ease. When students don't know where to start on a new project, I advise them to begin the flow sheet, confident that a potentially fruitful approach will be revealed by that exercise. Mechanics of flow sheet preparation are described in Chapter Three.

STEP 3   Equipment Design

Equipment cost is an important element in process economics. Partial design, at least, is necessary before such costs can be established. Estimating precision is dictated by the desired accuracy of an estimate. For predesign estimates, equipment must be specified quickly and without great detail. This is necessary because of the limited budget that can be devoted to the work. Chapter Four and Five describe rapid and approximate methods for determining equipment specifications and costs. Even with short-cut techniques, however, capital estimates generally compare within plus or minus 20 percent when executed by different engineers having equal competence. This is adequate for deciding whether to proceed with a project.

If the decision is positive, a detailed project control or contractor's estimate will follow. Techniques and precision required for these estimates are typical of those employed in unit operations or kinetics courses; but all details such as tube or tray layout, vessel thickness, and materials of construction must be specified to the extent that equipment can either be ordered from a vendor or fabricated in-house.

STEP 4   Economic Analysis

Most process feasibility studies lead to the same question; What return can be expected on the money invested? To answer this, raw material, labor, equipment, and other processing costs must be combined to provide an accurate economic forecast for the prospective manufacturing operation. The time value of money, inflation, taxes, and other factors influence profitability. These must be considered and evaluated in a manner that is meaningful to management. Although detailed manipulation of economic parameters is the province of economists, not engineers, the economist generally is not qualified to design equipment, define raw materials, and evaluate other processing costs. In practice, it is easier for an engineer to bridge the communication gap by learning elementary economic techniques than for an economist to learn engineering. These principles and techniques are presented in Chapters Five, Six, and Eight.

STEP 5   Optimization

A combination of economics and engineering, optimization is necessary in any engineering project for which alternate design possibilities exist. Since this is frequently the case, optimization is generally employed at several points in most design projects. Often, such as in determining pipe sizes, the optimum can be obtained from charts or nomographs prepared by others. In some situations, the optimum choice may be a simple matter of common sense if an engineer has had experience with similar alternative selections in the past. Occasionally, as is frequently true in reactor specification, the optimum must be determined uniquely for a particular process and configuration under examination.

STEP 6   Reporting

A design report may represent the only tangible product of months or years of effort. An effective report cannot be prepared from a poor engineering effort, but a poor or mediocre report can, and often does, obscure otherwise excellent engineering. This is another interface between engineering and humanity that must be crossed and crossed well by the engineer if his or her work is to be appropriately recognized and rewarded.

These six activities are prominent steps in the development of every modern chemical process. By their senior year, well-trained chemical engineering students have been exposed to most of the skills necessary to develop a successful design. The value, accuracy, and practicality of such a design depends on the diligence, creativity, and intellect devoted to it. The chapters that follow contain information in areas such as flow sheet techniques and economics that will not have been treated in previous courses. Some material will reinforce that which has already been learned. During the study of these chapters, you will execute segments of selected design problems and prepare the solution to a major case study. It is then that you will recognize the power of the tools you have gained.

OTHER BOOKS ON PROCESS DESIGN

Aerstin, Frank, and Gary Street, *Applied Chemical Process Design*, 294 pp., Plenum, New York (1978). Primarily a collection of charts and equations gleaned from the reference literature. Because of its brevity, this book is of limited value to the inexperienced designer.

Baasel, William D., *Preliminary Chemical Engineering Plant Design*, 490 pp., Elsevier, New York (1976). A complete and well-written treatise on the techniques of process design, based on the author's experience as a Ford Foundation resident at Dow Chemical Company. This guide to the details of design includes site selection, safety, layout, process control, construction and startup, planning tools, and pollution. More general techniques such as equipment design, energy and manpower definition, cost estimation, and optimization are also expounded. Although the book does not contain extensive economic data, it is an excellent source of design information.

Blackhurst, J. R., and J. H. Harker, *Process Plant Design*, 400 pp., Elsevier, New York (1973). Contains detailed equipment design information and considerable mechanical information; the treatment of economics, optimization, and flow sheet development is sparse, however. Some cost data are included.

Bodman, Samuel W., *The Industrial Practice of Chemical Process Engineering*, 231 pp., MIT Press, Cambridge, Mass. (1968). Employs case studies to illustrate solutions to problems in reactor design, optimization, and design of mass transfer and other commercial processes. Useful for background material on the individual projects described.

Evans, F. L., *Equipment Design Handbook*, Gulf, Houston, Tex. (Vol. 1, 1979; Vol. 2, 1980). For use in detailed equipment design, giving information beyond that necessary for preliminary design and cost estimation. Because of more extensive textual description and explanation, this set is recommended over the abbreviated treatment by Aerstin and Street.

Landau, Ralph (editor), and A. S. Cohan (assistant editor), *The Chemical Plant*, 327 pp., Reinhold, New York (1966). A qualitative discussion of the steps involved from conception to operation of a chemical plant, this book provides the background and flavor surrounding the evolution of a chemical plant. Organized according to chronological sequence, each chapter was prepared by an authority on that step.

Peters, Max S., and Klaus D. Timmerhaus, *Plant Design and Economics for Chemical Engineers*, 3rd edition, 973 pp., McGraw-Hill, New York (1980). Currently the most prominent U.S. chemical process design text, this treatise contains detailed information and extensive references to the journal literature. It also includes a substantial compilation of economic data. Some sections describe detailed procedures for design of heat and mass transfer equipment such as would be found in a book on unit operations or transport phenomena.

Resnick, William, *Process Analysis for Chemical Engineers*, 400 pp., McGraw-Hill, New York (1981). Contains extensive discussion of the creative element of process design and of techniques for generating and screening alternatives and making decisions. The text emphasizes principles and elements of thermodynamics and reactor design. An introduction to economic analysis is included along with substantial material on economic forecasting. Several selected case studies are thoroughly discussed.

Rudd, Dale F., and Charles C. Watson, *Strategy of Process Engineering*, 466 pp., Wiley, New York (1968). Contains useful guidance in defining and selecting process alternatives. Discussions of economic design and cost estimation are clear and concise. Optimization and simulation are emphasized and extensively discussed.

Sherwood, Thomas K., *A Course in Process Design*, 254 pp., MIT Press, Cambridge, Mass. (1963). Chapter 1 is an interesting discussion of process design as practiced commercially. Techniques are discussed briefly. The remaining chapters contain case studies of nine industrial operations to illustrate various design approaches and techniques.

Vilbrandt, F. C., and C. E. Dryden, *Chemical Engineering Plant Design*, 534 pp., McGraw-Hill, New York (1959). A classic U.S. text on the subject, this book provides rather complete information on such items as site selection and preparation, equipment specifications, and process auxiliaries. Although somewhat outdated, some portions of this book are useful for reference even today.

REFERENCES

1   Cronin, C.S., "Unleashing Innovation," *Chem. Eng.*, p. 5 (Dec. 3, 1979).

2   Nichols, W.T., *Ind. Eng. Chem.*, **43**, p. 2295 (1951). Also reproduced in Perry and Chilton [3], Sect. 25, p. 15.

3   Perry, J.H., and C.H. Chilton, *Chemical Engineers' Handbook*, 5th edition, McGraw-Hill, New York (1973).

4   Pikilik, A., and H.E. Diaz, "Cost Estimation for Major Process Equipment," *Chem. Eng.*, pp. 107–122 (Oct. 10, 1977).

Chapter Two
PROJECT CONCEPTION AND DEFINITION

Design projects take a number of forms. The executive vice president of a paper company may, for example, ask the engineering director to assess potential profits in making methanol from wood waste. Perhaps the research director in a chemical firm engages the engineering department to evaluate manufacturing costs for a promising laboratory product.

In a food processing operation, the plant manager may ask for a recommendation on how to reduce biological oxygen demand (BOD) of wastewater effluent. Municipal officials might approach a consulting firm for help in solving a refuse disposal problem. An importer and distributor of liquefied natural gas (LNG) may seek a method of recovering cryogenic energy of the LNG while reducing fuel used to gasify it before distribution.

In many assignments, such as the examples in the first paragraph, a product or process is identified. In other cases, like those in the second paragraph, any number of possibilities exist. Given these types of assignment, inexperienced engineers often waste time floundering in confusion over major and minor process details. The three steps in Table 2-1 are recommended to remedy this and to expedite the execution of a new assignment.

STEP 1   Understanding the Process

Since few companies are foolhardy enough to venture into unknown areas, there generally is a significant backlog of data and resource material in company files or in the minds of the employees. Such would logically result from exploratory research or other experience. If this is true of your project, exploitation of internal resources is obviously the first move. For neophytes such as engineering students in a design course and engineers working with nonfamiliar processes, other sources of information must be tapped. In industry, such cases arise when a plant is faced with processing an unusual by-product or solving a pollution problem.

TABLE 2-1
THREE STEPS IN CONCEIVING AND DEFINING
A DESIGN PROJECT

Action	Tools and Resources
1. Understand the process	Company files; *Kirk–Othmer Encyclopedia of Chemical Technology*; Shreve and Brink, *Chemical Process Industries*; *Chemical Abstracts*; *Engineering Index*; journals; monographs; meetings; consultants
2. Narrow the possibilities	Decision chart
3. Define conditions and capacities	Project statement, comon sense, intuition, experience

Of the many sources of information, the following are legal and usually helpful.

Kirk–Othmer Encyclopedia of Chemical Technology, 2nd edition, 22 volumes, Wiley, New York (1963–1972) [3rd edition, 25 volumes, in progress, 1978–1984]. An excellent compilation of detailed information on existing processes and products. Each section is written by an established expert. Information is complete and accurate. The Kirk–Othmer encyclopedia is a valuable source if a process has been in existence for 5 years or more.

Chemical Process Industries, 4th edition, R. Norris Shreve and Joseph A Brink, Jr., McGraw-Hill, New York (1977). A classic text containing surveys of all established processes in the chemical industry. Although much briefer than Kirk–Othmer, it is useful for learning about an unfamiliar chemical process or industry.

Chemical Abstracts, American Chemical Society, Columbus, Ohio, and *The Engineering Index*, Engineering Information, Inc., New York. Traditional guides to technical and research literature, these volumes cite published papers and articles associated with particular products or processes. Because of their specialized nature, many of the published articles may be of minor value at this stage of project definition. Computer search facilities are available in most corporate or university libraries.

Journals. Engineers familiar with the journals in a given discipline often can find useful information there. An effective search should begin by reviewing contents of the most recent issues first, working backward chronologically until annual indexes can be found and examined. The bibliography of a good recent article will contain references to pertinent prior literature. For chemical processing information, useful periodicals are *Chemical Engineering Progress*, *Chemical Engineering*, and *Hydrocarbon Processing*. The latter publishes a large catalog of flow sheets in its annual Petrochemical Handbook issue.

Monographs and Meeting Proceedings. Frequently, experts on a particular subject or process write monographs or small books. Often, publications result from technical meetings. The papers presented at a given meeting are frequently published in preprint form or in technical society journals. Attendance at meetings can also be valuable. Most engineers and scientists are willing to share information that is not secret or proprietary.

Consultants. If entering an unfamiliar area, use of a consultant should be considered. Engineering time saved through more intelligent direction will usually justify the fee.

STEP 2   Narrowing the Possibilities

If not narrowed, the number of design alternatives can be staggering. When a laboratory product or process is being scaled up, experience and earlier feasibility efforts will have pared the options somewhat. In any case, it is valuable, early in the project, to limit the options to those that deserve detailed analysis. These are not always obvious at the outset.

Take, for example, the problem of Durham, New Hampshire, a small university town faced with finding a suitable technique for trash disposal. In the distant past, waste had been buried, but soils were inadequate and leachate was polluting streams. Responsible town officials had promoted and constructed an incinerator, which resolved water pollution problems but was comparatively expensive. With passage of time, air pollution standards became more stringent and the town faced a federal mandate to reduce incinerator emissions. This was technically feasible but even more expensive. Meanwhile, fuel prices had tripled, and it seemed appropriate to consider the value of energy contained in the trash.

A chemical engineer assessing the situation first became familiar with waste disposal problems and technology. A particularly good introductory resource was found in a booklet published for semitechnical readers [1]. More advanced information was found in the American Chemical Society (ACS) periodical *Environmental Science and Technology* and in the *Proceedings of the Incinerator Division of the American Society of Mechanical Engineers* (ASME). These sources led to others that on the whole, provided sufficient background information for intelligent analysis. Next, a decision chart was prepared as shown in Table 2-2. This grid contains a list of alternatives in the first column and a tally of significant evaluation criteria as the top row. These criteria, gathered from the background survey and from notes, are significant to most municipal waste disposal operations. It had become apparent during literature research that capital costs and lack of a continuous energy customer were obstacles to most energy-from-waste projects. When the equipment is in place and the customer assured, however, such schemes are reasonable. A case in point is the Union Electric power generating station in St. Louis, where trash was used to supplement coal as fuel. In the case of Durham, New Hampshire these needs were answered by the presence of a university with its central heating plant and its continuous energy requirements (heat in winter and hot water in summer).

Table 2-2 was constructed using plus symbols to indicate criteria that are favorable to a particular process, minus symbols to denote those that are unfavorable, a question mark to indicate uncertainty, and a blank if there is no clear advantage or disadvantage.

Sanitary landfill, the first technique listed in Table 2-2, has been historically the cheapest and most common approach in small towns, but it has become more and more difficult to find suitable sites. This is a reflection of increased environmental awareness, stringent soil conditions, and public attitude. Even sites that can be employed tend to be remote. Hence, this alternative is unattractive by most of the criteria. Incineration, in Table 2-2, seems to be suitable by all standards except siting, cost, and wastage of resources. Recycling, despite a positive public image,

TABLE 2-2
DECISION CHART EMPLOYED TO IDENTIFY VIABLE SOLID-WASTE
DISPOSAL TECHNIQUES FOR A SMALL NEW ENGLAND UNIVERSITY TOWN

Alternate Disposal Techniques	Criteria						
	Resource Recovery	Public Image	Pollution	Transportation Costs	Siting	Total Cost	Volume Reduction
Sanitary landfill	−	−	−	−	−	+	−
Incineration	−	+	+	+	−	−	+
Recycling[a]	+	+	+	−	+	−	+
Burning in power plant boilers[b]	+	+	+	−	−	+	+
Incineration with heat recovery	+	+	+	+	+	?	+

Key

(−)   undesirable by this criterion   (?)   uncertain by this criterion
(+)   desirable by this criterion   ( )   neutral by this criterion

[a]Recycling is not an ultimate solution because it depends on other forms of disposal for unwanted residues.

[b]Limited to large population centers.

involves significant transportation costs in small towns. It is not an ultimate solution, in any event, because other modes of disposal are required for the unusable fraction, which represents a majority of the trash. (In fact, recycling can often be considered on its own merits along with other forms of ultimate disposal without affecting feasibility of the latter.)

Combustion of waste in an electric power plant meets some of the criteria but requires unique conditions (i.e., a nearby coal-burning power plant and appropriate matching of waste to coal) that were not present in this case. Incineration with heat recovery, on the other hand, is positive by most criteria. The exception is cost, which is somewhat uncertain, lying between landfill and incineration.

The decision chart makes it possible to carve out the alternatives that deserve more detailed attention. Table 2-2 suggests that an engineering analysis should be applied to three alternatives and possible combinations of the three—incineration, recycling, and heat recovery. In this particular case, these alternatives were studied by a consulting firm. To provide adequate waste volume, the consultants recommended a regional operation, involving 11 surrounding towns, whereby waste would be transported to the university and burned as fuel. This was to be combined with recycling in the individual towns as desired. The fact that approval was gained in a majority of the initial town meetings (in New Hampshire!) is evidence of clear financial advantages. Moreover, the plan would not have been viable economically before the rise in fuel prices that occurred in 1973.

In this illustration, the small number of alternatives does not require a formal decision chart for manageable analysis. Such can be accomplished through reasoning and conversation among a group of informed and interested parties. This more accurately describes the real selection process that occurred in Durham. The formal approach does, however, serve to aid communication even in such simple cases.

TABLE 2-3.

RANKING OF ALTERNATE TECHNIQUES FOR LNG REFRIGERANT ENERGY RECOVERY

Technique	(1) New Research and Development Expense	(2) Sophisticated Marketing	(3) Seasonal Compatibility	(4) Intermittent Operation	(5) Capacity
1.0 Miscellaneous					
1.1 Air conditioning	+		−	−	
1.2 Superconducting cable coolant	−	−	−	−	
2.0 Chemical processing					
2.1 Ethylene production	+	+	−	−	−
2.2 Hydrocarbon separations	+	+	−	−	?
2.3 Purification of synthetic natural gas					
2.4 Liquefaction and purification of hydrogen	?	?	−	−	?
2.5 Liquid air or nitrogen	+	+		−	−
2.6 Low temperature plastics processing	+	?	−	−	−
2.7 Extraction of cetane fractions	+	+	−	−	−
2.8 Cooling of heavy-wall molded polymers	−	−	−	−	−
2.9 Quench cooling of polyethylene or polyprophylene	−	−	−	−	−
2.10 Chemical reaction catalysis	−	−	−	−	−
3.0 Power generation					
3.1 Gas turbine precooling	−		−	−	
4.0 "Cold" utility					
4.1 Food freezing	+	−	−	−	−
4.2a Cold storage	+	−	+	−	−
4.2b Cold storage of whole blood	+	−	+	−	−
4.3 Metals treatment	+	−	−	−	−
4.4 "Cold packs" for truck transport	+	−	−	−	−
4.5 "Shrink fitting" of metals	+	−	−	−	−
4.6 Freeze drying of coffee	+	−	−	−	−
5.0 Cold grinding or pulverizing					
5.1 Grinding of refuse	−	−	−		
5.2 Grinding of polymers (except tires)	+	−	−	+	−
5.3 Grinding of rubber tires for recycling			+	+	?

The value of a formal decision chart is illustrated more forcefully when we consider an industrial example. Assume that a hypothetical company is importing LNG to the West Coast for use in meeting peak demands and supplementing winter usage. The fuel, stored at −100°C and modest positive pressure, must be vaporized and heated to ambient temperature before it is introduced into the distribution system. This not only requires energy but represents a serious loss of cryogenic energy invested in the LNG when it is liquefied.

(6)	(7)	(8)	(9)	
Exploitation of Low Temperature	Political Considerations	Social Acceptance	Economics	Notes
−	+	+		Serious seasonal incompatibility.
+	+	?	?	Technology and market too uncertain.
+	−	−	?	Insufficient refrigeration capacity and possible
	−	−		serious public objections.
+	−	−	?	Public objection likely to be substantial. Capacity greater than apparent market.
+	+	+	+	
+				Serious incompatibility of scale.
	−	−		Serious incompatibility of scale.
	+	+		Serious incompatibility of scale.
	+	+		Serious incompatibility of scale.
				Serious incompatibility of scale.
+	+	+	−	Questionable economic value.
−	+	+		
−	+	+		
−	+	+		Incompatibility of scale, need for sophisticated marketing, and seasonal discrepancies are serious
+	+	+		limitations. Any of these could be considered as options if a liquid air or nitrogen plant were
−	+	+		constructed.
+	+	+		
−	+	+		
+	+	−	−	Economic limitation is severe.
+	+	+		Same as items 4.1−4.6.
+	+	+	?	Questionable economics.

Discussion with technical staff and consultants resulted in a list of 23 possible alternatives. Nine criteria were applied and a chart was developed (Table 2-3). The problem was complicated by a number of factors, one being the amount of refrigeration capacity, which exceeds any reasonable local demand. Another was strong public concern about safety of the plant and consequent official scrutiny of any potential changes or additions at the site. Without going into detail, two alternatives—production of liquid air and cryogenic grinding of tires—emerge as clear candidates for further consideration. Most of the other alternatives can be dismissed. If one applies a numerical ranking, such as the percentage of criteria that are positive (excluding blanks or question marks), the favored alternatives have indices of 75 and 100, significantly larger than any others.

A more elaborate technique, which selects alternatives in a quantitative way, is one patterned after those promoted by Kepner and Tregoe [2] for improved decision making in business. Useful for such diverse decisions as choosing a mate or buying a car, the first step is division of criteria into "wants" and "musts." Alternatives that do not satisfy the essential criteria (musts) are automatically eliminated. Next, all criteria are given a priority rating, from 1 to 10. Finally, each alternative is assigned a numerical value according to its satisfaction of the criteria. With these modifications, Table 2-2 would appear as shown in Table 2-4. Because of the "musts" (required disposal of all waste and unavailability of an appropriate nearby facility), the recycling and power plant alternatives were eliminated. Each of the remaining was appropriately graded. To provide an overall numerical evaluation, numbers in each row were multiplied by the corresponding priority rating and summed to yield a total score. The process or technique with the highest score is, accordingly, considered to be superior to the others. As expected, alternatives are ranked numerically in the same order that was derived qualitatively from Table 2-2. However, quantitative comparison requires more careful discrimination to limit alternatives and to establish an index for each criterion. The total score also indicates not only which alternatives are better but the extent of their superiority.

Since qualitative and quantitative decision charts both depend on human judgment, one could argue with the selection of criteria and the relative importance of one factor or another. Often, as with informal decisions, human biases are reflected. Nevertheless, this powerful technique forces decision makers to identify and quantify the criteria employed. Equally important, it provides a comprehensive document that can be used to support or defend the decision and to serve as a basis for negotiations or discussions with others who may disagree.

STEP 3 Defining Conditions and Capacities

Raw materials, products, and processes have been identified in previous steps, though there may be several different processing options to be evaluated. Now, it is time to define primary variables for each alternative. This activity overlaps somewhat with flow sheet preparation as described in the next chapter.

TABLE 2-4.

"QUANTITATIVE" DECISION CHART APPLIED TO THE SOLID-WASTE DISPOSAL ANALYSIS OF TABLE 2-2[a]

	Criteria							
	Resource Recovery	Public Image	Pollution	Transportation Costs	Siting	Total Cost	Volume Reduction	Total Score
Priority Rating:	**2**	**7**	**10**	**5**	**6**	**8**	**4**	
Alternate Disposal Techniques								
Sanitary landfill	0	2	5	5	3	10	2	195
Incineration	0	5	8	6	4	4	10	241
Incineration with heat recovery	10	8	8	8	6	8	10	336

[a]The number at each intersection of a "Techniques" row and a "Criteria" column is an index of the positive value of the technique judged according to the particular criterion. A value of 10 indicates the technique is judged highly positive by the criterion in question.

In the incineration–heat recovery scheme explored in step 2, capacity was fixed either by the availability of refuse or by heating requirements of the university. This is typical of commercial processes, as well, which are limited in capacity either by raw material availability or by product demand. If limits are broad, analysis of several cases, using different capacities, may be necessary. In an approximate estimate typical of the predesign stage, it is rather easy to extrapolate costs from one capacity to another as discussed in Chapter Five.

With feed or effluent flows fixed, many of the remaining entering or leaving streams can be established by simple material balances. Next, temperatures, pressures, and other process properties must be defined. For entering streams, this often requires little more than common sense. For instance, nonreactive liquids and solids are normally stored at ambient temperature and pressure. For those that must be protected from exposure to the atmosphere, storage at 1 to 5 bara under an inert gas is common. Unless heated or cooled to retain certain properties, storage temperatures normally lie between 0 and 35°C. (An exception, among many, is number 6 fuel oil, which has the consistency of tar, and must be stored in heated tanks to be pumped.) Even if feedstock comes from another section of the process, it often is stored "in process" long enough to reach ambient temperature. Where ambient changes are significant, it is customary to employ the "worst case" in the process design. For example, in the northern United States, one might assume a feed temperature of −20°C if there is to be preheating. If cooling is required, 35°C would be appropriate. If energy costs for preheating or precooling are significant, consideration of seasonal variations may be necessary. If in doubt, most engineers make conservative assumptions, that is, those that assure a safe design.

Use of common sense and personal judgment to define temperatures, pressures, and other conditions is somewhat disconcerting at first, especially for students who are accustomed to ready-made homework problems. Students are even more uncomfortable if their assumptions differ substantially from those of the teacher. When such discrepancies are brought up in discussion, I find it upsets the students even more if I revise my figures to agree more closely with theirs. (Such apparent abandon and imprecision are acceptable to experienced designers, who recognize that some assumptions can vary within broad limits without having a significant impact on the final solution.) On the other hand, some students, caught up in the spirit, carry this to extremes, arbitrarily assigning parameters that are uniquely determined by nature. A common error, for example, is to assume the outlet temperature of an adiabatic reactor when it must be determined, instead, by executing an energy balance. With a little thought, it should become easy for mature students to identify situations that are arbitrary and those that are not.

Defining conditions and capacities contains an element of art. Thus, facility is developed with experience. In your first attempts to prepare an original flow sheet, some boldness will be required. As wisdom and judgment develop, this step in process design will become a natural one. Until you develop your own list of flow sheet boundary conditions, the rules of thumb in Appendix B should be useful.

Beginning the flowchart, at this point, may seem precipitous. Nevertheless, based on a literature survey, consultation, and efforts to define the process, you will be prepared to begin. The experience of preparing a flowchart will, in itself, identify areas in which more information is needed. It may even require repetition of the three "definition" steps in minor areas. The focus required for a flow sheet will promote speed and efficiency. Flow sheet practices and procedures are described in detail in Chapter Three.

REFERENCES

1   Glysson, E.A., J.R. Packard, and C.H. Barnes, *The Problem of Solid Waste Disposal*, University of Michigan Press, Ann Arbor (1972).

2   Kepner, C.H., and B.B. Tregoe, *The Rational Manager*, McGraw-Hill, New York (1965).

PROBLEMS

2-1   Decision Chart—College Calendar

Prepare a decision chart to determine the best alternate type of calendar for your university, that is, a quarter system, a semester system, or some other.

2-2   Decision Chart—General

Select a decision such as choosing a job, buying a car, electing a public official, or any other choice that is significant to you. Identify alternatives and criteria that apply. Then, employ the methods of this chapter to arrive at a decision. Show both qualitative and quantitative decision charts.

Chapter Three
FLOW SHEET PREPARATION

To most students, a flow sheet is a chart, printed in a book, containing a collection of symbols each crudely depicting an item of industrial equipment. Symbols are interconnected by straight line segments assembled in oblique, tortuous paths. The so-called qualitative flow sheet, described thus, is employed frequently and usefully to illustrate the general organization of a chemical process, but it has little value in industry.

To the practicing engineer, a flow sheet (diagram or chart) is quite different. Althouth it can be one of several types (process, mechanical, utility, etc.), in this book, *the* flow sheet, stated without qualification, denotes the process flow sheet—a key instrument for defining, refining, and documenting a chemical process.

The process flow diagram is the authorized process blueprint, the framework for cost estimation, and the source of specifications used in equipment designation and design. When "canonized" in its final form through exhaustive computation and vigorous discussion, it achieves the status of scripture; it is the single, authoritative document employed to define, construct, and operate the chemical process.

SKETCHING TECHNIQUES

Imitating a famous newspaper, a more humble publication printed the following slogan on its masthead: "All the news that fits the print." This, in many respects, describes the layout of an industrial flowchart. Unlike qualitative documents, which are often distorted, by necessity, to fit limitations of a printed page, the sheet size itself is expanded to accommodate the detail required. In conventional practice, oversized (typically 60 cm $\times$ 90 cm) flowcharts are folded and stowed in special pockets bound with the design report or "package." For less elaborate processes or reports and when illustration is more important than precision and detail, photoreduction and foldout sheets are employed. Diagrams of many large, complete plants exceed the capacity of a single, manageable sheet of paper and are represented as process segments or modules (each on a separate sheet) related to one another through an appropriate linking code. Some firms are said to fit the diagram on a single linear sheet even if it must be 6 meters long.

Beyond symbols and lines, what must a complete process flow diagram

TABLE 3-1

ELEMENTS OF A PROCESS FLOW DIAGRAM

Essential Constituents
    Equipment symbols
    Process stream flow lines
    Equipment numbers
    Equipment names
    Utilities designations
    Temperature and pressure designations on process lines
    Selected volumetric and molar flow rates
    A material balance table keyed to flow lines
Optional Constituents
    Energy exchange rates
    Major instruments
    Physical properties of process streams

contain? It must include equipment identification numbers and names, temperature and pressure designations, utility identifications, volumetric or molar flow rates for selected process streams, and a material balance table keyed to process lines. In addition, it may contain other information such as energy exchange rates and instrumentation. The "musts" and the "mays" are listed in Table 3-1, although the dogmatic emphasis may be somewhat overdrawn (qualitative variations do occur in industrial practice).

ILLUSTRATION 3-1

You have been asked to prepare a flow sheet for a small, oil-fired stream generator. For assistance, Table 3-2, containing stream designation symbols, Figure 3-1, giving flow sheet equipment illustrations, and Table 3-3, with its instructions for equipment designation, may be consulted.

Your project supervisor has identified the unit as a water-tube, "packaged" (i.e., shop-constructed) steam boiler. It is to produce 1.39 kg/s saturated steam at 20 barg (gage pressure in bar or 0.1 MPa). Eighty percent of the steam will be returned as condensate. The energy source will be number 6 (i.e., residual) fuel oil.

From past experience and background reading [2, 11, 13], you can identify the rough equipment inventory. The boiler itself consists of small tubes within a larger furnace structure. Some of the water-filled tubes (so-called water-wall or radiant tubes) line the wall of the combustion chamber where oil is atomized, mixed with air from the blower, and burned. Combustion products, through careful arrangement of baffles, flow past banks of closely spaced tubes (the convection section) wherein heat is transferred to boiling water inside the tubes. The process flow diagram, accordingly, must show an oil pump, a blower, heat exchangers, and other essential components. Since this is a "battery-limits" flow diagram, it does not include such auxiliary facilities as storage tank, administrative buildings, electrical substation, and the demineralized water plant. The equipment configuration is illustrated in a qualitative sketch comprising Figure 3-2.[1] It is not a flow sheet yet, but it already

[1]Appropriate templates for flow sheet sketching are Charrette No. 150, Koh-I-Nor No. 830050, Picket No. 1053, RapiDesign No. R-50, Timely No. 48, or equivalent.

TABLE 3-2

FLOW SHEET SYMBOLS FOR STREAM DESIGNATION AND
IDENTIFICATION

Symbol	Definition	Symbol	Definition
	Process Streams		*Utility Streams*
	Raw material influx (Identify stream by name)	A–	Compressed air [A–12, e.g., would denote 12 barg]
		CTW	Cooling water (tower)
		CTWR	Cooling water return (tower)
		CW	Cooling water (natural source)
	Finished product effluent (Identify stream by name)	CWD	Cooling water discharge (natural receiver)
		E–	Electricity [E–220, e.g., would denote 220 volts]
		HO–	Hot oil [HO–300, e.g., would denote 300°C]
		HW–	Hot water [HW–150, e.g., would denote 150°C]
B	Designates a skip or break in a process line on the sheet. "B" is a match letter to identify the stream at another place on the same or an adjoining sheet	HWR	Hot water return
		R	Refrigerant [R–10, e.g., would denote –10°C]
		S–	Steam [S–15, e.g., would denote 15 bara pressure]
	Stream designation for material balance point	SC	Steam condensate
	Temperature (°C)	WD	Demineralized water
	Pressure [barg (0.1 MPa gage)]	WM	Municipal water
	Pressure (kPa gage)	WP	Process water
	Pressure (kPa absolute)		
	Gas flow (std m^3/s; ideal gas at 273 K, 1 atm)		
	Liquid flow (liters/s)		
	Mass flow (kg/s)		
	Molar flow (kgmol/s)		

(Assembled from information provided by R. D. Beattie and from reference [1])

includes the following characteristics, which are common to almost all commercial flowcharts.

1 Flow is from left to right with raw materials entering from the left and finished products or waste streams leaving at the right.

2 The flow sheet is oriented horizontally with equipment elevations in the diagram resembling those in the real process. Some freedom is permitted with vertical spacing to allow adequate room for flow lines, to provide efficient use of paper, and to give visual balance. Horizontal spacing of equipment, on the other hand,

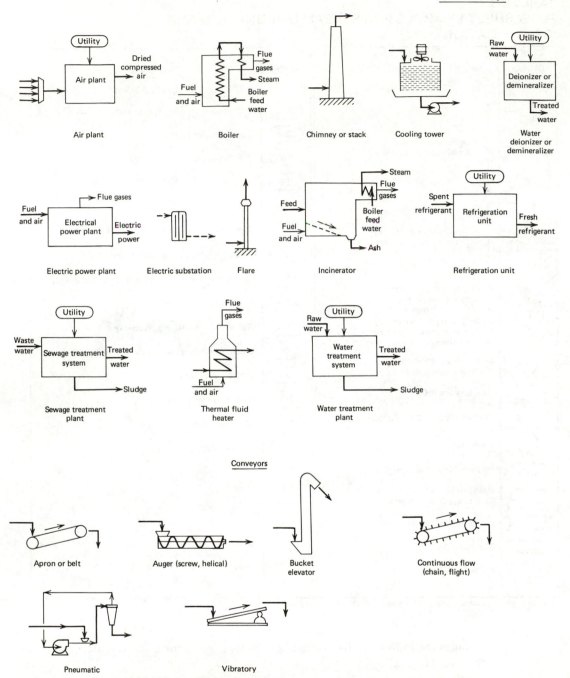

Figure 3-1 Equipment symbols for flowcharts. Symbols are organized according to the generic categories found in Table 4-1. (Assembled from information provided by R. D. Beattie and from Austin [1].)

Crushers, Mills, Grinders

Crushers

Jaw

Gyratory

Impact
(hammer, rotor, cage)

Roll

Mills and grinders

Rolling compression
(bowl, pan, ring–roll)

Disk
(Attrition)

Tumbling or vibrating
(rod, ball)

Hammer

Cutter

Steam or
compressed →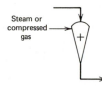
gas

Fluid energy

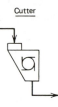

Rotary cutter
(chipper, dicer)

Drives and Power Recovery Machines

Electric motor or generator

Fuel
and air →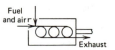
Exhaust

Internal combustion engine

Turbine or expander

Figure 3-1   (*Continued*)

Evaporators and Vaporizers

Evaporators

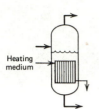

Short tube
(basket, calandria)

Once through

Long tube

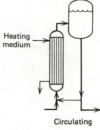

Circulating

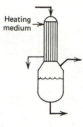

Falling–film

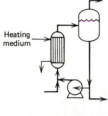

Vertical

Forced-circulation

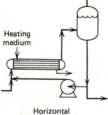

Horizontal

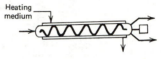

Agitated–film
(scraped–wall)

Vaporizers

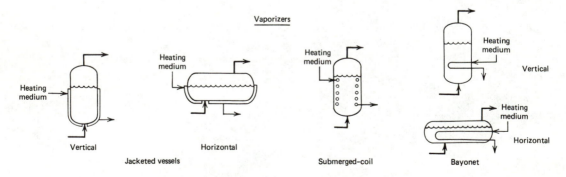

Vertical

Jacketed vessels

Horizontal

Submerged-coil

Vertical

Horizontal

Bayonet

Figure 3-1 (*Continued*)

<u>Furnaces</u>

<u>Boilers</u>

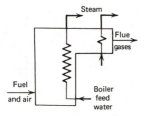

Water–tube and utility

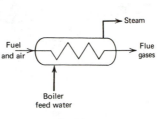

Fire–tube

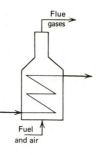

Thermal fluid system

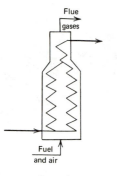

Process heater

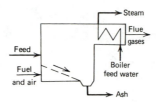

Incinerator

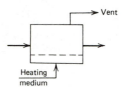

Oven

<u>Gas Movers, Compressors, Exhausters</u>

<u>Fans</u>

Centrifugal

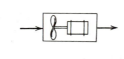

Axial

<u>Blowers and compressors</u>

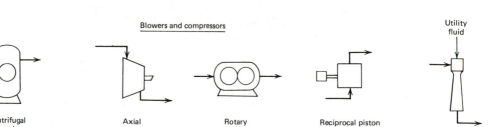

Centrifugal
or turbo

Axial

Rotary

Reciprocal piston

Ejector

Figure 3-1 (*Continued*)

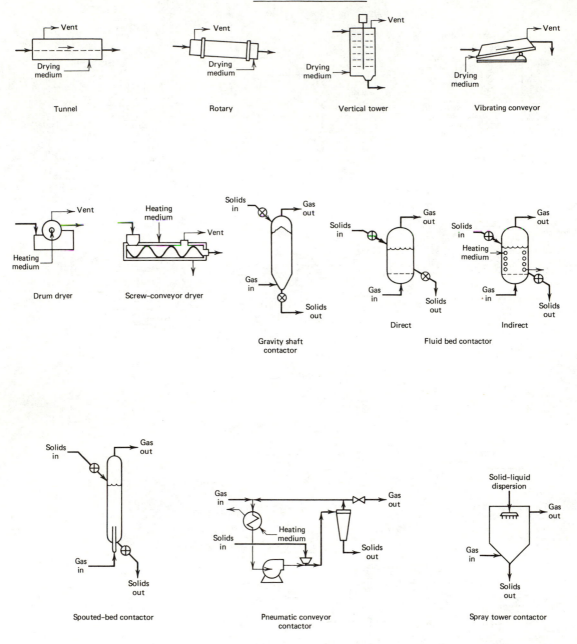

Gas-Solid Contacting Equipment

Tunnel

Rotary

Vertical tower

Vibrating conveyor

Drum dryer

Screw-conveyor dryer

Gravity shaft contactor

Direct

Indirect

Fluid bed contactor

Spouted-bed contactor

Pneumatic conveyor contactor

Spray tower contactor

Figure 3-1 *(Continued)*

Heat Exchangers

Conventional exchangers

Process fluid on
tube side

Process fluid on
shell side

Counterflow

Process fluid on
tube side

Process fluid on
shell side

Parallel flow

Condensers

Process fluid on
shell side

Process fluid on
tube side

Reboilers

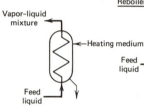

Thermosyphon

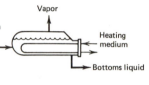

Kettle

Mixers

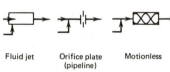

Fluid jet   Orifice plate   Motionless
              (pipeline)

Gas sparger   Pump   Propeller
                (agitated-line)

Axial   Turbine   Radial

Agitated tanks

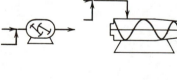

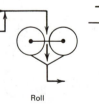

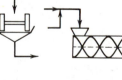

Kneader   Extruder   Roll   Muller   Rotor, ribbon

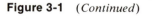

Figure 3-1 (*Continued*)

Process Vessels

Towers

 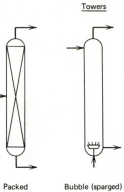  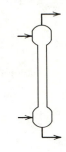

Tray　　　　　Packed　　　　Bubble (sparged)　　　Spray　　　　Proprietary

Venturi scrubber　　　　　　　　　　　　Drums

Holdup

Flash or knockout

Pumps

Axial flow　　　Centrifugal　　　Rotary　　　　Reciprocating　　　　Jet
　　　　　　　　　　　　(positive displacement)

Figure 3-1　(*Continued*)

34

Reactors
(See other equipment listed in Table 4–21)

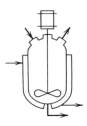

Separators

Centrifuge

Cyclone

Electrostatic precipitator
or bag filter

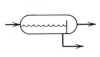

Clarifier, thickener

Decanter, settler

Cartridge filter

Sand filter

Continuous process
filter

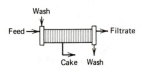

Plate and frame,
shell and leaf filter

Screen

Figure 3-1 (*Continued*)

Size–Enlargement Equipment

Tableting press

Roll–type press

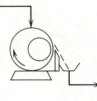

Pellet mill

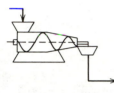

Extruder

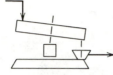

Disk agglomerator

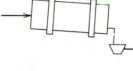

Drum agglomerator

Prilling tower

Storage Vessels

Fixed-roof
(conical)

Floating-roof

Gas holder

Bin

Pressure storage

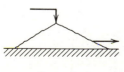

Open yard

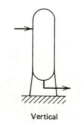

Vertical

Cylindrical

Horizontal

Spherical

Figure 3-1 (*Continued*)

TABLE 3-3

INSTRUCTIONS FOR DESIGNATING EQUIPMENT ON THE PROCESS FLOW SHEET[a]

Equipment Numbering System

1. Number each process area, starting with 100, 200, 300, etc.
2. Number major pieces of process equipment in each area, starting with 110, 120, 130, etc.
3. Number supporting pieces of equipment associated with a major process unit by starting with the next higher number than the major process unit, e.g., 111, 112, 113, 114 for supporting pieces associated with 110.
4. Attach a prefix letter to each equipment number from the following list to designate type. The equipment list might then appear thus:

F-110	B-120	D-210
G-111	G-121	E-211
E-112	G-122	G-212
E-113		E-213

5. The first nine numbers in each area are reserved for equipment servicing the entire area such as a packaged refrigerating unit (e.g., P-105). However, any packaged unit may be broken into its components and numbered as in steps 2 to 4.
6. Use letters following the equipment number to denote duplicates or spares (e.g., G-111A, B denotes two identical pumps in G-111 service).

Equipment Lettering System

A. Auxiliary facilities.
B. Gas–solids contacting equipment (calciners, dryers, kilns).
C. Crushers, mills, grinders.
D. Process vessels (distillation towers, absorption columns, scrubbers, strippers, spray towers).
E. Heat exchangers (coolers, condensers, heaters, reboilers).
F. Storage vessels (tanks, drums, receivers, bins, hoppers, silos).
G. Gas movers (fans, compressors, vacuum pumps, vacuum ejectors).
H. Separators (bag filters, rotary filters, cartridge filters, centrifuges, cyclones, settlers, precipitators, classifiers, extractors).
J. Conveyors (bucket elevators, augers, belts, pneumatic conveyors).
K. Instruments (control valves, transmitters, indicators, recorders, analyzers).
L. Pumps.
M. Agitators, mixers.
N. Motors, drives, turbines.
P. Package units (refrigeration, air units, steam generators, cooling towers, etc.).
Q. Furnaces, process heaters.
R. Reactors.
S. Size-enlargement equipment.
V. Vaporizers and evaporators.
X. Miscellaneous.

Source Assembled from information provided by R. D. Beattie.

[a]Numbering techniques vary from one organization to another. These guidelines are those preferred by the author and are applied consistently in this text.

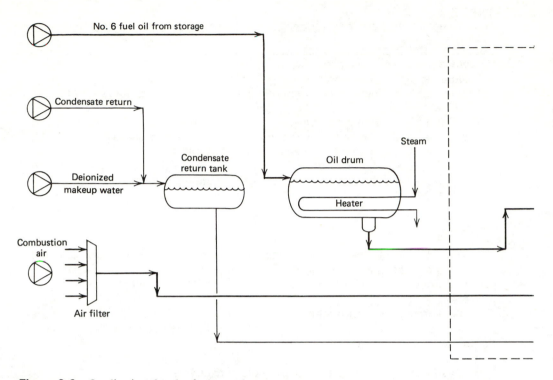

Figure 3-2   Qualitative sketch of water-tube steam generator.

departs much farther from physical reality to reinforce the left-to-right stream motion and to give adequate space for the material balance table, equipment information, and stream designations.

3   Process streams are designated with heavy lines. If streams cross without mixing, one line (usually the vertical one) is broken to allow space at the crossing point. Arrowheads are drawn at each line angle to indicate flow direction. To prevent ambiguity, arrowheads are drawn on all streams *entering* the intersection when flow lines merge.

EQUIPMENT NUMBERING

The next step toward a bona fide flowchart is completion of equipment naming and numbering. (Equipment items are already named in Figure 3-2.) In this case, the furnace is considered to be the central processing unit. It is given the number Q-110 (Q indicates that it is a process heater—see Table 3-3). Equipment items ancillary to the furnace are given numbers from 111 to 119. Tanks and drums carry the prefix F; heat exchangers, E; pumps, L; and blowers, G. The package unit is also numbered separately as P-101 in keeping with instruction 5 in Table 3-3. The oil feed drum and heater are considered to be a small separate subsection with numbers in the 120 series. The diagram now appears as shown in Figure 3-3.

Note that numbers are located in two places: on or near the item designated and

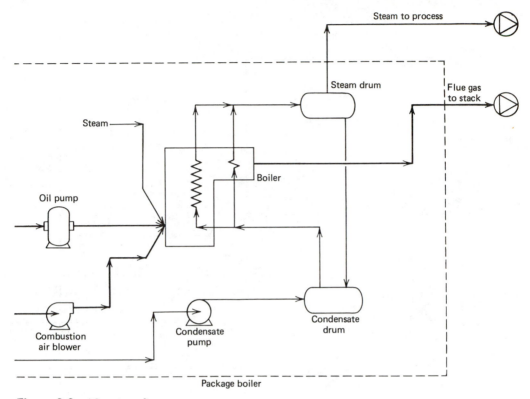

Figure 3-2 (*Continued*)

at the top or bottom of the sketch.[2] This, at first, may seem redundant. Designations are imposed on each unit to prevent ambiguity. The reason for numbers and names at the margins of the sketch will become apparent when equipment design or cost estimation is attempted. In the simple illustration shown, it would be difficult to overlook one of the components, *with* or *without* marginal names. In more complex flow sheets, an oversight is more likely. Such could cause serious errors in the design or cost estimate. It requires almost conscious effort to overlook any one item when they all are tabulated on the margins of the flow sheet.

At this point, readers may wonder whether they are studying engineering or drafting. The sketch prepared by the engineer will, in fact, be refined and its appearance improved later by a draftsperson. Nevertheless, it takes little more time to prepare a neat sketch than a sloppy one, and several important decisions and engineering discussions will be based on the rough sketch before it is drawn professionally.

Even equipment numbering should be taken seriously. The numbers will remain not only throughout the design sequence but also through construction and operation if the plant is built. A pump that is designated L-163 now will be referred to by the same number 6 years later when the plant engineer tells a maintenance foreman about suspicious noises or vibrations. In many respects, the process will be

---

[2]It is common, also, to impose the equipment *name* near the unit itself as well as in the margin, although this is optional.

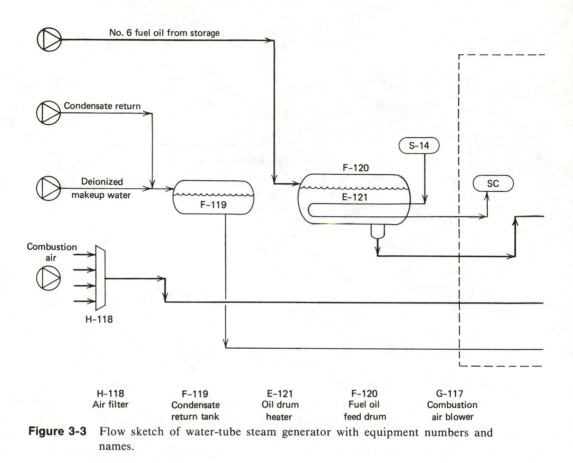

H-118	F-119	E-121	F-120	G-117
Air filter	Condensate return tank	Oil drum heater	Fuel oil feed drum	Combustion air blower

Figure 3-3   Flow sketch of water-tube steam generator with equipment numbers and names.

known as your "baby" for better or for worse. You are well advised to be careful in its conception and naming as well as in its development.

STREAM DESIGNATION

The next step, returning to Table 3-1, is designation of utilities. This is accomplished easily here, since only steam to heat the oil drum and to atomize the fuel must be specified. (It is true that the pumps require either electricity or steam for their motive energy, but utility consumption of pumps and blowers is usually assembled later during the equipment design step.) Steam is specified according to saturation pressure, as illustrated in Figure 3-3. In this case, as defined in Table 3-2, the steam is supplied at an absolute saturation pressure of 14 bar. The temperature of the oil tank is low enough to be heated adequately by "house" steam available at this pressure. (Such is also adequate for fuel atomization). In more critical situations, it may be necessary to postpone the specification of utilities until more information on process conditions is available.

Pressures and temperatures in some cases are defined loosely, using judgment and experience. Careful calculation is required in others. This flow sheet offers examples of both. According to convention, the streams have been characterized

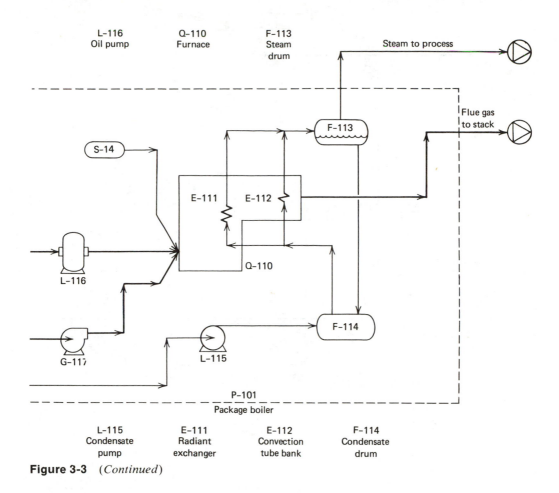

Figure 3-3 (*Continued*)

and identified with various symbols as prescribed in Table 3-2. Diamonds represent material balance flags, which are keyed to the table below the process sketch (Figure 3-4). Temperature and pressure symbols should be numerous enough to identify, unambiguously, conditions at each point on the diagram. Volumetric and molar flow rates are also shown by appropriate flags where such information will aid in specifying or designing pumps, blowers, or other equipment.

In numbering streams, no particular sequence is employed except to begin, as is natural, with the raw material streams entering from the left and then move toward the right until all have been designated. Information required to fill the blanks in Figure 3-4 is developed in the discussion that follows. As you read this material, you are encouraged to enter the data, as appropriate, on the flowchart. You may even wish to push ahead, entering the data first and then checking them against the text. This initial flowchart preparation exercise resembles one of filling in the blanks. Definition of the material balance table is much like working a crossword puzzle. Both require judgment and versatility, and both become easier with practice.

Several marginal temperatures and pressures can now be defined using the concepts outlined in Chapter Two. First, the steam pressure and condition were identified in the problem assignment. Accordingly, 20 barg saturated pressure corresponds to a temperature of 213°C at stream 7. The temperature of stream 1

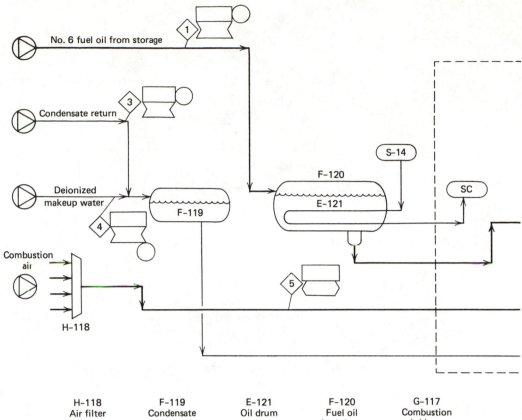

	H-118 Air filter	F-119 Condensate return tank	E-121 Oil drum heater	F-120 Fuel oil feed drum	G-117 Combustion air blower

Material Balance (g/s)

		Process Streams					
		Fuel Oil from Storage	Fuel to Boiler	Condensate Return	Makeup Water	Combustion Air	Boiler Feed Water
Components (mw)		◇1	◇2	◇3	◇4	◇5	◇6
C	(12)						
H	(1)						
N_2	(28)						
O_2	(32)						
S	(32.1)						
CO_2	(44)						
CO	(28)						
NO	(30)						
SO_2	(64.1)						
H_2O	(18)						
Ash							
Total							

Figure 3-4 Process sketch including blank material balance table.

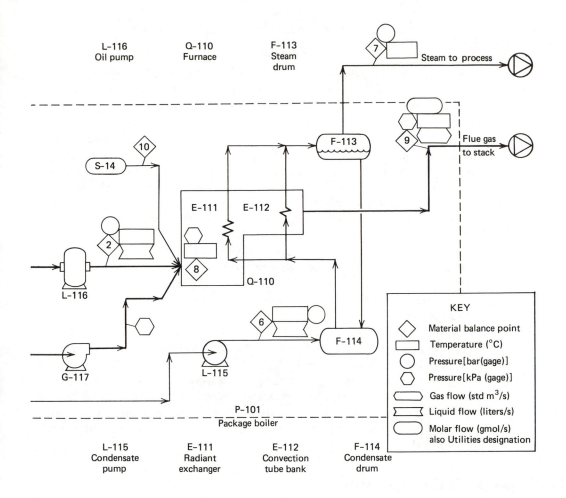

	Steam Product	Combustion Products	Flue Gases	Atomization Steam
	⟨7⟩	⟨8⟩	⟨9⟩	⟨10⟩

Figure 3-4 (*Continued*)

must be high enough for the fuel oil to be pumped easily. Based on references such as Nelson [7], this is approximately 95°C. Since the oil feed drum normally is under atmospheric pressure, the line pressure in stream 1 will not exceed that generated by conventional pumps (i.e., 2 to 5 barg). The condensate return and deionized water streams will be at ambient temperature except in cold climates, where special precautions must be taken to avoid freezing. Thus, 20°C is a logical temperature to assume. The pressure of these streams, as for fuel oil cited previously, will be 2 to 5 barg. Air will also enter at ambient conditions. Those who work with boiler design recognize that changes in ambient air temperature have little significant impact on the design. Thus, rather than employ the "worst-case" approach mentioned in Chapter Two, we will, using accepted boiler design practice, employ the traditional conditions of 27°C and 60 percent relative humidity [3].

At stream 6, the temperature will be the same as stream 3. The pressure will be that of stream 7 plus the difference in liquid head (approximately 10 m or 1 bar) minus friction loss in the recycle line, which is negligible. Thus, stream 6 is at 20°C and near 21 barg.

The condition of fuel oil at position 2 is that required for good atomization. This is approximately 100°C and 18 barg. The internal pressure of the furnace is approximately 20 kPa above atmospheric. Another 40 kPa is lost between the air blower and the burner, yielding 60 kPa at the blower outlet [2].

Temperatures in the combustion chamber and at the stack entrance must be determined from energy balances. It becomes natural in this case, as in many others, to combine specification of the remaining pressures and temperatures with material and energy balances.

MATERIAL AND ENERGY BALANCES

Since there is enough information to establish stream 2, this is the logical next step. Fuel oil consumption is related to steam rate, which, was specified in the problem definition as 1.39 kg/s (5000 kg/h).[3] This fixes column 7 of the material balance chart. (In an advanced flow sheet employed for final equipment design, trace elements in the steam would be included. In this preliminary chart, they are disregarded.) To specify the oil rate, some assumptions must be made concerning boiler operation. From experience or through background reading, one can expect a boiler of this type to operate at approximately 78 to 80 percent overall efficiency [9]. According to U.S. convention, this means that 78 to 80 percent of the gross or higher heating value of the fuel is usefully employed to generate steam. (An alternate design specification might have been to set the flue gas temperature. This, in some respects, is a more realistic approach, but calculations are more laborious. Flue gas temperature and efficiency are related and cannot be specified independently.) Further literature research reveals the composition and properties of a typical number 6 fuel oil (Table 3-4).

To calculate the quantity of oil, a simple overall energy balance is executed on the boiler system.

---

[3]The student might ask, at this point, if capacity should be increased to allow for operation above specified rates when necessary. This is not usually done on the flow sheet. In designing individual items of equipment, however, an overdesign factor is frequently applied to permit flexibility in operation.

TABLE 3-4
PROPERTIES OF NUMBER 6 RESIDUAL FUEL OIL [7, 8]

Composition (weight percent)	
Sulfur	0.7
Hydrogen	9.9
Carbon	87.0
Nitrogen	0.6
Oxygen	1.7
Ash	0.1
	100.0
Specific Gravity	0.97
Heating Value (*gross*)	−42.5 MJ/kg

Energy required to heat water from 20°C and to generate steam at 21 bara and 213°C

$$(1.39 \text{ kg/s})(2796 \text{ kJ/kg} - 84 \text{ kJ/kg}) = 3770 \text{ kJ/s}$$

Fuel Oil Required

$$\frac{3770 \text{ kJ/s}}{0.79 \text{(eff)}(42.5 \text{ kJ/g})} = 113 \text{ g/s}$$

At a specific gravity of 0.97, this corresponds to a volumetric flow rate of 0.116 liter/s.

From the mass flow rate of oil and its composition, column 2 in the material balance table can be completed easily. Logic indicates that stream 1 will flow intermittently as the feed drum is filled periodically from a plant storage tank. Normally, the transfer will occur automatically, controlled by a liquid-level instrument mounted on the drum. This could also be done manually by an operator. For purposes of illustration, assume it is done manually here. Consistent with the rule of thumb in Appendix B, we assume one transfer for each shift. This should be noted as a batch transfer in the table, with the word "Batch" written vertically in the column under stream 1. The volumetric flow rate of stream 1 will be adequate to complete the transfer in 15 min (another rule of thumb). At a continuous oil consumption rate of 0.116 liter/s, the transfer line must have a capacity of 3.9 liter/s.

Several other volumetric and mass rates can be readily established now. Employing guidelines for atomization steam, (i.e., 0.1 kg of steam per kilogram of oil), stream 10 can be specified at 11 g/s. Stream 6 is the same, on a mass basis, as stream 7. The volumetric equivalent, shown on the flow sketch, is 1.39 liters/s. Streams 3 and 4 must also equal 7. As stated at the outset, 80 percent of the steam is recovered as condensate and returned to the boiler. This requires 280 g/s of deionized makeup water added to the condensate return stream.

Computation of the air rate requires a conventional calculation of combustion stoichiometry. From material balance column 2, the gram-molar fuel rate is listed in Table 3-5. Typically, the ratio of actual to theoretically required air is approximately 1.1 (110 percent stoichiometric). Accordingly, the overall combustion

TABLE 3-5.

MATERIAL BALANCE VALUES FOR A 1.39 kg/s WATER-TUBE STEAM GENERATOR BASED ON 1-s OPERATION

Component	(mw)	Fuel Rate ⟨2⟩ g	Fuel Rate ⟨2⟩ gmol	Air Rate ⟨5⟩ gmol	Air Rate ⟨5⟩ g	Atomizing Steam ⟨10⟩ g	Atomizing Steam ⟨10⟩ gmol	Combustion Products ⟨8⟩ gmol	Combustion Products ⟨8⟩ g
C	(12)	98.3	8.19						
H	(1)	11.2	11.20						
N_2	(28)	0.7	0.03	$4.14x = 45.4$	1271			$(4.14x + 0.02)$ $= 45.4$	1272
O_2	(32)	1.9	0.06	$1.10x = 12.1$	386			$0.1x = 1.1$	35
S	(32.1)	0.8	0.02						
CO_2	(44)							8.19	360
CO	(28)								
NO	(30)							0.03	0.9
SO_2	(64.1)							0.02	1.3
H_2O	(18)			$0.11x = 1.2$	21.7	11.0	0.6	$(5.6 + 1.2$ $+ 0.6) = 7.4$	134
Ash		0.1	na					na	0.1
Total		113	19.5	$5.35x$ 58.7	1679	11	0.6	62.1	1803

reaction may be written[4]

$$C_{8.19}H_{11.2}N_{0.06}O_{0.12}S_{0.02} + 1.1x(O_2 + 3.76\,N_2) + 1.1x(4.76)\left(\frac{29}{18}\right)(0.013)\,H_2O \rightarrow$$

$$8.19\,CO_2 + (5.6 + 0.11x)\,H_2O + 0.03\,NO + 0.02\,SO_2 + 0.1x\,O_2 + (4.14x + 0.015)\,N_2$$

where x is the number of moles of added oxygen required to oxidize the carbon, hydrogen, and sulfur in the fuel to CO_2, H_2O, and SO_2 respectively. Thus

$$x + \frac{0.12}{2} = 8.19 + \frac{11.2}{4} + 0.02$$

or

$$x = 11.0 \text{ gmol}$$

This allows us to assemble material balance columns for air a flue gases as illustrated in Table 3-5. These values, entered in Figure 3-4, complete the material balance table in the flowchart.

One important characteristic of a tabular material balance is suggested by the temptation to compare rows and columns. This is a temptation you should always indulge. For example, the principle of mass conservation demands that totals in the columns be consistent. A simple illustration is streams 3 and 4: the totals in these columns must add up to the total for stream 6, which is, in turn, equal to the total for stream 7. In the case of chemical reaction, molecular totals will change, but total

---

[4]Based on results described by Blakeslee and Burback [4], it is reasonable to assume negligible combustibles in ash, negligible (less than 50 ppm) CO, and oxides of nitrogen consisting primarily of NO and corresponding to approximately half the nitrogen in the fuel. It is accepted practice in the boiler industry [3] to employ "standard air," which is defined as 79 mole percent nitrogen and 21 mole percent oxygen, containing 0.013 kg of moisture per kilogram of dry air (i.e., 90 percent relative humidity at 20°C or 60 percent relative humidity at 27°C).

system masses must add up. In Figure 3-4, for example, streams 2, 5, and 10 must give a sum equal to the total of stream 9. The fact that they do agree is strong evidence that the combustion calculations were executed correctly. A proficient supervisor, or teacher, will always check the material balance chart for internal consistency when examining a process flow sheet. An error here will undermine the credibility of the entire design effort.

Returning to Figure 3-4, since gas flow in standard units is directly proportional to molar flow (22.4 std m^3 = 1 kgmol), volumetric flow rates in streams 5 and 9 can be calculated from supporting data in Table 3-5.[5] (The molar flow in stream 9, 62 gmol/s, can be entered directly from the table.) Flue gas temperatures at positions 8, 9, and that between the radiant and convective heat exchangers are the only remaining undefined quantities in Figure 3-4. Energy balances are necessary to identify them. That procedure also illustrates, in this case, a useful technique for designing heat exchangers in series.

The temperature at position 8 is the so-called adiabatic flame temperature. This is the hypothetical temperature of combustion products, assuming no heat loss to surroundings. To review principles from thermodynamics, an energy balance on the flame, without heat loss, reveals that the enthalpy of the flame products is equal to the enthalpy of the reactants. Since enthalpy is a point function, any path can be taken for its evaluation. A convenient one is that illustrated in Figure 3-5. The path $abc \rightarrow d$, $d \rightarrow e$, $e \rightarrow f$, and $f \rightarrow g$ is equivalent to the direct actual path $abc \rightarrow g$. The enthalpy change of step $a \rightarrow d$ will be recognized as a reduction in fuel oil sensible heat.

$$\Delta \dot{H}_{a \rightarrow d} = \dot{m}_2 \int_{100}^{25} C_{p,2} \, dT \qquad (3-1)$$

where $\dot{m}_2$ is the mass flow of stream 2 on the flow sheet and $C_{p,2}$ is its specific heat. Similarly, for the air stream,

[5]In SI units, the standard molar unit is the kilogram mole having the symbol "mol" (see Appendix A). This is commonly confused with the gram mole which has, in the past, been designated by the same symbol. In this text, SI convention is followed, with "mol" meaning the kilogram mole. Often, for surety, I also employ "kgmol" to designate the same unit. The symbol "gmol" is employed whenever gram mole is intended.

Figure 3-5 Enthalpy path employed to calculate adiabatic flame temperature.

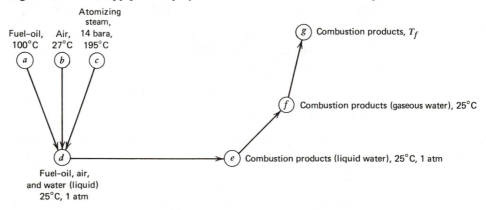

$$\Delta H_{b\rightarrow d} = \dot{m}_5 \int_{27}^{25} C_{p,5}\, dT + \dot{m}_5 \left(\frac{22}{1679}\right) \lambda \tag{3-2}$$

where the second term accounts for latent heat in water vapor carried with the air. For atomizing steam,

$$\Delta \dot{H}_{c\rightarrow d} = \dot{m}_{10}\, (h_{l,25°C} - h_{v,195°C,14\,bara}) \tag{3-3}$$

where $h_{l,25°C}$ and $h_{v,195°C,14\,bara}$ are the enthalpies of water and steam. Step $d\rightarrow e$ is simply the gross heat of combustion of the fuel multiplied by its mass flux.

$$\Delta \dot{H}_{d\rightarrow e} = \dot{m}_2\, \Delta H_c \tag{3-4}$$

Since the gross heat of combustion was employed, the transformation from water liquid to gas is:

$$\Delta \dot{H}_{e\rightarrow f} = \dot{m}_{H_2O} \lambda \tag{3-5}$$

where $\dot{m}_{H_2O}$ is the amount of water produced by reaction plus that from atomizing steam and in combustion air, and λ is the latent heat of water at 25°C. Step $f\rightarrow g$ represents the sensible heat of combustion products above 25°C.

$$\Delta \dot{H}_{f\rightarrow g} = \dot{m}_8 \int_{25}^{T_f} C_{p,8}\, dT \tag{3-6}$$

At this juncture, an approximation in the calculation should be noted. As a general rule, at temperatures above 1900 K, products of hydrocarbon combustion dissociate significantly into smaller molecules such as CO, monotomic species, and radicals [3]. These reactions are endothermic. Thus, a computation that fails to allow for dissociation will yield an artificially high value for T_f. Those familiar with combustion reactions will recognize that the adiabatic flame temperature for fuel oil combustion is substantially above 1900 K. To consider dissociation, however, requires computation of complex chemical equilibria, a problem beyond the scope of convenient manual calculation. The idealized calculation, in this case, however, is not greatly in error. As gases cool below 1900 K, enthalpy values and temperatures become rigorous because dissociated species recombine, releasing their energy to the process. With this in mind, the calculation is continued. The rigorous result is reported later for comparison.

Returning to the enthalpy evaluation, we recognize that

$$\Delta \dot{H}_{abc\rightarrow g} = 0 = \Delta \dot{H}_{a\rightarrow d} + \Delta \dot{H}_{b\rightarrow d} + \Delta \dot{H}_{c\rightarrow d} + \Delta \dot{H}_{d\rightarrow e} + \Delta \dot{H}_{e\rightarrow f} + \Delta \dot{H}_{f\rightarrow g} \tag{3-7}$$

The only unknown in Equation 3-7 is $\Delta \dot{H}_{f\rightarrow g}$, which depends on T_f.

Using the specific heat of 1.7 J/g · K taken from Perry and Chilton [10] for number 6 fuel oil, we have

$$\Delta \dot{H}_{a\rightarrow d} = (113\ g/s)(1.7\ J/g \cdot K)\,(-75\ K) = -14.4\ kJ/s$$

Similarly, for the air and steam [6]

$$\Delta \dot{H}_{b\rightarrow d} = (1679\ g/s)\,(1.0\ J/g \cdot K)\,(-2\ K) + (22\ g/s)\,(-2240\ J/g) = -57\ kJ/s$$

$$\Delta \dot{H}_{c\rightarrow d} = (11\ g/s)\,(105\ J/g - 2790\ J/g) = -29.5\ kJ/s$$

The gross heat of reaction, based on data reported in Table 3-4, is:

$$\Delta \dot{H}_{d\rightarrow e} = (113\ g/s)\,(-42.5\ kJ/g) = -4803\ kJ/s$$

Corrected for the small amount of energy absorbed in NO formation, $\Delta \dot{H}_{d \to e}$ is approximately -4800 kJ/s. Step $e \to f$, based on a latent heat of 2440 J/g, is:

$$\Delta \dot{H}_{e \to f} = (134 \text{ g/s}) (2440 \text{ J/g}) = 327 \text{ kJ/s}$$

Thus, according to Equation 3-7, T_f is the temperature at which $\Delta \dot{H}_{f \to g}$ is equal to the negative sum of the above, or:

$$\Delta \dot{H}_{f \to g} = \dot{m}_8 \int_{25}^{T_f} C_{p,8} \, dT = 4574 \text{ kJ/s}$$

This is a trial-and-error integration because of the unknown value of T_f. Looking ahead, we note that there is heat exchange between combustion products and other media. In such cases, experience suggests that an enthalpy diagram will be useful, especially if multiple exchangers, in series, are involved. To construct the diagram, temperatures are assumed at selected intervals, and the resultant values of $\Delta \dot{H}_{f \to g}$ are tabulated as shown in Table 3-6.

The enthalpy diagram, a plot of the sensible heat of combustion products versus temperature, is thus represented as illustrated in Figure 3-6. The adiabatic flame temperature is obtained merely by noting the temperature that corresponds to the required value $\Delta \dot{H}_{f \to g} = 4574$ kJ/s. As illustrated in Figure 3-6, $T_f = 2280$ K. Thus, the temperature corresponding to point 8 on the flow sheet is shown as $2000°$ C.

The results of a rigorous calculation, employing computer techniques [5], are listed in Table 3-7. The actual adiabatic flame temperature is 2190 K, 90 K less than the foregoing approximation. This discrepancy is caused by dissociation to the extent documented in Table 3-7. Convergence between rigorous and approximated solutions is evident at lower temperatures as illustrated by the two curves in Figure 3-6.

Now it is possible to determine the temperature at point 9 by an overall energy

TABLE 3-6

ENTHALPY TABULATION FOR STREAM 8 AT SELECTED TEMPERATURES

	Flue Gas Flows		$\Delta \dot{H}_{f \to g}$(kJ/s) at Trial Values of T_f shown				
Component	g/s	g mol/s	500 K	1000 K	1500 K	2000 K	2500 K
N_2	1272	45.4	268	971	1738	2541	3364
O_2	35	1.1	7	25	44	65	86
CO_2	360	8.19	68	273	504	747	996
NO	0.9	0.03	—	1	1	2	3
SO_2	1.3	0.02	—	1	2	2	3
H_2O	134	7.4	51	192	356	538	733
Ash	0.1	na	—	—	—	1	1
Total = $\Delta \dot{H}_{f \to g}$			394	1463	2645	3896	5186

Source *JANAF Thermochemical Tables* [14]. Selections from this reference are found in Appendix C. This is an important and convenient source of thermodynamic data for chemical engineers. Each figure in the chart is sensible heat of a particular component at the temperature shown relative to 298 K. Sensible heat integrals, that is, $\int_{298}^{T} C_p \, dT$, are tabulated in ref. 14. Taking nitrogen at 500 K as an example, the value, 268 kJ/s was computed from the JANAF integral $\int_{298}^{500} C_p \, dT = 5.192$ kJ/gmol as follows.

$$(45.4 \text{ g mol/s})(5.192 \text{ kJ/gmol}) = 268 \text{ kJ/s}$$

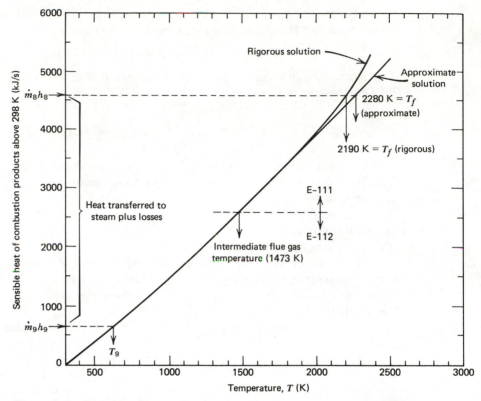

Figure 3-6 Enthalpy diagram for products of combustion in oil-fired boiler.

balance on the boiler. This simplifies to:

$$\dot{m}_8 (h_8 - h_9) = \dot{m}_6 (h_7 - h_6) + Q \tag{3-8}$$

where $\dot{Q}$ is so-called radiation, a traditional term describing heat losses through the walls of the boiler. For a boiler of this size, "radiation" represents approximately 2 percent of the gross heat input $\Delta \dot{H}_{d\rightarrow e}$. In this illustration, it is 100 kJ/s. Energy absorbed by the steam, $\dot{m}_6 (h_7 - h_6) = 3780$ kJ/s, was evaluated earlier in the process of calculating the oil rate. Thus, the left-hand side of Equation 3-8 is equal to 3880 kJ/s.

Since $\dot{m}_8 h_8$ is 4574 kJ/s as noted previously, $\dot{m}_9 h_9$ is the difference, or 694 kJ/s, as illustrated in Figure 3-6. The temperature at 9 is the corresponding value at the abscissa, 633 K or 360°C.

The intermediate flue gas temperature is arbitrary without more information about relative heat duties of E-111 and E-112. In practice, an optimum temperature between these exchangers is in the vicinity of 1200°C (1473 K), which is assumed here.

To appreciate the value of this enthalpy chart, consider an analysis of exchangers E-111 and E-112. By noting the ordinate that corresponds to 1473 K, the duties (including radiation) in E-111 and E-112 are easily evaluated as 1994 and 1886 kJ/s, respectively. Without Figure 3-6, a trial-and-error solution similar to that employed to calculate flame temperature would have been required.

With legend added, the preliminary flow sheet for the water-tube boiler (Figure

TABLE 3-7

ADIABATIC FLAME TEMPERATURE AND COMPOSITION OF PRODUCTS FROM
THE COMBUSTION OF NUMBER 6 FUEL OIL [5]

Ingredients	Mass (g/s)	Enthalpy (J/g)[a]
Fuel oil (100°C) $C_{8.19}H_{11.2}N_{0.06}O_{0.12}S_{0.02}$	113	−84
Ash (100°C) SiO_2	0.1	−13,943
Air (60% relative humidity, 27°C)		
$N_{36.4}O_{10.2}H$	1679	−174
Steam (Saturated, 195°C) H_2O	11	−13,206
Total mass	1803.1	

Overall Elemental Composition

Component	g atom/s	g/s
H	14.11	15
C	8.19	98
N_2	90.75	1271
O_2	26.13	418
Si	0.002	—
S	0.02	1
		1803

Adiabatic Flame Temperature   2190 K
Composition of Combustion Products

Component	gmol/s	g/s
N_2	45.23	1266.4
CO_2	7.74	340.6
H_2O	7.29	131.2
O_2	1.16	37.1
CO	0.45	12.6
NO[b]	0.30	9.0
OH	0.23	3.9
H_2	0.08	0.1
O	0.024	0.4
SO_2	0.02	1.2
SiO_2	0.0015	0.1
H	0.001	—
SiO	0.0005	—
NO_2	0.0001	—
	62.5	1802.6

[a]Based on the elements at 298 K.
[b]About 5000 ppm. In actual furnaces and boilers, flames do not burn adiabatically nor perfectly mixed. Equilibrium also shifts as the gases cool. Nitric oxide emissions are usually in the range of 100−200 ppm for process heaters.

3-7) is completed. This one document could serve as the basis for preliminary design of equipment (i.e., tanks, pumps, blowers, exchangers) and a predesign capital and operating cost estimate. In many design organizations, physical properties of each stream (viscosity, specific gravity, etc.) are also noted on process flow diagrams. Additional rows inscribed below the material balance table provide a convenient format for this.

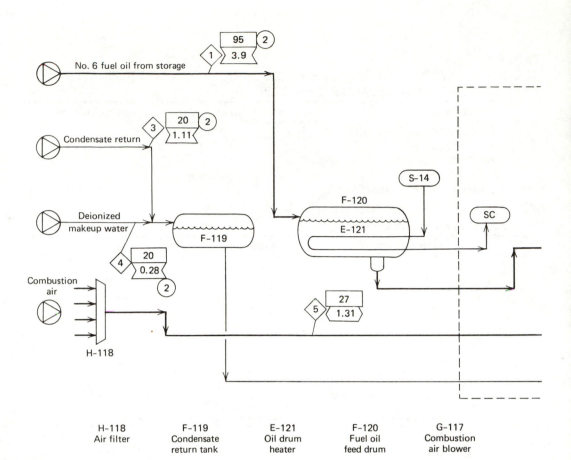

H-118	F-119	E-121	F-120	G-117
Air filter	Condensate return tank	Oil drum heater	Fuel oil feed drum	Combustion air blower

Material Balance (g/s)

				Process Streams			
Components (mw)	Fuel Oil from Storage ⟨1⟩	Fuel to Boiler ⟨2⟩	Condensate Return ⟨3⟩	Makeup Water ⟨4⟩	Combustion Air ⟨5⟩	Boiler Feed Water ⟨6⟩
C (12)		98.3				
H (1)		11.2				
N_2 (28)		0.7			1271	
O_2 (32)		1.9			386	
S (32.1)	Batch	0.8				
CO_2 (44)						
CO (28)						
NO (30)						
SO_2 (64.1)						
H_2O (18)			1110	280	22	1390
Ash		0.1				
Total		113	1110	280	1679	1390

Figure 3-7 Completed process flow diagram for 1.39 kg/s water-tube steam generator. (For most commercial processes, stream flows in kilograms per second are usually greater than unity. In this small-scale illustration, flows are given, for numerical convenience, in grams per second.

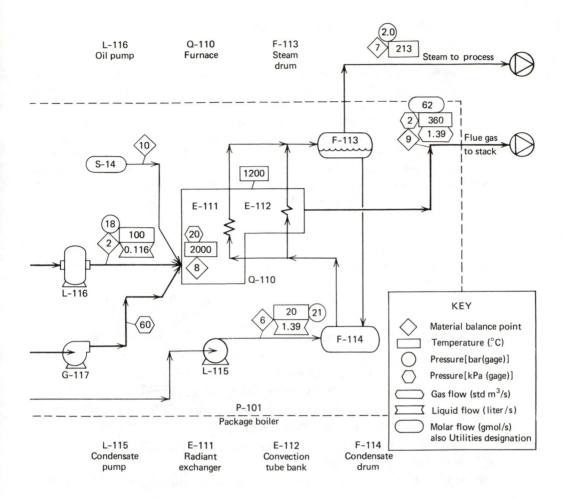

| | L-116
Oil pump | Q-110
Furnace | F-113
Steam
drum | | | |

KEY

◇ Material balance point
▭ Temperature (°C)
◯ Pressure [bar(gage)]
⬡ Pressure [kPa (gage)]
⬡ Gas flow (std m³/s)
⬡ Liquid flow (liter/s)
◯ Molar flow (gmol/s)
  also Utilities designation

| L-115
Condensate
pump | E-111
Radiant
exchanger | E-112
Convection
tube bank | F-114
Condensate
drum |

	Steam Product ◇7	Combustion Products ◇8	Flue Gases ◇9	Atomization Steam ◇10
		1272	1272	
		35	35	
		360	360	
		0.9	0.9	
		1.3	1.3	
	1390	134	134	11
		0.1	0.1	
	1390	1803	1803	11

Process Flow Diagram
Water — tube steam
generator
(1.39 kg/s steam rate)

By: *G. U.*   Date: *19 Jan 1981*

Figure 3-7 (*Continued*)

For definitive process design, transient behavior must be considered. To do this, imagine yourself starting up, shutting down, and operating the plant at any level within its design range. A list of useful questions one might consider is reported by Roodman [12]. Changes might occur, for example, as a catalyst ages over a period of months. Special equipment might be required to get the plant going or to turn it off. Traditionally, we design conservatively, that is, to satisfy the worst-case condition. Proficiency in designing for transient behavior or "operations" is best developed through actual operating or startup experience. For preliminary design, as emphasized in this text, steady-state operation normally is assumed.

REFERENCES

1   Austin, D.G., *Chemical Engineering Drawing Symbols*, George Godwin Ltd., London, and Wiley, New York (1979).

2   Babcock and Wilcox Company, *Steam*, Chapter 25, Babcock and Wilcox, New York (1978).

3   Babcock and Wilcox, *Steam*, Chapters 6, 7, and 25, Babcock and Wilcox, New York (1978).

4   Blakeslee, C.E., and H.E. Burback, "Controlling NOX Emissions from Steam Generators," *J. Air Pollut. Control Assoc.*, **23,** pp. 37–42 (January 1973)

5   Cruise, D.R., "Notes on the Rapid Computation of Chemical Equilibria," *J. Phys. Chem.*, **68,** pp. 3797–3803 (1964).

6   Irvine, T.F., and J.D. Hartnett, *Steam and Air Tables in SI Units*, Hemisphere, Washington, D.C. (1976).

7   Nelson, W.L., *Petroleum Refinery Engineering*, 4th edition, p. 72, McGraw-Hill, New York (1958).

8   Nelson, W.L., *Petroleum Refinery Engineering*, 4th edition, p. 416, McGraw-Hill, New York (1958).

9   Perry, J.H., and C.H. Chilton, *Chemical Engineers' Handbook*, 5th edition, p. 9–26, McGraw-Hill, New York (1973).

10   Perry, J.H., and C.H. Chilton, *Chemical Engineers' Handbook*, 5th edition, p. 3–133, McGraw-Hill, New York (1973).

11   Perry, J.H., C.H. Chilton, and S.D. Kirkpatrick, *Chemical Engineers' Handbook*, 4th edition, Section 9, McGraw-Hill, New York (1963).

12   Roodman, R.G., "Operations: A Critical Factor Often Neglected in Plant Design," *Chem. Eng.*, pp. 131–133 (May 17, 1982).

13   Singer, J.G., ed., *Combustion*, Combustion Engineering, Inc., Windsor, Conn. (1981).

14   Stull, D.R., and H. Prophet, *JANAF Thermochemical Tables*, 2nd edition, U.S. National Bureau of Standards, Washington, D.C. (June 1971).

PROBLEMS

3-1   Steam Boiler

With background information provided from the beginning of this chapter up to and including Table 3-4, fill in as many blank spaces as you can on the flow sheet of Figure 3-4. Do not refer to information given later in the chapter unless necessary.

3-2 Maple Syrup

The average farmer who processes maple syrup in the United States produces about 2000 liters of syrup per year. In a typical evaporation operation, sap flows by gravity from a storage tank across an open, flat-bottomed pan, which forms the ceiling of a brick-walled combustion chamber. Flow across the pan is directed by baffles into channels so that there is little mixing in the direction of flow. Steam escapes through a vent in the roof of the sugar house.

The concentration of sugar in the sap averages about 2.0 percent by weight, and the concentration of sugar in the syrup [specific gravity (sp gr) 1.32] is 65 percent by weight. The sugar is about 96 percent sucrose and 4 percent oligosaccharides. The sap contains about 0.02 percent mineral which becomes a residue (known as "sugar sand".) It is removed from the syrup by filtration. Although there are important chemical changes in the oligosaccharides that affect color and flavor, there are no significant chemical changes from a material balance point of view. Number 2 fuel oil is used at a rate of 45 liters/h to supply energy to the evaporator. Approximately 82 percent of the lower heating value of the oil is available to heat and boil the sap. The burner runs at 20 percent excess air. Using this information and reasonable assumptions:

1 Prepare a proper flow sheet and material balance for this process. Be certain to compute adiabatic flame temperature in the combustion chamber and exit temperature of the flue gases.

2 For a pan 5 m long and 1.7 m wide, calculate the average evaporation rate per square meter of surface. Estimate the overall heat transfer coefficient for the pan.

3 To conserve energy, this process is being modified so that vapor from the sap will be used to preheat liquid feed. This will be accomplished by placing over the pan a hood that contains tubes to preheat the sap plus a catch-pan and drain for condensate. Make a second flowchart (including, of course, a revised material balance) for the modified process.

3-3 Anhydrous Hydrogen Chloride

In a process for anhydrous hydrogen chloride (HCl) production, equal molar quantities of hydrogen and chlorine are burned in a combustion reactor. The reaction goes 99.5 percent to completion, and unreacted raw materials remain in the final product stream. Chlorine is stored as a pressurized, essentially pure liquid. Hydrogen is available from an adjoining process as a gas stream containing 2 percent nitrogen. The pressure of this stream is 10 barg. The reaction occurs in a water-jacketed vessel. Steam is generated in the jacket. The water level in the jacket is controlled to maintain a reactor effluent temperature of 1200°C. The steam is generated at a pressure of 40 barg. Effluent from the furnace–reactor passes through a second steam generator and then to a water cooler before exiting as a gas at 4 bar and 30°C to another part of the plant. The steam from the second generator is saturated at 16 barg. The temperature of the HCl stream leaving this unit is 315°C. The desired annual HCl capacity of the plant is 2500 metric tons, and the operating factor is 85 percent.

1   Prepare a flow diagram for this process.

2   As an emergency method for controlling the outlet temperature of the reactor, it is proposed that provision be made for recycling HCl from the outlet of the secondary steam generator. Assume that this is done without changing any of the existing equipment but by merely adding a recycle loop. To maintain the required reactor outlet temperature in the absence of water in the jacket, what must the recycle rate be? Prepare a revised flow sheet for the operation with recycle cooling. Note that the temperature and flow rate of the stream leaving the secondary boiler will change with recycle. Make all reasonable and necessary assumptions.

3-4   Multiple-Effect Evaporation of Maple Sap

To reduce the fuel consumption of the maple syrup process, the use of multiple-effect evaporators has been proposed. The first effect will use process steam at 6 barg, and the pressure in the last effect will be 7 kPa(abs). In this process, the fresh sap will be fed to the second effect and the liquid from the last effect will be pumped back to the first effect for finishing. Using the sap flow rates and concentrations of Problem 3-2, prepare a flow sheet for this case, assuming three effects. The heat transfer coefficients in each effect are the same at some reasonable value.

3-5   Coal-Fired Power Plant

A boiler in an electrical power generating plant consumes coal at a rate of 2700 metric tons per stream day. (A "stream day" is defined as a period during which the plant operates at design capacity, without interruption, for 24 h.) The boiler produces steam at 540°C and 174 barg. Steam is expanded through a turbine to 24 barg then reheated to 510°C and expanded further to the condenser pressure where it is condensed and recycled. Flue gases leaving the boiler are at 150°C.

1   Using your best judgment and employing reasonable assumptions, prepare a flow sheet for the unit.

2   Estimate the amount of heat transfer surface in the boiler and the condenser.

3   Estimate electrical power output and overall efficiency.

Note: The coal is West Virginia medium volatile bituminous, having a higher heating value of 33.2 MJ/kg (as delivered). It contains 7 percent ash, 4 percent moisture, 80 percent carbon, 5.5 percent hydrogen, 1 percent nitrogen, 1 percent oxygen, and 1.5 percent sulfur.

3-6   Nitric Acid Production

A process employing air and ammonia produces 454 Mg/d of nitric acid. The ammonia is oxidized with air on a platinum gauze catalyst. Air is pressurized with a steam- and gas-turbine-driven compressor, to 9.3 bar(abs). It is preheated by exchange with product gases to 260°C. Ammonia, supplied as a liquid raw material, is vaporized and preheated to the same temperature with steam. The two streams are then mixed and fed to the adiabatic reactor. The feed ratio is 10 mole percent ammonia and 90 mole percent air. Product gases, at chemical equilibrium, pass through a steam boiler, the air preheater, a tail-gas heater, and a fiberglass filter before entering a water cooler–condenser. The ratio of nitric oxide to nitrogen dioxide (NO/NO_2) shifts throughout the process to maintain equilibrium. Products leave the cooler–condenser at 38°C, partially condensed as weak acid. The liquid is separated from the gases and fed to the middle tray of a bubble cap absorption tower having interstage cooling coils to maintain the temperature at 38°C. Process gases plus additional air enter the bottom of the tower where they react to produce more NO_2, which is absorbed in a countercurrent water stream. The product, 58 percent nitric acid, is discharged from the bottom of the tower. Tail gas from the top of the tower is preheated with steam, reheated by exchange with reactor products, and expanded through a second turbine on the air compressor shaft. After expansion, the gases are vented to the atmosphere.

Prepare a flowchart for this process.

3-7 Synthesis Gas Manufacture

A process for synthesis gas manufacture (*Hydrocarbon Processing*, **44**, p. 273, 1965) involves partial combustion of heavy fuel oil with a gas stream containing 95 percent oxygen and 5 percent nitrogen. This fuel oil contains 84.7 percent carbon, 11.3 percent hydrogen, 3.4 percent sulfur, 0.13 percent oxygen, 0.4 percent nitrogen, and 0.07 percent ash. The product stream, on a dry basis, has the following analysis: 46.1 percent H_2, 46.9 percent CO, 4.3 percent CO_2, 0.4 percent CH_4, 1.5 percent N_2 and A, 0.35 percent H_2S, and 0.45 percent COS. Gross heating value of the fuel is 41.5 kJ/g. Fuel oil, preheated to 250°C, is atomized with steam (0.4 kg of steam per kilogram of oil) and burned with the oxygen-rich stream, which has been preheated to 250°C. The reactor is well insulated. The pressure in the reactor is 9 barg. Products from the reactor are cooled to 260°C in a waste heat boiler, which generates saturated steam at 38 barg. Three percent of the carbon in the oil is converted to carbon particles in the reactor. These are removed in a water spray tower or carbon catcher downstream of the boiler. The carbon–water slurry exits from the base of the carbon catcher to a carbon separator. The carbon sludge exits from the process, and the carbon-free water is recycled to the spray tower. Gaseous products from the spray tower enter the base of a packed tower, where they are cooled by a circulating water stream. Water from the bottom of the packed tower is pumped through a heat exchanger and returned to the top of the tower. Excess water condensed in the packed tower is returned to the carbon separator.

Prepare a flowchart for a plant of this type having a production capacity of 2.6×10^8 std m³/yr dry synthesis gas. Assume an operating factor of 90 percent.

3-8 Kraft Pulping Process (Courtesy of R.J. Willey)

Prepare a process flow sheet for the continuous Kraft pulping operation of a medium-sized paper mill. A block diagram of the process is shown in Figure P3-8-1. Assume that operation consumes 500 cords per day of softwood averaging 1360 kg/cord and containing 12 percent average moisture and 0.3 percent ash. The plant operating factor is 90 percent. Assume that 95 percent of the cellulose in the wood is found in the pulp product. The wood itself consists of 50 percent cellulose. The balance, 25 percent hemicellulose and 25 percent lignin, is digested by the digesting liquor. You can disregard the volatile organics produced in the digester and assume that all organics remain in the black liquor.

Assume that the liquors have the following dry compositions (kilograms per 100 kg of dry wood fed, 16 percent effective alkali).

	White Liquor	*Black Liquor*	*Green Liquor*
NaOH	17.1	0.4	0.0
Na_2S	6.8	0.2	6.8
Na_2CO_3	7.5	7.5	Balance
		Plus	of sodium
		Na–organic	
		groups	

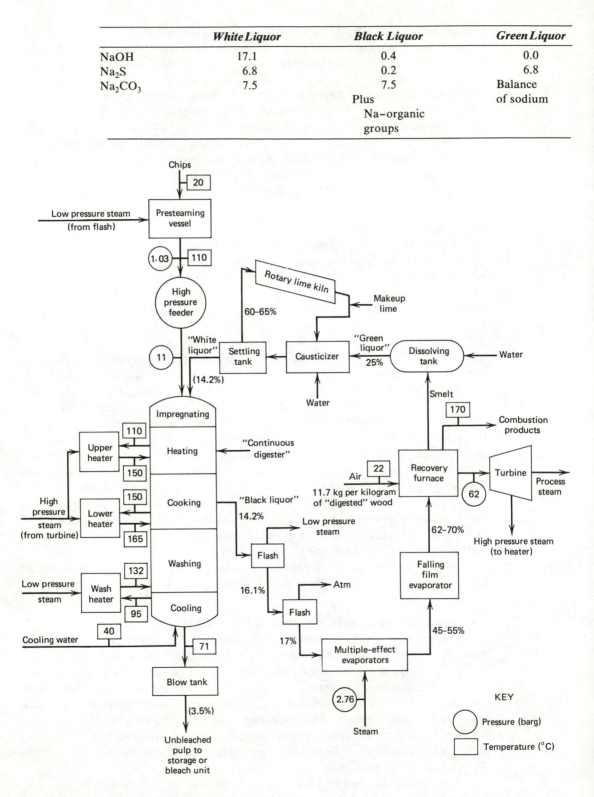

Figure P3-8-1   Sketch of Kraft pulping operation.

Typical concentrations are shown in Figure P3-8-1, where water is diluent. Temperature and pressure requirements are also shown where needed for energy balances. The higher heat of combustion of lignin and hemicellulose may be taken as 15,350 kJ/kg. Assume that pulping liquor has the same heat capacity as water. NaOH is reformed by reacting Na_2CO_3 with lime in the causticizer. Lime is formed by burning $CaCO_3$, generated in the previous reaction, in a lime kiln. Assume that the upper and lower heaters are designed to heat 110 liters/s of liquor (sp gr 1.2) and that the wash heater heats 13 liters/s of liquor. All heaters are shell and tube heat exchangers. The turbine exhausts process steam at 3.8 barg and the multiple-effect evaporators require 0.2 kg of 2.8 barg steam per kilogram of water evaporated. The high pressure steam is at 11 barg and the low pressure steam is at 1.0 barg.

3-9   Synthesis-Gas from Coal

One method of manufacturing liquid fuels from coal is gasification of the coal with steam and oxygen (e.g., Lurgi and Koppers–Hulsh processes) followed by purification of the gaseous products and conversion to a hydrocarbon liquid by a Fischer–Tropsch synthesis. In the process considered here, the coal to be gasified is an Illinois number 6 type having the following characteristics.

Bulk density, (kg/m³)	720
Proximate analysis (weight percent)	
Moisture	1.5
Ash	10.3
Volatile matter	35.5
Fixed carbon	52.7
Total	100.0
Ultimate analysis (weight percent)	
Carbon	70.2
Hydrogen	4.6
Nitrogen	1.0
Sulfur	3.6
Oxygen	10.5
Ash	10.1
Total	100.0

The reactor pressure is 15 atm(abs) and the reactor temperature must be 1250°C to assure complete conversion of carbon in a brick-lined vessel. Coal is ground to minus 100 mesh and conveyed to one of two lock-hopper feeders, which can be alternately pressurized with CO_2 to allow feeding into the pressurized reactor. The powdered coal is metered through a screw conveyor into the preheated and premixed oxygen–steam gases at the burner nozzle. The ratio of steam to carbon is 1.6 mol/atom and the ratio of oxygen (O_2) to carbon is 0.30 mol/atom. Concentrations of major gaseous species in the product stream are as follows.

	mole percent
H_2O	33.5
H_2	28.0
CO	23.0

CO$_2$	14.0
CH$_4$	1.5
Total	100.0

Sulfur in the coal can be assumed to react completely to H$_2$S, and the nitrogen is converted to gaseous nitrogen.

Prepare a process flow sheet for a battery-limits plant which, beginning with mine-run coal and liquid oxygen, will generate a gas stream of the composition given and separate hydrogen and carbon monoxide (less than 0.5 mole percent other compounds) as a mixed gas for delivery to a Fischer–Tropsch unit.

Chapter Four
SPECIFICATION AND DESIGN OF EQUIPMENT

Once the engineer has prepared a flow diagram as outlined in Chapter Three, the next logical step is to specify equipment that appears on it. In this regard, engineers might be classified in two categories: buyers and builders. To illustrate the difference, consider two people who wish to acquire the same material object, such as an automobile. One might choose to build from scratch, the other to buy a vehicle manufactured by someone else. In both cases, intelligence, experience and judgment are valuable, but the levels of expertise and detailed knowledge required are vastly different.

Consider this analogy applied to a processing plant. If you were constructing the plant, you would need detailed design specifications and fabrication drawings for all equipment, including instruments, wiring, piping, and auxiliary equipment. All mechanical details would need to be specified completely and unambiguously. (This, by the way, is the degree of detail necessary for a definitive cost estimate, where, as discussed in Chapter One, such engineering itself often accounts for 10 to 20 percent of the cost of a plant.) Any well-trained engineer, given enough time, can prepare a definitive design. Most courses in the undergraduate curriculum are geared to this level of accuracy. In this regard, students are good builders but poor buyers. Well trained in the workings of the "carburetor," they usually know little about the "tires," "upholstery," and "optional equipment."

Few proposed processes ever actually materialize, and money spent to design the discarded ones is lost. Thus, an engineer must be able to identify appropriate equipment costs quickly so that the economic promise of future efforts can be predicted. This requires short-cut design techniques that can be employed to obtain inexpensive answers having reasonable accuracy but limited precision. Such tools or techniques are discussed extensively in this chapter. Not only are they useful for tentative decisions and cost estimates, but they are invaluable for rapidly checking your own or another's detailed design calculations and as a source of background data to support negotiations with a specialist or vendor.

Short-cut or approximate techniques are normally no substitute for the more rigorous designs required to actually build equipment. They are, however, adequate for many cost estimates. Equipment design is a step toward assessing the capital cost (i.e., purchase price) of a manufacturing plant. Since the plant will operate (and its cost will be distributed) over a number of years, an error in the capital estimate is frequently less significant than one might suspect. For reasons discussed later, less than 30 percent of the total capital cost is reflected in annual manufacturing expenses. If one estimates the capital with an accuracy of plus or minus 25 percent, the uncertainty in annual expenses is much less, only about 6 percent of fixed capital. In many cases, this is less than the error in other estimated manufacturing costs and is certainly tolerable for preliminary decision making. In a coal-fueled electric power plant, an uncertainty of 25 percent in the capital estimate is equivalent to one quarter of a cent per kilowatt-hour. This is less than 6 percent of the total manufacturing cost and is insignificant when compared with the fuel cost alone.

Although short-cut techniques must be used with wisdom, there is no cause to discredit them because of their rather unsophisticated nature. An automobile buyer, caring little about carburetion and combustion, can still make an intelligent purchase. Inexperienced engineers are often appalled to find that their mentors can design equipment in one tenth the time that they require and with better accuracy. Such newcomers come to appreciate the value of abbreviated methods when applied by those who have developed superior judgment through intimate and lengthy applications of more tedious and rigorous techniques.

ESSENTIALS OF SHORT-CUT EQUIPMENT DESIGN

Students, in their years of undergraduate training, see photographs of such standard items as heat exchangers, towers, and compressors. They even conduct experiments with the equipment, as isolated units, in the laboratory. But they are shocked, on their first plant visit, to realize that these items are almost unrecognizable when in use, integrated and shrouded, as they are with instruments, controls, insulation, utility lines, piping, valves, and other peripheral equipment. It is reassuring to know that costs of peripheral and auxiliary items can be estimated easily using factors developed by cost engineers through prior experience. Thus, only the costs of the major process components shown on the flow sheet must be assessed in detail. Cost estimation techniques are explained in Chapter Five. In this chapter, rapid methods for designing and specifying major equipment are described.

Categories of Process Equipment

In reality, cookers, coolers, economizers, kettles, preheaters, reboilers, super-heaters, and thermosyphons are all heat exchangers, often identical in design. Similarly, fume collection, stripping, scrubbing, humidification, and fractionation are usually accomplished in packed or tray towers. These are the same process vessels that chemical engineering students encounter as distillation or absorption columns in a mass transfer or unit operations course. These examples suggest that much of the mystery in process design can be resolved with better communication. To aid definition and minimize the jargon, I have listed in Table 4-1 more than 140

TABLE 4-1
SPECIFIC TYPES OF PROCESS EQUIPMENT, CATEGORIZED GENERICALLY AND CITED BY PAGE NUMBER

Generic Equipment Type

Specific Equipment Type	Auxiliary facilities	Conveyors	Crushers, Mills, Grinders	Drives and Power-Recovery Machines	Evaporators-Vaporizers	Furnaces	Gas Movers, Compressors, Exhausters	Gas-Solids Contacting Equipment	Heat Exchangers	Mixers	Process Vessels	Pumps	Reactors	Separators	Size-Enlargement Equipment	Storage Vessels
Absorption towers											193					
Accumulators											186					250
Adsorption towers											195					
Agitators										174						
Agglomerators															243	
Air classifiers														233		
Air coolers									156							
Air plants	69															
Augers		71														
Autoclaves					95						186		212			
Auxiliary facilities	69					108										245
Bag filters														234		
Ball mills			81													
Belt conveyors		72														
Bins																246
Blenders										167						
Blowers							115									
Boilers	69					110			160							
Bucket elevators		72														
Calciners								129								
Centrifuges														224		
Chillers									157							
Chimneys	69															
Clarifiers														232		
Classifiers														233		
Coalescers														229		
Coils					95											
Collectors														218		
Columns											190					
Compressors							116									
Concentrators					94									238		
Condensers									158							
Cookers					95											
Coolers								129	159							
Cooling towers	69															
Crushers			74													
Crystallizers														232		
Cyclones														224		
Decanters											202			227		
Dehumidifiers											193					
Deionizers	69															
Delumpers			74													
Demineralizers	69															
Dewaterers														218		
Digesters														232		
Distillation towers											190					
Drives				83												
Drums											190			227		247
Dryers								129								

63

TABLE 4-1
(Continued)

Generic Equipment Type

Specific Equipment Type	Auxiliary facilities	Conveyors	Crushers, Mills, Grinders	Drives and Power-Recovery Machines	Evaporators-Vaporizers	Furnaces	Gas Movers, Compressors, Exhausters	Gas-Solids Contacting Equipment	Heat Exchangers	Mixers	Process Vessels	Pumps	Reactors	Separators	Size-Enlargement Equipment	Storage Vessels
Dust collectors														218		
Economizers									159							
Ejectors							119					211				
Electrostatic precipitators														226		
Electric generating plants	69															
Electric substations	69															
Elevators		72														
Engines				86												
Evaporators					97											
Exhausters							115									
Expanders				89												
Extractors														224		
Extruders															242	
Fans							115									
Feeders		70														
Filters														234		
Flares	69															
Flocculators														232		
Fluid energy mills			82													
Fluidized beds								137					217			
Fractionators											190					
Fume collectors											201					
Furnaces						108										
Gas movers and compressors							115									
Generators, electric	69															
Generators steam	69					110										
Grinders			74													
Grit separators														233		
Heat exchangers									145							
Heaters						112			145							
Hoppers																246
Humidifiers											193					
Incinerators	69					113										
Internals											199					
Ion exchangers											195					
Jets (ejectors)							119					211				
Kettles									160		186					
Kilns								129								
Kneaders										179						
Knockout drums											202			227		
Leaching equipment											200					
Lump breakers			74													
Mills			74												240	
Mixers										167						
Motors				84												
Mullers			80							180						
Ovens						114										
Packed towers											195		217			
Pelletizers															243	

Table 4-1
(Continued)

Specific Equipment Type	Generic Equipment Type															
	Auxiliary facilities	Conveyors	Crushers, Mills, Grinders	Drives and Power-Recovery Machines	Evaporators-Vaporizers	Furnaces	Gas Movers, Compressors, Exhausters	Gas-Solids Contacting Equipment	Heat Exchangers	Mixers	Process Vessels	Pumps	Reactors	Separators	Size-Enlargement Equipment	Storage Vessels
Plate columns											195					
Precipitators, electrostatic														226		
Preheaters									159							
Presses														239	240	
Pressure vessels											187					247
Prilling towers															244	
Process vessels											182		212			
Pug mills		71								181						
Pulverizers			74													
Pumps												204				
Purifiers														218		
Reactors											182		212			
Reboilers					95				160							
Refrigeration systems	69															
Roasters						129										
Rotary kilns						131										
Screens														239		
Screw conveyors		71														
Scrubbers											193					
Separators														218		
Settlers														227		
Sewage treatment systems	69															
Shredders			74													
Sifters														239		
Silos																246
Stacks	69															
Steam generators	69				110				160							
Stills											190					
Strainers														239		
Stripping towers											193					
Superheaters									160							
Syphons												211				
Tanks																245
Tableting press															240	
Thermal liquid heaters	69				112				159							
Thermosyphons									160							
Thickeners														232		
Trays											199					
Turbines				86												
Vacuum pumps							128									
Vaporizers					95				160		186					
Ventilators							115									
Vessels											187					245
Washers											193					
Water treatment plants	69															

65

equipment items one might find in a chemical equipment catalog and have combined them into 16 generic groups.[1]

In describing design of specific equipment items, I first introduce general approaches applicable to each generic group as a whole. Then, the peculiarities of the respective individual items are considered.

Parameters for Equipment Specification

For each generic category in Table 4-1, certain information must be specified before equipment can be designed. These design data, listed in Table 4-2, will be familiar to most from past experience in fluid mechanics, heat transfer, mass transfer, and reactor design courses. Their use is illustrated in the example that follows.

ILLUSTRATION 4-1   PUMP SELECTION

Describe the procedure and derive equations for preliminary design of a pump.

On inspecting a commercial catalog, the prospective buyer recognizes a need to know the medium being pumped, the flow rate, the pressure rise, the temperature, the inlet pressure, and the power consumption (all shown in Table 4-2). In addition, a purchaser must specify materials of construction and the pump type. The first five quantities can be either extracted directly or derived from information in the diagram and material balance that comprise the flowchart. Definition of construction materials and pump type usually demands experience. For the novice seeking a tentative result, however, guidelines are provided Tables 4-28 and 4-20.

Power consumption can be computed from other known parameters using the following generalized mechanical energy equation (Perry 5-17; McCabe and Smith 86; Peters and Timmerhaus 509).

$$\frac{\dot{w}_f}{\dot{m}} = \int_{p_1}^{p_2} \frac{dp}{\rho} + \frac{\Delta u^2}{2} + g\,\Delta z + \frac{u^2}{2}\left(\frac{4fL}{D} + \sum_i K_i\right) \qquad (4\text{-}1)$$

(This equation and nomenclature will be familiar to the advanced chemical engineering student. In this section and throughout, equations and terminology typical of Perry's *Handbook*, 5th edition [37], are employed. References to specific items in this and other prominent chemical engineering design texts are designated by author and page number as above. The usual form of citation—reference number in brackets—is retained for more general cases.)

Applying Equation 4-1 specifically to the pump boundaries, there is a negligible change in velocity and elevation, and the terms involving Δu^2 and Δz drop out. Friction and fitting losses (terms in parentheses on the right-hand side of Equation 4-1) are complex functions of fluid motion within the pump casing. This sophisticated subject, the province of pump specialists and researchers, is translated

---

[1]Table 4-1 serves as a ready index to design information in this chapter. To locate in the text the discussion of an individual item, find it in the first column and refer to the generic category or categories indexed in the table. The same generic sequence was employed to organize the cost data in Chapter Five. If you look carefully, you will also note a similar structure in Figure 3-1, which contains flowchart symbols.

TABLE 4-2
DATA FOR THE DESIGN OF PROCESS EQUIPMENT[a]

Generic Equipment Type	Flow Sheet Letter Designation	Temperature In	Temperature Out	Pressure In	Pressure Out	Concentration In	Concentration Out	Heat Duty	Utility Consumption (electricity, steam, fuel, etc.)	Flow Rate or Capacity	Other Types of Required or Useful Data
Auxiliary facilities	A	(See Table 4-3)									
Conveyors	J								C	√	Particle size and bulk density
Crushers, mills, grinders	C								C	√	Particle size
Drives and power recovery machines	N								√		
Evaporators, vaporizers	V	√	√	√	√	√	√	C	C	√	Heat transfer coefficients
Furnaces	Q	√	√	√	√			C	C	√	
Gas movers, compressors, exhausters	G	√		√	√			S	C	√	
Gas–solid contacting equipment	B	√	√			√	√	C	C	√	Equilibria and rates
Heat exchangers	E	√	√	√				C	S	√	Heat transfer coefficients
Mixers	M	√		√		√	√	S	C	√	
Process vessels	D	√	C	√	C	√	√	C	C	√	Equilibria and rates
Pumps	L	√		√	√				C	√	
Reactors	R	√	C	√	C	√	√	C	S	√	Equilibria and rates
Separators	H	√		√		√	√		C	√	Pressure drop sedimentation rate
Size-enlargement equipment	S								C	√	Particle size
Storage vessels	F	√		√						√	

[a]Items marked with a check are those normally required; those designated "C" are usually calculated in the design process; "S" denotes information that is sometimes required.

by them to an intrinsic efficiency factor ϵ_i, which can be employed simply and directly by the process engineer. With these considerations and the knowledge that liquid densities vary little with pressure, Equation 4-1 can be reduced to

$$\frac{\epsilon_i \dot{w}_s}{\dot{m}} = \frac{\Delta p}{\rho} \tag{4-2}$$

which is usually rearranged to a more convenient form for pump design.

$$\dot{w}_s = \frac{\dot{m}\,\Delta p}{\rho\,\epsilon_i} \tag{4-3}$$

In this case, efficiency was specified for the system excluding the driver, which is normally an electric motor and sometimes a steam turbine. Thus, $\dot{w}_s$ is the so-called shaft or brake power, that which enters the system through a moving shaft. (Note that $\dot{w}_s$, a positive quantity here, is opposite to thermodynamic convention where work done by the system is positive.)

To compute consumption of a utility (electricity, steam, compressed air, or fuel), one must consider energy losses in the drive motor, engine, or turbine itself. This involves a drive efficiency ϵ_d. Overall power consumption is consequently related to shaft power and other variables by the following expression.

$$P = \frac{\dot{w}_s}{\epsilon_d} = \frac{\dot{m}\,\Delta p}{\epsilon_d \epsilon_i \rho} = \frac{\dot{m}\,\Delta p}{\epsilon_o \rho} \tag{4-4}$$

Overall efficiency, obviously, is the product of intrinsic and drive efficiencies. Representative values of ϵ_i are discussed later; they are generally in the range of 40 to 85 percent for pumps.

Given flow rate, pressure difference, density, and efficiency, an engineer can easily specify a pump with enough accuracy to identify it in a vendor's collection. In this step, one final factor must be considered—the net positive suction head (NPSH). This is, in effect, the pressure at the inlet to the pump. One of the most destructive effects in centrifugal pump service is cavitation, that is, vaporization and recondensation of a fluid as it experiences rapid acceleration and deceleration. To avoid cavitation, pressure at the pump entrance must be appreciably greater than the vapor pressure of the fluid. This exact minimum pressure NPSH is specified by the pump manufacturer and must be available at the pump inlet. In preliminary design, it is usually not necessary to specify NPSH, but you should be aware of this limitation so that you can place the pumps physically at low elevations or conceptually in the flow diagram where adequate inlet pressure is always available. (Information on various types of pump and their selection as well as materials of construction is presented later in this chapter.)

Returning to the general principles of equipment design, five conceptual steps taken in the preceding illustration will be applied again and again to the design of other equipment (the order of application may differ from case to case).

1   Identification of parameters that must be specified.
2   Application of fundamental underlying theoretical equations or concepts.
3   Enumeration, explanation, and application of simplifying assumptions.
4   Employment of a nonideal correction factor or efficiency.
5   Enumeration of other factors that must be considered for adequate specification.

Because this book is designed for novices, I have attempted to provide basic, dependable techniques appropriately balanced with science and art, which will permit the readers to competently identify equipment items and estimate their costs. Specialists, no doubt, can use standard repositories with facility and will have no need for this chapter. On the other hand, most experienced engineers will recall that the process of locating, identifying, and applying material from scattered reference locations was not trivial the first time. Hence the inclusion of specific design techniques in this admittedly general text.

SPECIFIC EQUIPMENT DESIGN METHODS

Intended as a guide and reference for short-cut equipment design, this section, except for the discussion of auxiliary facilities, does not demand or deserve reader engagement at the level of a mystery novel. Rather it is a catalog to be used when and if the need for insight or guidance in design of a particular equipment type is required. Topics are organized alphabetically according to the generic categories listed in Tables 4-1 and 4-2.

A major factor in all equipment design, defining the specific fabrication material (i.e., metalic, ceramic, polymeric, or other) is discussed at the conclusion of this chapter.

AUXILIARY OR "OFFSITE" FACILITIES

In a definitive plant design or construction project, auxiliary facilities represent a substantial expenditure of engineering resources. Within this broad grouping lie two subcategories: utility and service. Utility facilities include plants to produce steam, electricity, refrigerants or cooling water, and equipment for sewage treatment, pollution control, and waste disposal, to name a few. Service facilities, also popularly known as yard facilities, include roads, railroad sidings, site excavation and preparation, cafeteria, and administrative offices, plus other auxiliary buildings and site amenities.

Most utility plants are of standard design and are frequently purchased as "package" or prebuilt units. Such are designed, constructed (often in the vendor's own shop), transported, and installed by the supplier at the plant site and connected to the main process or "battery-limits" plant. (Facilities thus obtained are also known as turnkey or ready-to-use plants.)

Precise characterization of utility plants would require a complete process design and economic evaluation in each instance. Thus, it is impractical to treat them in great detail here. Such depth is usually unnecessary because utility plants are routinely supplied and continuously operated to the extent that units can be purchased "off the shelf" with the assurance of reliable, predictable, long-term service. In construction of a grass-roots plant, auxiliary facilities are purchased much as an automobile is bought by an individual.

Later, in assessing economic parameters for battery-limits processes, the costs of various utilities provided by auxiliary facilities must be defined. In this book, they are generally evaluated as direct fees per unit, such as cents per kilowatt-hour, as if purchased directly from a separate utility company. In some situations, such as the purchase of a package plant, it is necessary to identify the utility facilities in more detail. To do so requires specification of several key parameters. Most common types of packaged utility units or variants are listed in Table 4-3, which also defines parameters required for adequate and unambiguous designation. In a predesign economic evaluation, more information than that listed is seldom required. Even in a definitive estimate, these data may be adequate for purchase of a package plant from a vendor. If a unique utility process is involved, it can be evaluated as a separate process module by developing flow charts, material and energy balances, and individual equipment specifications.

Other auxiliary amenities, general, and service or yard facilities are also listed in Table 4-3. In a definitive estimate, these are usually designed in detail by civil engineers working for the contractor. They are outside the scope of typical chemical process engineering and are considered further only as they relate to total capital costs as outlined in Chapter Five.

TABLE 4-3
COMMON AUXILIARY (OFFSITE) FACILITIES AND DATA REQUIRED FOR THEIR SPECIFICATION[a]

Facilities	Utility or Service Provided	Temperature		Pressure		Concentration		Heat Duty	Utility Consumption	Flow Rate or Capacity	Other
		In	Delivered[b]	In	Delivered[b]	In	Delivered[b]				
Utility Facilities											
Air plants	Pressurized, dried air for process or instrument use		C		√		√	C	C	√	
Boilers	Process steam	C	√	C	√	C		C	C	√	Fuel type
Chimneys or stacks	Atmospheric dispersion of waste gases	C				√				√	
Cooling towers	Process cooling water	√	√	C	C	C	C	C	C	√	Air wet bulb temperature
Deionizers and demineralizers	High purity water for process use, boiler makeup, or cooling tower feed water				√	√	√		C	√	
Electrical generating plants	Electricity									√	
Electrical substations	Conversion and distribution of purchased electricity									√	
Flares	Disposal of combustible gases and vapors	√		√		√		C	C	√	
Incinerators	Disposal of liquid or solid waste					√	C	C	C	√	Properties of waste
Refrigeration systems	Ultra-cold fluid for extreme cooling service	√	√	C	C			√	C	√	
Sewage treatment systems	Cleansing of discharged water	√	C			√	√		C	√	
Thermal fluid heaters	Ultra-hot special fluids for process heating	C	√	C	C			C	C	√	
Water treatment plants	Purification of fresh, brackish, or salty water for drinking, cooling, or process use	√	C		C	√	√		C	√	

Service (yard) Facilities

Buildings (nonprocess)	Offices and auxiliary structures
Change house	Pipelines (outside battery limits)
Cafeteria	Roads, walks, and railroad sidings
Communications network	Storage and loading of products and by-products
Emergency power system	Storage and unloading of raw materials and supplies
Fire protection system	Storm drain network
Fuel receiving, blending, and storage facility	Yard lighting
Landscaping, fencing, and site improvements	

[a]Items checked are data normally required. Those designated by "C" are usually specified by the vendor. Columns marked "S" denote information that is sometimes required. Many specifications (e.g., instrument air pressure) are standard and constant throughout industry.)

[b]To the stream leaving the facility.

CONVEYORS (FEEDERS)

Although liquid and gas pumps and pipelines are conveyors of one sort, their design is considered elsewhere. This section pertains to the unique and specialized equipment employed to convey solids.

Solids conveying and feeding equipment, available in numerous modifications and variations, can be classified into seven broad categories: apron, auger,

TABLE 4-4
CRITERIA AND DATA FOR THE RAPID DESIGN OF SOLIDS-CONVEYING
EQUIPMENT

	Type of Conveyor						
	Apron	Auger (screw, helical)	Belt	Bucket Elevator	Continuous Flow (chain, flight)	Pneumatic	Vibratory
Range of Common Equipment Sizes							
Diameter or width, *D* (m)	0.5–2	0.15–0.50	0.3–2.0	0.15–0.5	0.2–1.0	0.1–0.3	0.15–3.0
Length, *L* (m)	10–50	5–25	10–50	8–25	10–50	10–100	5–30
Maximum solids bulk capacity (m³/s)	0.06	0.007–0.08^a	1.0	0.02	0.01	0.03	0.06
Normal conveying speeds (m/s)	0.1–0.3		1–2	1–2	0.5–1	15–50	—
Simultaneous actions		Mixing, heating, cooling, drying				Drying, calcining	Drying, sieving heating, cooling
Compatibility							
Dusty solids	B	A	B	B	A	C	B
Lumpy materials	A	B	A	B	B	X	B
Fibrous solids	B	A	A	B	B	B	D
Abrasives	D	C	A	A	D	D	A
Corrosive materials	D	C	C	C	C	C	C
Sticky and gummy solids	D	A	D	E	E	D	E
Controlled environments	D	C	D	D	B	A	C
Vacuum conveying	E	C	E	E	C	D	C
Conveying up or down an incline	B	A	B	A	B	A	D
Vertical lifting	X	C	X	A	A	A	X
Versatility of path	E	X	E	X	A	A	X
Limiting angle of incline	30°	None	30°	None	None	None	5°
Normal temperature limit (°C, carbon steel)	300	400	100	200	400	450	300
Relative annual cost	Moderate	High	Low	Moderate	Moderate	Moderate	Moderate
Power consumption (kW)^b	$0.006\dot{m}^{0.82}L$	$0.07\dot{m}^{0.85}L$	$0.0027\dot{m}^{0.82}L$	$0.07\dot{m}L^{0.63}$	$0.07\dot{m}^{0.85}L$	See nomograph, Perry 7-18	$0.02\dot{m}^{0.72}L$

KEY

A   excellent or no limitations
B   modest limitations or problems
C   special units available at higher cost to minimize problems
D   limited in this regard
E   severely limited in this regard
F   unacceptable

^{a}For nonabrasive solids, 0.08 m³/s; for mildly abrasive materials, 0.025 m³/s; for highly abrasive solids, 0.007 m³/s (see Perry, Table 7-5).

^{b}Except for bucket conveyors, this applies to horizontal conveying. If elevation is involved, add or subtract power for the change in potential energy: $0.012\,m\,\Delta h$.

belt, bucket, continuous flow (chain, flight), pneumatic, and vibratory. Each is listed with its respective advantages and disadvantages in Table 4-4. A brief description of each type follows (for evaluating the properties of numerous solid materials, Tables 7-3 and 7-4 in Perry, p. 7-4 and 7-5 are useful.)

Auger or Screw Conveyor

This device, also known as a helical conveyor, employs a rotating screw to transport solids through the enclosing trough or duct. Most readers would recognize it as the

element that forces food through the blades in an old-fashioned manual domestic food grinder.

Available in numerous sizes and configurations, the auger is probably the most flexible of conveyors. It can transport sticky, gummy solids, under a variety of conditions, in controlled atmospheres and with the simultaneous transfer of heat. Screw conveyors are, however, limited in diameter. Capacity is also controlled by the size and abrasiveness of conveyed materials. Rapid design can be executed using the information in Table 4-4 in combination with the capacity chart (Table 7-5, Perry 7-7). For simultaneous mixing and conveying of gummy materials (e.g., for feeding some dryers, where dried solids are recycled to make gummy or sticky materials tractable), the flight may be segmented to form a propellerlike screw. This is known also as a paddle conveyor or pug mill.

Belt Conveyor

(Perry 7-7)

For high capacity, conventional conveying in noncritical situations, a belt conveyor usually is the most economical choice. It consists of a continuous flexible belt

passing over rollers or idlers and driven through sets of power rollers; belt conveyors exist that travel at speeds of several meters per second and transport materials such as minerals or ores over distances of several kilometers. These conveyors can transport a rather broad variety of materials except those that adhere to the belt. Vertical lifting is limited by a maximum belt incline of 30 degrees with few applications exceeding 20 degrees. Changes in direction are limited with belt conveyors, requiring special design or multiple conveyors. Rapid design can be performed using the information in Table 4-4.

Apron Conveyor

(Perry 7-16)

Apron conveyors are similar to belt units except that the conveying element is made of overlapping segmented plates connected to form a continuous chain. The apron conveyor is more expensive than a belt conveyor and is perferred only when the nature of the conveyed solid or the temperature is incompatible with feasible belt materials.

Bucket Elevator

(Perry 7-11)

A bucket elevator is the common conveyor chosen for vertical lifting of noncritical, nonsticky solids. It can be visualized as a beltlike conveyor with the belt replaced by a series of metal buckets. These are linked to form a continuous chain that moves up and down between rotating top and bottom sprocket wheels. Bucket elevators are rugged and dependable, capable of lifting a broad range of materials. They can be loosely enclosed for dust control but cannot easily be made totally leakfree for operation with controlled environments or under reduced pressure. Bucket elevators are common in mineral processing where vertical lifting of abrasive, lumpy materials is necessary. Specifications, power consumption, and other data required for preliminary design are listed in Table 4-4.

Continuous Flow Conveyor

(Flight, Chain, Closed-Belt: Perry 7-15)

This flexible, though relatively expensive, conveyor is available in numerous configurations and modifications. The conveying element is a moving chain or belt with protruding rakes or "flights," which drag solids along.

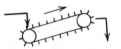

Dual channels are required, one operating full to convey and one for return of the belt or chain. Because of flexibility in the continuous element, these conveyors are compact and can be made in numerous configurations to follow tortuous horizontal and vertical flow paths. Because of the enclosure, highly abrasive, lumpy, or sticky materials tend to jam the unit and cannot be handled easily. On the other hand, enclosure permits operation in controlled or vacuum environments. Convenient self-feeding and self-discharge characteristics of continuous flow conveyors contribute to their flexibility of use. Thus, they are frequently chosen for relatively low capacity operations where convenience and versatility justify a higher price.

Pneumatic Conveyor

(Perry 7-16)

When conditions and material characteristics permit, pneumatic conveyors are popular for high volume transport of solids through substantial horizontal and

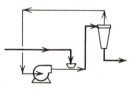

vertical distances. These units employ a high velocity gas stream to entrain solids and conduct them through a duct much as a fluid would be pumped. They are flexible, compact, and inexpensive and require relatively little maintenance. Gas-tight design is, of course, necessary, allowing operation with controlled atmospheres or under pressure. Use of vacuum is limited because a modestly dense gas phase is essential to convey solids. Large lumps and abrasive solids are not compatible with pneumatic units. Dust control is integral and necessary in this conveyor. Explosive gas–solid mixtures must, of course, be handled with special care. Pertinent information for design of pneumatic conveyors is included in Table 4-4, although for complete information, you are referred to information and the nomograph in Perry (pp. 7-18 and 7-21), which provides excellent guidance for determining equipment size.

Vibratory Conveyor

(Perry 7-13)

This device, also known as an oscillatory conveyor, employs a rapidly vibrating and oscillating pan to throw the solid particles, moving them in the desired direction.

Transportability is strongly influenced by characteristics of the solid itself. It must not be slippery or sticky and should not aerate easily. Because of simple pan geometry, vacuum or controlled atmospheres can be employed. With jacketed pans or screens, other operations such as drying, sieving, heating, and cooling can be conducted simultaneously. The smooth fluidlike flow is advantageous in certain applications. Simultaneous elevation, although possible, is limited to an incline of about 5 degrees.

CRUSHERS, MILLS, GRINDERS

Among the most ancient of processing operations, milling (size reduction, or comminution) was developed for the conversion of grains to flour for bread and other foodstuffs. Indeed, many entire processing plants are known to the workers as "mills," illustrating the significance of comminution in mining, mineral processing, cement manufacturing, metallurgy, paper production, food processing, and a multitude of chemical production operations.

Types of equipment are almost as varied as their numerous applications, but underlying mechanisms of size reduction fall within four categories. These are compression, impaction, attrition (or rubbing), and cutting. McCabe and Smith [30] mention the nut cracker, the hammer, the file, and the shears as familiar manual devices that comminute solids in these respective ways.

The words "crushing" and "grinding," in popular usage, suggest a type of action or motion. In process terminology, they divide machines that crush large lumps or rocks from those that pulverize smaller grains and powders. This somewhat arbitrary distinction is employed in Table 4-5, where types of comminution equipment are listed in order of decreasing feed dimensions. Jaw and gyratory units, known as primary crushers, can accept large quantities of friable materials of up to 2 m in diameter. Impact and roll devices, generally designed for lumps of 0.5 m or less, are examples of secondary crushers. Pan, bowl, ring-roll, and attrition mills are usually employed for fine crushing or coarse grinding, whereas ball (Figure 4-1), rod, and high speed hammer mills are medium to fine grinders. Fluid energy mills perform fine and ultra-fine grinding to produce particles as small as a fraction of 1 micrometer (μm) in diameter.

Comminution technology is so advanced that the theoretical energy required to fracture a given solid can be accurately predicted. Unfortunately, less than 2 percent of the power supplied to the mill is consumed in this way. The balance is lost as

Figure 4-1 Ball mill used to grind clinker in the manufacture of portland cement. (Ideal Basic Industries, by permission.)

friction and dissipated as heat. Thus, an accurate prediction of power consumption requires expert attention, often combined with laboratory tests.

One theoretical concept is obvious. Since the intrinsic energy required for comminution is related to the increased surface area that results, specific power consumption increases as the particles become smaller. Thus, fine grinders consume many more kilowatts per kilogram of product than primary crushers. Hence, machine capacities become smaller and smaller as particle size decreases.

Power consumption also depends on relative hardness of the feed. Since no devices are suitable for all types of solid, the power relationship shown in Table 4-5, for a given mill, applies throughout the practical range of materials. These relationships are approximate, to be used only for preliminary design and in the absence of specific data.

The presence of fines reduces efficiency of a mill, so most modern units are designed for closed-circuit operation. That is, high flows are promoted with much oversize material passing through. The composite product is then separated by screens, cyclones, air classifiers, settlers, or filters. Oversized solids are returned to the feed. Several pulverizers such as the rolling-compression grinder with air classification and the fluid energy mill integrate separation into the basic design. Information in Table 4-5 and cost data of Chapter Five are for package mills in closed-circuit operation.

In most situations, efficiency is higher if fine solids are ground in a liquid suspension. Such is generally employed when moisture is not detrimental to the product or separation is relatively inexpensive. If the desired final product is a liquid

TABLE 4-5
CRITERIA AND DATA FOR THE PRELIMINARY SPECIFICATION OF CRUSHING AND GRINDING EQUIPMENT

	Equipment Type					
	Crushers				Grinders	
	Jaw	Gyratory	Impact (hammer, rotor, cage)	Roll	Rolling Compression (bowl, pan, ring-roll)	Disk (attrition mill)
Maximum feed particle of lump diameter, D (m)	2	2	0.3	0.7	0.08	0.5
Typical maximum reduction ratio, R	8	8	35	4	15	10
Maximum capacity, $\dot{m}$ (kg/s)	200	1000 (coarse) 100 (intermediate)	400	125	15	5
Performance Characteristics						
Narrowness of size distribution	A	A	D	B	A	A
Compatibility						
Hard solids	A	A	E	D	D	E
Abrasive materials	A	A	D	C	D	E
Sticky or cohesive solids	D	E	B	B	C	A
Soft materials	D	E	A	A	A	A
Resilient substances	E	E	B	C	E	A
Suitability						
Wet grinding	E	E	D	C	A	A
Controlled atmospheres	E	E	E	C	D	C
Heating or cooling	E	E	E	C	E	B
Other types of simultaneous processing					Blending, kneading	Blending, fluffing

Crushing and Grinding Size Ranges (arrows indicate grinding range) Particle Diameter

Size range	Particle Diameter	Jaw	Gyratory	Impact	Roll	Rolling Compression	Disk
	1 m	A ↑	A ↑	B ↑	D	D	X
Coarse crushing	0.1 m (10 cm)	B ↓	A	A	B ↑	B ↑	E
Intermediate crushing	1 cm	B ↓	B ↓	B	A ↓	A	A ↑
Coarse grinding	1 mm	D	D	B	B ↓	A	A ↕
Intermediate grinding	0.1 mm (100 μm)	E	E	E	E	B ↓	C
Fine grinding	0.01 mm (10 μm)	X	X	X	X	X	X
Extra-fine grinding	0.001 mm (1 μm)	X	X	X	X	X	X
Ultra-fine grinding	0.1 μm						

Power Consumption (kW)[a]	Jaw	Gyratory	Impact	Roll	Rolling Compression	Disk
Hard materials (8–10 Moh)	$3.0\dot{m}^{0.88}R$	$2.5\dot{m}^{0.88}R$		$0.75\dot{m}R$		
Medium (4–7 Moh)				$0.30\dot{m}R$	$0.30\dot{m}R$	
Soft (1–3 Moh)			$1.0\dot{m}^{0.88}R$			$10\dot{m}$
Leathery, tenacious materials			$4.0\dot{m}^{0.88}R$			$50\dot{m}$

Typical Materials Processed	Jaw	Gyratory	Impact	Roll	Rolling Compression	Disk
Asbestos			X			X
Bone			X			X
Carbon black						X
Cereals and grains				X		X
Cement	X	X	X			
Charcoal						
Clay					X	
Coal	X	X		X	X	
Coke				X	X	
Feldspar						
Filter cake			X			X
Foods				X		X
Fuller's earth						
Graphite				X	X	
Gypsum				X	X	
Leather						X
Lime			X	X	X	
Limestone	X	X	X	X	X	
Mica			X			
Minerals and ores	X	X	X			
Organic solids						
Phosphates				X	X	
Pigments						X
Plastics						
Refractories	X	X		X	X	
Resins						
Rubber						
Salts			X			
Scrap iron			X			
Slags	X	X				
Soaps						
Soapstone				X	X	
Silica	X	X				
Sodium bicarbonate						
Solid waste (municipal)			X			
Sulfur				X	X	
Talc				X	X	
Wood						X

KEY
A   excellent or no limitations
B   modest limitations
C   special units available at higher cost to minimize problems
D   limited in this regard
E   severely limited in this regard
X   unacceptable

TABLE 4-5
(Continued)

		Equipment Type				
		Grinders				Cutters
		Tumbling Rod Mill	Tumbling Ball (pebble) Mill	Vibrating or Stirred Ball Mill	High Speed Hammer (pin)	Fluid Energy	Rotary Cutter Chipper, or Dicer
Maximum feed particle of lump diameter, D (m)							0.5
Typical maximum reduction ratio, R		15	20	30	50	50	50
Maximum capacity, $\dot{m}$ (kg/s)		50	15	0.1	2	1	50
Performance Characteristics							
Narrowness of size distribution		B	A	A	B	A	A
Compatibility							
Hard solids		A	A	A	X	A	X
Abrasive materials		A	A	A	X	A	X
Sticky or cohesive solids		B	D	D	A	D	A
Soft materials		C	D	E	A	D	A
Resilient substances		E	E	E	A	E	A
Suitability							
Wet grinding		A	A	A	A	X	E
Controlled atmospheres		C	C	A	A	A	E
Heating or cooling		A	A	A	A	A	E
Other types of simultaneous processing		Drying, dispersing	Drying, dispersing	Dispersing	Drying	Classification	

Crushing and Grinding Size Ranges (arrows indicate grinding range)	Particle Diameter						
	1 m	E	E	X	X	X	X
Coarse crushing	0.1 m (10 cm)	C	C	X	X	X	A
Intermediate crushing	1 cm	A	A	X	B	D	A
Coarse grinding	1 mm	A	A	E	A	B	A
Intermediate grinding	0.1 mm (100 μm)	D	A	B	A	A	D
Fine grinding	0.01 mm (10 μm)	E	B	A	B	A	X
Extra-fine grinding	0.001 mm (1 μm)	X	D	B	D	B	X
Ultra-fine grinding	0.1 μm						

Power Consumption (kW) [a]							
Hard materials (8–10 Moh)		$0.007\dot{m}/D_p$	$0.008\dot{m}/D_p$	$40\dot{m}/D_p^{0.3}$		$1–10$ [b]	
Medium (4–7 Moh)							
Soft (1–3 Moh)					$40\dot{m}\ln R$		$100\dot{m}$
Leathery, tenacious materials					$40\dot{m}\ln R$		$500\dot{m}$

Typical Materials Processed							
Asbestos					X		
Bone					X		
Carbon black					X		
Cereals and grains					X		
Cement		X	X	X			
Charcoal					X		
Clay						X	
Coal		X	X	X			
Coke		X	X	X			
Feldspar		X	X	X			
Filter cake					X		
Foods							
Fuller's earth					X		
Graphite		X	X	X			
Gypsum		X	X	X			
Leather							X
Lime		X	X	X	X		
Limestone		X	X	X			
Mica							
Minerals and ores		X	X	X		X	
Organic solids							X
Phosphates		X	X	X			
Pigments		X	X	X	X	X	
Plastics					X		X
Refractories		X	X	X			
Resins					X		X
Rubber							X
Salts					X		
Scrap iron							
Slags		X	X	X			
Soaps					X		
Soapstone							
Silica		X	X	X		X	
Sodium bicarbonate							
Solid waste (municipal)							
Sulfur							
Talc					X	X	
Wood							X

[a] Mass flow rate $\dot{m}$ is in kilograms per second; reduction ratio R is dimensionless; D_p is final particle diameter in meters.

[b] Fluid energy grinding power is reported as kilograms of compressed air or steam consumed per kilogram of solids. See Perry (p. 8-43) for more specific data. Steam or air pressure is normally about 8 bara.

dispersion, the mill is often a mixer as well as a grinder. Power consumption figures in Table 4-5 are for dry grinding. Thus, they will yield conservative values for wet-grinding situations.

Other simultaneous operations are possible in some machines as indicated in Table 4-5. Heating or cooling and drying are typical examples. In fact, because of low grinding efficiencies, substantial natural heating may occur. This may require external cooling of heat-sensitive and hazardous materials. In grinding organic or combustible materials, designers should always be aware of the potential for dust explosions.

Detailed information on specific comminution equipment is included in the paragraphs that follow and is outlined in Table 4-5. Perry (pp. 8-44 to 8-55) discusses a number of specific commercial products and the equipment employed for crushing and grinding them.

Jaw Crusher

(Perry 8-16; McCabe and Smith 831)

This device, simple in concept, is composed of two plates that form a vee-shaped chute. In the most common (Blake) type, one plate is pivoted at the top and

oscillates at the bottom, compressing and crushing solid chunks and boulders until they are small enough to drop through the opening at the bottom of the vee. These massive crushers are capable of accepting material up to 2 m in diameter. Since the jaw crusher is strictly a compression device, it is limited to friable, relatively nonsticky materials, but it can break even the strongest and most abrasive solids that will fracture under compression. Guidance for selection and preliminary design of jaw crushers is provided in Table 4-5.

Gyratory Crusher

(Perry 8-18; McCabe and Smith 832)

This machine is composed of a funnel-shaped mortar with a tapered pestle passing through the hole in its base. The pestle is suspended, spiderlike, small diameter up,

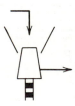

from a pivot at the top of the bowl. The base of the pestle gyrates or rotates eccentrically, causing the pinch space at the crusher discharge to oscillate. In effect, the action of the gyratory crusher is almost identical to that of the jaw crusher.

Capable of processing essentially the same types and sizes of solid, it is more prone than the jaw crusher to clog with some materials. Because of its continuous discharge, slightly better efficiency, and lower maintenance, the gyratory crusher has become the mainstay for hard ore and mineral crushing. Single units having capacities up to 1000 kg/s are larger than any other type of coarse crusher.

Impact Breakers

(Hammer, Rotor, Cage: Perry 8-22; McCabe and Smith 836)

Capable of breaking, cutting, and tearing cohesive and tenacious materials, these devices contain a cylinder that rotates in a stationary housing. In a *hammer mill*,

centrifugal force causes a number of pivoted hammers, mounted on the preiphery of the rotor, to swing within the housing. Stationary anvil bars or breaker plates are mounted on the shell, forming an impact and cutting surface. Solids are broken by impact of the high speed hammers combined with shearing and attrition between hammers and anvils.

Rotor impactors, with rigid bars or ribbons attached to the outside of a high speed rotating cylinder, do not employ breaker plates and depend solely on impact for size reduction. *Cage mills* employ concentric cagelike cylinders rotating in opposite directions to create impaction and attrition forces.

Impact crushers are not suitable for hard and abrasive solids but are excellent for tough, fibrous, or sticky materials like scrap iron, leather, or caked clays. Rotor impactors consume less power than hammer crushers, but they are limited to nonabrasive solids that fracture easily by impaction. Cage mills are employed for medium-scale operations with rather uniform feeds and where relatively large reduction ratios are required.

Power consumed by impact breakers varies considerably depending on the nature of the feed and the reduction ratio. Representative limits are shown in Table 4-5. In applications where either compression or impaction crushers will serve, the latter are generally chosen because capital costs are one half to one third those of gyratory or jaw crushers.

Roll Mills

(Perry 8-19; McCabe and Smith 833)

In their simplest form, roll mills consist of a horizontal rotating cylinder that crushes material either against a vertical wall or against a similar opposing cylinder.

Smooth rolls are limited in reduction ratio because oversized particles cannot be drawn into or "nipped" by the crusher. Hard and abrasive materials are better comminuted by gyratory or jaw crushers, but roll crushers are superior for processing soft, cohesive feeds.

Rolls can be modified in many ways for specific applications. For example, opposed rolls having fine serrations and rotating at different speeds are used extensively to produce flours from wheat and other grains. Here, the action is attrition rather than compression. Toothed rolls in various configurations can be used for some resilient materials. Tooth mills of special design are capable of processing more abrasive materials than smooth roll mills. Roll mills, in general, produce a continuous discharge of reasonably uniform product.

Rolling Compression Grinders

(Pan, Ring-Roll, Bowl: Perry 8-24: 8-33; McCabe and Smith 837)

In general, these machines contain a number of cylindrical rollers that crush solids against a flatter surface. This surface can be shaped like a flat pan, a bowl, or a ring,

with the rollers or mullers oriented accordingly. Generally, the pan, bowl, or ring moves against grinding wheels that are mounted on stationary axles.

Rolling compression machines are suitable for both coarse and fine grinding of medium-hard nonabrasive solids such as coal, cement clinker, limestone, clay, shale, cinder, and soft minerals. Because of their low maintenance, high efficiency, and large reduction ratios, they are often employed in operations that require economy and versatility such as pulverizing of coal for direct feed to a boiler. The ability to blend or knead a mixture of solids and grind simultaneously is an attractive asset for some applications.

Disk or Attrition Mills

(Dispersion, Colloid: Perry 8-41; McCabe and Smith 837)

Modern attrition mills are counterparts to the buhrstone mill, used for centuries to grind wheat and other grains. In the original, solids were ground by attrition

between the faces of stones. Grain was fed at the center of the top rotating stone, becoming pulverized to flour as it migrated to a discharge bin at the rim. Modern mills often have metal disks mounted on horizontal shafts. Serrations or channels in the disks provide abrasion suited for the particular application. Although their efficiency is relatively low, disk attrition mills are excellent for reducing tough or resilient particles such as leather, rubber, rags, seeds, and grains. They can be cooled or heated and serve, in some applications, for blending as well.

Disk mills are used for some processes that are not strictly comminution operations such as curling feathers and fluffing asbestos. Another commercially significant operation is dispersion of fine solids or liquids in a suspending liquid. This is accomplished in *colloid* or *dispersion mills*. Numerous food syrups, sauces, purees, pastes, and pulps are prepared in rotating-disk colloid mills. Paints, medicines, and other clinical products are processed by such mills to break down agglomerates or emulsify liquids.

Tumbling Mills

(Media, Ball, Pebble, Rod, Tube, Nonrotary Ball or Bead: Perry 8-25; McCabe and Smith 839)

For fine and extra-fine grinding of hard and abrasive powders, tumbling mills are superior. Also known as media mills, these devices commonly consist of rotating

horizontal cylinders that contain grinding media such as metal balls, rods, or pebbles. Centrifugal force causes the media to rotate near the top of the mill and then fall to the base, crushing solid grains against the wall or other balls, rods, or pebbles. These mills can be operated either wet or dry. Wet grinding, when permissible, is more efficient and yields a finer powder because grinding surfaces are kept free of cushioning powder layers. A *tube mill* is merely a longer *ball mill* that uses smaller balls and produces finer powder. Perforated partitions may be placed in a tube or ball mill, to provide zones where balls segregate according to size and permit a larger overall reduction ratio. One device employs a cone containing a distribution of ball sizes rather than a cylinder. This produces natural zones, since smaller balls migrate to the narrower end of the mill.

Rod mills are not efficient for fine grinding but are excellent for coarse and intermediate grinding. They are frequently employed upstream of a ball mill. *Pebble mills* are so named because they contain stones rather than metal balls. Some media mills do not rotate. Instead, balls or beads are stirred with a slowly moving armature or by rapid vibrations. These devices are most advantageous for fine and ultra-fine grinding where they have the ability to produce particles as small as 1 μm in diameter.

Media mills are relatively inexpensive to purchase and operate. Their efficiency is good, but because of the fine particles processed, specific power consumption is large. This is reflected by the inverse particle diameter dependence shown in Table 4-5.

Hammer Mill

(Pin: Perry 8-35; McCabe and Smith 836)

Similar in many respects to the heavy-duty hammer impactor mentioned earlier, this grinder features pivoting hammers pinned to the periphery of a high speed revolving disk or cylinder. Clearance between fixed blades in the housing and moving hammers can be adjusted to provide the desired particle fineness. Grinding is caused by impaction and attrition in the highly sheared environment. As with impact crushers, high speed hammer mills cannot accept abrasive materials without excessive wear. As long as the Moh hardness is 1.5 or less, however, hammer mills are more versatile and efficient than other available types. The domestic garbage disposal grinder employed to grind food scraps is a well-known high speed hammer mill. Typical applications include grinding of sugar, carbon black, phermaceuticals, plastics, dyestuffs, pigments, and cosmetics. Some grinds are finer than 50 μm.

Another grinder that produces similar effects with a different mechanical configuration is the *pin mill*. Rather than hammers, this machine has one pin-studded disk that rotates facing another disk with similar pins, designed to produce large impaction and attrition forces. Uses and characteristics of pin mills are similar to those of hammer mills.

Fluid Energy or Jet Mills

(Perry 8-43; McCabe and Smith 843)

Rather than depend on contact with an external surface, fluid energy mills employ sonic gas jets to accelerate particles and create intense turbulence. Particles collide

at high enough velocities to abrade and fracture one another. Fluid energy grinders are mechanically simple and maintenance-free. Nonsticky feeds ranging up to 0.2 mm in diameter can be reduced to a fraction of a micrometer. Since steam or compressed gas is used rather than mechanical energy, power consumption is expressed in terms of steam or gas pressure and quantity. Some typical values for grinding various media are shown in Perry Table 8-30 (p. 8-43).

Cutters

(Perry 8-55; McCabe and Smith 845)

For soft and fibrous materials such as rubber, leather, paper, cloth, and wood, cutting is generally the most economical means of size reduction. The device most commonly employed for this is the rotary knife cutter. In concept, it is similar to a hammer mill except that hammers are replaced with rigid "flying" knives and the anvils with fixed or "bed" knives. Rotary cutters, with razor-sharp alloy blades, are capable of rapidly reducing the most tenacious of nonabrasive materials. Observing a rotary cutter consuming logs 30 to 50 cm in diameter in a paper mill is an awesome and somewhat fearsome experience. With appropriate mechanical design, cutters can be used to cut strands or rods into pellets or to dice sheets into cubes.

DRIVES AND POWER RECOVERY MACHINES

For many types of process equipment such as crushers, grinders, dryers, and kilns, motors or drives are standard components, included in the design package and integrated into the purchase price. In other types such as blowers and compressors, several alternative drives may be practical. In these cases, identifying the driving device and specifying its size and cost are separate steps for the designer.

Recovering power from pressurized gases or liquids is another situation that requires knowledge of motors, engines, and turbines. An experienced hiker will travel long distances along a ridge rather than go from peak to peak by the more direct path crossing a valley. This behavior, designed to conserve energy, has an analogy in fluid pumping. Once a gas or liquid is pressurized, you, as an engineer, should carefully evaluate any step that includes a drastic drop in pressure, avoiding it if possible. When a drop in pressure is necessary, recovery of the potential energy through an expander or turbine is often advisable. Equipment employed for this purpose is described in this section.

Selection of Drives

In most process applications, particularly those involving less than 100 kW of power, an electric motor is the drive of choice. Low capital cost, extremely low maintenance, and almost perfect reliability contribute to the attractiveness of this modern workhorse. Historically, many factories and mills were located near rivers and reservoirs to tap water power. Eventually, with the advent of modern central power stations, these machines were converted to use electricity. Even with recently escalating costs of electricity, these plants will undoubtedly continue to employ electric power rather than the seasonably variable and less convenient water flowing by.

In some situations, where fuel or high pressure steam is available from a process, it is tempting to use this energy source either to generate electricity internally or to drive other equipment. In general, this approach is practical only if the following conditions exist [12].

1 Fuel cannot be sold externally at a reasonable market price or cannot be used efficiently as a heat source or boiler fuel.

2 High pressure steam is in excess of high temperature heating needs, and low pressure steam can be used for low temperature duty somewhere in the process.

Reasons for these rather limited conditions stem from the high efficiency of central station power plants. They, of course, like any other process for converting heat to work, are restricted by the Carnot limit. With temperature–pressure constraints on modern turbines and boilers, a large power plant is limited to an overall efficiency of about 40 percent, where the balance of the energy is exhausted as low grade heat to the environment. On the other hand, smaller power generating systems are even more severely restricted and operate at lower efficiencies. Thus, even considering transmission losses, purchased electricity will be as cheap as self-generated electricity if internal fuels and energy sources can be sold or employed at market value elsewhere. Considering convenience and the absence of a generator and its maintenance, purchased electricity is usually an automatic choice.

If fuels or other sources of energy cannot be employed at market value (e.g., because of remoteness), self-generation of power for internal use or export is attractive. Internal conversion is also attractive in case 2, where high pressure steam is abundant and low pressure steam is needed. Here, low pressure steam still contains most of the original energy and is suitable for process heating at moderate temperatures. Thus, low grade energy normally exhausted to the environment in a central power station is employed productively, and premium excess energy of the high pressure steam is converted to bonus electricity.

An extension of this situation exists when fuel is available on site and low pressure steam is needed in the process. In the past, the fuel normally was used to generate process steam in a low pressure boiler. Because of current and projected energy prices, however, cogeneration has become attractive. In this scheme, the fuel is used to generate steam at higher pressure, which is then used to drive compressors, pumps, and other large equipment, or to generate electricity. Exhaust steam is subsequently employed for process use. In prospective new plants having single drives rated at 1000 kW or larger and a matching need for low pressure steam, this choice is economically attractive.

Many steam pressures are possible. In most new plants, however, the high pressure steam is at 45 bara pressure, superheated to 400°C, and the low pressure steam is saturated at 4.5 bara and 150°C.

Machines to drive process equipment or recover power from a potential energy source can be identified and evaluated with the aid of information contained in Table 4-6. A more detailed discussion of each is found in paragraphs that follow. Because of improvements in other machines, those artifacts of the Industrial Revolution, reciprocating steam engines, have become obsolete and are not included in this discussion.

Electric Motors

(Perry 24-3)

The availability factor shown in Table 4-6 indicates why electric motors are used almost exclusively where the shaft power is below 100 kW. This asset plus high

efficiency explains why they predominate in most other situations. Electric motors having power outputs up to 10,000 kW are commercially available. This is near the

TABLE 4-6
CRITERIA AND DATA FOR THE PRELIMINARY SELECTION OF DRIVES AND POWER RECOVERY MACHINES

| | Drives | | | | | Power Recovery Machines | | |
| | Electric Motors | Internal Combustion Engines | Steam Turbines (noncondensing) | Combustion Gas Turbines | Air Expanders | Gas Expansion Turbines | | Liquid Radial Expanders |
						Axial	Radial	
Maximum capacity, P (kW)	10,000	15,000	15,000	15,000	—	5000	1000	1000
Compatibility								
Outdoor environments	C	A	A	A	A	A	A	A
Corrosive or dirty atmospheres or fluids	C	B	D	D	A	D	B	C
Explosive atmospheres	C	C	B	X	A	B	B	A
Mobile use	E	A	E	A	E	E	E	E
Remote locations	D	A	E	A	D	E	E	E
Fuel flexibility		D	A	B	—	B	B	—
Normal feed temperature (°C)		25	400	750	25	<500	<550	25
Normal feed pressure (bara)		1.1	45	6	4	<175	<175	—[a]
Normal exhaust temperature (°C)		200	150	300		100	Various	25
Normal exhaust pressure (bara)		1.0	4.5	1.1	1.1	Various	Various	1.1
Turndown ratio	0.1	0.2	0.6	0.7	0.5	0.8	0.7	0.6
Compatible simultaneous processes							Cryogenics and refrigeration	
Maximum liquid in discharge (percent)			<20			<20	<20	
Energy available for process heat (percent of input)		30% at 75°C 15% at 175°C	90% at 150°C	20% at 200°C				
Efficiency (percent)	See Figure 4-2	See Figure 4-2	See Figure 4-2[b]	30–34		See Figure 4-2[b]	75–88	50–60
Availability (percent)	>99	>95	>95	>90	>98	>90	>98	>98

KEY
A   excellent or no limitations
B   modest limitations
C   special units available at higher cost to minimize problems
D   limited in this regard
E   severely limited in this regard
X   unacceptable

[a]Liquid expanders can tolerate any pressure that can be contained by a centrifugal pump.

[b]If condensate is present in the exhaust from gas expanders, efficiency is reduced. A corrected value is obtained by multiplying the efficiency taken from Figure 4-2 by the weight fraction of vapor in the exhaust system.

limiting size of most drives, electric or otherwise. As depicted in Figure 4-2, efficiencies range from 70 percent for fractional kilowatt motors to greater than 95 percent for the largest units. Motors having numerous shaft speeds, wattages, and mechanical designs are available. For the generalist, it is necessary to know only shaft power and the application. Shaft power determines the motor size. Electrical power consumed by a motor can be calculated simply by dividing the shaft power by the efficiency. Efficiencies shown in Figure 4-2 are representative of modern motors except that they should be decreased by about 5 or 10 percent if speed reduction or variation is necessary [22].

Mechanical designs are available in three major types. The open, dripproof configuration is standard and is employed for most indoor and some shielded outdoor applications. Weather-protected units are designed for outdoor use with their ventilating passages arranged to shield against wind-driven dirt and moisture. In corrosive or explosive environments, totally enclosed motors are employed. They can be cooled with water or air and are sometimes purged with inert gas or instrument air. If energy is available from the process, as in a downhill conveyor, a motor can be electronically reversed and used as a brake to regenerate rather than consume electricity.

Internal Combustion Engines

(Perry 24-13)

Although superb as portable power sources, the higher maintenance and capital costs of internal combustion engines make them inferior to electric motors for most

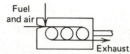

stationary process situations. For remote or mobile situations, these familiar engines can be specified directly according to required shaft power. Fuel consumption is determined from lower heating value and the appropriate efficiency curve in Figure 4-2. Choosing between a gasoline or diesel engine usually requires help from a specialist. The efficiency curve in Figure 4-2 applies to either, depending on which is optimum at a given power level. Some of the heat from engine cooling or in the exhaust gases is available for process use where desired. Quantities and temperatures are indicated in Table 4-6.

Steam Turbines

(Perry 24-16; Neerken [34])

When high pressure steam is abundant and low pressure steam is required, noncondensing turbines provide an attractive way to recover energy that otherwise would be degraded. As described in the introductory remarks, expansion to

subatmospheric pressures and subsequent condensation is accomplished more efficiently in central power stations. The turbogenerators employed there are massive units, sometimes hundreds of times larger than those that drive process equipment.

To the nonspecialist, a turbine—gas, steam, or otherwise—is essentially identical in appearance to an axial compressor (Figure 4-3). Steam turbines offer excellent reliability at a reasonable capital cost. They are somewhat more expensive

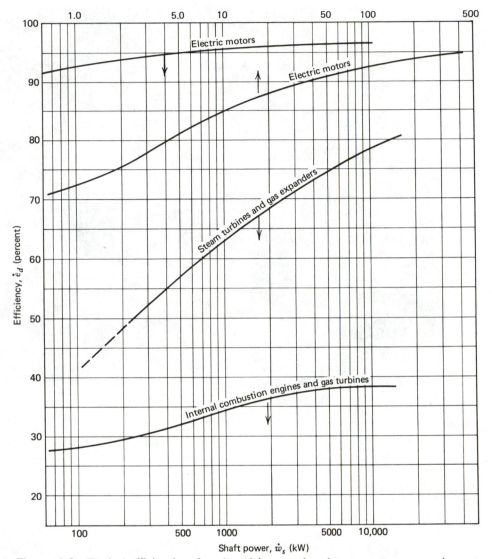

Figure 4-2 Typical efficiencies of modern drives employed to power process equipment. (Efficiencies of internal combustion engines are based on the lower heating value of the fuel, others are based on the theoretical performance of ideal machines.)

than electric motors but cheaper than internal combustion engines. Because of limited turndown ratio, turbines are seldom used in applications where broad variations in capacity or speed are encountered. They are superior, however, for driving centrifugal pumps, compressors, or generators where speeds and demands are relatively constant. Turbines are discussed in more detail in the section on power recovery machines.

Cost and design constraints limit steam turbines for service below 100 kW. As indicated in Figure 4-2, the efficiency also improves markedly with size. Thus, they are often used for high capacity duty. Gas compression in the manufacture of ammonia is a classic example. As mentioned previously, the large amount of low

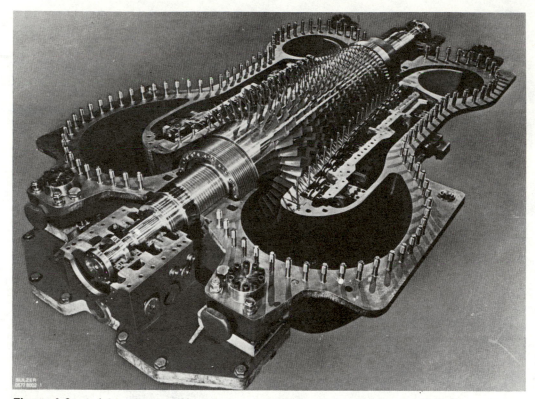

Figure 4-3   Axial compressor with rotor exposed. (Sulzer Bros., Inc., by permission.)

temperature heat available from noncondensing turbine exhaust is another incentive to employ one whenever conditions warrant.

Combustion Gas Turbines

(*Perry 24-28*)

Combustion gas turbines, like steam turbines, are employed for high speed, high capacity, and fixed-load service. In most process applications, gas turbines are less economical than steam units and are not chosen except for remote, mobile, or other nonconventional applications. Because of their high energy density and versatility, they find employment in aircraft (turboprop engines) and electric power plants (as load-matching generators to respond to rapid changes in consumer demand).

Part of the gross power from a turbine is consumed by a compressor, which is often mounted on the same shaft and is required to pressurize air for combustion. For preliminary design, one can use thermal efficiencies characteristic of internal combustion engines as shown in Figure 4-1.

Because of materials limitations, inlet gas temperatures are normally not allowed to exceed 750°C. Although electric motors and steam turbines are superior to combustion gas turbines in most conventional situations, current research is focused on developing turbine components to withstand higher inlet temperatures. If these efforts are successful, gas turbines will become attractive as topping devices to extract power from high temperature combustion gases. The exhaust, still relatively hot, will be employed to generate steam and power as is practiced now in conventional boilers and turbines.

Air Expanders

In some applications, particularly those requiring small high speed drives in hazardous or critical environments, air expansion motors are employed. (The modern dentist's drill is a rather miniature illustration.) Air expanders can be designed by the same techniques employed for other gas turbines or expanders. However, for preliminary purposes, it is much easier and quicker to designate and price an electric motor. This is recommended, since the difference in cost will be insignificant for decision-making purposes. The final selection of air or electric drive can be reserved for a specialist.

Power Recovery Machines

In principle, power recovery machines do not differ from the drivers just discussed. In fact, steam and gas turbines similar to those for drivers are employed as often for power recovery as they are for driving. These are the so-called axial turbines, which resemble axial compressors. They are used where flow rates, inlet temperatures, or energy drops are high.

In practice, another more versatile expander is employed for recovery of power from numerous high pressure process gases and liquids. It is the radial flow or turboexpander. Originally developed for efficient recovery of energy in cryogenic service, it is useful in any application where inlet temperatures do not exceed 550° C. Radial flow expanders can be visualized as turbocompressors (in the case of gases) or centrifugal pumps (for liquids) operating in reverse. In fact, tests have shown "that a good centrifugal pump generally makes a good hydraulic turbine" (Perry 24-38) [7].

In an interesting commercial operation, one device functions as both pump and expander. To schedule the capacity of electrical power plants more effectively, hydroelectric storage basins are sometimes used. During nighttime hours when demand is low, electricity from the regional power grid energizes huge electric motors that in turn drive centrifugal pumps to move water from a lake or river at low elevation to a large elevated pond. During daytime hours, the water flows in reverse, passing backward through the pump and driving the motor as a generator to produce electricity for the grid.

In the past, when power was plentiful and cheap, expanders were not often employed because a simple, inexpensive letdown throttling valve can be used to drop the pressure of a process stream. (The writer in Perry 24-37 points out, perceptively, that there is no similar cheap, foolproof alternative for the reverse process, i.e., pumping or compressing a fluid.) In energy-intensive applications such as refrigeration and gas liquefaction, expanders have been employed extensively, and they will be even more widely used in other process applications as energy prices rise.

Design procedures for expanders are the same as those outlined below for turbines. In detailed process design, much attention is devoted to mating the expander with a driven unit. This requires rather complex matching of loads and duties. For our purposes, it is important only that the expander provide a continuous steady source of power. In these rapid predesign evaluations, unless the driver–driven combination is obvious, it is appropriate to design the expander to drive an electrical generator. This will reflect a reasonable credit for power recovery in the economics. More refined decisions can come later.

Expanders, like centrifugal pumps, do not operate efficiently at reduced capacity. In fact, they do not operate at all below 50 percent of design flow and pressure. Thus, the limiting turndown is 60 to 80 percent of design. Radial flow gas expanders are capable of high efficiencies (75 to 80 percent). Liquid units are slightly less efficient (50 to 60 percent) than pumps. As with axial turbines, gas expanders cannot tolerate more than 20 percent condensate in the exhaust; when condensate is present to that extent, efficiencies are reduced (see design discussion below). Liquid expanders, unless so designed, are hampered if gas evolution of cavitation occurs.

For the same reasons that compressors consume more power than pumps, power recovery from process liquids is generally not as attractive as that from gases. To be practical, the power available should exceed 100 kW. The potential can be estimated quickly from Figure 24-41 in Perry (p. 24–37) or simply from

$$P = \frac{-\epsilon_o \dot{m} \Delta p}{\rho} \tag{4-5}$$

Design of Turbines and Expanders

The design of gas expansion equipment uses the same theoretical principles as the design of compressors. Either the simplified mechanical energy equation from fluid mechanics or an overall energy balance can be applied. Disregarding potential and kinetic energy effects, the steady-flow energy balance is

$$\dot{m}(h_1 - h_2) + \dot{Q} - \dot{w}_s = 0 \tag{4-6}$$

where h_2 is the specific enthalpy of the leaving steam and h_1 that of the entering gas. The adiabatic assumption is even more legitimate here than it is for compressors, which are often intentionally cooled. Thus, Equation 4-6 simplifies to a familiar form:

$$\dot{w}_s = \dot{m}(h_1 - h_2) \tag{4-7}$$

where *delivered* shaft power is positive as defined in Equation 4-6. The usual impediment to applying Equation 4-7 is an unknown value for h_2, the effluent enthalpy. The usual method of attack is to evaluate ideal power produced by a reversible adiabatic or isentropic expansion:

$$\dot{w}_1 = \dot{m}(h_1 - h_{2,s}) \tag{4-8}$$

where $h_{2,s}$ is enthalpy at the outlet pressure and the inlet entropy. Since

$$\dot{w}_s = \dot{w}_i \epsilon_i \tag{4-9}$$

then

$$\dot{w}_s = \dot{m} \epsilon_i (h_1 - h_{2,s}) = \dot{m} (h_1 - h_2) \tag{4-10}$$

and actual exit enthalpy can be determined from the ideal enthalpy change and the efficiency. Shaft power follows directly from Equation 4-10. This is especially convenient if an enthalpy–entropy (Mollier) diagram is available. The following example illustrates this procedure.

ILLUSTRATION 4-2   POWER PRODUCTION FROM A STEAM TURBINE

A paper mill has 0.5 kg/s of steam available at 45 bara and 400° C, which it currently throttles to 4.5 bara for use in a paper dryer. If this steam were passed through a turbine first, how much power could be recovered?

From a Mollier diagram or the steam charts, we find that the inlet enthalpy and entropy are 3202 kJ/kg and 6.71 kJ/kg · K. The enthalpy at 4.5 bara and the same entropy is 2680 kJ/kg. Thus, ideal isentropic work is

$$\dot{w}_i = (0.5 \text{ kg/s}) [(3202 - 2680) \text{ kJ/kg}] = 261 \text{ kW}$$

From Figure 4-2, the efficiency is estimated to be 42 percent. Shaft power is consequently 110 kW. The actual exit enthalpy can now be calculated from a rearranged form of Equation 4-10.

$$h_2 = h_1 - \epsilon_i (h_1 - h_{2,s})$$

$$= 3202 - 0.42(3202 - 2680)$$

$$= 2983 \text{ kJ/kg}$$

With true enthalpy and pressure known, the actual exit conditions are fixed, that is, $T = 263°$ C and $S = 7.37$.

An alternate analysis can be conducted using the mechanical energy balance, Equation 4-1. To calculate ideal work, only the pressure, density, and mass flow rate are involved.

$$\dot{w}_i = -\dot{m} \int_{p_1}^{p_2} \frac{dp}{\rho} \tag{4-11}$$

An efficiency factor compensates for irreversibilities within the expander to give:

$$\dot{w}_s = -\epsilon_i \dot{m} \int_{p_1}^{p_2} \frac{dp}{\rho} \tag{4-12}$$

where $\dot{w}$, is power delivered by the expander shaft. For isentropic expansion of an ideal gas, pressure and density are related by $p\rho^{-k} = \text{const}$, where k is the specific heat ratio. Substitution of this into Equation 4-12 and integration yields a result for this idealized case.

$$\dot{w}_s = \frac{\epsilon_i \dot{m} R T_1 k}{k - 1} \left[1 - \left(\frac{p_2}{p_1} \right)^{(k-1)/k} \right] \tag{4-13}$$

(The ideal gas equation, $\rho = pM/RT$, was also employed to eliminate density.) The relation between pressure and temperature, in this instance, is represented by a familiar equation from thermodynamics.

$$\frac{T_2}{T_1} = \left(\frac{p_2}{p_1} \right)^{(k-1)/k} \tag{4-14}$$

For nonisentropic (polytropic) expansion of a real gas, a different exponent n, the polytropic "constant," is employed. With this substitution plus the equation of state for a real gas, $\rho = pM/zRT$, Equation 4-12 becomes:

$$\dot{w}_s = -\frac{\epsilon_i \dot{m} z_1 RT_1}{p_1 M} \int_{p_1}^{p_2} \left(\frac{p_1}{p}\right)^{1/n} dp = \frac{\epsilon_i \dot{m} z_1 RT_1 \bar{n}}{\bar{n}-1}\left[1 - \left(\frac{p_2}{p_1}\right)^{(\bar{n}-1)/\bar{n}}\right] \quad (4\text{-}15)$$

The polytropic constant depends on both efficiency and the extent of departure from gas ideality. In some situations, n can be determined from past experience. For nonpolar gases, compressibility data have been used to correlate n with critical constants. In polytropic expansion, the final temperature is related to the pressure ratio by

$$T_2 = T_1 \left(\frac{p_2}{p_1}\right)^{\bar{m}} \quad (4\text{-}16)$$

where $\bar{m}$ is the average of another "constant" dependent on expander efficiency, specific heat, and compressibility of the gas. Values of n and m can be calculated from Equations 4-17 and 4-18:

$$n = [Y - m(1 - X)]^{-1} \quad (4\text{-}17)$$

and

$$m = \frac{zR}{C_p}(\epsilon_i + X) \quad (4\text{-}18)$$

where X and Y are presented in Figures 24-39 and 24-40 in Perry (pp. 24–35 and 24–36).

ILLUSTRATION 4-3   POWER PRODUCTION FROM A STEAM TURBINE (ALTERNATE METHOD)

Calculate power generated by the steam turbine in Illustration 4-2, but use Equation 4-15.
Known data are as follows.

$$T_c = 647 \text{ K}, \; p_c = 221 \text{ bara}, \; \dot{m} = 0.5 \text{ kg/s}, \; \epsilon_i = 0.42, \; C_p = 2.0 \text{ kJ/kg} \cdot \text{K}$$

	Inlet	Outlet		Inlet	Outlet
p (bara)	45	4.5	X	0.2	~0.02
T (K)	673	?	Y	1.06	~1.01
p_r	0.203	0.02	n	1.05	1.09
T_r	1.04	?	m	0.136	0.097
z	0.95	1.0			

Outlet temperature is unknown, but we are in the region of the compressibility charts where it will make little difference in defining key parameters. Thus, the other properties are evaluated and estimated as shown.

The shaft power computed from Equation 4-15 is:

$$\dot{w}_s = \frac{(0.42)(0.5 \text{ kg/s})(0.95)(0.0832 \text{ m}^2 \cdot \text{bar/mol} \cdot \text{K})(673 \text{ K})(1.07)[1 - (4.5/45)^{0.07/1.07}]}{(18 \text{ kg/mol})(0.07)(10^{-5} \text{ bar} \cdot \text{m}^3/\text{J})}$$

$$= 133 \text{ kW}$$

Using an average value for m of 0.116, the outlet temperature is:

$$T_2 = T_1(0.1)^{0.116} = 673(0.765) = 515 \text{ K} = 242° \text{C}$$

Agreement with the preceding more rigorous analysis is quite good considering that water vapor, being a polar gas, is not precisely represented by the generalized compressibility charts. (This technique involves trial and error if outlet temperature is guessed incorrectly or if the efficiency, found from Figure 4-2, is not assumed correctly in the first calculation.)

One modification to the efficiency is required if condensation occurs. As a rule of thumb, efficiency, with condensate present, is equal to the condensate-free value multiplied by vapor fraction. Thus, if 10 percent of the steam had condensed in the steam turbine, of the preceding example, efficiency would be 0.42(0.90) = 0.37.

Machines powered by high pressure liquids cannot strictly be called expanders because for all practical purposes, the specific volume does not change appreciably. Power generated in a liquid power recovery turbine can be simply estimated from Equation 4-2 integrated with constant density.

$$\dot{w}_s = \frac{\epsilon_i \dot{m}(p_1 - p_2)}{\rho} \tag{4-19}$$

ILLUSTRATION 4-4 POWER PRODUCTION FROM A LIQUID EXPANDER

Calculate the power generated in a liquid expander accepting 0.5 kg/s of water at an inlet pressure of 45 bara and 25° C, discharging it at 4.5 bara.

Assuming an efficiency of 55 percent, the shaft power, from Equation 4-19, is:

$$\dot{w}_s = \frac{(0.55)(0.5 \text{ kg/s})(40.5 \text{ bar})(1 \times 10^5 \text{ J/bar} \cdot \text{m}^3)}{1 \times 10^3 \text{ kg/m}^3}$$

$$= 1.1 \text{ kW}$$

Exit temperature can be estimated from the energy balance on an adiabatic system:

$$\dot{m}(h_1 - h_2) = \dot{w}_s = \dot{m} C_p(T_1 - T_2)$$

or

$$T_1 - T_2 = \frac{1.1 \text{ kW}}{(0.5 \text{ kg/s})(4.19 \text{ kJ/kg} \cdot \text{K})} = 0.5 \text{ K}$$

a temperature drop of 0.5° C. This example shows that the energy stored in a compressed liquid is many times smaller than that stored in the same mass of gas at the same pressure. The safety implications of that principle explain why new pressure vessels are tested hydrostatically, with compressed water, rather than with compressed air.

Liquid expanders, where a substantial fraction of the liquid vaporizes during expansion, require expert analysis and should be evaluated by a specialist.

EVAPORATORS AND VAPORIZERS

As denoted by their names, evaporators and vaporizers transform liquids to vapors by application of heat. Thus, they are closely related to heat exchangers and some types of process vessels.

In conventional process technology, evaporators are considered to be liquid dryers or concentrators, devised to separate solvents from solutes by evaporation.

TABLE 4-7
CRITERIA AND DATA FOR THE RAPID DESIGN AND SELECTION OF VAPORIZERS AND EVAPORATORS

	Type of Vaporizer or Evaporator									
	Vaporizers		Evaporators							
			Short-Tube (basket, calandria)	Long-Tube		Falling-Film		Force-Circulation		Agitated-Film (scraped-wall)
	Jacketed Vessel or "Pot"	Vessel with Submerged Coil or Bayonet		Once-Through	Circulating	Once-Through	Circulating	Vertical	Horizontal	
Maximum Vessel or Reservoir Size										
Diameter, D (m)	4	4	4	4	4	4	4	4	4	—
Height, L (m)	16	16	12	8	8	4	4	8	8	—
Maximum heating surface, A (m²)	$3D^{0.33}L^{0.67}$	$4D^{1.33}L^{0.67}$	30–300	100–10,000	100–10,000	30–300	30–300	20–2000	20–2000	2–20
Velocity through tubes, (m/s)	—	—	0.3–1	1–3	1–3			2–6	2–6	
Maximum tolerable viscosity (Pa · s)	0.01	0.01	0.01	1.0	1.0	1.0	1.0	2	2	100
Compatibility										
Low viscosity liquids	A	A	A	A	A	A	A	A	A	X
High viscosity liquids	D	D	D	B	B	B	B	A	A	A
Slurries	X	X	D	B	B	D	D	A	A	A
Scaling or salting liquids	E	E	C	D	E	D	E	B	B	A
Corrosive liquids	C, A	C, A	C, E	C, A	C, A	C, A	C, B	C, B	C, B	C, D
Crystal-forming liquids	E	E	D	B	B	E	E	B	B	A
Foaming liquids	B	B	D	B	B	E	E	B	B	D
Heat-sensitive liquids	A	A	D	A	D	A	D	B	B	B
Sticky or gummy liquids	D	D	X	X	X	X	X	E	E	B
Suitability										
High capacity	D	B	D	A	A	B	B	A	A	E
Multiple-effect use	E	E	A	A	B	A	B	A	A	E
Process feed vaporization	A	A	B	B	D	D	D	D	D	D
Restricted vertical space	A	A	A	E	E	E	E	D	A	D
Small ΔT's	E	B	D	D	D	A	A	D	D	A
Other Criteria										
Purchase cost	B	A	A	A	A	B	B	B	B	D
Power consumption	A	A	A	A	A	A	A	B	B	(1–100 kW/m²)
Heat transfer efficiency	B	B	D	B	B	B	B	A	A	A
Entrainment	A	B	B	B	B	B	B	B	B	A
Ease of cleaning	A	D	A	B	B	B	B	B	B	B
Typical heat transfer coefficients, U (J/m²·s·k)	100–500	100–500	100–2000	100–10,000	100–10,000	100–2000	100–2000			
Pressure drop through exchanger, ΔP (bar)	—	—	—	—	—	—	—	0.2–0.5	0.2–0.7	—
Typical fluids processed	Organic liquids for supplying vapors to process vessels and reactors		Sugar syrups	Kraft liquor	Evaporated milk, foods	Fruit juices	Foods			

KEY

A	excellent or no limitations	D	limited in this regard
B	modest limitations or problems	E	severely limited in this regard
C	special units available at higher cost to minimize problems	X	unacceptable

In almost all applications, the solvent of water. In most instances, the solute is the more valuable product. Water desalination, however, is a prominent exception. Evaporation, as a chemical engineering unit operation, is distinct from distillation in that the solute is nonvolatile, and complete separation can be accomplished in one stage.

Vaporizers, although resembling evaporators physically, are employed to transform a pure liquid or a mixture of volatile liquids to a vapor without causing separation. They are employed where it is necessary to convert a liquid feed to a vapor for subsequent transport or processing. Vaporization of LNG for introduction into a pipeline is one example.

In essence, an evaporator or vaporizer is merely a heat exchanger attached to a vessel of one sort or another. A distillation reboiler, by this definition, is certainly also a vaporizer. A distillation column is the process vessel in this case. Although such reboilers could be included in this section, in keeping with other chemical engineering literature, they are treated in the discussion of heat exchangers.

Dilution of a process stream, like the descent from a mountain peak or the expansion of a gas, is accompanied by an increase in entropy. Reversing the process by evaporation requires an inordinate quantity of energy. In conceiving a process, one should avoid dilution or minimize it whenever possible. If a stream must be concentrated, techniques such as precipitation and filtration or reverse osmosis, which do not require a heat cycle, should be considered. If there is no practical alternative to evaporation, one of the conventional designs listed in Table 4-7 should be adequate. In some situations, particularly in the food and detergent industry, essentially all the solvent is removed. A spray dryer (such as that used to produce powdered milk) often serves as the finishing stage in these cases. Spray dryers are described in the section on gas–solid contactors.

Since vaporization is energy intensive, efficiency is a major consideration. The most effective and successful means of improving efficiency is to reuse the vapors that leave one unit for further service elsewhere. In practice, individual evaporators are staged or joined together such that each unit operates at a lower pressure than the preceding one. Hence, vapor from one stage can be employed as a source of heat to evaporate liquid in the next. The first stage must, of course, be supplied with independent energy such as process steam. In most cases, vapor leaving the last stage is condensed by cooling water. This allows operation at subatmospheric pressure, permitting a broader temperature range and recovery of energy from low grade vapor.

Vaporizers

The conventional feed vaporizer is merely a cylindrical process vessel having a heating surface in contact with the boiling liquid. In relatively clean service with modest heating rates, a steam jacket or hollow wall around the lower portion of the vessel is adequate. If more heat is required than can be conveniently transferred through the vessel wall, vaporizers are designed with internal steam coils or bayonet heaters rather than jackets. Costs of jacketed and coil vaporizers are comparable. Heat exchangers in the latter, although difficult to clean in place, can be removed for easier access. Recirculation and volumetric heat release are less intense in jacketed units, making them less likely to experience problems with foaming and entrainment.

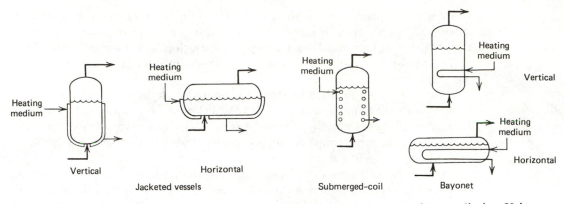

Vaporizer vessels are designed on the basis of entrainment limits. Using an approach similar to that employed to determine the diameter of a distillation tower, an equation such as the well-known Souders-Brown expression is used:

$$u = K_{SB} \left(\frac{\sigma}{0.020} \right)^{0.2} \left(\frac{\rho_l - \rho_g}{\rho_g} \right)^{0.5} \tag{4-20}$$

where u is allowable gas velocity (m/s) and σ is the surface tension (N/m). The reference value for σ of 0.020 is near the minimum for a majority of conventional liquid–vapor mixtures. Maximum values seldom exceed 0.10. With its fractional exponent, the surface tension correction rarely exceeds 1.4 and it is not sensitive to temperature. This correction is insignificant for most vaporizers, and Equation 4-20 can be rewritten as follows,

$$u = 0.06 \left(\frac{\rho_l - \rho_g}{\rho_g} \right)^{0.5} \tag{4-21}$$

where 0.06 m/s is the recommended value for K_{SB}. For water vapor at atmospheric pressure, this yields a superficial vapor velocity of 2.4 m/s, which will not entrain droplets larger than 100 μm in diameter. It is common practice to install across the top of a vaporizer a fiber or mesh pad that with less than 2 kPa pressure drop, will remove most particles larger than 2 μm in diameter. The steps in preliminary vaporizer design should be quite obvious.

Step 1   Determine superficial vapor velocity from Equation 4-20 or 4-21.

Step 2   Calculate the cross-sectional area and diameter of a vertical, cylindrical vaporizer using the velocity obtained in step 1 and the mass flow rate of vapor required. (If the diameter is larger than 4 m, it will be necessary to employ a horizontal pressure vessel, multiple vertical vessels, or use special entrainment separation.)

Step 3   Calculate the heat duty of the vaporizer.

Step 4   Employing an overall heat transfer coefficient, taken from Table 4-7 or Tables 4-15, calculate the heat transfer area required in a jacketed vessel. (The temperature inside the vaporizer is known. Steam, the most likely heat source, is normally available saturated at standard pressures of 4.5, 9, and 17 bara or superheated to 400° C at 45 bara.

Step 5.   From the known heat transfer area, calculate the height of jacket required. [The area of the dished vessel base ($1.2\pi D^2/4$) is also available for heat transfer.]

Step 6. Allowing one additional diameter for vapor disengagement, determine the total height of the vessel. If calculated height is more than four times the diameter, an internal heating coil or bayonet will be preferable. Typical coefficients from Table 4-7 or Tables 4-15 can be used to estimate coil or bayonet size. The maximum heat transfer area of an internal coil is approximately

$$A = 4V^{2/3} \tag{4-22}$$

where V is vaporizer vessel volume. If more area is required, a bayonet or an external heat exchanger is recommended.

Evaporators and Vaporizers

Employed for many years to concentrate solutions and suspensions, evaporators have been constructed in numerous configurations to suit a variety of applications. The most prominent evaporator types are listed and characterized in Table 4-7. In the discussion that follows, each type of evaporator is first described qualitatively. Design methods common to all types are mentioned next, and multiple-stage or multiple-effect design is discussed last. (Excellent and more detailed treatments can be found in Perry 10-32, 11-27, McCabe and Smith 427, and Foust 494).

NATURAL CIRCULATION EVAPORATORS (Perry 11-28; McCabe and Smith 430; Foust 494)

Short-Tube, Calandria, Basket

One of the earliest types still in commmon use, the short-tube evaporator contains a heat exchanger, sometimes called the "basket," which is a vertical bundle

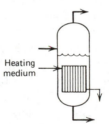

of relatively large (5 to 8 cm diameter) tubes, open at both ends. It is fully immersed in liquid near the bottom of a vertically oriented process vessel. A heating medium, usually steam, enters the shell of the bundle. Open space, consisting of either a large tube at the bundle center or an annulus between bundle and vessel wall, allows liquid to flow downward. Liquid vaporizes and flows upward inside the individual tubes at relatively high velocity, causing substantial internal circulation. This, coupled with the high heat transfer coefficient for nucleate boiling, produces respectable overall coefficients.

A short-tube evaporator is compact, relatively inexpensive (when constructed of conventional materials), and efficient. It performs well with conventional, rather low viscosity liquids. It still serves faithfully in the sugar refining industry. Designed to allow a person (invariably a man in the early days) to enter through a "manhole" and push a brush or rod through the tubes to clean them, this device was once very popular. With the development of long-tube and forced-circulation evaporators, which in many applications have higher heat transfer coefficients and require less manual cleaning, the popularity of short-tube units has declined. Calandrias can be equipped with propeller agitators to alleviate some problems. In general, however, they are not suitable for slurries or solutions that deposit a scale on heated surfaces. Highly viscous, foam-producing or heat-sensitive liquids are processed more efficiently in other evaporator types. In new processes, the short-tube unit would normally be specified only when head space is limited, capacity is small, or flexibility, including the potential for batch operation, is desired. Since the vessel itself is rather massive, short-tube evaporators can become exceptionally expensive if constructed from special alloys for corrosive service.

Long-Tube Vertical

To provide greater circulation rates, long-tube vertical evaporators were developed. They are composed basically of a conventional single-pass shell and tube

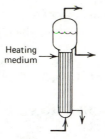

heat exchanger, oriented vertically and attached to a relatively small reservoir for vapor–liquid disengagement. Since exchangers are 6 to 12 m long, they cannot be installed where vertical space is restricted. On the other hand, with their much smaller reservoir size, they require less floor space than short-tube evaporators. Circulation at high velocities is caused by large buoyancy differences between the vertical liquid-filled pipe, which connects the reservoir to the base of the exchanger, and the heated liquid and vapor–liquid mixture inside the exchanger tubes. These high circulation rates not only create large heat transfer coefficients, but they permit operation with fairly viscous (up to 1 Pa·s) liquids, suspensions, and slurries. Impingement baffles are effective in processing liquids that are prone to foam. With short contact times, long-tube units are more suitable for heat-sensitive liquids than are calandrias. Scaling can create severe problems in long-tube units as well as short, and long-tube evaporators are somewhat more difficult to clean. Nevertheless, cleaning the insides of heat exchanger tubes is a standard common practice and is acceptable if it is not required too frequently. Since shell-side cleaning is more difficult, the process fluid, with rare exception, flows inside the tubes. The area exposed to process fluid is relatively small; thus long-tube evaporators are a natural choice for corrosive service.

Because of their high efficiency and the potential for substantial vaporization in a single passage through the exchanger, long-tube units can be designed for either once-through or recirculating operation. Single-pass units are generally less prone to foul and scale, but they are not as flexible. Thus, single-pass, long-tube evaporators often are assembled in multiple effects connected to a finishing unit of a different type, designed to handle the more difficult final liquid.

Because of hydrostatic pressure, liquid at the entrance to a long-tube heat exchanger is below its boiling point. Rising in the tube, it first becomes superheated, then flashes, as the pressure decreases rapidly approaching the discharge. Thus, the

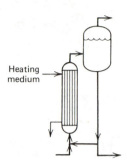

true ΔT varies with position along the tubes. The temperature difference employed for design, on the other hand, is constant, based on the boiling point of the liquid in the reservoir and the condensing temperature of vapor in the exchanger shell. The approximation is valid in most instances unless design temperature differences become less than about 6°C.

Because of compact size and high efficiency, the long-tube evaporator is the conventional choice for large-capacity service where scaling is not severe. A common application of single-pass designs is in the concentration of black liquor in paper pulp processing. Recirculating long-tube evaporators are used to produce evaporated milk.

Falling-Film

Almost identical in construction to the long-tube (sometimes called rising-film) evaporator just described, falling-film devices are designed so the liquid flows as a

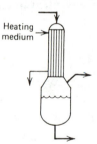

film, downward inside the tubes. This modification eliminates the problem of superheat. This permits operation when temperature differences are small but with a substantial sacrifice in capacity. Falling-film units are severely limited in their ability to concentrate slurries and viscous and scale-producing liquids, but they are superb for processing heat-sensitive fluids at low temperatures and with a small temperature differential. Fruit juices are commonly concentrated in this type. Falling-film units may be designed for once-through or circulating modes in either single- or multiple-effect service, but a pump is necessary in the circulating mode.

FORCED-CIRCULATION EVAPORATORS

Vertical and Horizontal

To alleviate problems with scale-forming and highly viscous fluids, a pump may be employed to create adequate circulation, keep exchanger tubes clean, and provide efficient heat transfer. Otherwise, vertical forced-circulation evaporators are very similar to long-tube units. Unlike to the latter, however, because of the need

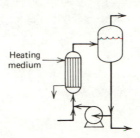

for recirculation, forced-circulation devices are seldom operated in the once-through mode. They experience the same problems with superheat and small temperature differentials attributed to long-tube natural circulation evaporators.

Since a pump eliminates the need for convective flow, forced-circulation units can operate with either vertical or horizontal heat exchangers. Vertical exchangers

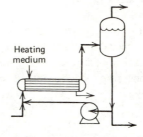

do require somewhat less pumping power and can, in a pinch, operate at reduced efficiency without a pump. However, they also have head space limitations and are somewhat more inconvenient to clean. All things considered, a horizontal design normally is specified for new construction where floor space is adequate. (Such detail, although interesting, is, I suppose, totally unnecessary for preliminary design.)

Agitated-Film (Scraped-Wall)

In concentrating viscous, sticky, and gummy liquids, where other types of evaporators cannot succeed, agitated-film devices are often employed. They are

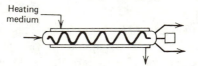

mechanically sophisticated, double-pipe exchangers having agitators and scrapers that rotate inside the core tube to keep its surface clean. Capacity is, of course, somewhat restricted, but an agitated-film unit can proces fluids having viscosities up to 100 Pa·s. If processing to complete dryness is desired, an engineer should consider spray or roll drying as described in the section on solid–gas contacting equipment.

Design of Single-Effect Evaporators

Evaporator design employs material and energy balances on process and heating fluids plus the traditional rate equation for heat transfer.

$$\dot{Q} = UA \ \Delta T \tag{4-23}$$

For economic analysis and preliminary design, the type of evaporator, its heat transfer surface, material of construction, and power consumption, if any, are the specifications necessary.

One precaution must be observed in detailed design. This pertains to noncondensable gases and is, perhaps, more significant to the operator than to the designer. Inert or noncondensing gases such as air entrained in vapor or leaking into systems that are below atmospheric pressure can almost totally block an evaporator. Such gases convert the steam-side coefficient from a high value typical of pure condensing vapors to a very much lower value controlled by diffusion through an inert gas barrier that forms around the tube surface. Normally, venting or evacuating noncondensable gases from the system is necessary to avoid this blockage.

A second precaution, important to both preliminary designer and specialist, stems from an elevation of solution boiling point caused by the solute. In contrast to many other liquid systems, concentrations are often high in evaporators, and boiling point elevation is frequently significant. This is especially true for highly ionized solutes such as inorganic salts, acids, and bases. For suspensions and dissolved organic compounds such as those found in sugar solutions, numerous food fluids, and Kraft liquor, boiling point elevation is negligible. In solutions exhibiting significant boiling point elevation, the heat of solution is also substantial and should be considered in the energy balance. Note also that steam generated from the liquor will be superheated by the boiling point elevation. Its true effective temperature in subsequent condensation is not the superheat temperature but, as is typical in condensers, that corresponding to saturation at the true pressure. This represents a loss in thermal potential that is especially significant in multiple-effect evaporators and is discussed further. A useful nomograph for determining boiling point elevation is found in Perry (Figure 11-18, p. 11-31).

The phenomenon of superheat in long-tube and forced-convection evaporators is different from boiling point elevation. Correction for this is built into heat transfer coefficients so that the conventional definition of ΔT applies for these units in Equation 4-23.

With these precautions is mind, you should be able to formulate the appropriate balance equations and rate expressions. If help or comfort is needed, the approach is outlined with considerable care in McCabe and Smith (p. 441). Specialists calculate heat transfer coefficients from correlations available in the literature. In practice, however, the range of coefficients is rather narrow. An optimum evaporator for a given application generally is one that has a coefficient greater than $500 \, J/m^2 \cdot s \cdot K$. For preliminary design, coefficients can be selected from Figure 4-4 or 4-5. These are valid only for water-based systems. Power consumption in forced-circulation and agitated-film units can be assessed using information from Figure 4-5.

Design of Multiple-Effect Evaporators

Multiple-effect evaporation is a classical operation often mentioned in chemical engineering courses to illustrate economic optimization. To improve energy efficiency, single evaporators are connected in series, each operating at successively lower pressures (see Figure 4-6). Process steam is provided to the first effect. Water vapor from this effect is used in turn to heat the second, and so on through the last,

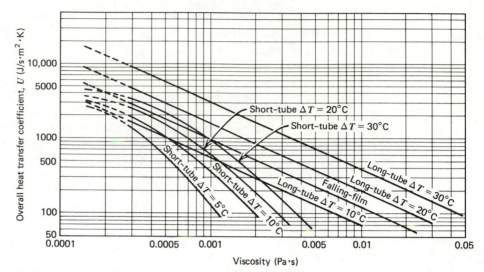

Figure 4-4   Overall heat transfer coefficients for preliminary design of natural convection evaporators (water-based systems).

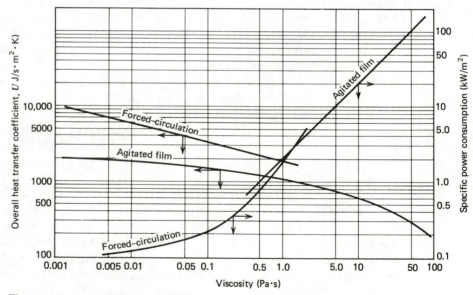

Figure 4-5   Overall heat transfer coefficients and specific power consumption for preliminary design of forced-circulation evaporators (water-based systems).

where vapor effluent is discharged to a condenser. Since the condenser is usually cooled by air or water, its temperature is near ambient and the steam pressure there (equal, in essence, to that of the last effect) is subatmospheric. Increasing the number of effects obviously increases the productivity of a kilogram of process steam. But, since the total available temperature driving force (that of process steam minus that of cooling medium) must be divided among individual effects, the vaporization capacity of each is reduced. Thus, in essence, a single evaporator has about the same vapor and liquid capacity as a group of similar units connected in series. This provides the classical situation for an optimum: decreasing operating costs opposed by increasing capital costs.

For tentative design of a multistage system, one decision must be made to assemble the flow diagram. This concerns liquid flow scheme. Forward flow is the easiest from an operations viewpoint. In this arrangement, shown in Figure 4-6, feed enters the first or high pressure effect, the same one heated by process steam. It can then flow from effect to effect driven by the natural pressure gradient. For the designer, there is one major disadvantage of this scheme: temperature in the last effect is lower. Consequently, for viscous products, the heat transfer coefficient may be much lower in the last effect than it would be if final evaporation were to occur in the first effect. Thus, it is common to employ a backward feed scheme where fresh liquid enters the last effect and is subsequently pumped, countercurrent with the vapor steam, from stage to stage. The first effect, in this scheme, yields final concentrated product. There are, of course, many possible alternate feed routes. Occasionally, one even finds parallel or cross flow, where the process fluid is divided into separate streams, each passing through only one evaporator but with the vapor flowing from stage to stage as before.

A rapid decision regarding feed flow can be made by assuming forward feed, tentatively estimating temperatures in the first and last effects, and determining heat transfer coefficients from Figure 4-4 or 4-5. If the coefficient in the first effect is greater than that in the last by no more than 50 percent, forward feed is acceptable. Otherwise mixed feed is preferable. You must use your own judgment to provide an arrangement having large coefficients and minimal pumping and instrumentation requirements. Since liquid residence time is greatest in the finishing stage, forward feed may be necessary anyway where exposure of the materials to high temperatures must be limited.

Selecting the number of effects, to be rigorously done, requires an optimization calculation similar to those described in Chapter Seven. If time is pressing, you may wish to employ guidelines from current practice. Based on temperatures and costs of steam and cooling water in a typical aqueous-based process, four to six units will normally be optimum where boiling point elevations are significant. Six to ten stages are typical otherwise. These are for medium- to large-scale continuous operations functioning twenty-four hours per day throughout most of the year. Otherwise the optimum number of stages will be lower. For reassurance while scribbling on the back of an envelope, you might recall, from detailed optimization work, that the cost curve is rather flat near the minimum. Thus, an error of one or two stages one way or the other should not harm the validity of your work.

Having established the number of stages, one can employ short-cut procedures described below to determine the stage size, intermediate flow rates, pressures, and temperatures. In an optimization calculation, this procedure is used for several cases involving different numbers of stages. As you will note, in many cases, equations can be expressed with the number of stages a variable that can be retained algebraically until the final solution. For guidance in a rigorous but tedious hand calculation, see McCabe and Smith (p. 448). Before proceeding with a rigorous analysis, however, short-cut calculations are recommended to provide a beginning point and a check on the final result. Ultimate design is invariably performed by specialists, usually with the aid of proprietary computer programs.

The following steps are recommended in short-cut design of multiple-effect evaporators. They are based on several assumptions common to industrial practice.

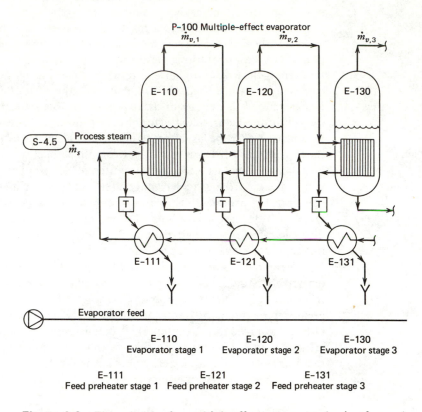

Figure 4-6   Flow sketch of a multiple-effect evaporator having forward feed.

Some of these are discussed at each step, but certain general assumptions are specified here. First, all evaporator stages are of identical design, having the same heat transfer area. Unless practical limitations dictate otherwise (e.g., an "impossible" final product, which must be finished in an agitated-film evaporation), this is a good assumption. Second, the number of stages has been assigned arbitrarily or according to the foregoing discussion. Third, terminal conditions and flow rates are fixed. (Process stream conditions normally appear on the flow diagram.) Finally, it is assumed that hot product and condensate liquids are employed to preheat the feed so that it enters near its saturation temperature.

Step 1.   Identify evaporator type. Selection criteria in Table 4-7 will be useful in this step.

Step 2.   Estimate the vapor generated in each effect. Considering sensible heat losses and the increase in latent heat with reduced pressure, 1 kg of vapor from a stage normally yields 0.80 to 0.90 kg of vapor in the one following. The quantity of vapor $\dot{m}_v$ removed from the first effect is, thus

$$\dot{m}_v = \dot{m}_s \, y \tag{4-24}$$

where $\dot{m}_s$ is steam rate and y is vapor yield. The vapor from the second effect is

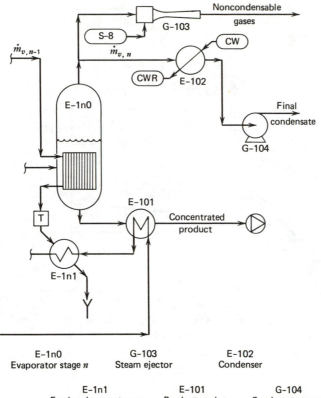

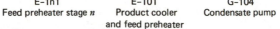

| E–1n0 | G–103 | E–102 |
| Evaporator stage n | Steam ejector | Condenser |

| E–1n1 | E–101 | G–104 |
| Feed preheater stage n | Product cooler and feed preheater | Condensate pump |

Figure 4-6 (Continued)

$$\dot{m}_{v,2} = \dot{m}_{v,1} \, y = \dot{m}_s \, y^2 \tag{4-25}$$

or the vapor from any effect is

$$\dot{m}_{v,i} = \dot{m}_s \, y^i \tag{4-26}$$

where i is the stage number. If n represents total number of effects, the combined vapor removed from the flow stream in all effects is:

$$\dot{m}_{v,\Sigma} = \sum_{i=1}^{n} \dot{m}_{v,i} = \dot{m}_s \sum_{i=1}^{n} y^i \tag{4-27}$$

The value of the summation in Equation 4-27 is shown in Figure 4-7 for up to 10 stages and for yields between 0.65 and 1.0. Thus, from total vapor rate and yield, the process steam and individual vapor rates can be calculated from Equations 4-27 and 4-26. In the absence of boiling point elevation, y will be approximately 0.85. With boiling point elevation, it will be smaller. For a more precise number, an energy balance in a typical effect will suffice. (For purposes of algebraic manipulation, curves in Figure 4-7 can be represented by

$$\sum_{i=1}^{n} y^i = n \exp \left(y^{2/y} \right) \tag{4-28}$$

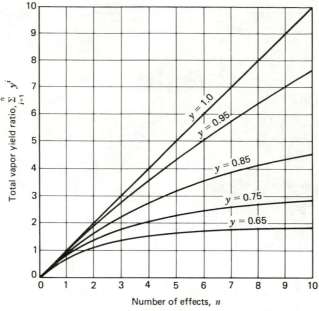

Figure 4-7   Total vapor yield from a multiple-effect evaporator as a function of number of stages and single-effect vapor yield.

with an accuracy better than 10 percent throughout the range of interest.)

Step 3   Calculate the concentrations of intermediate liquid streams. With vapor, feed, and product rates known, this step is a series of straightforward and simple material balance calculations.

Step 4   Estimate temperatures in the system. This step is not so straightforward, and the estimate must be adjusted later. For a first trial, one can take the difference between process steam and cooling water temperatures, subtract the boiling point elevations, summed over all effects, and divide the result by $n + 1$. This is equivalent to assuming that the decrease in vapor load with each effect is balanced by a decrease in heat transfer coefficient and an increase in latent heat. To amplify, consider the standard transfer equations:

$$\dot{Q}_2 = U_1 A \ \Delta T_1 = \dot{m}_s \lambda_s$$

$$\dot{Q}_2 = U_2 A \ \Delta T_2 = \dot{m}_{v,1} (\lambda_1 - S_1) \tag{4-29}$$

$$\dot{Q}_i = U_i A \ \Delta T_i = \dot{m}_{v,i} (\lambda_i - S_i)$$

where λ is latent heat, S is the net heat of solution per kilogram of water, and the other terms are known. (Note that S_i is not total heat of solution but the difference between those of liquids entering and leaving an effect.) Alternate forms of Equation 4-29 can be combined to give

$$\frac{\Delta T_1}{\Delta T_2} = \frac{\dot{m}_s}{\dot{m}_{v,1}} \frac{\lambda_s}{\lambda_1 - S_1} \frac{U_2}{U_1} = \frac{1}{y} \frac{\lambda_s}{\lambda_1 - S_1} \frac{U_2}{U_1} \tag{4-30}$$

where

$$\frac{\dot{m}_{v,1}}{\dot{m}_s} = \frac{\dot{m}_{v,i}}{\dot{m}_{v,i+1}} = y \qquad (4\text{-}31)$$

The term containing latent and solution heats in Equation 4-30 will usually be less than unity. The ratio of heat transfer coefficients will also be less than one, especially in feed-forward operation. These will balance the inverse yield term, making ΔT_1 approximately equal to ΔT_2.

To understand the influence of boiling point elevation, consider a single stage, say the next to last effect in Figure 4-6. We find the liquid temperature equal to its boiling point at the pressure of that effect. Assume, for example, that the liquor is a 30 weight percent sodium hydroxide solution and the pressure is 1.01 bara. Although the boiling point of water at that pressure is 100°C, the actual temperature, because of boiling point elevation, is found from the Perry nomograph (p. 11–31) to be 115°C. Although water vapor leaves this stage superheated at 115°C, essentially all its heat is available at 100°C, the saturation temperature. Hence, boiling point elevation, 15°C, is useless for heat transfer purposes. The ΔT's in Equation 4-29 are based on temperatures thus corrected. In calculating the average ΔT_i, coefficient and area (or their product) are the same in the condenser as in each evaporator.

Step 5 Obtain heat transfer coefficients. From the temperatures now defined, a heat transfer coefficient can be estimated for each stage from viscosity data and Figure 4-4 or 4-5. All coefficients will normally exceed 500 J/m^2·s·K. If not, selection of another evaporator type or rearrangement of the feed sequence should be considered.

Step 6 Repeat step 4 with the improved heat transfer coefficients. This is a modified trial-and-error procedure, controlled by Equation 4-30. In essence, since y and the heat terms are related, we have:

$$\frac{\Delta T_i}{\Delta T_{i+1}} \cong \frac{U_{i+1}}{U_i} \qquad (4\text{-}32)$$

This restriction, and the requirement that the sum of the ΔT's and boiling point elevations equal the difference between steam and cooling water temperatures, establish a unique temperature profile. Steps 5 and 6 are repeated until temperatures and coefficients remain fixed. This may require one or two iterations but seldom more.

Step 7 Calculate the area of an evaporator stage from Equation 4-29. This is readily done with data at hand.

Step 8 Determine power consumption. If pumps have been added for mixed-feed operation, they can be evaluated by conventional techniques described at the beginning of this chapter. If forced-circulation or scraped-surface evaporators are in use, power consumption can be calculated using known surface areas and data provided in Figure 4-5.

The foregoing sequence of steps will yield enough information to complete the flow sheet and obtain equipment costs.

As mentioned earlier, the optimum number of effects is always lower for liquids having significant boiling point elevation than for ideal solutions. A lower available temperature differential is one reason. Another is related to yield. An energy balance reveals that yield in a given effect is approximately equal to the latent heat minus heat of solution in that effect divided into the same quantity for the downstream effect:

$$y_i \cong \frac{(\lambda - S)_{i-1}}{(\lambda - S)_i}$$

(4-33)

If solutions are nonideal, having large negative heats of solution, $\lambda - S$ will increase substantially with i, and the yield will decrease. The impact of yield on total vapor produced is clearly revealed in Figure 4-7. If y is 0.85 or greater, substantial vaporization continues into the tenth effect. If y is 0.65, productivity becomes minimal beyond three or four effects. The detrimental effect of small values of y is compounded in multiple-effect evaporators. Thus, special care must be taken to conserve heat and prevent losses in multiple-effect systems.

FURNACES

Fired process equipment or furnaces is a category that includes boilers, heaters, incinerators, ovens, and stream generators. Such equipment is employed prominently in most chemical plants to provide heat conveniently, efficiently, and at the temperature level required. Occasionally, heat is applied by burning fuel directly in a process stream or vessel (direct-fired heaters). In these situations, combustion gases blend with the process stream and must be compatible with it. Design of direct-fired systems is a custom job. For preliminary purposes, an engineer can consider the direct-fired heater simply as a process vessel (having an integral burner) that provides 10 to 60 s residence time for the process and combustion fluids.

Indirect-fired furnaces, where heating media are separated from process streams, represent the most common configuration employed in chemical processing. Various types and their characteristics are listed in Table 4-8. Unless stated otherwise, the terms "boilers" and "steam generators" are used synonymously. Occasionally, liquids other than water are vaporized in boilers. Some boilers, for example, vaporize special heat transfer fluids. The most prominent of these is called a "Dowtherm" vaporizer.

Indirect-fired furnaces are often employed to vaporize or heat process streams directly. The reboiler in a crude oil fractionation column is heated, in most instances, by combustion. Some reactions, especially homogeneous hydrocarbon cracking reactions, occur inside metal tubes suspended in a furnace. Many viscous, high temperature liquids are preheated by passage through a furnace. Furnaces employed for these varied process functions are quite similar and are known generally as process heaters.

Incinerators are, as the name implies, furnaces designed to dispose of unwanted wastes. Currently, because of the value of energy, they are usually fitted with auxiliary steam or water coils to recover heat that otherwise would be wasted.

Ovens are enclosures, heated either by combustion or by electricity. Objects such as ceramics or metal bodies are heated, tempered, or sintered therein.

Furnaces are almost always purchased as packaged units from a vendor. If size permits, they are shop fabricated, shipped intact to a construction site, and erected

TABLE 4-8
CRITERIA AND DATA FOR THE PRELIMINARY DESIGN OF INDIRECT-FIRED FURNACES, BOILERS, AND OVENS

	Boilers			Thermal Fluid Systems					Process Heaters			
	Industrial		Utility	Hot Water	Diphenyls	Fused-Salt	Mineral Oil	Silicon Oil	Reactive	Nonreactive	Incinerators	Ovens
	Fire-Tube	Water-Tube										
Maximum steam rate, m_s (kg/s)	25	150	1000	—	—	—	—	—	—	—		
Maximum heating duty (kJ/s)[a]	60,000	360,000	2,600,000	20,000	20,000	20,000	20,000	20,000	150,000	150,000	700,000	15,000
Fuel efficiency (percent of LHV of fuel transmitted to utility fluid)	75–80	85–90	90–93	85–90	80–85	80–85	80–85	80–85	80–85	90–92	60–70	—[b]
Electric power efficiency (percent of LHV of fuel transformed to electric power)			35–40									
Utility Fluid	Steam	Steam	Steam	Water	Various organics[c]	Inorganic salts	Mineral oil	Silicon fluids	Process fluid	Process fluid	Steam	Air or combustion gases
Maximum temperature (°C)	210	400	540	200	400	590	320	315	600	600	375	2000
Maximum pressure (bara)	18	45	175	16	11	<1	<1	<1	200	200	45	1.01
Freezing temperature (°C)	0	0	0	0	10	150	−50	−40	—	—	0	—
Flammability	A	A	A	A	D	A	D	D	D	D	A	A
Toxicity	A	A	A	A	D	B	B	B	B	B	A	A
Fuel Type												
Nuclear	X	X	A	X	X	X	X	X	X	X	X	X
Coal (10–15% excess air)	E	B	A	B	D	D	D	D	D	D	X	X
Wood Bark (15–20% excess air)	E	B	A	B	D	D	D	D	D	D	A	X
Residual oil (10–15% excess air)	B	A	A	B	C	C	C	C	C	C	C	X
Light oil (5–10% excess air)	A	A	D	A	A	A	A	A	A	A	A	B
Gas (5–10% excess air)	A	A	D	A	A	A	A	A	A	A	A	A
By-product fuels (5–20% excess air)	A	A	D	A	B	B	B	B	B	B	A	B
Solid waste (30–50% excess air)	X	E	C	X	X	X	X	X	X	X	A	X
Shop-fabricated	A	A	E	A	A	A	A	A	A	A	A	A
Field-erected	X	A	A	B	D	D	D	D	B	B	A	A
Tube Construction Material												
Carbon steel	√	√	√	√	√	√	√	√			√	
Chromium–molybdenum steel		√				√			√	√		
Stainless steel									√	√		
Chromium–nickel alloys									√	√		

KEY

A   excellent or no limitations
B   modest limitations
C   special units available at higher cost to minimize problems
D   limited in this regard
E   severely limited in this regard
X   unacceptable

[a] This represents heat absorbed by the utility fluid except for ovens, where it denotes heat absorbed in the process plus losses.

[b] This depends on the application. Oven heat losses can be expressed approximately by $\dot{Q}\,(\text{J/s}) = 0.5\,V^{2/3}\,T^{5/3}$, where V is internal oven volume (m^3) and T is internal temperature (°C). Heat loss plus heat absorbed by the process represents 70 percent of the fuel lower heating value (LHV) for oil or gas firing and 100 percent of the power in an electric oven.

[c] Mixtures of diphenyl derivatives. These are known by various trade names, the most common being Dowtherm.

by the manufacturer. Very large furnaces and boilers must, of course, be field erected. Because of its specialized nature, the general designer merely characterizes a furnace by such macroscopic criteria as type, process fluid, pressure–temperature limits, and heat duty. This is adequate to determine equipment cost, fuel, and air rate. As you will recall from Table 4-3, furnaces are prominent auxiliary facilities and, even when part of a process module, they can often be purchased, like auxiliaries, as package units. If more detail is required, a furnace can be represented by a flow sheet and analyzed like any other processing operation. The steam generator employed in Chapter Three to illustrate flowchart formulation is an example of this. Such analysis, from the perspective of the "builder" rather than the "buyer" is usually unnecessary. For most purposes, basic specifications mentioned above are adequate for predesign purposes.

In this section, furnaces of various types and their application are described briefly. For more detailed information, books published by Babcock and Wilcox [1] and Combustion Engineering [44], two firms that design and fabricate furnaces and boilers, may be helpful.

Boilers and Steam Generators

(Perry 9-34, 9-40)

Steam, because of its low cost, cleanliness, noncorrosiveness, high energy content, and high heat transfer rate, is by far the most common medium for utility heating in process plants. To minimize corrosion, boiler feed water is normally demineralized and treated. Operating economy dictates, therefore, that condensate be recycled whenever possible. Seventy to ninety percent recycle is common in process boiler loops.

Industrial boilers are of two types. *Fire-tube* units are similar to shell and tube heat exchangers with combustion gases flowing through the tubes. The center tube

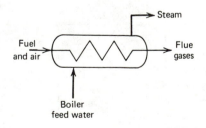

of the bundle, much larger than the rest, comprises the combustion chamber. Flow reverses at the end of the bundle and passes back through numerous smaller outer tubes. Efficient and compact, fire-tube boilers are always shop fabricated. Steam pressures are limited by the strength of the large cylindrical shell. These, of course, are less than could be contained in smaller tubes. Thus, fire-tube furnaces are employed primarily for generating modest amounts of low pressure saturated steam. Because of geometry, the combustion chamber and flue gas tubes are not compatible with continuous cleaning. This, plus a limited combustion residence time, restrict fire-tube boilers to fuels no dirtier or less convenient than residual oil. Additional data for preliminary specification and design can be found in Table 9-8.

Water-tube boilers contain steam within the tubes while combustion occurs in a boxlike open chamber. In large boilers, hundreds to thousands of tubes, usually 7

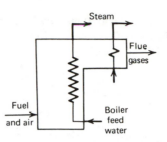

to 12 cm in diameter, are installed side by side, forming the walls of the combustion chamber and of baffles that control flow of and remove heat from combustion gases. In the combustion area, known as the radiant section, gas temperatures drop from about 2200 K to 1300 K. After combustion products have been thus cooled by radiation to wall tubes, they pass at high velocity through slots between more tubes suspended as large banks in the gas stream. This is known as the convection section. In the radiant section, such direct exposure to higher temperature gases would damage the tube metal. Gas entering the convection section at about 1300 K leaves near 600 K. Tubes in the radiant section are normally filled with circulating, boiling liquid to avoid hot spots. When superheating is desired, this occurs in the hot end of the convection system.

Since small tubes are capable of much higher pressures than is the large shell of a fire-tube boiler, elevated steam pressures as well as superheat are common in water-tube furnaces. Steam at 45 bara pressure superheated to 400°C is a typical maximum. Saturated process steam is also commonly generated at pressures of 17 and 33 bara in water-tube boilers. Pressures lower than this are impractical because of distribution piping costs. If lower pressure process steam is needed in substantial quantity (i.e., greater than 5 kg/s), it will probably prove practical to generate high pressure steam at 45 bara and 400°C, pass it through an expansion turbine to recover cheap power, and employ the exhaust for process needs. This is known as cogeneration. (See the section on drives and power recovery machines for more details.)

Because of the large, open combustion chambers, coal and wood fueling is common in water-tube furnaces. Fly ash and soot are cleaned from convection tubes by automatic "soot blowers," which direct high velocity steam or air jets against outer surfaces of tubes while the boiler is operating. Water-tube boilers can be shop fabricated with heating duties up to 100,000 kJ/s. Modern units burning coal and wood or residual oil are fitted with dust collectors for fly-ash removal.

Utility power boilers are designed for the sole purpose of generating electricity from nuclear or fossil fuels at high efficiency. To overcome, as far as possible, Carnot limitations, temperatures and pressures are the highest that reasonably can be tolerated by modern construction materials in the boiler itself and in the turbine. Condensate temperatures are controlled by the ambient cooling medium. To achieve high efficiencies and to dissipate copious quantities of low temperature heat, thermal–electrical power plants are located near oceans or large bodies of water where possible. Otherwise massive cooling towers are employed.

Electrical power stations are usually an order of magnitude larger than industrial boilers. As a consequence, they can burn lower quality fuels such as coal, which most now do. Modern plants not only include dust collectors but some employ scrubbers or other gas-cleaning devices as well to remove sulfur compounds. The effluent from such plants matches the cleanliness of a modern oil-fired boiler.

Because of size, nuclear fuels can be employed efficiently in large central power stations. Additional data pertaining to electric utility power boilers are contained in Table 4-8. More extensive information can be found in references 1 and 44, cited at the beginning of this section.

Thermal Fluid Heaters

(*Perry 9-4*)

If a process, application requires heat in the 100–250°C range, steam is the ideal working fluid. For reasons of safety or convenience, liquid water is often employed

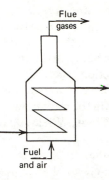

instead of steam. In a closed system, corrosion is low, and, with high velocity pumping, heat transfer coefficients are large. Hot water systems are simple and easy to control. They often are selected rather than steam when heating loads are light and temperatures are low (up to 200°C). Above 200°C, the pressure required to prevent vaporization, 16 bara, is too high to justify this system in preference to steam or others that are available. Above 250°C, steam pressures became excessive, and above the critical point, 374°C, there is no latent heat. Thus, it becomes necessary to consider other media for high temperature heat exchanger. In practice, several fluids are commonly used.

In the range from 250 to 400°C, one commonly used heat transfer fluid is Dowtherm A, a mixture of diphenyl and diphenyl oxide. This special medium was developed by Dow Chemical Company for high temperature transfer. Since it has a low vapor pressure, heaters and exchangers can be designed to contain the fluid safely. Dowtherm is readily vaporized as well and serves with the efficiency of steam in higher temperature service. At its maximum service temperature, 400°C, Dowtherm exerts a vapor pressure of only 11 bara. Decomposition prohibits its use at temperatures greater than this. Dowtherm, being noncorrosive and stable, is compatible with carbon steel and other conventional materials. Its major drawback is the need for absolute system sealing, since leakage produces economic loss and a potential health, fire, and explosion hazard. Leakage is more difficult to eliminate with this ultra-hot, low viscosity liquid than with conventional fluids. Data on Dowtherm and other established heat transfer liquids can be found in Perry (Table 9-31, p. 9-40, and Table 9-33, p. 9-42), and Singh [45].

Alternate heat transfer fluids include mineral oils, chlorinated diphenyls, silicon oils, fused salts, and liquid metals. Unlike Dowtherm, these liquids, except for the molten metals, do not exert appreciable vapor pressures. Thus, high pressure sealing is not as difficult, but the advantages of a condensing medium are also absent. All fluids except mercury and sodium–potassium (NaK) alloys can be contained in carbon steel equipment. The molten salt medium is limited by its freezing point, which prevents frequent or unattended shutting down of the system.

Heaters for thermal liquids are similar in design and cost to water-tube boilers. However, most of them are vertical cylinders rather than boxes and the tubing is often coiled in a helix around the furnace wall. Since these systems are seldom chosen when steam will serve, they are usually smaller than the boilers described above.

Process Heaters

(*Wimpress* [*53*])

Frequently, the need arises for process heat at temperatures above those available from the systems already described. In these situations and even where an

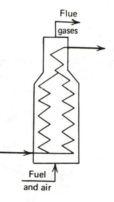

intermediate medium can be used, the process fluid itself is passed through tube coils in a fired furnace. The process system may be reactive, as with pyrolysis furnaces, which have been used extensively to thermally crack hydrocarbons for ethylene and propylene manufacture. The process stream may be nonreactive as well. Such is the case when a fired furnace is used as a reboiler in the distillation of heavy petroleum liquids.

Configurations and costs of process heaters are comparable to those for thermal liquid systems and steam boilers. Because of corrosion and danger due to leakage of process fluid into the combustion chamber, selection of construction material is somewhat more conservative. Thus, tubing from premium steels or chromium–nickel alloys is routinely specified for process heaters.

Incinerators

(*Perry 9-35*)

Before energy prices skyrocketed in the 1970s, waste materials were disposed of in the United States by incineration, with little incentive for heat recovery. Since then, most new incinerators are fitted with heat transfer coils to generate steam for useful purposes. Because the fuel is generally heterogeneous, bulky, and variable, incinerators require special feed mechanisms. With solid waste, reciprocating or rotary grates are employed to gently agitate the feed as it migrates through the

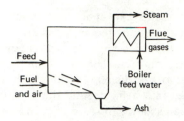

furnace. Liquid and gaseous wastes require similar custom injection devices depending on the nature of the waste and its tendency to cause slagging and fouling. Modern units usually require gas-cleaning equipment to maintain responsible emission limits. To assure complete combustion, more air is required than with traditional fuels. This causes the lower efficiency indicated in Table 4-8.

Ovens

(*Perry 20-25*)

Used more often for batch processing, ovens are enclosures constructed of firebrick and designed to maintain solid objects at high temperatures for extended periods.

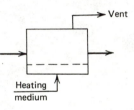

Fitted with conveyors or moving cars, they have been adapted for continuous processing. Heat is provided either by direct combustion of fuel in the enclosure or by radiation from electrical resistance elements. In fuel firing, approximately 10 to 20 percent of the fuel energy is transferred to the treated object. With electrical heating, efficiencies are 30 to 50 percent. This makes electrical ovens generally competitive with direct-fired ones, especially for small-scale operations.

Design

Selection of furnace types should be quite obvious based on process specifications, data provided in Table 4-8, and the foregoing discussion. As a general rule, steam generators or boilers are chosen above all others if they can perform the needed service. Once the type has been established and the fuel chosen, a furnace is characterized by its heating duty. This, divided by the efficiency, specifies the fuel energy required. Based on typical excess air levels and flow stream quantities, temperatures and compositions can be defined. Equipment cost, as discussed in Chapter Five, is dictated by heat duty and other known parameters. Please note that cost data for some categories of furnaces and boilers (particularly boilers) are in alternate locations in Chapter Five because of the dual role of such equipment as process and auxiliary facilities.

GAS MOVERS AND COMPRESSORS

Gas movers (including blowers, compressors, ejectors, exhausters, fans, vacuum pumps, and ventilators) are mechanically and theoretically analogous to liquid pumps. Primary differences stem from the much larger specific volumes and much lower viscosities of gases. As a consequence, gas movers are larger, have closer tolerances, operate at higher speeds, require more maintenance, consume more power, and are many times more expensive than liquid pumps operating at similar mass flow rates and differential pressures. Indeed, one often finds that pumps represent a minor fraction of purchase and operating costs in a process, whereas blowers or compressors represent a major portion of these expenses.

Reference to Table 4-2 yields the parameters required for specification of a gas mover. These, you will note, are the same as those denoted for pump analysis. Similarly, design of gas-moving equipment (excluding fluid jet ejectors, which are discussed later) is founded on the same basic mechanical energy balance, Equation 4-1, which was employed for pump design.

As with pumps, potential and kinetic energy changes through a gas mover are minimal, and fluid friction is included in an intrinsic efficiency factor. However, with gases, the density is a variable, and the equation must be applied in its integral form.

$$\frac{\epsilon_i \dot{w}_s}{\dot{m}} = \int_{p_1}^{p_2} \frac{dp}{\rho} \tag{4-34}$$

Substitution for ρ from the generalized real gas equation ($\rho = p/zRT$) yields:

$$\frac{\epsilon_i \dot{w}_s}{\dot{m}} = \int_{p_1}^{p_2} \frac{zRT \, dp}{p} \tag{4-35}$$

Fans

At this point, it is convenient to separate the analysis according to the type of gas mover. Fans operate, as a rule, near atmospheric pressure, and pressure differentials are generally less than 15 kPa. (Exhausters and ventilators are fans employed for the specific functions described by their respective names.) Thus, the variation in density or specific volume will be less than 10 percent. Within the limits of engineering accuracy then, Equation 4-34 can be reduced to the same form as that used to rate liquid pumps:

$$\dot{w}_s = \frac{\dot{m} \Delta p}{\bar{\rho} \, \epsilon_i} \tag{4-36}$$

where ρ is the average gas density.

Blowers

Blowers, which are more sophisticated and expensive than fans, operate with higher pressure differentials, from about 3 kPa to 5 bar. The distinction between blowers and compressors is not terribly precise because these names are used interchangeably for most machines that produce discharge pressures greater than 2 bara. To be more definitive, compressors generally have discharge pressures greater than 5 bara and require special provisions for cooling and containing the high pressure gas.

The theoretical analysis is similar for both blowers and compressors. Equation 4-34 still applies, but, as differential pressures increase, density can no longer be considered to be constant. For adiabatic, reversible compression of an ideal gas, *absolute* temperature and *absolute* pressure ratios are related by:

$$\frac{T}{T_1} = \left(\frac{p}{p_1} \right)^{(k-1)/k}$$

(4-37)

From thermodynamics, we remember that k is the ratio of specific heats, C_p/C_v. Using Equation 4-37 to eliminate T, Equation 4-35, can be integrated directly to yield:

$$\frac{\epsilon_i \dot{w}_s}{\dot{m}} = \frac{RT_1 \, k}{M(k-1)} \left[\left(\frac{p_2}{p_1} \right)^{(k-1)/k} - 1 \right]$$

(4-38)

where the compressibility z is unity for an ideal gas and k is assumed constant.

Equation 4-38 applies to isentropic (i.e., reversible, adiabatic) compression of an ideal gas. Corrections for friction in the machinery and turbulent losses in the gas are lumped within the intrinsic efficiency factor. In blowers, where the heat removed is small relative to the difference in gas enthalpies, the adiabatic assumption is legitimate. If the gas is ideal, Equation 4-38 may be used with confidence. If the gas is not ideal, a modified form is valid for the relatively low pressures encountered in blowers:

$$\frac{\epsilon_i \dot{w}_s}{\dot{m}} = \frac{\bar{z} R T_1 \, \bar{k}}{M(\bar{k}-1)} \left[\left(\frac{p_2}{p_1} \right)^{(\bar{k}-1)/\bar{k}} - 1 \right]$$

(4-39)

where $\bar{z}$ represents compressibility and $\bar{k}$ is the specific heat ratio, both averaged between inlet and outlet conditions of compressor or blower.

Compressors

(Pressure ratios less than 4:1)

Compressors, as opposed to blowers, operate in pressure ranges and with compression ratios that often require external cooling to prevent damage to sensitive seals and metal surfaces. This physical situation falls between the isentropic and isothermal extremes. Such is called polytropic compression where, instead of using k in Equation 4-38, one would use a polytropic constant n, smaller than k and greater than 1. For large compressors where gas–surface contact is limited, n approaches k. Evaluation of n in a specific application requires experience with the process or information from a specialist. For small units with extensive cooling surface or in many vacuum pumps, operation is nearly isothermal and n approaches unity. This limit can be evaluated by integrating Equation 4-35 with constant temperature but variable pressure.

$$\frac{\epsilon_i \dot{w}_s}{\dot{m}} = \bar{z} R T_1 \ln \frac{p_2}{p_1}$$

(4-40)

The work required in isentropic compression is always greater than that for the polytropic or isothermal. Thus, for a preliminary analysis of predesign accuracy, if the system is not isothermal and n is unknown, Equation 4-39 yields a conservative result.

Unfortunately, nonideal behavior is often synonymous with gas compression. It is clear that nonideality will be most extreme for compression of a vapor. In the vicinity of the saturation curve, especially if entropy data are available, it is more accurate to use the first law of thermodynamics directly rather than Equation 4-39. Applied to a steady-flow, adiabatic system, an energy balance reduces to:

$$\frac{\epsilon_i \dot{w}_s}{\dot{m}} = h_{2,s} - h_1 \tag{4-41}$$

Here, h_1 is fluid enthalpy at the inlet or "suction" port and $h_{2,s}$ is enthalpy at the pressure of the exit or "discharge" port and the same entropy as the inlet. The right-hand side of Equation 4-41 is an expression for adiabatic-reversible or isentropic work of compression. Real compressors, of course, are not reversible. However heat transfer, in large water-cooled or smaller air-cooled units, is often negligible compared with the enthalpy change, and they are essentially adiabatic. Thus, another first-law expression can be written based on actual rather than isentropic exit enthalpy.

$$\frac{\dot{w}_s}{\dot{m}} = h_2 - h_1 \tag{4-42}$$

The analysis assumes that friction-generated heat at sliding surfaces inside the compressor is transferred to the gas.

To evaluate a compressor, Equation 4-41 is employed first, and the isentropic power consumption $\epsilon_i \dot{w}_s$ is calculated from the relation:

$$\epsilon_i \dot{w}_s = \dot{m}(h_{2,s} - h_1) \tag{4-43}$$

Next, isentropic efficiency[2] is employed to calculate true shaft power $\dot{w}_s$. True exit enthalpy is calculated using a rearrangement of Equation 4-42. With true enthalpy and pressure known, other properties of the effluent compressed stream can be specified and the analysis completed. (This approach is basically the inverse of that for designing turbines and expanders.)

If the compressor is small, or, as with vacuum units, mass flow rates are low, exhausted heat $\dot{Q}$ may be significant when compared with the gas enthalpy change. In this situation, the first law can be written as follows:

$$\dot{w}_s = \dot{m}(h_2 - h_1) + \dot{Q} \tag{4-44}$$

Evaluation of shaft power from Equation 4-44 requires a value for $\dot{Q}$ or more information about the specific compressor under study. For definitive design, or in predesign situations where compression costs are dominant, consult with a vendor or someone familiar with the specific application. In less critical situations, it is safe to assume that the compressor is adiabatic—calculate the power from Equation 4-43 and employ an isentropic or intrinsic efficiency from Table 4-9. As long as heat is lost and not added, the adiabatic calculation always yields an overestimate of power requirements. An efficiency of about 70 percent is representative in most

---

[2]"Isentropic efficiency" has the same meaning as "intrinsic efficiency," but it is a particular term applied to systems that are analyzed by an adiabatic, reversible first-law analysis.

normal situations. One exception is vacuum pumping at discharge pressures well below atmospheric. Here mechanical friction is more significant, and efficiencies may be considerably lower.

If one has detailed information on compression path and adequate property data, the integral in Equation 4-34 can be evaluated directly (graphically or otherwise) to yield a valid result.

An alternate technique, useful when entropy data are not accessible, is that developed by Shultz [43] and applied earlier to polytropic expansion in turbines and expanders. An equation analogous to Equation 4-39 is derived from the integration of Equation 4-34 for a real gas undergoing polytropic compression.

$$\dot{w}_s = \frac{\dot{m} z_1 R T_1 \bar{n}}{\epsilon_i (\bar{n} - 1)} \left[\left(\frac{p_2}{p_1} \right)^{(\bar{n} - 1)/\bar{n}} - 1 \right] \qquad (4\text{-}45)$$

As before, the temperature ratio is given by

$$T_2 = T_1 \left(\frac{p_2}{p_1} \right)^{\bar{m}} \qquad (4\text{-}46)$$

where $\bar{m}$ is the average at inlet and outlet, given by

$$m = \frac{z R}{C_p} \left(\frac{1}{\epsilon_1} + X \right) \qquad (4\text{-}47)$$

(The effect of efficiency is inverted compared with the expander treatment in Equation 4-15.) The definition of polytropic constant is unchanged, and it can be calculated by the following relation,

$$n = [Y - m(1 - X)]^{-1} \qquad (4\text{-}48)$$

where X and Y are taken from Figures 24-39 and 24-40 in Perry (p. 24-34 and 24-36). If condensate is present at any stage of compression, this approach is invalid, and a first-law analysis is recommended.

Staged Compressors

(Compression ratios greater than 4:1)

For ultra-high pressures, the adiabatic temperature rise is so great that compression units must be staged. To cool them, discharge gases pass through heat exchangers (intercoolers) between compression stages. Even though numerous stages and intercoolers may be combined in the same packaged unit, the system is analyzed as though each stage, with its intercooler, were a separate compressor. In this case, power consumed per stage can be determined as outlined in preceding paragraphs.

If gas enters at ambient temperature and the pressure ratio is 4 or 5, the isentropic discharge temperature approaches 200° C for a diatomic ideal gas. Temperatures higher than this can cause serious damage to lubricants, seals, and other sensitive materials. Thus, as a rule of thumb, if the pressure ratio is greater than 4:1, multiple stages are employed (Perry 6-25). Detailed analysis reveals that total work is minimized if each stage does the same amount. Since gases are generally cooled to the same suction temperature, compression ratios, according to Equation 4-45, must be approximately the same in each stage. To find the number of stages required under these circumstances, one can, by simple logic and algebra, derive the expression:

$$(r_{p,i})^q = r_{p,T} \qquad (4\text{-}49)$$

where q is the number of stages, $r_{p,i}$ is the compression ratio per stage, and $r_{p,T}$ is the ratio of absolute discharge to suction pressures. Thus, for example, if air were being compressed from atmospheric pressures to 400 barg, the number of stages required would be

$$q = \frac{\ln (401/1.01)}{\ln (3.7)} = 5 \qquad (4\text{-}50)$$

where 3.7 is the compression ratio per stage, nearest 4, which yields a whole number for q.

Ejectors

Used widely for vacuum service in corrosive and noncontinuous service, ejectors employ the momentum of a high velocity utility stream to pump another fluid.

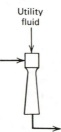

Utility
fluid

Ejectors can be used to pump liquids, but their most common application is in the transport of gases. Steam is the motive fluid employed in most cases. Steam (8 bara or higher in pressure) enters through a jet nozzle, as shown in Figure 4-8, where it is expanded to supersonic velocity. This entrains process gases, and both are decelerated in the diverging diffuser, creating a higher pressure in accordance with Bernoulli's principle. For design, the procedure presented in Perry 6-29 is recommended.

In terms of utility consumption, ejectors are much less efficient than mechanical compressors. On the other hand, they are simple, have no moving parts, and are inexpensive and almost maintenance free. They can also be staged to increase pressure or vacuum range.

A major limitation occurs if it is undesirable to dilute the process stream with a motive fluid. In steam ejectors, heat exchangers are frequently installed downstream of each unit to condense and remove the water vapor as in Figure 4-8.

Ejectors are ideal choices for intermittent or pilot plant service with corrosive gases where dependability and freedom from maintenance are more important than utility consumption. They are used primarily in vacuum applications where the discharge pressure is near 1 atm.

Equipment Details

General equipment descriptions follow. Criteria to aid selection and design are shown in Table 4-9.

TABLE 4-9

CRITERIA AND DATA FOR THE PRELIMINARY SPECIFICATION OF GAS MOVERS AND COMPRESSORS

	Type of Gas Mover or Compressor					
	Fans				Centrifugal or Turbo	
	Centrifugal		Axial			
	Radial (paddle-wheel)	Backward-Curved (squirrel cage)	Tube	Vane	Single	Staged
Absolute pressure range (atm)	Near 1	Near 1	Near 1	Near 1	0.1–2	0.1–700
Maximum differential pressure or compression ratio per stage	15 kPa	10 kPa	1 kPa	5 kPa	1.4	1.2
Maximum stages per casing						8
Maximum capacity of stock equipment, $\dot{q}$ (std m^3/s)	300	500	300	300	80	200
Typical efficiency, ϵ_i (percent)	65–70	75–80	60–65	60–70	70–80 (50–70)[a]	70–80 (50–70)[a]
Relative Costs						
Purchase price	Moderate	Low	Low	Moderate	Moderate	Moderate
Installation	Moderate	Moderate	Low	Low	Moderate	Moderate
Maintenance	Low	Low	Low	Low	Low	Low
Utilities	Moderate	Low	Moderate	Moderate to low	Low	Moderate
Compatibility						
Corrosive gases	C	C	C	C	C	C
High temperature gases	C	C	D	D	D	D
Particle-laden (dusty) gases						
Abrasive	A	C	D	D	C	C
Sticky	B	D	E	E	D	D
Vacuum Service	X	X	X	X	C	C
Variable pressure service	E	E	E	E	D	D
Variable capacity service	A	A	A	A	C	C
Common Construction Materials						
Carbon steel	√	√	√	√	√	√
Stainless steel	√	√	√	√	√	√
Plastics	√	√	√	√		
Special alloys	√	√	√	√	√	√
Performance Problems						
Lubricant contamination	A	A	A	A	A	A
Flow pulsations	A	A	A	A	A	A
Noise	B	B	D	D	D	D
Vibration	B	B	A	A	D	D
Explosion hazards	B	B	B	B	D	D
Other advantages or disadvantages	_b	_b	Flow direction is easily reversed[b]			

KEY

A   excellent or no limitations
B   modest limitations
C   special units available at higher cost to minimize problems
D   limited in this regard
E   severely limited in this regard
X   unacceptable

[a]Vacuum operation

[b]For more definitive distinctions among fans, see J.E. Thompson and C.J. Trickler, "Fans and Fan Systems" *Chemical Engineering*, pp. 46–63 (March 24, 1983).

TABLE 4-9
(Continued)

				Type of Gas Mover of Compressor			
				Blowers and Compressors			
Axial		Rotary				Reciprocal–Piston, Single and Staged	Ejector, Single and Staged
Single	Staged	Twin-Lobe, Single and Staged	Liquid-Ring, Single and Staged	Rotary-Screw, Single and Staged	Sliding-Vane, Single and Staged		
0.1–2	0.1–14	0.3–2	0.01–6	1–10	0.1–10	0.01–3000	0.01–5
Seldom used without staging	1.4	2.0	4.0	4.0	4.0	4.0	—
	15	1	1	1	1	8	5
	300	20	6	15	0.8	1.5	—
80–85 (50–70)[a]	80–85 (50–70)[a]	60–80 (40–60)[a]	60–80 (50–70)[a]	60–80 (40–60)[a]	60–80 (40–60)[a]	60–80	25–30 (see Perry 11-35)
Moderate	High	Moderate	Moderate	Moderate	Moderate	High	Very low
Moderate	High	Moderate	Moderate	Moderate	Moderate	High	Low
Moderate	Moderate	Moderate	Moderate	Moderate	Moderate	High	Very Low
Low	Moderate	Low to moderate	Low to moderate	Low to moderate	Low to moderate	High	High
E	E	E	C	D	E	D	A
D	D	D	D	D	D	E	A
E	E	E	C	D	E	X	A
E	E	E	C	D	E	D	A
C	C	B	A	B	B	A	A
E	E	A	A	A	A	A	A
E	E	C	B	C	C	D	A
√	√	√	√	√	√	√	√
√	√		√	√			√
			√				√
							√
A	A	C	E	C	C	C	D
A	A	C	C	B	B	C	C
D	D	B	B	C	B	D	B
D	D	A	A	A	A	B	A
D	D	B	B	B	B	E	A
							Process fluid is contaminated by motive fluid

FANS

Industrial fans, designed to move large volumes of gases at low pressure differentials, are of two general types. More common is the centrifugal type (Figure

Centrifugal

Axial

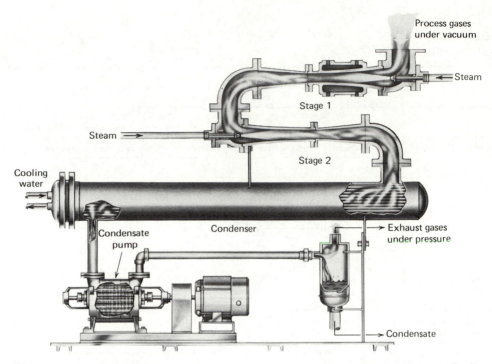

Figure 4-8   Two-stage steam ejector with condenser and condensate pump. (Croll-Reynolds, Inc., by permission).

4-9) which, like a centrifugal liquid pump, employs a rotating, circular impeller to move a fluid. This rotor is enclosed by the volute, a spiral-, scroll- or snail-shaped casing, which is connected to process piping or ducting.

Another fan, the axial flow type (Figure 4-10), is much like a common household room fan except that motor and blades are in a cylindrical duct. More than one set of propellerlike blades may be mounted on the shaft to provide multiple stages for higher pressures. Both centrifugal and axial types are characterized by rather large clearances. Because of this, internal recirculation occurs at flow rates below design values. Thus, as illustrated by characteristic pressure–volume curves (Perry 6-21), the pressure is almost constant for flows varying from zero to the design value. (You can "feel" this by closing the outlet of a common personal hair dryer or other small fan with your hand.) For specific design characteristics and efficiencies, see Table 4-9.

Centrifugal fans are most commonly constructed with one of two different impeller designs, radial curved or backward curved. Axial fans are also found in two oft-used designs. The tube-axial is cheaper, whereas the vane-axial type can achieve higher pressures. These four fan types are illustrated in Figures 4-9 and 4-10. Advantages and disadvantages of each are outlined in Table 4-9 and discussed at length by Summerell [49].

BLOWERS

The numerous blower configurations can be classified quite neatly into two general types: centrifugal-axial and rotary positive displacement. Centrifugal (also known as turbo) and axial blowers are analogous to centrifugal and axial fans except that to achieve higher pressures, machining is much more precise, rotation velocities are higher, and they are in order of magnitude more expensive. Turboblowers are shaped like centrifugal liquid pumps except the volute is larger (Figure 4-11). Compression ratio is limited by current machine technology to a value less than about 1.4 (with air at room temperature and pressure). To achieve higher pressures, up to eight stages are combined in one unit. (Encased in a single cylindrical casting, the identifying scroll-like shape is not as evident; see Figure 4-12.) If higher pressures are required, multistage turboblowers are connected in series, separated by intercoolers.

The axial blower, a sophisticated relative of the simple propellar fan, has many precisely machined blades attached to the periphery of a high speed rotor (Figure

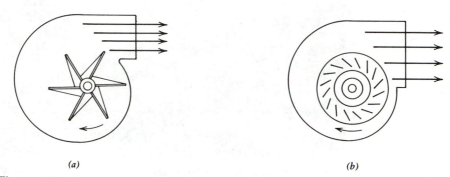

(a)   (b)

Figure 4-9   Two common types of centrifugal fan: (*a*) radial (paddle-wheel) impeller and (*b*) backward curved (squirrel cage) impeller. [Excerpted by special permission from CHEMICAL ENGINEERING (June 1, 1981). Copyright © 1981 by McGraw-Hill, Inc., New York, N.Y. 10020.]

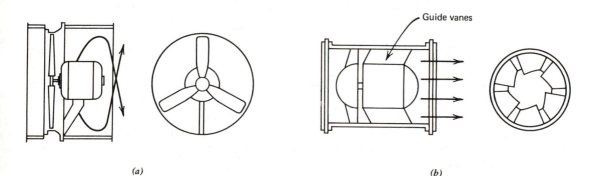

(a)   (b)

Figure 4-10   Two designs for axial flow fans: (*a*) tube-axial and (*b*) vane-axial. [Excerpted by special permission from CHEMICAL ENGINEERING (June 1, 1981). Copyright © 1981, by McGraw-Hill, Inc., New York, N.Y. 10020.]

Figure 4-11   Single stage centrifugal (turbo) compressor. (Allis-Chalmers Corporation, by permission.)

4-13). It resembles the more common stream or gas turbine. High efficiency and capacity, combined with relatively low weight, motivated the development of these units for combustion air compression in aircraft jet engines. They accommodate changing loads by variations in rotational speed. This explains why they are frequently coupled with turbine drives on the same shaft. These advantages are offset, in some chemical plant applications, by high cost, inflexibility, and intolerance to adverse environments. Pressures up to 7 barg are possible in single machines.

Like fans, turbo and axial blowers are essentially constant pressure devices. Unlike the less sophisticated fans, they become unstable when flows are constricted. This "surge" occurs at about 50 percent of capacity for turboblowers and much nearer the design rate for axial blowers.

A second general group of blowers is known as the rotary positive displacement type. ("positive displacement" to distinguish them from centrifugal-axial units and "rotary" to separate them from reciprocating compressors discussed below. They are termed simply "rotary" in this book.) Many models in this category are similar

to positive displacement liquid pumps. In particular, sliding-vane and screw-type units operate on the same principle and have the same appearance as the analogous liquid pumps shown and discussed later in this chapter. The lobe type (also known as the Roots compressor after the original patentee) consists of two mating impellers that trap and compress the transported gas (Figure 4-14). A similar design is employed for liquid pumping but with less prominence.

Another rotary, positive displacement blower, but one unique to gas pumping, is the so-called liquid-piston or liquid-ring (Figure 4-15) type. Commonly employed

Figure 4-12 Multistage turbocompressor. (Sulzer Bros., Inc., New York, by permission.)

in laboratories and small shops for vacuum pumping, it consists of an impeller with vanes, similar to those in centrifugal pumps. This impeller, on the other hand, rotates within an elliptical or eccentric housing. The housing is partially filled with liquid, which, under the influence of centrifugal force, forms a ring while operating. The liquid moves back and forth inside each vaned compartment in a pistonlike motion. Inlet and outlet ports are positioned to take advantage of this reciprocal motion and pump the gas. Liquid-ring vacuum pumps are being used more and more frequently in industry because of their high energy efficiency.

In contrast to turbo and axial blowers, rotary types are of the positive displacement type; that is. the discharge pressure increases if outlet flow is restricted (see Perry, Figure 6-48, p. 6-24). Since sealing is not perfect in rotary blowers, discharge pressures are limited to about 8 bara for single-stage, screw-type units and are lower for the others. In general, rotary blowers are useful in small-scale specialty or multipurpose applications. Their efficiencies are somewhat lower than those of centrifugal or axial devices. Some rotary designs, however, are superior in the transport of corrosive gases.

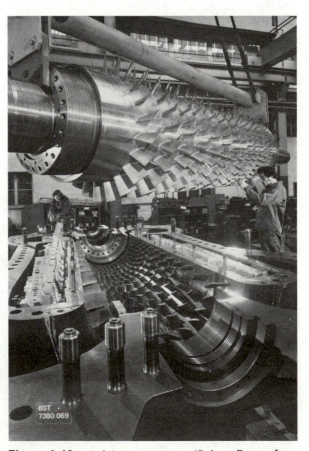

Figure 4-13   Axial compressor. (Sulzer Bros., Inc., New York, by permission.)

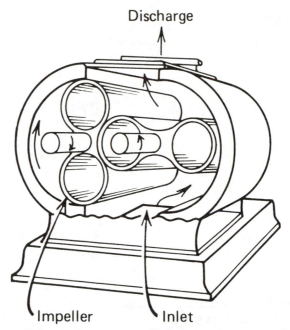

Figure 4-14 Rotary lobe-type compressor. (From *Unit Operations of Chemical Engineering*, 3rd edition, by McCabe, W. L., and J. C. Smith. Copyright © 1976. Used with the permission of McGraw-Hill Book Company.)

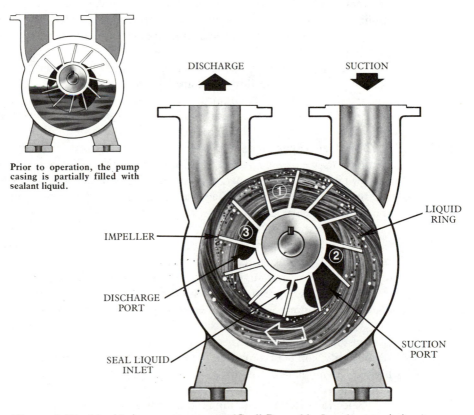

Figure 4-15 Liquid-ring vacuum pump. (Croll-Reynolds, Inc., by permission.)

COMPRESSORS

Within the definition mentioned earlier (discharge pressures greater than 2 bara), most of the blowers described above can serve as compressors when specifically

designed or staged to develop large compression ratios. When ultra-high pressures are required, reciprocal compressors are the most popular choice in the chemical industry. These contain a piston moving back and forth in a cylinder. Valves synchronized with piston movement control flow of gases in and out of the cylinder. These machines are available in many designs and configurations (see Perry 6-25 to 6-29). Generally, they are lubricated with oil, but special units are available constructed of materials that can be operated dry.

Selection and Specification of Fans, Blowers, and Compressors

Information to aid in the design and selection of gas-moving equipment is contained in Table 4-9. In general, if pressures are near ambient and the required pressure differential is less than 5 kPa, fans are the preferred type of gas mover. In ventilating applications where the gas is air, either centrifugal or axial fans are commonly employed. In moderately corrosive or demanding applications, centrigual fans are usually preferred.

Rotary blowers or compressors are often used when pressures range from low vacuum to 10 bara and flow rates are less than 15 std m³/s. They are especially preferred where delivery volumes must be maintained constant against variable downstream pressures. Selection of a particular type of machine for given process conditions is aided by the compatibility ratings and criteria outlined in Table 4-9.

For high capacity and constant delivery pressure, centrifugal blowers or compressors are usually preferred. Modest cost, high efficiency, and low maintenance, which characterize centrifugal machines, favor their selection when discharge pressures are below 15 atm. Because of high rotation speeds, staged centrifugal compressors are somewhat smaller than reciprocal units having the same capacity. Furthermore, they can be staged rather simply and inexpensively, accounting for their advantages at moderately high pressure and ultra-high capacity. In the low capacity, low pressure range, rotary blowers are often competitive because of the precise machining and alignment necessary in centrifugal units.

At ultra-high pressures and modest capacities, reciprocating positive displacement compressors retain an edge despite their mechanical complexity.

Vacuum operation, because of condensation, corrosion, and other unique circumstances, requires a specialist for confident formal design. The review by Ryans and Croll [41], however, is an excellent resource for the newcomer and is highly recommended for anyone designing or evaluating commercial vacuum equipment.

Selection of a construction material for gas-moving equipment coincides with general guidelines at the end of this chapter. One should, however, pay specific attention to the damaging effects of trace impurities and moisture. These are very likely to be present in a compressor.

Another serious consideration is safety. The high pressures and temperatures characteristic of gas compression can be disastrous when explosive mixtures develop as the result of entrained oil mist or polymer seals exposed to or contained in a high pressure, hot, oxidizing gas.

In many process design efforts, several gas-moving units are found in a single process module. Chapter Five provides an equipment tabulation sheet that will be helpful in the design and pricing of fans, blowers, and compressors.

GAS–SOLID CONTACTING EQUIPMENT

In terms of equipment included under one generic category (calciners, coolers, dryers, kilns, fluidized beds, roasters), the gas–solid contacting equipment grouping is probably the broadest, most diverse and, as might first appear, illogical of those listed in Table 4-1. Combining of equipment that cools, dries, calcines, and promotes reaction into one category is, of course, a simplification. On the other hand, since gas–solid interaction is often the major consideration in each of these operations, the mechanical devices are quite similar. For example, a rotary drum dryer used to process starch may be almost identical mechanically to machines used to calcine gypsum, cure titanium dioxide pigments, or roast copper ore.

Absence of a uniform, reliable, consistent design technique is another characteristic shared by this group. Not only must each type of equipment be evaluated separately, but every application of each type, as well. Thus, even the experienced process designer will defer to a specialist when seeking a reliable equipment specification or definitive cost estimate. The specialist, in turn, will often require pilot plant or laboratory testing of the solid in question before framing a response. With these complications, one wonders how the novice can presume to prepare a design, even a tentative one. Fortunately, gas–solid contactors are among the most ancient of unit operations. Thus, despite their variety and complexity, guidelines are available to establish limits on equipment size and energy consumption (quantities necessary for a predesign evaluation and cost estimate). Even if we select a type of unit that differs from that ultimately constructed, costs for alternate equipment that could serve the application are near enough that conclusions drawn from our work would normally not be altered.

The variety of mechanical equipment available for gas–solid service is staggering. For this discussion, batch devices have been omitted.[3] Some continuous equipment types designed for unique and rarely encountered situations have also been excluded.

Gas–solid processing devices can be subdivided into two groups, those that depend on mechanical means for gas–solid contact and those that depend on fluid motion. Within each, there is a second subdivision: indirect versus direct heating or cooling. Direct heating occurs, for instance, when hot gases or combustion products come in *direct* contact with the solid being processed. In indirect service, heat is transferred through a conductive surface that separates the process stream

---

[3]Batch contactors are employed more extensively in the manufacture of small-volume specialty chemicals, pharmaceuticals, and mechanical–metallurgical components. Engineers concerned with small-scale, labor-intensive applications such as these should review Perry, Section 20.

from the heating or cooling medium. Categories and equipment types are listed in Table 4-10. Strengths and weaknesses of each device are indicated for the particular processing constraints that one commonly encounters.

Before considering individual contactors, some comments about the group as a whole are in order. In the processing of slurries or solutions, successful devices either provide a transition zone at the entrance, atomize the fluid, or premix it with recycled solids to make it flow freely. From an energy consumption standpoint, total evaporation of liquid from a slurry or solution should be employed only when other separating or concentrating techniques are impractical or inappropriate.

For gummy pastes, a pretreatment or *preforming* step is often necessary to convert the paste to a briquet or pellet that can be processed in the normal way. (For more information on preforming, see Perry 20-28).

With dusty solids, special equipment implied in Table 4-10 usually includes an add-on dust collector (bag filter, cyclone, electrostatic precipitator, or scrubber).

Heat-sensitive solids require contactors having precise temperature control and well-defined gas motion. Vacuum applications demand gas-tight containers and easily sealed feed and discharge components. Ideally, the vacuum-tolerant system will be small and compact. If large, it must be structurally designed to withstand the external pressure.

In applications exposing mechanical components to heat, service temperatures are more strictly limited. If only the vessel surface or a static component is exposed, temperatures are restricted by oxidation or mechanical limits of construction materials alone. For carbon steel, this maximum service temperature is about 425° C. For stainless steel, it is near 650° C, and for more expensive metal alloys, up to 1100° C. By nature, surface temperatures of indirectly heated equipment exceed those of the bulk solids. Thus, for these units, processing temperatures are about 200° C below those just mentioned. To obtain higher temperatures, some types of gas–solid contactors can be lined with firebrick and heated directly by fuel–oxygen combustion or other exothermic reactions.

The unique and qualitative characteristics of each gas–solid contactor, as listed in Table 4-10, are discussed next. Preliminary design or rating procedures are discussed for the group as a whole in the section that follows.

Mechanically Aided Equipment

TUNNEL CONTACTORS   (Perry 20-26; McCabe and Smith 773; Treybal 687)

Solid material is passed through these simple heated tunnels for drying, calcining, or converting. Tunnel units evolved logically from a common form of batch dryer, the

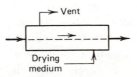

tray type. In one configuration (tray-tunnel contactors), stacks of trays, known as trucks, move on a continuous rail or chain through the tunnel. The trays are loaded and unloaded manually, and each contains solids, generally at a depth of 3 to 5 cm. If characteristics of the solid permit, trays may have porous bottoms so that gas can circulate through the bed for more efficient contact. Temperatures can be varied along the tunnel length to provide a programmed exposure. It is rather obvious from Table 4-10 that continuous tunnel–tray dryers can be made to process almost any solid material. This flexibility accounts for their popularity in small-scale, multipurpose applications. On the other hand, the high labor requirements associated with loading and unloading prohibit their use in most large-scale manufacturing operations. Temperatures are limited only by materials of construction. The motion of discrete mechanical trucks or trays in and out of the tunnel limits its use for atmospheres other than air. Vacuum conditions are impractical in continuous tunnel–tray equipment of any size. For processing small loads under vacuum conditions, batch units often are chosen.

To eliminate much of the manual labor required in tray dryers, tunnels have been fitted with continuous conveyors including automatic feeding and discharge attachments (through-circulation equipment). In most applications, a conveying screen is used so that gas can pass through the solid bed. Through-circulation, screen-conveyor tunnel units preserve the flexibility of atmospheric control typical of tray units and improve the efficiency of gas–solid contact. Sealing the tunnel from air is also more easily accomplished. On the negative side, dusty, fine-particle materials may fall through the mesh. Also, mechanical components within the heated zone limit maximum temperatures to values below those of tray tunnels. Structural design makes vacuum operation of both types impractical.

ROTARY CONTACTORS   (Perry 20-30; McCabe and Smith 774; Treybal 689)

The rotary contactor has been an industrial workhorse since the dawn of continuous mineral and chemical gas–solid processing. In contrast to more ancient

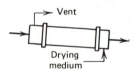

batch techniques, it permits continuous flow of both solids and gases, can be operated at high temperatures and, first impressions notwithstanding, runs with reasonably high efficiencies.

Consisting of long cylindrical drums, supported on rollers, and rotated slowly with their axes slightly inclined from horizontal, these units are known by several names: rotary dryers, rotary calciners, rotary kilns. Because of their size, they dominate the landscape in many mining or milling applications. Even the casual visitor to a cement plant would be impressed by the rotary kiln (Figure 4-16).

In all rotary contactors, feed enters at the elevated end and slowly migrates to the discharge point. Gas flow can be either countercurrent or cocurrent, depending on the exposure conditions desired. Tumbling and mixing of solid with gas is often promoted by "flights" or shelves welded to the inside of the cylinder. Slurries and gummy solids can be accommodated without preforming in some instances where hanging chains or fixed scrapers are capable of agitating the mass until it dries enough to become free-flowing.

TABLE 4-10

CRITERIA AND DATA FOR THE PRELIMINARY DESIGN OF GAS–SOLIDS CONTACTING EQUIPMENT

	Type of Gas–Solids Contacting Equipment						
	Tunnel		*Mechanically Aided*				
			Rotary				
	Tray (direct)	Through-Circulation (direct)	Dryer (direct)	Dryer or Calciner (indirect)	Kiln or Calciner (direct)	Vacuum (indirect)	Vertical Tower (direct)
Range of Common Equipment Sizes							
Diameter or width, D (m)	0.3–4	0.3–4	1–3	1–3	1–4	0.5–3	2–10
Length or height, L (m)	5–20	5–20	4–20	6–30	10–160	1.5–12	2–20
Length-diameter ratio, L/D			4–6	6–10	10–40	3–4	1–2
Other							6–1800 m² tray area
Solids Flow, $\dot{m}_s$ (kg/s)					$0.002\,LD^2$–$0.006\,LD^2$		$0.002A$–$0.01A^{a}$
Average velocity, u_s (m/s)	0.006–0.2		0.02–0.06		$2\times10^{-5}L$ to $6\times10^{-5}L$		
Bed or tray depth (cm)	3–10	3–15					3–10
Percentage of cross section occupied by solids	10–20	10	10–15	10–15	3–12	50–65	
Particle size range, D_p (mm)	>0.5	>0.5					
Typical residence time, θ (s)		300–7000		70 L/D 70–90^{c}		Batch	1×10^5 to 4×10^5
Percentage void volume							
Gas Flow							
Average superficial velocity, u_g (m/s)	2–5	1–2	0.3–1.0				0.6–3
Average mass flux, G (kg/s·m²)			0.5–5.0	0.05–0.5	0.5–5.0	Batch	
Maximum inlet temperature (°C)			800				
Pressure drop							
Maximum Solids Temperature (°C)							
Carbon steel construction	300	300	450	400	450		350
Stainless steel	300	300	750	650	750		350
Nickel-based alloy	300	300	1200	1100	1200		350
Brick-lined	—	—	—	—	1500		—
Mean Heat Transfer Coefficients							
Area (J/s·m²·K)	(Perry 20-29) 5–50^{d}	30–300^{d}	$60G^{0.67\,e}$	15–$85^{e,f}$	$24G^{0.67\,e}$	40–60^{e}	40–70^{d}
Based on internal volume (J/s·m³·K)			$240\,G^{0.67}/D$				
Thermal efficiency (percent)	20–50	55–75	55–75	30–55	65–75	60–80	55–75
Typical water evaporation rates (kg/s·kg solid)	0.0001–0.001						
Equipment power requirements (kW)		$0.5A$–$1.0A^{d}$	$8.0D^2$	$8.0D^2$	$0.15V^{h}$	$8.0D^2$	$0.2V^{h}$
Processing Compatibility							
Slurries or solutions	D	X	D, C	D, C	D, C	E, C	C
Gummy pastes	C	C	C	C	C	D	D
Friable sludges	A	A	C	C	C	B	B
Fine, free-flowing solids	A	D	A	A	A	A	A
Granular or fibrous solids	A	B	A	A	A	B	D
Dusty materials	A	D	C	C	C	B	B
Large solids, special-shapes	A	A	X	X	X	X	X
Heat-sensitive materials	A	A	D	B	E	A	A
Controlled-atmosphere applications	D	B	C	C	C	A	B
Vacuum applications	E	E	E	E	X	A	C
Ease of dust recovery	C	B	C	B	C	B	A

KEY—A excellent or no limitations, B modest limitations, C special units available at higher cost to minimize problems, D limited in this regard, E severely limited in this regard, X unacceptable

TABLE 4-10
(Continued)

Type of Gas–Solids Contacting Equipment

	Mechanically Aided			Fluid-Activated					
	Vibrating Conveyor (direct)	**Drum Dryer (indirect)**	**Screw Conveyor (indirect)**	**Gravity Shaft (direct)**	**Fluid Bed** Direct	**Fluid Bed** Indirect	**Spouted Bed (direct)**	**Pneumatic Conveyor (direct)**	**Spray Tower (direct)**
	0.3–2	0.6–3	0.1–0.7	0.5–5	1–10		0.2–1	0.2–1	2–10
	3–20	0.6–5	1–6	3–30	0.3–15		0.5–4	10–30	8–25
	3–10	1–8	3–10		0.3–15		1–10	>50	4–5
					0.4D freeboard		1.0 freeboard		
	2–15	$0.001A$–$0.01A$[b]						$0.6u_g$	See Perry
	3–10		0.01D–0.1D	80L	30–1500		30–300		(p. 20-63,
			40	40–60	20–30		20–30	5–10	Fig. 20-72)
	>0.15			>10	0.01–3		>1	<0.4	
		6–15						<4	<30
				40–60	60–80		60–80	60–95	
	Perry (Table 20-29)				0.1–2		Perry 20-54 (Eq. 20-62)	25	0.2–3
			0.05–0.5					150–700	150–800
				Perry (Fig. 20-61)	1.1 (bed weight) (area)		0.5 (bed weight) (area)		
	300	300	400	450	450	400	450	450	450
	300	300	650	750	750	650	750	750	750
	300	300	700	1200	1200	1100	1200	1200	
	—	—	—	1500	1500	—	1500	1500	
		500–2000[b]	15–60			400–800[g]			
	$1000G^{0.67}$				$1000G^{0.67}$		$300G^{0.67}$	2000	
			30–60		60–80	50–70	50–80	55–75	40–60
	$0.4A$[d]		$1.0V$[h]		$1.0V$[h]	$1.0V$[h]	$5V$[h]	$10V$[h]	
	X	A	D, C	X	X	X	X	E, C	A
	D	D	C	X	E	E	E	D, C	X
	B	X	B	E	C	C	D	C	X
	A	X	A	E	A	A	E	A	X
	B	X	B	D	B	B	A	B	X
	C	X	A	D	C	C	E	C	X
	E	X	X	A	X	X	D	X	X
	B	B	B	A	B	B	B	A	A
	B	C	A, C	A	A	A	A	A	A
	D	C	A, C	D, C	D	D	D	D	C
	C	B	A	B	A	A	A	A	A

[a] Based on total tower or tube cross-sectional area (m²). [b] Based on drum or plane surface (m²). [c] Steam-tube dryers. [d] Based on bed exposed surface area (m²). [e] Based on internal dryer or kiln contact area (m²). [f] Based on tube area in steam-tube dryers (m²). [g] Based on surface of exchanger (m²). [h] Based on internal volume (m³).

Figure 4-16   Rotary kilns in a cement plant. These twin units are 122 m long and 3 m in diameter. (Ideal Cement Company, Devil's Slide, Utah, by permission.)

Many rotary driers, calciners, or kilns employ direct heat transfer from combustion gases. Where the internal atmosphere must be more carefully controlled, the cylinder is shrouded with a combustion hood or external gas enclosure, and heat is transferred indirectly through the wall. One common form of indirect rotary contactor is the so-called rotary steam-tube dryer. This consists of a cylinder containing rows of longitudinal tubes adjacent to the inside wall and rotating with it. Steam or cooling water is circulated inside the tubes to dry or cool the solid. Another variant of the direct type is the *Roto-Louvre*® dryer, which forces a gas stream to pass through a porous, cylindrical shell, upward through the moving solids bed, and out through the center of the rotating drum.

With appropriate modifications, rotary contactors can be employed to process most conventional solids. This virtue, combined with minimal labor requirements, large capacities, flexibility, and moderately high efficiency has, historically, compensated for high capital cost, making these devices the contactor of choice in many solid–gas processes. Sealing of rotating surfaces is adequate for controlling gas composition when internal pressures are near atmospheric. Vacuum operation has not been practical except for popular vacuum-rotary driers (Perry 20-45). These, although similar in design and cost to continuous steam-tube driers, are operated batchwise.

TOWER CONTACTORS   (Perry 20-47; McCabe and Smith 776; Treybal 687)

In some respects, tower contactors resemble gas–liquid process vessels such as distillation towers where liquid flows by gravity, stage to stage, from top to bottom.

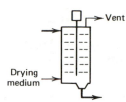

The gas–solid contactors are, of course, more complex mechanically than the conventional liquid–gas tray tower. Stirrers and wipers are required to mix, expose, and transport the solid, tray to tray, from top to bottom of the tower. Various heating and circulating arrangements are possible to provide programmed exposure, if necessary. Since the shell is stationary, gas-tight sealing is possible, making operation with controlled atomspheres easy and with vacuum, at least, feasible. Because of a relatively complex internal mechanism, maximum temperatures do not normally exceed 300° C. This, plus high capital expense and a capacity limited by mechanical–structural constraints, prevents the vertical tray contactor from displacing rotary equipment in large-volume production.

VIBRATING CONVEYOR CONTACTORS   (Perry 20-54)

An enclosed vibrating conveyor could be called a hybrid of mechanically agitated and fluid-agitated contactors. An upward-flowing gas stream, with the aid of

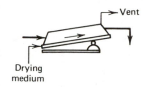

mechanical vibration, agitates solids in fluidized suspension. This provides the excellent contacting efficiency of a fluid bed with some solids that cannot be suspended in a conventional fluid bed.

DRUM DRYERS   (Perry 20-30; McCabe and Smith 778; Treybal 694)

Employed with slurries or solutions that are impractical to concentrate otherwise, drum driers, heated from inside, rotate through a pool of liquid to form a film. As

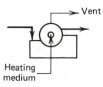

the film moves with the outer surface of the drum, water evaporates, the film becomes solid, is scraped with a fixed blade, and deposits in a discharge hopper. Only materials that are fluid enough to spread on a surface can be processed. Commonly, drums are mounted in pairs and are known as double-drum driers. Because of the relatively large amount of water removed per kilogram of solid, energy consumption is high. Thus, this type of contactor is used only when more efficient techniques will not work. This equipment can be enclosed for use with controlled atmospheres and, with small units, can be operated in a vacuum chamber if necessary.

SCREW CONVEYOR   (McCabe and Smith 775)

Patterned after the horizontal rotating screw or auger conveyor, this device, suitable for indirect heating or cooling only, has a jacketed conveyor trough. Relatively

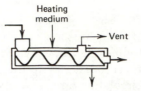

limited in capacity, it is, nevertheless, a popular choice for solids that are gummy, sticky, or otherwise difficult to process. Screw conveyor dryers or coolers can be enclosed rather easily for use with controlled atmospheres or moderate vacuum.

Fluid-Activated Contactors

From blast furnaces and lime kilns to catalytic crackers, much of modern as well as ancient chemical technology is and was based on fluid-activated gas–solid

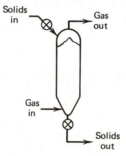

contactors. Included in this cateogry are all devices such as the classic reactors named earlier, which depend on a fluid medium rather than mechanical devices for transfer of heat and mass. These are discussed in their order of appearance in Table 4-10. With one or two exceptions (left for the reader to identify), solid particle size decreases and gas velocity increases as we progress through this discussion.

GRAVITY SHAFT CONTACTOR   (Perry 20-50)

The blast furnace (shaft furnace), employed to manufacture pig iron, is among the most ancient and familiar of gravity contactors. Other classic examples include the limestone calciner, the phosphate kiln, and the Lurgi coal gasifier. As a group, these devices are characterized by high efficiency, ease of control, and large capacity. They are most severely limited by the character of the solid feed or reagent. If it is too soft, friable, or dusty, pores in the packed bed will close, and gas flow will become choked. Plugging of the Lurgi gasifier when operated with a slagging coal is a pertinent example. Because of a relatively static bed and lack of solids mixing, hot spots and pockets can develop in highly exothermic operations. Construction simplicity, durability, large capacity, and broad temperature limits are characteristics that favor use of this equipment when the nature of the solid permits.

Processing under controlled atmospheres is relatively easy and common in gravity contactors. Vacuum and pressure operation are, on the other hand, more difficult, since solids in these units, by nature, are relatively large and abrasive, hampering their feeding and discharge under gas-tight conditions.

For drying applications, modifications of the gravity contactor that have been developed permit its use with less cooperative solids. One type includes a hollow shaft for gas collection with solids flowing downward in an annular space. In another, a mechanical conveyor lifts the solids, allowing them to cascade by gravity through the gas stream. (This device, in fact, might have been included more appropriately under the category of mechanically aided equipment.)

FLUID BED   *(Perry 20-65; McCabe and Smith 778; Treybal 697)*

Fluid beds were developed in the United States for the catalytic cracking of petroleum. The first commercial fluid bed cracker was a wartime effort put into

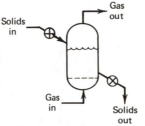

operation in 1942. Characterized by intimate gas–solids contact, high capacity, mechanical simplicity, low cost, and uniform internal temperatures, fluid beds are used for numerous reaction and drying applications. In contrast to gravity contactors, finely divided solids are required for efficient fluidization. Gummy materials can be handled rarely and only after preforming. Friable or fragile materials are limited because of their tendency to fracture in the highly agitated bed, generating dust that escapes with the gas. Even with stable solids, dustiness is a problem, and cyclones, filters, or scrubbers are normally required.

With characteristic flexibility, fluid beds are well suited for controlled atmospheres. Reduced pressure operation is possible, but there must be enough differential gas pressure for solids fluidization. Research and development has concentrated and continues to focus on methods for introducing and discharging solids efficiently. By nature, cocurrent flow is difficult in fluid beds.

SPOUTED BED   *(Perry 20-53; Treybal 697)*

On a microscale, the spouted bed is a combination of fluid bed and gravity contactor. Activated by a high velocity gas jet, the solid is lifted in the center of the

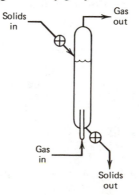

jet and then recirculated by gravity in the surrounding annular space. On a macroscale, the spouted bed can be considered to be a well-stirred gas–solid mixture and is treated, for preliminary design, much like a fluid bed. In reviewing the performance characteristics listed in Table 4-10, the spouted bed is a complement to the fluid bed, usable when solids are too large for efficient fluidization. Spouted beds are frequently used for drying of wheat, beans, and other uniform granular solids. With a solid containing both large and small particles, fluidization is usually possible, and fluid beds are therefore normally employed.

PNEUMATIC CONVEYOR (FLASH) CONTACTOR   *(Perry 20-55; McCabe and Smith 776; Treybal 698)*

Named after a common conveying mode where entrained solids are carried in a high velocity gas stream, this type of contactor is used prominently for drying, cooling,

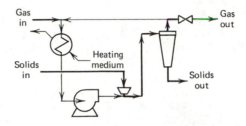

and reacting. The flash dryer is an example of the former. Pulverized-coal combustors and Koppers–Toltzek coal gasification units are examples of reactors featuring this type of contact. A pneumatic contactor would be the natural result if one placed a fine-particle or dusty solid in a spouted bed. As with a fluid bed, gas–solid contact is efficient, but unlike the case of a fluid bed, solid and gas elements remain together over most of the processing time. Also, because of limited equipment size, contact times are very brief (in the range of one second). Thus, materials that can react or dry in short times are required.

Reactors, with unique characteristics for each application, vary widely in design. Dryers, on the other hand, are more conventional. They contain a feeding device that injects solids into a high velocity gas stream. Drying occurs as the dusty fluid flows through a residence tube or duct and ceases when the solid is disengaged in a separator (usually of the cyclone type). By nature, only fine materials can be processed. Some problem solids can be handled by either preforming them, recycling a large fraction of dried product, or employing a mill in the feed line. Sealing of feed and discharge lines is relatively simple, allowing for convenient control of the atmosphere. Because of the large gas momentum required, operation under vacuum is limited. Another limitation is unique to pneumatic contactors: since each solid particle is exposed to a limited volume of gas, the degree of drying or reaction is bounded by enthalpy, moisture, and reagent capacity of the gas. To remedy this and to provide longer average residence times as well, most pneumatic dryers recycle a large fraction of the solids.

SPRAY DRYER-COOLER   [*Perry 20-58, 8-64 (prilling); McCabe and Smith 780; Treybal 695*]

Because spraying is a form of liquid dispersion, this device is as much a gas–liquid as a gas–solid contactor. However, since the final product leaves as a solid, spray

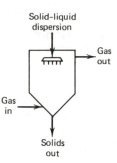

dryers–coolers are included here. Such equipment consists quite simply of an atomizer that disperses a solution, slurry, or melt into a larger gas-filled chamber. As with pneumatic units, the degree of drying or cooling is limited, although not as severely, by thermal capacity of the surrounding gas. Unlike pneumatic conveyors, residence times can be quite large as dictated by chamber size, gas velocity (both magnitude and direction), and particle settling rates.

In drying, this operation is attractive because of its mechanical simplicity, large capacity, and low capital cost. These advantages, however, are offset by the excessive energy required for total evaporation of the liquid. For this reason, spray drying is attractive only when liquid cannot be removed by less expensive methods, when special characteristics are desired in the final product, or when very sensitive materials are involved. Numerous food products fall in the latter category and are spray dried. Because of the unique spherical or hollow-sphere nature of the solid and the capacity to control size, spray drying is employed to manufacture detergents and other consumer products where dispersibility, pourability, and ease of handling are important. Some slimy, sticky fluids cannot be efficiently dried by other means.

Atmospheres are easily controlled in spray units. Vacuum operation is limited by the thermal capacity of the gas and the strength of the large residence chamber. There is no inherent limit on temperature except that imposed by materials of construction or sensitivity of the solid.

Procedures for Approximate Design and Rating of Gas–Solids Contactors

In essence, this equipment must promote intimate contact between solids and gases so that heat and/or matter can be transferred from one to another. As indicated earlier, each application is a special case, often with unique circumstances. Consequently, it is impossible to give general guidelines that will apply in each case. In fact, design of gas–solid contactors illustrates the interaction of several phenomena such as fluid transport, heat and mass transfer, and chemical kinetics, plus wisdom and judgment. As with most chemical process problems, analysis can be pursued directly with the aid of conventional material balance, energy balance, and rate equations.

As a general rule, the following equations and approaches are employed to design gas–solid equipment.

MASS FLOW RELATIONSHIP

In most devices, such as rotary driers, kilns, calciners, tunnel equipment, and vertical towers, average solids mass velocities fall within rather narrow bounds. Gas flows, temperature, and composition, on the other hand, are easily varied and can be adjusted for particular needs.

The average solids flow rate for a longitudinal flow contactor can be expressed simply by the continuity equation, so familiar from fluid mechanics:

$$\dot{m}_s = \rho_s u_s A_s \tag{4-51}$$

where ρ_s is the solid density, u_s is its average flow velocity, and A_s the average cross-sectional area through which is flows.

MASS BALANCE

From terminal solid conditions, which are normally dictated by process requirements, the mass transferred from one medium can be calculated by traditional techniques.

ENERGY BALANCE

Solid terminal conditions also define the amount of energy crossing the gas–solid interface. In evaluating this, a designer must remember basic concepts such as the need to carefully define the system to which the balance applies and the necessity of a consistent datum for enthalpy data. In reactors, heats of reaction can be derived by conventional techniques, using a convenient path. In most dryers, solvents are only physically bound. Thus, heats of solution or adsorption can be disregarded and only latent heats considered. Heat losses normally range from 5 to 15 percent of that supplied to the process. In most cases, mechanical work, in the energy balance equation, is negligible compared with heat and enthalpy fluxes. (The work must, however, be defined for estimating utility costs.) During this step, the gas flow rate and inlet temperature must be fixed to define its inlet and outlet enthalpy. For the first trial, a representative value may be selected from Table 4-10. This can be revised and the calculation repeated later to refine the result.

RESIDENCE TIME

Not only must a contactor have adequate flow capacity, it must maintain the solid and gas in proximity for long enough to accomplish the degree of drying, cooling, or reacting necessary.

In some reactors, *gas* residence time is a critical variable. In dryers, coolers, and a majority of reactors, the *solid* residence time is critical. This can be expressed by:

$$\theta = \frac{V f_s \rho_s}{\dot{m}_s} \tag{4-52}$$

where V is the volume of the reactor and f_s is the fraction of the contactor occupied by the solid.

Residence time for a linear unit can be expressed simply in terms of contact length and mean solids flow rate.

$$\theta = \frac{L}{u_s} \tag{4-53}$$

For well-stirred units such as fluid beds, Equation 4-52 can be used directly. Thus, from solids density, average flow velocity, and required residence time, equipment volume can be determined.

TABLE 4-11
REPRESENTATIVE APPLICATIONS AND OPERATING DATA FOR
ROTARY KILNS

Application	Temperature Range (°C)	Residence times (h)
Alumina calcination	800–1100	1.5–3.0
Clay calcination	800–1100	1.5–3.0
Diatomaceous earth revivification	550–750	1.0–1.5
Fluxing of silica	1000–1350	1.0–1.5
Gypsum calcination	500–650	0.5–1.0
Petroleum coke calcination	1100–1200	1.0–1.5
Phosphate rock calcination	800–1000	1.0–1.5
Roasting cinnabar ore (mercury production)	550	1.0–1.5
Roasting ores (gold, silver, iron)	600–1200	0.75–1.0
Sodium aluminum sulfite calcination	550	1.0–1.5
Titanium dioxide calcination	1000	1.0–1.5
Zinc ore calcination	600–800	1.0–1.5

In reactors, residence times are evaluated uniquely for each situation. An engineer must obtain them from theoretical analyses, the literature, or laboratory tests. Processing data for some special cases, taken from Perry, Section 20, and other sources, are shown in Tables 4-10 and 4-11. For drying calculations, residence times are controlled by both diffusion and heat transfer rates. In tunnel, vertical tower, and drum dryers, the rate depends on solids porosity, temperature difference, moisture content, and bed thickness. This can be expressed by the equation:

$$\frac{dX}{d\theta} = \frac{k_d X (T_g - T_{wb})}{t} \qquad (4\text{-}54)$$

where X is the ratio of liquid to dry solid (kg/kg), T_g is gas dry bulb temperature, T_{wb} is gas wet bulb temperature, and t is bed thickness. The drying rate constant k_d has units of meters per degree-seconds (m/°C · s). Integration of Equation 4-54 yields an expression for drying or residence time:

$$\theta = \frac{t \ln(X_o/X_i)}{k_d (T_g - T_{wb})} \qquad (4\text{-}55)$$

where an arithmetic average of terminal temperature differences is used.

For materials that are normally dried in static bed units, values of k_d (calculated from data in Perry, Tables 20-10 and 20-12) vary from about 1×10^{-6} m/°C · s for dense, nonporous materials to approximately 5×10^{-6} m/°C · s for more granular, fibrous beds.

The molecular process of drying is complex. Thus Equation 4-55 should be used with skepticism and as a last resort. Pilot plant or laboratory drying data, if available, are much more dependable.

For the dryers listed in Table 4-10 that have agitated beds or intimate gas–solid contact, heat transfer can be assumed to limit vapor transport. Residence time can be estimated from an energy balance on the solids stream plus a heat transfer relationship

$$\dot{Q} = UA \, \Delta T_{lm} \qquad (4\text{-}56)$$

or

$$\dot{Q} = U'V\,\Delta T_{lm} \tag{4-57}$$

where U is an overall heat transfer coefficient based on area and U' that based on volume as reported in Table 4-10. The logarithmic–mean temperature difference is based on differences between gas dry bulb and wet bulb values at inlet and outlet of the dryer.[4]

With area or volume determined from Equation 4-56 or 4-57, residence time can be calculated from Equation 4-52 or 4-53.

HEAT TRANSFER RATE

For a given residence time and a particular equipment geometry, several sets of terminal temperatures and flow rates will satisfy mass and energy balances, but a unique set is dictated by the heat transfer characteristics of the contactor. If not already completed in step 4, the traditional heat transfer relation, Equation 4-56 or 4-57, must be employed to find this set. As with other heat exchangers, the logarithmic–mean driving force applies. As noted above, solid temperatures are nearly constant at the wet bulb temperature of the gas throughout most of the drying period. For this situation, as with evaporators or condensers, that temperature is employed for the logarithmic–mean calculation rather than true terminal temperatures, which may change sharply because of superheating or subcooling at inlet and outlet of the contactor.

SATURATION

Because of some arbitrary assumptions, the analysis normally is overspecified. One implicit assumption, for example, is absence of saturation in the gas stream. With most parameters fixed, it is important to check the exit gas concentrations to guarantee that saturation or near-saturation has not occurred.

REEVALUATION

In the preceding steps, certain characteristics of the process may seem to be overly confining and expensive. At this point, it is important to stand back and look at the whole picture. Terminal gas properties may require reevaluation. Perhaps other assumptions are justified or other equipment alternatives should be considered.

ILLUSTRATION 4-5   DRYING OF SILICA GEL

Application of the preceding steps is illustrated by considering the drying of silica gel. A given gel contains 0.5 kg of water per kilogram of dry solids and enters the dryer at 20° C. It is dried with air that has been preheated by steam. Final desired moisture content X_o, is 0.05 kg water per kilogram of dry solids. The air was initially at 20° C and 60 percent relative humidity. Steam is available, saturated at 10 bara (180° C). Wet solids enter the dryer at a rate of 0.02 kg/s with a density of 0.8 kg/m³. Determine the size of dryer required, the air flow rate, and its temperature. As a first trial, consider a direct rotary dryer.

---

[4]For hygroscopic materials (i.e., those containing bound moisture), the saturation temperature rather than wet bulb value should be employed. With water at 1 atm adsorbed on calcium chloride, for example, 100° C should be substituted for T_{wb} to yield a conservative estimate.

STEP 1 Equipment Size

Employ the mass flow rate to define equipment size. From data given in Table 4-10, a solids velocity of 0.05 m/s is selected. The solids cross-sectional flow area is given by a rearrangement of Equation 4-51.

$$A_s = \frac{\dot{m}_s}{\rho_s u_s} = \frac{0.02 \text{ kg/s}}{(0.8 \text{ kg/m}^3)(0.05 \text{ m/s})} = 0.5 \text{ m}^2$$

From guidelines suggested in Table 4-10, we can assume that 12.5 percent of the drum cross section is filled with solids. The total drum cross section is given, accordingly, by the following relation.

$$A_t = \frac{\pi D^2}{4} = \frac{A_s}{0.125} = 4 \text{ m}^2$$

The diameter is thus found to be 2.3 m.

STEP 2 Mass Balance

Water evaporated from the gel is found from the following equation.

$$\dot{m}_s = (0.02 \text{ kg/s}) \frac{1.0 \text{ kg dry solids}}{1.5 \text{ kg wet solids}} \left(0.5 - 0.05 \frac{\text{kg water}}{\text{kg dry solids}} \right)$$

$$= 0.006 \text{ kg water/s}$$

STEP 3 Energy Balance

Assuming an outlet solids temperature of $100°$ C, the total heat transferred from the gas to the wet solid is determined by an energy balance on the solid stream.

$$\dot{Q}_s = \dot{m}_{s,i} \frac{1}{1 + X_i} C_{p,s} (100 - 20° \text{C}) + \dot{m}_{s,i} \frac{X_i - X_o}{1 + X_i} (h_{v,T_{g,o}} - h_{l,20° \text{C}})$$

$$\text{(4-58)}$$

$$+ \dot{m}_{s,i} \frac{X_o}{1 + X_i} C_{p,l} (100 - 20° \text{C})$$

Substituting data from the problem and using a heat capacity of 2.0 kJ/kg · °C for the solid, this equation reduces to:

$$\dot{Q}_s = (0.02) \left[\left(\frac{1}{1.5} \right) \left(2.0 \right) \left(80 \right) + \left(\frac{0.45}{1.5} \right) \left(h_{v,T_{g,o}} - 84 \right) + \left(\frac{0.05}{1.5} \right) \left(4.19 \right) \left(80 \right) \right]$$

or

$$\dot{Q}_s = 2.4 + 0.006 (h_{v,T_{g,o}} - 84) \text{ kJ/s} \qquad \text{(4-59)}$$

where $h_{v,T_{g,o}}$ designates enthalpy of water vapor at the exhaust temperature. The heat $\dot{Q}_s$ must be provided from sensible heat contained in the air. Allowing for 10 percent heat loss through dryer walls, we have

$$\dot{Q}_s = 0.9 \dot{Q}_g = 0.9 \dot{m}_g C_{p,g} (T_{g,i} - T_{g,o}) \qquad \text{(4-60)}$$

STEPS 4 AND 5 Residence Time and Heat Transfer Rate

The silica-gel surface is agitated and exposed during passage through the rotary dryer. Thus, residence time is assumed to be controlled by heat transfer rather than by mass transfer. From Table 4-10, the overall heat transfer coefficient from gas to solid can be approximated by $U = 60\,G^{0.67}$ (J/s · m^2 · K). This is based on internal drum surface, where G is the gas mass flux in kilograms per second-square meter (kg/s · m^2). The heat transfer rate is accordingly written as follows.

$$Q_g = \frac{60 G^{0.67}\, \pi DL\, \Delta T_{lm}}{1000}$$

$$= 0.060 \left(\frac{4 \dot{m}_g}{0.875\, \pi D^2} \right)^{0.67} \pi DL\, \Delta T_{lm} \tag{4-61}$$

$$= 0.18 L\, \dot{m}_g^{0.67}\, \Delta T_{lm}\ \text{kJ/s}$$

Equations 4-59, 4-60, and 4-61 can be manipulated to give the following:

$$\dot{Q}_s = 2.4 + 0.006\,(h_{v,T_{g,o}} - 84) = 0.9 \dot{m}_g C_{p,g}(T_{g,i} - T_{g,o})$$

$$= 0.90\,(0.18)\,L \dot{m}_g^{0.67}\, \Delta T_{lm} \tag{4-62}$$

which is, in essence, two equations in three fundamental unknowns $\dot{m}_g$, $T_{g,i}$, and $T_{g,o}$. Since the amount of water evaporated is relatively small, we will set $\dot{m}_g$ according to the lower value of G recommended for rotary dryers.

$$m_g = (0.50\ \text{kg/m}^2\text{s}) \left[\frac{\pi}{4}\,(2.3\ \text{m})^2 \right] (0.875)$$

$$= 1.8\ \text{kg/s}$$

Assuming a gas outlet temperature of 100° C, $h_{v,T_{g,o}}$ (from the stream tables) is 2676 kJ/kg and $\dot{Q}_s = 18$ kJ/s as computed from Equation 4-59. Using $C_{p,g} = 1.0$ kJ/kg · °C, we have

$$T_{g,i} = \frac{18\ \text{kJ/s}}{(0.4)\,(1.8\ \text{kg/s})\,(1.0\ \text{kJ/kg} \cdot °\text{C})} + 100°\text{C} = 111°\text{C}$$

Based on 60 percent relative humidity of 20° C, the incoming air contains 0.009 kg of water per kilogram of dry air. From a material balance, the leaving air contains 0.012 kg of water per kilogram of dry air. Using a psychometric chart such as that in Perry, Figure 20-11, the inlet and outlet wet bulb temperatures are found to be approximately equal at 38° C. Thus

$$\Delta T_{lm} = \frac{(118 - 38) - (100 - 38)}{\ln \left(\dfrac{111 - 38}{100 - 38} \right)} = 67°\text{C}$$

From Equation 4-62, the dryer length, as calculated, is:

$$L = \frac{18.0}{(0.90)\,(0.18)\,(1.8)^{0.67}\,(67)} = 1.1\ \text{m}$$

STEP 6 Saturation

A check for saturation reveals that water concentrations in the air are well below levels that would limit mass transfer.

STEP 7   Reevaluation

It is clearly impractical to have a length smaller than the diameter. Air temperature and flow rate are substantially higher than are required for this situation. Calculations could be repeated, setting length equal to a minimum value of four times the diameter. The corresponding air conditions would be computed by repeating the procedure just done. In performing this analysis, it becomes obvious that the equipment, as dictated by solids flow rate, is oversized according to heat and mass transfer potential. For instance, the equipment size as far as heat transfer and mass transfer are concerned, may be reduced substantially, if higher air temperatures are employed. Since the solid is not heat sensitive, the only limitation on air temperature is that imposed by steam pressure. Because solids flow rate dictates an impractically large contactor, our selection of equipment was not appropriate. If the feed had been a slurry or sludge containing 5 kg of water per kilogram of solid rather than 0.5 kg, the rotary dryer would have been more suitable. For the conditions specified, however, a tunnel, fluid bed, or even a transport dryer would have been more suitable.

These seven steps apply to the design of any particular type of gas–solid contactor. Individual steps may be emphasized or repeated more or less depending on the particular equipment or problem. In most cases, the value of the result will be controlled by the accuracy of residence time calculations. If true, this should be acknowledged by the designer in his or her report and, where necessary, more laboratory or pilot plant data requested. Some improvement in assumptions can probably be accomplished by reviewing Perry 20-1 to 20-16, McCabe and Smith Chapter 29, and Treybal Chapter 12.

HEAT EXCHANGERS

(*Perry 11-3; McCabe and Smith 398; Foust 327*)

Fundamental instruments in energy transfer and conservation, heat exchangers are prominent and ubiquitous throughout the chemical process industry. Heat exchange is not only a major unit operation in isolation, it is an important element of many others. Its role in evaporation and gas–solid contacting, for example, is obvious. Heat exchangers are also necessary for distillation and important or necessary for numerous other process steps such as reaction, gas compression, and absorption. Even though applications and duties are legion, heat exchanger types are limited and quite standard, allowing for easy description and characterization.

Most advanced chemical engineering students can manipulate equations and combine resistances to determine an overall heat transfer coefficient with high precision. Ironically, they often employ it with careless abandon in the simple heat transfer equation:

$$\dot{Q} = UA \, \Delta T_m \qquad (4\text{-}63)$$

This often leads to serious errors in equipment design. To illustrate, consider the terms in Equation 4-63. The heat transfer rate $\dot{Q}$ is straightforward and can be calculated directly from an energy balance on the exchanger. Entering and exiting temperatures flow rates and other characteristics of at least one stream will almost always be available from the flow sheet. The overall heat transfer coefficient U is a composite term. It includes inside and outside film coefficients (h_i and h_o), the resistance of the separating, usually metal, wall and fouling coefficients ($h_{i,f}$ and $h_{o,f}$). These quantities are added according to the classical equation

$$\frac{1}{U} = \frac{1}{h_o} + \frac{1}{h_{o,f}} + \frac{D_o \ln(D_o/D_i)}{2k} + \frac{D_o}{D_i h_i} + \frac{D_o}{D_i h_{i,f}} \qquad (4\text{-}64)$$

which pertains to a tubular separator. The terms and their significance should be familiar to all chemical engineers and most are capable of quickly deriving the equation.

In Equation 4-63, the area A is traditionally the outside bare-tube surface and is the area corresponding to U in Equation 4-64. If an exchanger contains finned tubes, A is the outside area of the tubes as though fins were absent. Reported heat transfer coefficients, in this case, are, by convention, adjusted accordingly.

The most misapplied term in Equation 4-63 is mean temperature difference (MTD) or ΔT_m, which, most often, is the logarithmic mean. All design engineers should be able to derive Equation 4-63 by applying an energy balance to a differential length of exchanger surface along the flow path. Several conditions must be met for the derived result to require an MTD equal to the logarithmic mean.

1 The system must be at steady state.
2 The overall heat transfer coefficient must be constant throughout the exchanger.
3 There must be no phase change.

Temperature profiles for a counterflow exchanger meeting these criteria are illustrated in Figure 4-17. The energy balance applied to a differential length of exchanger is:

$$d\dot{Q} = U(T_h - T_c)\,dA = \dot{m}_h C_{p,h}\,dT_h = \dot{m}_c C_{p,c}\,dT_c \qquad (4\text{-}65)$$

If all the terms except T_h and T_c in Equation 4-65 are constant, integration yields Equation 4-66, where ΔT_m is the logarithmic mean of the hot- and cold-end "approach" temperatures.

$$\text{MTD} = \Delta T_m = \frac{\Delta T_h - \Delta T_c}{\ln(\Delta T_h / \Delta T_c)} = \text{LMTD} \qquad (4\text{-}66)$$

Equation 4-66 is easily applied except when curves happen to be parallel and the approach ΔT's are equal. LMTD in this case becomes simply an arithmetic mean or the value at either end (although Equation 4-66 is indeterminant in this trivial case).

$$\text{MTD} = \Delta T_m = \Delta T_h = \Delta T_c \qquad (4\text{-}67)$$

The hot and cold streams could flow cocurrently rather than countercurrently. With both streams entering from the same end, the temperature profile would be as in Figure 4-18.

Since the temperature achievable by the cold stream can never exceed that of the exiting hot stream, cocurrent heating or cooling is employed less often than counterflow exchange. The latter generates less entropy and allows the cold fluid

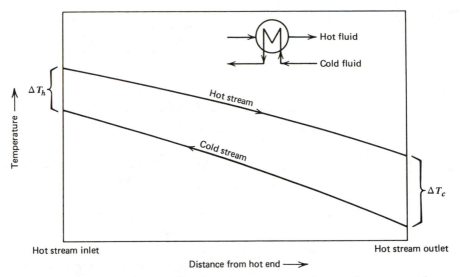

Figure 4-17   Temperature profiles for a countercurrent heat exchanger, no phase change.

temperature to approach that of the *entering* hot fluid. Cocurrent exchange is used with heat-sensitive or other materials where a limiting outlet temperature is desired. It is also useful when quenching or a rapid initial change in temperature is important.

For cocurrent flow, Equation 4-65 is integrated, with appropriate changes in sign, to yield the same result as before (i.e., Equation 4-66).

The pitfall in using Equation 4-63 occurs when there is phase change. Consider, for example, the use of superheated steam in the reboiler of a distillation column or in the first stage of an evaporator. The temperature profile in this situation will resemble that shown in Figure 4-19. Note that the steam temperature drops rapidly to its saturation value because of a small sensible heat content. It remains constant

Figure 4-18   Temperature profiles for a cocurrent heat exchanger, no phase change.

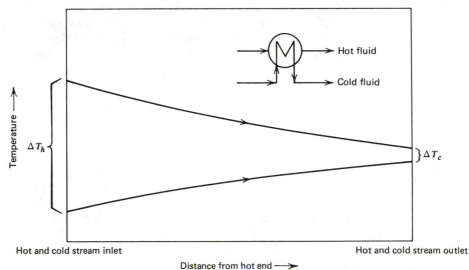

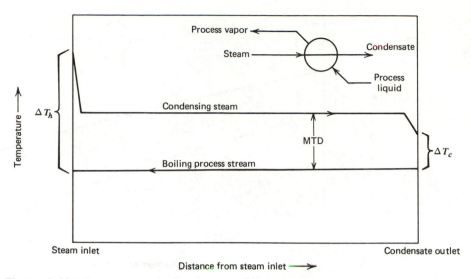

Figure 4-19 Temperature profiles in a reboiler or vaporizer with phase change: super-heated steam in, subcooled condensate out.

during condensation as the large quantity of latent heat is released. Finally, the temperature drops again as condensate becomes subcooled. There are obviously discontinuities in the steam temperature profile. Use of a logarithmic mean of hot- and cold-end approach temperatures most certainly is not appropriate. One could treat the unit as a system of three exchangers with desuperheating, condensation, and subcooling occurring in sequence. The LMTD, applied to each case, would yield a correct result. From a practical point of view, however, surface areas devoted to desuperheating and subcooling are generally negligible compared with that of condensation. Thus, for preliminary design, the difference in saturation temperatures or the MTD shown in Figure 4-19 may be employed in Equation 4-63 to yield a satisfactory result. The significant temperatures are constant in this case, and the direction of flow, co- or counter-, is irrelevant.

In phase changes involving mixtures rather than pure compounds, temperature profiles might appear as in Figure 4-20. Here, boiling and condensing temperatures change with concentration and location in the exchanger. Counterflow or cocurrent flow, in this situation, obviously affects the result. It should also be apparent that the appropriate MTD for this exchanger is the logarithmic mean of the pseudo-approach temperatures

$$\text{MTD} = \frac{\Delta T_h' - \Delta T_c'}{\ln (\Delta T_h' / \Delta T_c')} \qquad (4\text{-}68)$$

where $\Delta T_h'$ is the difference in dew points and $\Delta T_c'$ the difference in bubble points of the two streams indicated in Figure 4-20.

The appropriate MTD should be obvious in most situations. If it is not, a designer should construct the true temperature profile, apply Equation 4-65 to the differential area, and integrate it accordingly. Independent analysis may be necessary when immiscible liquids or other systems having unique temperature profiles are involved. One common situation is the vertical thermosyphon or forced-circulation reboiler or evaporator. In this exchanger, process liquid enters at

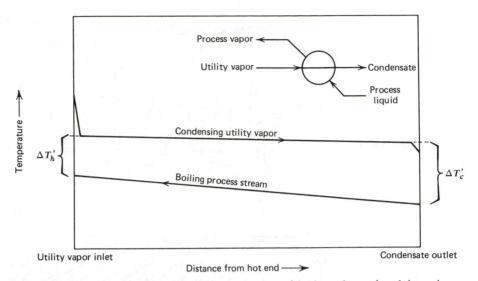

Figure 4-20 Temperature profiles in an exchanger with phase change involving mixtures.

elevated pressure. It becomes superheated in the flowing through the first section and then begins to vaporize. As it progresses, the pressure drops, causing the saturation temperature to decrease as well toward the exit. A typical profile might be quite complex, as in Figure 4-21. For rapid design, analysis is simplified by using the MTD defined by Equation 4-68, where the approaches are as shown in Figure 4-21. In most cases, this pseudo-MTD will be simply the difference in saturation temperatures of the two fluids. Compensation for such nonrigorous assumptions is included within the recommended overall heat transfer coefficients, which have been adjusted accordingly.

Figure 4-21 Temperature profiles characteristic of thermosyphon reboilers or forced-circulation evaporator exchangers.

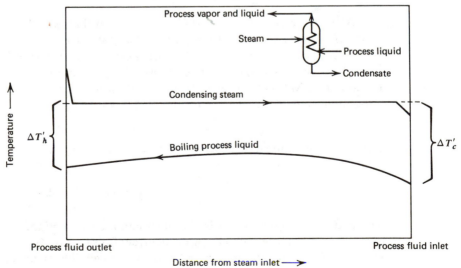

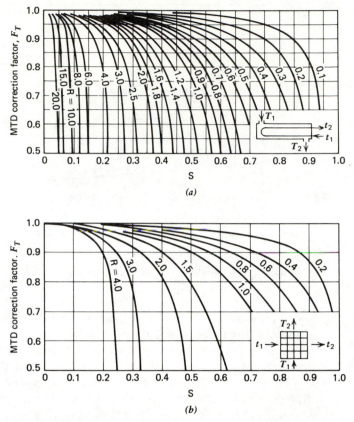

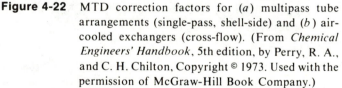

Figure 4-22 MTD correction factors for (a) multipass tube arrangements (single-pass, shell-side) and (b) air-cooled exchangers (cross-flow). (From *Chemical Engineers' Handbook*, 5th edition, by Perry, R. A., and C. H. Chilton, Copyright © 1973. Used with the permission of McGraw-Hill Book Company.)

A final correction to the MTD occurs in exchangers that may be neither counterflow nor cocurrent. These, for mechanical convenience or necessity, have mixed or cross-flow arrangments. A typical example, the so-called 1-2 exchanger, is a U-tube bundle inserted into a shell (illustrated by the insert in Figure 4-22a). Here, the apparent MTD determined from Equation 4-66 or 4-68 is multiplied by a correction factor F_T, which can be obtained from the figure.

$$\Delta T_m = \text{MTD} \times F_T \qquad (4\text{-}69)$$

Parameters in Figure 4-22 depend on terminal temperatures as shown in the insert, where

$$R = \frac{T_1 - T_2}{t_2 - t_1} \quad \text{and} \quad S = \frac{t_2 - t_1}{T_1 - t_1} \qquad (4\text{-}70)$$

The pattern in Figure 4-22a is the most common exception, in shell and tube exchangers, from true counterflow or cocurrent flow. Figure 4-22a applies as well if shell-side flow direction is reversed and, with negligible error, to an exchanger having one shell pass and any number of tube passes. For more unusual and complex flow arrangements, or when multiple exchangers are connected in various series–parallel flow patterns, correction factors are given in Figure 10-14, Perry 10-24. Air-cooled exchangers employ a true cross-flow pattern. Correction factors for this are presented in Figure 4-22b. It is obvious, in application, that correction factors are unity for most situations involving condensation or boiling applications regardless of the flow arrangement. As a practical rule of thumb, correction factors less than 0.8 are seldom employed. Rather, the flow pattern is redesigned to provide a more efficient configuration.

With ΔT_m appropriately defined, only U must be known before Equation 4-63 can be employed to compute the required heat transfer area. (As illustrated in Chapter Five, surface area is invariably the basis for cost estimation.) Contrary to the tradition in heat transfer courses where much time is invested in assessing overall coefficients, representative approximate values are adequate for preliminary design.[5] These are presented later in the discussion of design procedure.

Selection of Heat Exchangers

DOUBLE-PIPE

Conceptually and mechanically, the double-pipe heat exchanger is undoubtedly the most simple and familiar. It consists of a tube inserted concentrically inside another larger one (the shell). One fluid passes through the annulus (shell side) while another passes through the central tube (tube side). Heat is exchanged by transfer through the wall of the inner tube. (The outer tube is usually insulated to prevent losses.) In commercial use, double-pipe exchangers become excessively cumbersome and costly as their size increases. Thus, they are limited to about 10 m^2 surface area (see Table 4-12) [8].

SHELL AND TUBE

A logical alternative for increasing heat transfer surface is to place a multitude of smaller tubes inside a single shell. This, of course, is the most common heat exchanger configuration, the so-called shell and tube exchanger. Although other types (to be discussed later) are used in certain applications, shell and tube exchangers of one design or another are and can be used almost universally. Their prevalence justifies a rather detailed discussion of their characteristics. A complete and comprehensive discussion of mechanical details is found in Perry (pp. 11-3 to 11-20). A well-educated chemical engineer should be familiar with this material.

Several inherent properties are obvious from the mechanical design of a shell and tube heat exchanger. For example, high pressures are more cheaply contained in small-diameter tubes. Thus, a high pressure stream can be accommodated quite easily on the tube side of a shell and tube unit. Since a corrosive fluid inside the tubes is exposed only to the tubes and the end chambers (bonnet or channel and tube sheet, see Perry, Figure 11-2, pp. 11-6 to 11-8), these can be constructed of special alloys, whereas the shell can be made of a conventional inexpensive material. If deposits or scale accumulate on the tube side, end covers or bonnets can be easily removed and cleaning rods pushed through to clean the individual tubes. If a tube

---

[5] A sound, fundamental theoretical knowledge is still essential for intelligent employment of the approximate coefficients or for perceptive analysis of a unit that is not performing adequately.

TABLE 4-12

CRITERIA AND DATA FOR THE PRELIMINARY SPECIFICATION OF HEAT EXCHANGERS

	Type of Heat Exchanger			
	Shell and Tube			
	Double-Pipe	*Fixed Tube Sheet*	*U-Tube*	*Bayonet*
Maximum surface area per unit,[a] *A (m²)*	10	800	800	100
Typical number of passes (shell/tube)	1/1	1–2/1–4	1–2/2–4	2/1
Maximum operating pressures of stock units (bara), (shell/tube)	1000/1000	140/140	140/140	140/140
Maximum operating temperatures of stock units (°C)	150	150	350	350
Minimum practical ΔT of approach (°C)	5	5	5	5
Maximum Flow Capacity, $\dot{q}$ Liquid (m³/s, shell/tube) Gas (std m³/s, shell/tube)	1/1	15/15	15/15	2/2
Typical Mean Flow Velocity, u (m/s) Liquid (shell/tube) Gas (shell tube)	2–3/2–3 10–20/10–20	1–2/2–3 5–10/10–20	1–2/2–3 5–10/10–20	1–2/2–3 5–10/10–20
Compatibility Fouling service (shell/tube) Cleanability (shell/tube) In-service tube replacement Differential thermal expansion Thermal shock Toxic or hazardous fluids (shell/tube) Condensing service (shell/tube) Evaporative service (shell/tube) Viscous liquids (shell/tube) Maintenance Alloy construction (shell/tube) Heat transfer efficiency	D/B D/B B B B E/A D/D D/D B/A D C/C B	E/B E/B A C E A/A A/B A/A E/B B D/C B	D/D D/D D A A A/A B/B A/D E/B D D/C B	A/D B/E A A A A/A A/B A/D D/D B D/C D
Relative cost (1 = low, 4 = high)	1	1	1	4
Pressure Drop (bar) shell tube	0.5–1.0 0.2–0.6	0.2–0.6 0.2–0.6	0.2–0.6 0.2–0.6	0.2–0.6 0.4–1.0
Power consumption (kW/m²)				
Typical Service Chilling Condensing Cooling Exchanging Heating Vaporizing Superheating	√ √ √ √ √	 √ √ √	√ √ √ √ √ √	√ √ √ √ √ √ √
Typical Fluids Aqueous Organics Gases	√ √ √	√ √ √	√ √ √	√ √
Other service characteristics	For small-scale applications			Excellent under thermal stress and in corrosive environments

KEY—A   excellent or no limitations, B   modest limitations, C   special units available at higher cost to minimize problems, D   limited in this regard, E   severely limited in this regard, X   unacceptable

TABLE 4-12
(Continued)

			Type of Heat Exchanger					
Shell and Tube					Plate–Spiral			
Floating Head								Air-Cooled (fin-fan)
Packed Tube Sheet	Internal Clamped Ring	Internal Bolted Head	Teflon Tube	Scraped-Surface	Flat Plate	Spiral Plate	Spiral Tube	Air-Cooled (fin-fan)
1000	800	1000	75	10	1500	200	50	2000
1-2/1-4	1-4/2-6	1-4/2-6	1/1	1/1	1/1	1/1	1/1	1/2
140/140	140/140	140/140	8/16	140/14	20/20	20/20	500/500	–/140
350	350	350	175	150	260	260	350	260
5	5	5	5	5	1	3	2	5
15/15	15/15	15/15	0.1/0.1 —	0.1/0.03 —	0.7/0.7	3/3	1/1	–/15
1-2/2-3	1-2/2-3	1-2/2-3	1-2/1-2	1-2/1-2	—	1-2/1-2	2-3/2-3	–/2-3
5-10/10-20	5-10/10-20	5-10/10-20	5-10/5-10	—	—	5-10/5-10	5-10/5-10	3-6/10-20
B/B	B/B	B/B	D/B	D/A	A/A	B/B	B/D	–/A
B/B	B/B	B/B	D/D	D/B	A/A	B/B	B/D	–/A
A	A	A	E	D	A	D	C	A
B	A	A	A	B	A	A	A	A
D	B	B	A	B	A	A	A	A
X/A	B/A	A/A	A/A	D/D	B/B	B/B	A/A	–/A
A/B	A/B	A/B	B/E	C/D	E/E	D/D	A/D	–/B
A/A	A/A	A/A	D/E	E/B	E/E	D/D	A/D	–/E
D/B	D/B	D/B	D/A	D/A	B/B	B/B	B/B	–/B
B	D	B	B	D	A	A	B	A
D/C	D/C	D/C	D/—	C/C	C/C	D/C	C/C	–/C
B	B	B	B	B	A	A	A	B
2	2	3	4	4	4	3	3	2
0.2-0.6	0.2-0.6	0.2-0.6	0.4-1.0	0.2-0.6	0.5-1.5	0.4-0.8	0.2-0.6	0.0012
0.2-0.6	0.2-0.6	0.2-0.6	0.6-1.5	0.2-0.6	0.5-1.5	0.4-0.8	0.2-0.6	0.2-0.6
				0.5-150[b]				0.1-0.15
√	√	√	√	√	√	√	√	
√	√	√					√	
√	√	√		√			√	√
√	√	√	√		√	√	√	
√	√	√	√		√			
			√		√			
			√					
√	√	√	√	√	√	√	√	√
√	√	√	√	√	√	√	√	√
			√				√	√
For rugged process service, heating or cooling of chemicals, condensing of toxic or corrosive vapors			For corrosive, viscous, fouling liquids	For highly viscous liquids	For viscous, corrosive liquids			

[a]By convention, overall heat transfer coefficients are based on the outside bare-tube surface area. In exchangers having fins, the same equivalent area is used in the heat transfer equation even though the actual area of fluid contact is much greater than that of the equivalent bare tubes. The value of the heat transfer coefficient is corrected to compensate. [b]See Figure 4-5.

ruptures or otherwise fails, it can be plugged or removed and replaced rather simply with only the end covers removed.

Fixed Tube

Individual exchanger characteristics depend on the type of mechanical construction employed. Various types are compared in Table 4-12. For example, if the shell-side fluid is such that mechanical cleaning is unnecessary, the perforated plates to which tube ends are sealed or welded (tube sheets) often are welded directly to the shell. This (fixed tube sheet) is among the cheapest and is the most commonly employed construction (Perry Figure 11-26). One serious limitation, other than that the shell side cannot be exposed for cleaning, is stress that may result from differential expansion of shell and tubes. In practice, stresses are usually relieved by bellows or expansion joints welded into the shell.

U-Tube

Another way to avoid thermal stress is to bend tubes in the shape of a "U" and seal both ends to the same sheet. This forms a bundle, which is inserted from one end into the shell (Perry Figure 11-2d). Because they are free to expand and contract, U-tubes have excellent resistance to thermal shock created by sudden temperature changes. Bundles can also be removed completely from the shell for external washing or brushing. Tube bundles are frequently inserted into vessels or more complex shells to serve as tank heaters or reboilers. U-Tube exchangers are inexpensive, but because of the bent tubes, they cannot be thoroughly cleaned by reaming, and individual tube replacement is seldom possible.

Bayonet

This alternative to the U-tube exchanger overcomes some of its disadvantages but at higher cost. The bayonet employs a series of straight tubes or fingers, each sealed at one end and fixed to a tube sheet at the other. This bundle is inserted into the shell. A second series of smaller tubes, each open at both ends but fastened at only one end in another sheet, forms a second bundle. This is inserted into the first bundle with tube sheets separated by a spoollike spacer. Fluid can then flow inside the inner tubes, through the annuli and out. Thermal expansion is completely unhindered and tube replacement is possible. DeVore et al. [8] term it the "most underused type," the "only perfect solution to the thermal expansion problem." However, only a fraction of total tube surface can transfer heat to the shell-side fluid, and pressure drops and capital costs are relatively large.

Floating Head

The most flexible shell and tube exchanger is designed so that one end of the tube bundle, the floating head, can move freely within the shell. This is accomplished by a sliding seal between tube bundle and shell (packed tube sheet type, Perry Figure 11-2c,f) or by a separate internal cover attached to the floating head (internal clamp ring or internal bolted head, Perry Figure 11-2a). Packed units are employed where the tube side is exposed to corrosive, toxic, or hazardous materials at high temperatures and pressures, with a more innocuous fluid such as water on the shell side. Internally clamped or bolted heads are designed for severe, rugged service with corrosive, toxic, or hazardous fluids on both shell and tube sides. These are premium units that can be cleaned inside and out and repaired with minimum effort.

Teflon Tube

A relatively modern variation in shell and tube construction is an exchanger that has a conventional shell packed with a bundle of small-diameter Teflon tubes. At first blush, this seems illogical because of the relatively low thermal conductivity of teflon. On the other hand, in service with fouling or highly viscous fluids, higher tube-wall thermal resistance often is insignificant. Another compensation is the much larger heat transfer area that can be packed in a given shell volume with small-diameter tubes. Since Teflon is inert and has a low coefficient or friction, its tendency to foul or scale is less than that of other surfaces. Tube replacement, because of the excellent corrosion resistance of Teflon is seldom necessary. Thus, for small-scale applications with fouling, corrosive, or highly viscous fluids, Teflon tube exchangers may offer distinct advantages over regular shell and tube units. The major problem seems to be temperature and pressure limits (15 barg at 20° C, dropping to 3.5 barg at 150° C), which are inadequate for many commercial applications. Tube cleaning, if necessary, is rather difficult.

Scraped-Surface

For heating or cooling of sticky, gummy fluids, designers often resort to exchangers having mechanically scraped walls. Similar to scraped-wall evaporators, these units are double-pipe exchangers having rotating blades in the center tubes. Design criteria in Table 4-12 are similar to those pertaining to evaporators in Table 4-7. As a rule of thumb, one should consider a scraped-surface exchanger over a nonscraped design when liquid viscosity exceeds 1 Pa · s or when there is heavy fouling or deposition. Commercial uses are common in crystallization and other applications involving slurries.

PLATE AND SPIRAL

Flat Plate

For moderately viscous fluids, higher heat transfer coefficients can be achieved by forcing the fluid through spaces between parallel corrugated plates. Plates are mounted and spaced much like the elements in a plate and frame filter and thus, can be cleaned quite easily by hand. Flat plate exchangers are economically attractive when both fluids dictate exotic, expensive materials of construction. Because of the limited strength of large flat surfaces, they are not suitable for high pressure service.

Spiral Plate

For reduced cost and mechanical simplicity, plates can be coiled to form countercurrent flow passages in a special configuration. End plates can be removed for easy access as well as cleaning, and corrosion-resistant metals can be used quite economically in such units. Spiral plate exchangers are attractive for small-capacity service with viscous, corrosive, fouling, and scaling fluids, but they have pressure limitations similar to those of flat plate units.

Spiral Tube

One adaptation that overcomes the pressure limitations of plate exchangers employs a spiral of adjacent tubes. Adjacent tubes touch one another so that shell-side fluid cannot pass between individual tubes but must flow in the spiral gap between turns of the tube coil. This exchanger is extremely compact and can be adapted to a variety of fluids and services. The shell side is accessible for cleaning merely through removal of a flange cover. Because of the disklike shell, capacities are somewhat limited. Another major drawback is inability to ream and replace individual tubes, although the bundle itself is easily replaced. In small-scale applications where such limitations are not prohibitive, spiral tube exchangers are attractive selections.

AIR-COOLED EXCHANGERS   (*Brown* [4])

With the growing scarcity and cost of cooling water, air-cooled exchangers have become increasingly prevalent for ambient temperature service. They have been termed "fin-fan" units because the tubes have external fins and fans are employed to force air through tube banks in a cross-flow arrangement (Figure 4-23). At first, it might appear that cost of the fan, low air-side coefficients and power consumption would make these units impractical. On the other hand, with fins, overall heat transfer coefficients of fin-fan exchangers rival those of water-cooled devices. In addition, there is no shell required and no cooling water (including pumps and cooling tower) necessary. Shell-side fouling is not generally a problem, and internal tube cleaning is relatively simple. High pressure service is practical and common. Fin-fans, on the other hand, are obviously impractical for heat recovery or transfer to a fluid other than ambient air.

Design of Heat Exchangers

Steps in preliminary design of exchangers include selection, specification of duty, calculation of the differential temperature driving force, assessment of the overall heat transfer coefficient, calculation of heat transfer area, and identification of construction materials. Some of these procedures have been touched on. The most difficult step for the novice is often the selection process, which is now discussed at length, followed by a review of the other steps.

Figure 4-23   Air-cooled (fin-fan) heat exchanger. [Excerpted by special permission from CHEMICAL ENGINEERING (March 27, 1978). Copyright © 1978, by McGraw-Hill, Inc., New York, N.Y. 10020.]

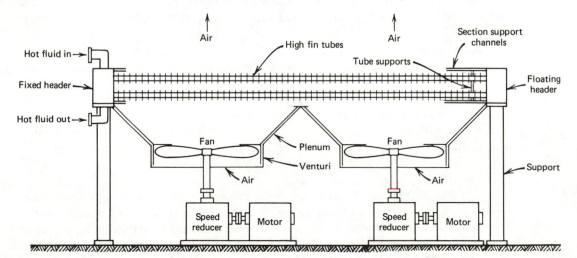

STEP 1   Selection of Exchanger Type

Considerable guidance in the selection of an exchanger is provided by data in Table 4-12. Although exceptions exist, some additional rules for selecting the flow path through an exchanger are given below.

Corrosive Fluids   In shell and tube exchangers, corrosive fluids are normally routed through the tube side. Expensive, corrosion-resistant materials can be used for tubes without requiring the same of the shell, an economical arrangement. This is particularly obvious with Teflon units. The same can be said regarding spiral tube exchangers except that shells are smaller and the cost penalty of an alloy shell is smaller. Selection of flow path for corrosive fluids should be obvious for other exchanger types.

High Pressure Fluids (> *10 barg*)   Because of the superior ability of small-diameter cylinders to contain force, high pressure fluids should, where possible, be on the tube side. Plate exchangers, flat or spiral, are limited to moderate service pressures.

Fouling or Scaling Fluids   In a fixed-tube exchanger, dirty fluid must be inside the tubes, which can be easily cleaned. For deposits that can be removed by high velocity steam or water jets, the fouling fluid might be passed through the shell of a removable bundle unit. During cleaning then, the bundle is removed from the shell and flushed. If a hard scale or deposit is formed, tube-side cleaning is generally quicker and easier. In spiral tube exchangers, internal tube cleaning can be done only by chemical means. Either side of a spiral plate exchanger can be cleaned easily of a nonadherent deposit. Flat plate exchangers are especially attractive in fouling service because they are easily dismantled for cleaning.

Highly Viscous Fluids   When a heat transfer coefficient is marginal and no factors dictate otherwise, the more viscous fluid should be placed on the shell side, where film coefficients are generally higher for a given pressure drop. If the viscosity is greater than 1 Pa · s, scraped-wall exchangers should be considered. Plate exchangers offer higher coefficients and more efficient service than shell and tube units where viscous liquids are involved.

Common heat transfer applications are discussed below. Using the rules of thumb just given, plus pressure and temperature differentials, exchanger selection and stream specification can be readily accomplished based on data provided in Tables 4-12 and 4-13.

Specific Applications
Chilling   This term is applied to applications requiring refrigerant or subcooled brine to reduce the process temperature below that of cooling water or air. Either a conventional shell and tube, spiral plate, or spiral coil exchanger can be employed. Refrigerant, which is normally nonfouling, should pass through the shell side of a fixed tube sheet unit and the tube side of a spiral tube exchanger. If a corrosive cooling brine is being employed, the most economical design might call for a removable tube bundle to permit the use of alloy tubes, with cleaning of the shell or process side still possible.

TABLE 4-13
PRELIMINARY DESIGN AND FLOW SHEET DATA FOR VARIOUS HEAT EXCHANGE APPLICATIONS

		Flow Sheet Heat Exchanger Designation								
		Condenser		Cooler				Reboiler or Vaporizer		
	Chiller	**Water-Cooled**	**Air-Cooled**	**Water-Cooled**	**Air-Cooled**	**Exchanger**	**Heater**	**Kettle**	**Thermosyphon**	**Superheater**
Normal orientation	Horizontal	Horizontal or vertical	Inclined	Horizontal		Horizontal	Horizontal	Horizontal	Vertical	Horizontal
Normal Pressure Drop (bar)										
Shell side	0.2–0.6	0.1	0.0012	0.2–0.6	0.0012	0.2–0.6	0.2–0.6	0.1	0.1	0.05–0.6
Tube side	0.2–0.6	0.1	0.1	0.2–0.6	0.2–0.6	0.2–0.6	0.2–0.6	0.1	0.2–0.6	0.05–0.6
Common range of approach, ΔT's (°C)	10–50	10–50	5–50	5–50		10–50	10–50	10–50	20–60	50–100
Process Stream Temperature (°C)										
In	<20	<500	<500	<150		150–500	<500	<500	<500	<500
Out	Various	~T_{in}	~T_{in}	<T_{in}		Various	>T_{in}	~T_{in}	~T_{in}	>T_{in}
Fraction vapor or gas										
In	0	1.0	1.0	0 or 1.0		0 or 1.0	1.0	0	0	1.0
Out	0	Various	Various	Same as entering		Same as entering	1.0	1.0	0.05–0.10	1.0
Utility Stream	Brine, Ammonia, Freon	Cooling Water	Air	Cooling Water	Air	Another Process Fluid	Steam, Dowtherm (liquid or vapor), Hot Water, Flue Gases, etc.[a]			
Temperature (°C)										
In	<10	30[b]	35	30[b]	35	Various	Various	>100	>100	>100
Out	Various	45[b]	40–60[c]	45[b]	40–60[c]	Various	Various	>100	>100	>100
Fraction vapor or gas										
In	0	0	1.0	0	1.0	0 or 1.0	Various	Various	Various	Various
Out	0–1.0	0	1.0	0	1.0	Same as entering	Usually 0	Usually 0	Usually 0	Various
Power consumption (kW/m²)			0.10–0.15	—	0.10–0.15		—	—		

[a] As discussed in the section on furnaces and boilers, steam is normally available saturated at 8, 16, and 32 bara. Dowtherm is available at up to 400°C. Fused-salt utility fluids can be provided as hot at 590°C. See Table 4-8 for more information.

[b] Except for closed-circuit water cooling, where inlet and outlet temperatures are approximately 40 and 60°C, respectively.

[c] The optimum temperature rise in an air-cooled unit requires an economic evaluation for rigorous determination. To a close approximation, Brown [4] suggests $t_2 - t_1 = 0.0009U \left[(T_2 + T_1)/2 - t_1 \right]$, where T_2 and T_1 are the leaving and entering temperatures of the process fluid, t_1 is the entering air temperature, and U is the overall coefficient.

Condensation   Under normal conditions, a shell and tube unit is employed with condensing liquid on the shell side so that liquid films do not grow too large and reduce efficiency. In some applications, liquid subcooling is desirable. This is accomplished by orienting a shell and tube exchanger vertically and allowing vapor to condense on the outside of tubes and subcool while descending as a film. Occasionally horizontal shell and tube or spiral tube units create subcooling by maintaining an appreciable level of liquid in the shell. In applications such as a distillation tower, partial condensation may be employed to enhance separation efficiency. This is commonly done by passing vapors through the tubes of a vertical shell and tube unit. Condensation surface is adequate for condensation of only part of the total stream. In all condensation service, the presence of inert substances should be avoided because noncondensable gases will blanket the condensing surface and create a diffusion barrier. In extreme cases, heat transfer coefficients are reduced from normally large numbers to very low values characteristic of transfer to a gas. If noncondensable gases are unavoidably present, the heat transfer coefficient must be adjusted accordingly.

Cooling   This service implies discharge of sensible heat from a process fluid to the natural surroundings. Traditionally, cold water from a natural source or from a cooling tower was employed. More recently, air coolers have become popular. The best choice in a given situation depends on a number of factors. As far as the exchanger itself is concerned, the rules of thumb in this chapter and data in Tables 4-12 and 4-13 will provide guidance. A decision chart is helpful in making the choice of air versus water cooling. Employing the techniques discussed in Chapter Two, the qualitative chart in Table 4-14 was assembled.

TABLE 4-14
DECISION CHART FOR SELECTION BETWEEN AIR AND WATER COOLING

| Alternatives | Criteria | | | | | | | | | Costs | | | |
	Corrosion	Temperature Control	Space	Freeze-up	Water Pollution (direct)	Water Pollution (thermal)	Noise	Water Consumption	Achievable Cooling Temperature	Power	Capital	Maintenance	Piping
Air cooling	+	−	−	−	+	+	−	+	−	−		+	+
Water cooling (cooling tower)	−	+	+	+	−	+	+	−	+	+	−	−	−
Water cooling (closed-circuit)	+	+	+	+	+	+	+	+	−	+	−	+	−

Closed-circuit water cooling is actually air cooling where water is circulated between the process and a large water–air exchanger. Conventional water cooling involves evaporation of water directly to the atmosphere in a cooling tower. Water consumption is significant, and some flushing of water is necessary to maintain acceptable mineral concentrations. The discharge water, containing treatment chemicals, can be a pollution problem as well. Because of evaporation, cooling water from a tower can be below the ambient dry bulb temperature, whereas that of closed-circuit water and air cannot. In typical designs, water that has been cooled by evaporation is available at 30° C and is returned to the tower at 45° C. Closed-circuit water and air under worst-case design conditions are typically available at 40° C with a practical temperature rise in the cooler to about 50 or 60° C. Generally, noneconomic factors are about balanced. Thus, selection ultimately is made on the basis of cost, which depends on the particular plant situation. For tentative design, it is reassuring to know that the difference in cost between these alternatives seldom has a deciding impact on overall process economics. Thus, either may be selected for approximate design purposes. In a final design, a quantitative decision chart is helpful when economic parameters have been precisely determined.

Exchange   An unqualified heat exchanger is one that conserves heat given off in the cooling of a hot fluid by absorbing it usefully through the heating of a cooler stream. Criteria for the selection of exchanger type have already been expounded (see also Table 4-13).

Heating   This term applies to the adding of sensible heat to a liquid. Such heat is provided by condensing steam, Dowtherm, or some other utility stream. Design criteria for Dowtherm heaters are similar to those for conventional exchangers. Additional useful data can be found in Table 4-8. With steam, the exchanger acts much like a condenser, and factors identical to those enumerated earlier for condensers dictate selection of an exchanger type.

Reboiling or Vaporization   (Treybal 392)   A vital operation in distillation and evaporation practice, energy, normally from condensing steam or Dowtherm, is transmitted to evaporate a process fluid. Criteria applied to condensers and heaters also apply to the utility side of reboilers and vaporizers. The process side, on the other hand, is quite different, and devices of numerous types are employed. In the past, it was common to encouter kettle evaporators or reboilers comprised of a tube bundle inserted into a specially designed reboiler shell or directly into the base of a distillation tower. Utility vapor condenses inside the tubes while process vapor is released from the liquid pool. Traditional kettle reboilers have oversized shells to allow vapor disengagement. Wiers are employed to assure adequate liquid level (Figure 11-2e, Perry, p. 11-7). The combination of condensing vapor on one side and boiling liquid on the other is an extremely efficient heat transfer combination as long as fouling is minimal.

In recent years, because of superior fouling resistance, vertical thermosyphon or forced-circulation reboilers have become popular. Identical to exchangers employed in long-tube evaporators, thermosyphon reboilers are merely shell and tube exchangers oriented vertically with circulating liquid boiling inside the tubes. Rapid flow, caused by buoyancy forces, creates high film coefficients and reduces fouling. To maintain long-term efficiency, only 5 to 10 percent of the liquid is normally vaporized per pass through the tubes, the balance recirculating through the exchanger. Optimum differential approach temperatures usually lie between 20 to 60° C [47].

In vacuum distillation and vaporizing of heat-sensitive materials, falling-film reboilers are often used. Kettle and falling-film reboilers provide separation equivalent to one theoretical stage. In principal, a thermosyphon reboiler should do the same, but it is common practice to assume no separation.

Superheating   This designates the heating of vapor above its dew point. A common application is in steam generators or boilers. There steam is superheated in the first convection bank of the furnace for more efficient power generation in a turbine. Heat transfer coefficients in superheaters are characteristically low. Fins may improve performance unless, as in a boiler, both film coefficients are low and service conditions are too dirty for efficient use of fins.

STEP 2   Duty Specification

Determining exchanger duty is a direct application of information on the flow sheet. As dictated by other aspects of the process, mass flow rate, terminal temperatures, pressures, and states of one fluid passing through an exchanger will be defined. The energy balance is usually simple. Applied at steady state to one of the fluid streams, with no work involved, it is:

$$\dot{m}(h_2 - h_1) = \dot{Q} \qquad (4\text{-}71)$$

Here, $\dot{Q}$ is heat transferred through the wall of the exchanger per unit time, or its "duty." If the fluid to which this balance is applied is on the shell side, losses should be considered. At maximum temperatures, near 400°C, about 10 percent of its enthalpy difference will be lost from the shell of a well-insulated unit. This declines linearly to zero as shell temperature approaches ambient.

STEP 3   Determination of Temperature Driving Force

Based on guidelines outlined in Table 4-13, the type of utility fluid and its condition should be relatively obvious. If not, alternatives must be compared, as air versus water cooling was evaluated above. Approach temperatures listed in Table 4-13 define an optimum economic range and can be used to establish exit temperatures of exchanger fluids. The quantity of exchange or utility fluid can be calculated as in step 2 and the appropriate MTD calculated by Equation 4-66, 4-67, or 4-68. (Be certain to avoid the pitfalls with discontinuous temperature profiles noted at the beginning of this section.) If flow in the exchanger is not true co- or countercurrent, an appropriate correction factor must be applied according to Equation 4-69.

STEP 4   Overall Heat Transfer Coefficients

Calculation of heat transfer coefficients, a tedious step in definitive design, is avoided in predesign evaluations where approximate values are more than adequate. Considering uncertainties caused by fouling, coefficients based on experience are appropriate even for many final designs. An extensive tabulation of typical overall coefficients, based on industrial practice, is found in Perry (pp. 10-39 to 10-42). Values from other sources and those from Perry, converted to SI units, appear in Tables 4-15. These values should be adequate for most preliminary purposes.

STEP 5   Calculation of Heat Transfer Area

This is the easiest step of all. It is accomplished by employing data from the preceding steps in a rearranged form of Equation 4-63.

$$A = \frac{\dot{Q}}{U \, \Delta T_m} \qquad (4\text{-}72)$$

STEP 6   Materials of Construction

Adequate preliminary specification requires knowledge of the area, pressures, temperatures, and materials of construction. General guidelines for this last step are provided at the conclusion of the chapter.

TABLE 4-15

TYPICAL OVERALL HEAT TRANSFER COEFFICIENTS *U* FOR VARIOUS TYPES OF SERVICE (J/m² · s − K)

TABLE 4-15a   Coefficients for Shell and Tube Heat Exchangers[a]

Shell Side	Tube Side	U (J/m² · s · K)
Liquid−Liquid Media		
Aroclor 1248	Jet fuels	570−850
Cutback asphalt	Water	57−110
Demineralized water	Water	1700−2840
Ethanol amine (MEA or DEA) 10−25% solutions	Water or DEA or MEA solutions	790−1140
Fuel oil	Water	85−140
Fuel oil	Oil	57−85
Gasoline	Water	340−570
Heavy oils	Heavy oils	57−230
Heavy oils	Water	85−280
Hydrogen-rich reformer stream	Hydrogen-rich reformer stream	510−680
Kerosene or gas oil	Water	140−280
Kerosene or gas oil	Oil	110−200
Kerosene or jet fuels	Trichlorothylene	230−280
Jacket water	Water	1310−1700
Lube oil (low viscosity)	Water	140−280
Lube oil (high viscosity)	Water	230−450
Lube oil	Oil	62−110
Naphtha	Water	280−400
Naphtha	Oil	140−200
Organic solvents	Water	280−850
Organic solvents	Brine	200−510
Organic solvents	Organic solvents	110−340
Tall oil derivatives, vegetable oil, etc.	Water	110−280
Water	Caustic soda solutions (10−30%)	570−1420
Water	Water	1140−1420
Wax distillate	Water	85−140
Wax distillate	Oil	74−130
Condensing Vapor−Liquid Media		
Alcohol vapor	Water	570−1140
Asphalt (230°C)	Dowtherm vapor	230−340
Dowtherm vapor	Tall oil and derivatives	340−450
Dowtherm vapor	Dowtherm liquid	450−680
Gas plant tar	Steam	230−280
High boiling hydrocarbons V	Water	110−280
Low boiling hydrocarbons A	Water	450−1140
Hydrocarbon vapors (partial condensers)	Oil	140−230
Organic solvents A	Water	570−1140
Organic solvents high NC, A	Water or brine	110−340
Organic solvents low NC, V	Water or brine	280−680
Kerosene	Water	170−370
Kerosene	Oil	110−170

TABLE 4-15a (Continued)

Shell Side	Tube Side	U (J/m² · s · K)
Naphtha	Water	280–425
Naphtha	Oil	110–170
Stabilizer reflux vapors	Water	450–680
Steam	Feed water	2270–5700
Steam	Number 6 fuel oil	85–140
Steam	Number 2 fuel oil	340–510
Sulfur dioxide	Water	850–1140
Tall oil derivatives, vegetable oils (vapor)	Water	110–280
Water	Aromatic vapor stream azeotrope	230–450
Gas–Liquid Media		
Air, nitrogen, etc. (compressed)	Water or brine	230–450
Air, nitrogen, etc., 1 atm	Water or brine	57–280
Water or brine	Air, nitrogen (compressed)	110–230
Water or brine	Air, nitrogen, etc., 1 atm	28–110
Water	Hydrogen-containing natural gas mixtures	450–710
Vaporizers		
Anhydrous ammonia	Steam condensing	850–1700
Chlorine	Steam condensing	850–1700
Chlorine	Light heat transfer oil	230–340
Propane, butane, etc.	Steam condensing	1140–1700
Water	Steam condensing	1420–2270

Source: Perry, R. H., and C. H. Chilton, *Chemical Engineers' Handbook*, 5th edition, Table 10-10, McGraw-Hill.

KEY
NC noncondensable gas present   V vacuum   A atmospheric pressure

TABLE 4-15b Coefficients for Shell and Tube Exchangers in Refinery Service (J/m² · s · K)[a]

Heating or Cooling Stream	Fluid	Specific Gravity	Reboiler, Steam-Heated	Condenser, Water-Cooled[b]	Exchangers, Liquid to Liquid (tube-side fluids C, G, and H)			Reboiler (heating liquids C, G, and K)			Condenser (cooling liquids, D, F, G, and J)			
					C	G	H	C	G[c]	K	D	F	G	J
A	Propane		910	540	480	480	450	620	540	200				
B	Butane		880	510	450	430	430	500	510	200	450	310	230	170
C	93°C end-point gasoline	0.68	680	450	400	370	340	370	280	170				
D	Virgin light naphtha	0.60	790	480	400	310	310	430	340	200	430			
E	Virgin heavy naphtha	0.80	540	430	370	310	280	310	260	170	400	280	200	170
F	Kerosene	0.83	480	340	340	310	280		260	140		280	200	170
G	Light gas oil	0.88	400	280	340	280	280		230	140	400	260	170	170
H	Heavy gas oil	0.92	340	260	310	280	260	280	230	110	400	230	170	110
J	Reduced crude	0.95			310	260	230							
K	Heavy fuel oil (tar)	1.00			280	230	200							

Source: Perry, R. H., and C. H. Chilton, *Chemical Engineers' Handbook*, 5th edition, Table 10-10, McGraw-Hill, New York (1973), by permission.
[a]All coefficients are based on outside bare-tube area and include fouling factors typical of the service given.
[b]Cooler, water-cooled, coefficients are about 5 percent lower.
[c]With heavy gas oil (H) as heating medium, coefficients are about 5 percent lower.

TABLE 4-15 (Continued)

TABLE 4-15c Coefficients for Air-Cooled (Fin-Fan) Heat Exchangers[a]

Liquid Coolers[b]	
Material	*U* $(J/m^2 \cdot s \cdot K)$
Oils (sp gr = 0.93)	
95°C	57–90
150°C	74–125
200°C	170–230
Oils (sp gr = 0.875)	
65°C	68–130
95°C	140–200
150°C	260–310
200°C	280–340
Oils (sp gr = 0.825)	
65°C	140–200
95°C	280–340
150°C	310–370
200°C	340–400
Heavy oils (sp gr = 0.97–1.02)	
150°C	34–57
200°C	57–90
Diesel oil	260–310
Kerosene	310–340
Heavy naphtha	340–370
Light naphtha	370–400
Gasoline	400–430
Light hydrocarbons	430–450
Alcohols and most organic solvents	400–430
Ammonia	570–680
Brine, 75% water	510–620
Water	680–790
50% ethylene glycol and water	570–680
Condensers	
Steam	790–850
Steam	
10% noncondensables	570–620
20% noncondensables	540–570
40% noncondensables	400–430
Pure light hydrocarbons	450–480
Mixed light hydrocarbons	370–430
Gasoline	340–430
Gasoline–steam mixtures	400–430
Medium hydrocarbons	260–280
Medium hydrocarbons, steam	310–340
Pure organic solvents	425–450
Ammonia	570–620
Freon-12	380–420

Table 4-15c (Continued)

Vapor Coolers

Material	Stream Pressure				
	0.7 barg	**3.5 barg**	**7.0 barg**	**21 barg**	**35 barg**
Light hydrocarbons	85–110	170–200	260–280	370–400	400–430
Medium hydrocarbons and organic solvents	85–110	200–230	260–280	370–400	400–430
Light inorganic vapors	57–85	85–110	170–200	260–280	280–310
Air	45–57	85–110	140–170	230–260	260–280
Ammonia	57–85	85–110	170–200	260–280	310–340
Steam	57–85	85–110	140–170	260–280	310–340
Hydrogen					
100%	110–170	260–280	370–400	480–540	540–570
75 vol %	100–160	230–260	340–370	450–480	480–510
50 vol %	85–140	200–230	310–340	430–450	480–510
25 vol %	68–130	170–200	260–280	370–400	450–480

Source Brown [4, Table 1].

[a]All coefficients based in outside bare-tube area and including fouling factors typical of the given service. Actual area of finned tubes is approximately 15 to 20 times that of bare tubes.

[b]Average temperatures are given.

TABLE 4-15d Coefficients for Coils Immersed in Liquids[a,b]

Hot Side	Cold Side	U ($J/m^2 \cdot s \cdot K$)	
		No Agitation	**Agitation**
Heating Applications			
Steam	Watery solution	570–1140	850–1560
Steam	Light oils	230–260	340–620
Steam	Medium lube oil	200–230	280–570
Steam	Bunker C or number 6 fuel oil	85–170	340–450
Steam	Tar or asphalt	85–140	230–340
Steam	Molten sulfur	110–200	200–260
Steam	Molten paraffin	140–200	230–280
Steam	Air or gases	6–17	22–45
Steam	Molasses or corn syrup	85–170	340–450
High temperature hot water	Watery solutions	400–570	620–910
High temperature heat transfer oil	Tar or asphalt	55–110	170–280
Dowtherm or Aroclor	Tar or asphalt	70–110	170–280
Cooling Applications			
Water	Watery solution	370–540	600–880
Water	Quench oil	40–60	85–140
Water	Medium lube oil	30–45	55–110
Water	Molasses or corn syrup	23–40	45–85
Water	Air or gases	6–17	22–45
Freon or ammonia	Watery solution	110–200	230–340
Calcium or sodium brine	Watery solution	280–430	450–710

Source: Tranter, Inc., by permission.

[a]All coefficients are based on outside bare-tube area and include fouling factors typical in the given service.

[b]More explicit guidelines for detail design are provided by F. Bondy and S. Lippa, "Heat Transfer for Agitated Vessels," *Chemical Engineering* pp. 62–71 (April 4, 1983).

TABLE 4-15   (Continued)

TABLE 4-15e   Coefficients for Jacketed Vessels[a,b]

Fluid Inside Jacket	Fluid in Vessel	Wall Material	Agitation	U (J/m²·s·K)
Steam	Water	Enameled cast iron	0–6 rps	550–680
Steam	Milk	Enameled cast iron	None	1140
Steam	Milk	Enameled cast iron	Stirring	1700
Steam	Milk, boiling	Enameled cast iron	None	2840
Steam	Milk	Enameled cast iron	3 rps	490
Steam	Fruit slurry	Enameled cast iron	None	190–510
Steam	Fruit slurry	Enameled cast iron	Stirring	870
Steam	Water	Cast iron and loose lead lining	Agitated	25–50
Steam	Water	Cast iron and loose lead lining	None	17
Steam	Boiling sulfur dioxide	Steel	None	340
Steam	Boiling water	Steel	None	1060
Hot water	Warm water	Enameled cast iron	None	400
Cold water	Cold water	Enameled cast iron	None	245
Ice water	Cold water	Stoneware	Agitated	40
Ice water	Cold water	Stoneware	None	28
Brine, low velocity	Nitration slurry		0.5–1 rps	180–340
Water	Sodium alcoholate solution	"Frederking" (cast-in-coil)	Agitated, baffled	450
Steam	Evaporating water	Copper		2160
Steam	Evaporating water	Enamelware		210
Steam	Water	Copper	None	840
Steam	Water	Copper	Simple stirring	1390
Steam	Boiling water	Copper	None	1420
Steam	Paraffin wax	Copper	None	155
Steam	Paraffin wax	Cast iron	Scraper	610
Water	Paraffin wax	Copper	None	140
Water	Paraffin wax	Cast iron	Scraper	410
Steam	Solution	Cast iron	Double scrapers	990–1190
Steam	Slurry	Cast iron	Double scrapers	910–990
Steam	Paste	Cast iron	Double scrapers	710–850
Steam	Lumpy mass	Cast iron	Double scrapers	430–550
Steam	Powder (5% moisture)	Cast iron	Double scrapers	230–290

Source: Perry, R.H., and C.H. Chilton, *Chemical Engineers' Handbook*, 5th edition, Table 10-14, McGraw-Hill, New York (1973), by permission.

[a]All coefficients are based on outside bare-tube area and include fouling factors typical in the given service.

[b]More explicit guidelines for detailed design are provided by F. Bondy and S. Lippa, "Heat Transfer in Agitated Vessels," *Chemical Engineering*, pp. 62–71 (April 4, 1983).

MIXERS

(*Agitators, Blenders, Kneaders, Mullers: Perry 5-19, 19-3, 21-30; McCabe and Smith 221, 895; Treybal 139, 521, 726*)

In the minds of many, agitators and mixers are synonymous. McCabe and Smith, however, point out that agitators perform numerous additional functions. They suspend solid particles in fluids for leaching or reaction, disperse gases, as bubbles, in liquids, emulsify one liquid as droplets within another, and promote heat transfer between a fluid and a solid surface. Mixing is the subdivision and blending of separate compounds by macroscopic means so that microscopic diffusion or shear will lead to more complete homogeneity.

Equipment design itself depends not so much on the function but on the nature of the fluid or solid being processed. Devices for agitation and mixing are divided, accordingly, in Table 4-16. Mixing of low viscosity miscible fluids is done quite

TABLE 4-15f Coefficients for Vessels Wrapped with External Coils[a,b]

| Type of Coil | Coil Spacing (cm) | Fluid in Coil | Fluid in Vessel | Temperature Range (°C) | U (J/m² · s · K) | |
					Without Cement[c]	With Heat Transfer Cement
1 cm o.d. copper	5	0.35–3.5 barg steam	Water under light	70–100	6–30	240–260
tubing	8		agitation	70–100	6–30	280–300
attached with	16			70–100	6–30	340–360
bands	32 or greater			70–100	6–30	390–410
	5	3.5 barg steam	Number 6 fuel oil under	70–125	6–30	110–170
	8		light agitation	70–125	6–30	140–220
	16			70–115	6–30	170–230
	32 or greater			70–115	6–30	200–260
Panel coils		3.5 barg steam	Boiling water	100	160	270–310
		Water	Water	70–100	45–170	110–270
		Water	Number 6 fuel oil	110–135	35–85	135–320
		Water	Water	55–65	40	85
		Water	Number 6 fuel oil	55–65	20	50–110

Source: Perry, R. H., and C. H. Chilton, *Chemical Engineers' Handbook*, 5th edition, Table 10-15, McGraw-Hill, New York (1973), by permission.

[a]All coefficients are based on outside bare-tube area and include fouling factors typical in the given service.

[b]Heat transfer area is the total external surface of tubing or the side of panel coil facing tank.

[c]For tubing, the coefficients depend more on tightness of the coil against the tank than on either fluid. The low end of the range is recommended.

efficiently by simple devices that create turbulence by relative fluid motion or by passage through flow constrictors. As viscosity and/or immiscibility increases, mechanical energy in various forms and at growing intensity is added.

Suspension of solid particles requires a degree of agitation of the same order as that required for dispersion of immiscible liquids in extraction or reaction vessels. Because of their controllability and flexibility, propeller or turbine agitators are almost universally employed for this service. Pumps or other pipeline agitators are sometimes used, but they do not have the same ability to sustain a suspension or control the size of dispersed liquid droplets so that they can be easily settled at the appropriate time. Rather, pumps tend to produce stable, unsettlable emulsions that sometimes, although rarely, are desired.

For mixing of highly viscous liquids, pastes, and solid powders, more energy-intensive mechanical mixers are required. As illustrated in Table 4-16, the type of device depends on the characteristics of a given feed. This equipment, in several cases, resembles that used for conveying, crushing and grinding of solids.

TABLE 4-16

CRITERIA AND DATA FOR THE PRELIMINARY DESIGN OF AGITATORS AND MIXERS

| | Type of Mixer | | | | |
| | Fluid-Agitated | | | | |
	Fluid Jet	Orifice Plate (pipeline)	Motionless Mixer	Gas Sparger	Pump or Agitated-Line Mixer
Range of Equipment Sizes					
Vessel diamenter, D_t (m)	$30\,D_a$	0.005–0.5	0.003–2.0	0.01–5	0.01–0.5
Vessel length or height, L (m)	$100\,D_a$	$50\,D_t$	0.03–80	0.03–5	0.3–2
Agitator diameter, D_a (m)	0.001–0.1	$0.2\,D_t$–$0.5\,D_t$			0.05–0.5
Vessel volume, V(m^3)					
Mixed Fluid Flow Rate, m (kg/s)				—[a]	
Gases	0.001–100	0.03–300	0.001–100		
Liquids	0.1–10,000	0.16–16	0.01–1000		0.1–400
Typical Residence Time, θ (s)					
Mixing	0.1–200	0.16–16	0.02–5.0	—[a]	0.15–1[b]
Liquid–Liquid extraction					
Solids leaching					
Chemical reaction					
Viscosity range (Pa·s)	0.0–0.01	0–0.1	0–1000	0–1.0	0–1.0
Volume fraction of dispersed medium, ϕ	<0.4	<0.4		<0.1	<0.8
Suitability					
Gas–gas mixing	A	A	A	X	D
Gas–Liquid mixing	E	D	B	A	B
Liquid–liquid mixing (miscible)	A	A	A	B	A
Liquid–liquid dispersion (immiscible)	B	B	D	D	B
Liquid–solid suspension	B	B	B	D	E
Paste–paste mixing	X	X	A	X	X
Solid–solid mixing	X	X	D	—[d]	X
Heat–transfer enhancement	A	B	D	B	D
Chemical reaction	A	B	D	A	B
Liquid–solid mixing	D	D	D	B	D
Mixing of sticky materials	E	E	A	X	D
Pressure Differential, Δp (bar)					
Gases	0.3–1.0	0.0002–0.001	6×10^{-6}	—[e]	
Liquids	1–3	0.05–0.3	0.006–0.6	—[e]	
Power Consumption, P (kW) [f]					
Gas–Gas	$1.5\dot{m}$–$5\dot{m}$	$0.03\dot{m}$–$0.15\dot{m}$	$0.001\dot{m}$		$0.15\dot{m}$–$1.5\dot{m}$
Gas–Liquid		$0.007\dot{m}$–$0.04\dot{m}$	$0.001\dot{m}$	—[e]	$0.04\dot{m}$
Liquid–Liquid					
Mild	$0.007\dot{m}$	$0.007\dot{m}$	$0.001\dot{m}$	—[e]	$0.007\dot{m}$
Vigorous	$0.01\dot{m}$	$0.02\dot{m}$			$0.02\dot{m}$
Intense	$0.02\dot{m}$	$0.04\dot{m}$			$0.04\dot{m}$
Liquid–Solid	$0.007\dot{m}$–$0.02\dot{m}$	$0.007\dot{m}$–$0.04\dot{m}$	$0.001\dot{m}$–$0.1\dot{m}$	—[e]	$0.007\dot{m}$–$0.04\dot{m}$
Paste–Paste			$0.1\dot{m}$		
Solid–Solid					
Typical overall heat transfer coefficients, U (J/s·m²·K)					

KEY—A excellent or no limitations, B modest limitations, C special units available at higher cost to minimize problems, D limited in this regard, E severely limited in this regard, X unacceptable

[a]Gas fluxes typically range from 0.004 to 0.06 m^3 per square meter of vessel cross section. Bubble rise velocities normally fall between 0.15 and 0.30 m/s. For detailed design procedures, see J. N. Tilton and T. W. F. Tussell, "Designing Gas-Sparged Vessels for Mass Transfer," *Chemical Engineering*, pp. 61–68 (November 29, 1982). [b]This is residence time within the pump or line mixer itself. For a pump installed in an external pipe loop, the time required for

TABLE 4-16
(Continued)

	Type of Mixer									
	Mechanically Agitated									
	Turbine						Single, and Twin-Rotor (porcupine)	Hammer Cage, and Attrition-Mills		Drum, Vibratory, Pebble, and Jet Mills
Propeller	**Axial**	**Radial**	**Kneader**	**Extruder**	**Roll**	**Muller**			**Ribbon**	
<50	<20									
<20	<40									
<1.5	<5									
<40,000	<1200									
$12{,}000\,(\mu V/P)^{-1/2}\,V^{0.2}$			20–200	20–200	2–10	100–200	20–200	—[c]	20–200	—[c]
60–20,000	60–1000									
60	60									
	From Kinetic Analysis									
0–5	0–500		200–2000	200–10,000	500–2000	500–5000	100–1000	—[c]		
<0.8	<0.8									
E	E	E	X	X	X	X	X	X	X	X
D	D	A	X	X	X	X	X	X	X	X
B	A	B	X	X	X	X	X	X	X	X
B	A	B	X	X	X	X	X	X	X	X
B	A	B	X	X	X	X	X	X	X	X
X	C	X	A	A	D	A	B	D	E	E
X	X	X	X	X	X	D	B	B	A	A
B	A	B	B	A	D	D	A	E	C	B
B	A	B	B	B	E	E	A	E	D	B
B	A	B	A	D	B	B	A	B	D	D
E	D	B	B	A	E	E	D	B	E	E
1.0V–3.0V		1.0V–2.0V								
0.17V–0.3V	$0.1V^{0.8}$–$0.2V^{0.8}$									
0.7V–1.0V	$0.4V^{0.8}$–$0.8V^{0.8}$									
1.0V–3.0V	$0.8V^{0.8}$–$2.0V^{0.8}$									
1.0V–3.0V	$0.8V^{0.8}$–$2.0V^{0.8}$									
			←	See Figure 4-23			→	—[c]		
							1$\dot m$–5$\dot m$	—[c]	1$\dot m$–6$\dot m$	—[c]
	See Table 4-15		50–500	50–500			20–300[g]			

circulating a volume equal to the tank contents is considered to be adequate for mixing of vessel contents. [c]See Table 4-5. [d]See fluid beds. [e]Power consumption can be calculated for a sparger from $P = m_g \epsilon_o \Delta p/\rho_g$, where m_g is the gas flow rate, ρ_g is the density, ϵ_o compressor overall efficiency, and Δp the compressor differential pressure. To determine the latter, assume 0.1 to 0.3 bar for pressure drop through the sparger and add the static pressure exerted by the fluid in the vessel. [f]All values include efficiency losses in drive and gears and correspond to the direct electricity or utility consumption. [g]V. W. Uhl and W. L. Root, "Heat Transfer to Granular Solids in Agitated Units," *Chem. Eng. Prog.* **63,** pp. 81–92 (1967).

Fluid Jet

(Perry 5-19, 18-74; McCabe and Smith 244)

Gases, because of low viscosity and high molecular and eddy transfer, are easily mixed through differential fluid motion or injection of one stream into another. For

high intensity mixing of gases, injection of one stream at sonic velocity into a second stagnant or low velocity gas is a common practice. This requires an absolute pressure in the motive gas about twice that of the mixing chamber. For a 1 atm mixture, this is a pressure differential of approximately 1 bar. Moderate turbulence is created with 0.66 bar and mild with 0.33 bar differential pressure. An examination of Reynolds numbers reveals that comparable turbulence intensity is possible in liquids flowing at velocities somewhat less than those for gas jets. Yet, from the mechanical energy equation, one can easily demonstrate that pressures required to achieve comparable velocities are considerably higher in liquid systems. Thus, pressure differentials of about 1, 2, and 3 bar are required for mild, moderate, and intense jet mixing of liquids having a viscosity of 0.001 Pa · s. With more viscous liquids, the pressure differential for a given intensity of mixing increases linearly with the square of viscosity. Thus, high intensity jet mixing at a viscosity of 0.01 Pa · s requires a pressure differential of 300 bar, which is near the practical limit.

Perry (Table 5-6, 5-20) lists some parameters useful in the preliminary design of jet mixers. The angle of divergence of a jet issuing from a circular tube is typically 20 degrees for gas systems and 15 degrees for liquid systems. Hence, a cylindrical element $100 D_o$ long and about $30 D_o$ in diameter is effectively mixed by a single nozzle of diameter D_o. Most intense mixing is near the jet itself, which rapidly entrains fluid from the region surrounding the nozzle. There is reverse flow near the outside of the cylindrical mixing volume as material from downstream is recycled to replace entrained fluid. The total volumetric flow of a jet increases with distance according to the relations

$$\frac{q}{q_o} = 0.3 \frac{x}{D_o} \text{ (air)} \tag{4-73}$$

$$\frac{q}{q_o} = 0.2 \frac{x}{D_o} \text{ (water)} \tag{4-74}$$

where x is distance downstream of the nozzle and q_o is the volumetric nozzle flow at the chamber pressure. A jet is capable of entraining and mixing about 20 to 30 times its own fluid volume before it becomes weakened at about 100 nozzle diameters. Ribbon or slot jets behave similarly with parameters somewhat different from those noted. In general, they entrain about one third the fluid of a circular jet at the same downstream distance, but they persist from much larger distances. (See Perry Table 5-6 for more specific details.)

Fluid jet mixing is universally found in combustors, reactors, and other equipment where gas uniformity is necessary. Because of its simplicity, the cost and maintenance of a jet mixer usually are negligible compared with other equipment.

The relative power consumption of jet mixers, based on mixed fluid flow rate, is indicated in Table 4-16. The values, which cover a range from mild to intense agitation, represent the power consumption of pumps or compressors employed to pressurize the motive fluid. On a flow sheet, the pump or compressor usually appears separately. In this case, the values shown are not needed. They are merely included in the table for comparison with other forms of agitation.

Orifice Plate

[*Pipeline Disperser (McCabe and Smith 260)*]

Another device employed to promote agitation of gases and nonviscous liquids is a constriction or partially closed valve in a pipe or flow channel. This induces

turbulence and recirculation; much like the fluid jet above. To accomplish moderate to intense mixing, the ratio of orifice to pipe diameters will range from 0.5 to 0.2. This creates a pressure drop of 0.05 to 0.3 bar in typical liquid pipelines. Turbulence created by an orifice is capable of dispersing one immiscible liquid in another. Droplet size of the dispersed phase can be predicted from McCabe and Smith (Equation 9-56, p. 260). With gases or miscible liquids having viscosities less than 0.1 Pa · s, mixing can be accomplished efficiently with an orifice plate followed by a length of pipe equal to 50 pipe diameters. The range of power consumption for mixing or dispersion is shown in Table 4-16.

Motionless Mixer

(*Perry 19-22; McCabe and Smith 249*)

Motionless mixers are ingenious pipeline devices that subdivide and recombine filaments of viscous liquids, slurries, or pastes. The mixer is a series of twisted metal

ribbons or similar elements inserted inside a section of pipe. Each element divides the fluid, rotates it, and delivers it to the following element, which subdivides and twists the fluid further. Two to twenty of these steps are enough to completely mix even highly viscous materials (see Figure 4-24). The pressure drop is low, only about four times greater than that of straight pipe. The typical pressure drop in a motionless mixer generally is in the range of 0.01 bar for most liquids, increasing to about 1 bar for 100 Pa·s pastes, which flow only with difficulty in pipes. Corresponding ranges of power consumption are shown in Table 4-16.

Motionless mixers were developed for viscous fluid mixing under conditions where turbulence is not necessary or desired. Because of low capital, maintenance, and operating costs, they have become increasingly popular in other applications. These include heat transfer enhancement (expecially for highly viscous flow), gas–gas mixing, liquid–liquid suspension, gas–liquid dispersion, and other applications listed in Table 4-16. With appropriate modification, they can even be employed for solid–solid mixing. The positive mixing characteristics and low pressure drops of motionless mixers make them superior to orifice plates in almost all applications.

Gas Sparger, Bubble Column

(*Perry 18-67, 19-12, 21-11; McCabe and Smith 255; Treybal 140*)

In extremely corrosive service or situations requiring gas–liquid contact for brief periods and with mild agitation, spargers are commonly employed. These are merely perforated tubes or porous elements immersed near the bottom of a tank

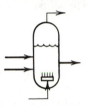

through which gases bubble and rise to the surface. In the process, they agitate the liquid and, in some cases, react or exchange mass or heat with it.

For critical design, equations in the references cited above are available to define most situations with reasonable precision. For approximate analysis, the range of commercial practice is rather narrow, allowing us to employ some rules of thumb. Depending on sparger design and gas flow rate, bubbles normally range from 2 to 6 mm in diameter, rising through thin liquids at a rate of 0.15 to 0.30 m/s. The recommended gas rate is 0.004 m^3/s per square meter of tank cross section for mild agitation, 0.008 for relatively complete agitation, and 0.02 for violent motion. This yields gas volume fractions varying from 2 to 10 percent. Sparging is possible for liquids as viscous as 1 Pa·s.

Power consumption is that required by the compressor, which supplies gas to the sparger (see note e, Table 4-16). The pressure differential can be calculated from liquid depth plus 0.1 to 0.3 bar added to force the gas through holes in the sparger.

Sparged vessels are commonly employed for ore flotation and reactors requiring gas–liquid–solid contact. Since gas volume fraction and residence time rae narrowly limited in sparged vessels, mechanical agitators are often added to provide more intimate contact and longer residence times.

Pump or Agitated-Line Mixer

(*Perry 19-9, 21-5*)

Agitators that depend on relative fluid motion are limited in versatility and in duration of contact. To overcome these shortcomings, mechanical energy can be

applied directly through various devices. The simplest of these is a centrifugal pump. For mixing, dispersion, or emulsification, it can be installed directly in the pipeline.

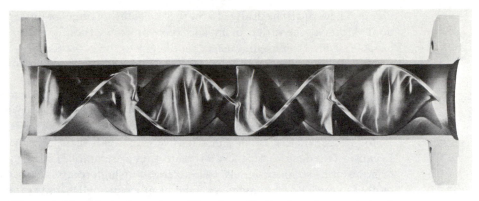

Figure 4-24 Motionless mixer. (Chemineer-Kenics, Inc., by permission.)

For agitation within a vessel, a pump or blower can be installed in an external pipe loop, which withdraws liquid or gas from the tank and reinjects it through a nozzle. Installed directly in a pipeline, the pump behaves somewhat like an orifice or motionless mixer except that pump mixers require additional motive power. A vessel fitted with a pump or blower has the agitation characteristics of a fluid jet or sparger, depending on design of the reinjection nozzle. Pressure drops, mixing intensities, and power consumptions for various types of pump agitation are similar to those for the nonmechanical modes that they simulate. Representative values are listed in Table 4-16. As a rule of thumb, a fluid volume equal to that inside the tank should be pumped through the external loop to adequately mix a vessel fitted with a pump. This rule can be employed to determine mixing times in batch operations or residence times in continuous ones.

To improve on pump mixers, which admittedly were designed for a different purpose, certain proprietary pumplike units, installed directly in a pipeline, have been developed for special applications. Their power consumption and capital costs are similar to those of the centrifugal pumps they replace.

Although simple to install, maintain, and operate, pump or agitated-line mixers are much less versatile than propeller or turbine mixers, which represent the majority choice for mechanical agitation of low to medium viscosity fluids.

Propeller and Turbine Agitators

(Perry 18-75, 19-4; McCabe and Smith 221; Treybal 146, 521, 726, 732; Gates et al. [18]; Oldshue [36])

Similar to mechanical mixers and blenders found in laboratories and kitchens, these devices are, by far, the predominant type of agitator employed in chemical process

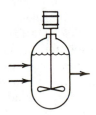

plants. They consist basically of a motorized rotating impeller immersed in a liquid pool. Although the variety of impeller types and vessel configurations is large, only propeller and turbine agitators are significant to the generalist.

Propeller agitators (Figure 4-25) resemble three-bladed marine propellers employed for ship propulsion except that in commercial service, the vessel remains stationary while the fluid moves. Propellers cause fluid to flow parallel with the rotating shaft (axial flow). Installed vertically in a tank, fluid circulates in one direction along the axis and in the reverse direction along the walls (see Perry, p. 19-6). Propellers are employed extensively in small-scale, flexible applications. Because of the deep vortex they can create, they are employed often to disperse gases or nonwetting solids in liquids. Characterized by high rotational speeds, propellers seldom exceed 1.5 m in diameter. One or more propeller units often are inserted through the side, to agitate large storage tanks or vessels. They range in power from those of laboratory size to 50 kW. An interesting application of a propeller mixer is that combined in a heat exchanger shell to serve as a well-stirred alkylation reactor for petroleum processing (illustrated in Perry 21-7).

Turbine impellers are mounted on shafts like propellers, but they are usually much larger and rotate at lower speeds. Turbines, available in a variety of impeller

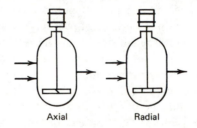

Axial   Radial

designs, are more flexible and more efficient than propellers in a number of critical applications. Most common turbine impellers fall in either the radial flow or axial flow category. Radial impellers can be visualized as flat-bladed stars (see Figure 4-26a). Like impellers in centrifugal pumps, they discharge liquid at high velocity in the radial direction. This acts like a jet mixer, entraining surrounding fluid, while setting up two circulation systems. One is above the impeller, the other below. Liquid flowing outward separates at the wall, part flows upward to the surface and returns to the eye of the impeller along the shaft. The other stream flows downward along the wall, across the vessel bottom, and back to the center of the impeller (see Figure 4-26a and Perry 19-6). Axial impellers are similar to the radial ones, except blades are pitched, usually at about a 45-degree angle. This, as illustrated in Figure 4-26b, causes flow to move downward parallel to the shaft and then upward along the vessel wall. Because of uniformity and control of circulation, axial flow turbines are superb for suspension of solids, dispersion of immiscible liquids, heat transfer enhancement, and promotion of chemical reaction. Radial impellers, on the other hand, are superior for gas dispersion.

Practical turbine impellers are limited to about 5 m in diameter, restricting vessels to the size indicated in Table 4-16. If larger capacities are required, multiple vessels or multiple agitators are recommended.

Except when gases or solids are to be entrained from the surface, vortex formation represents a loss of mixing energy and is undesirable. To avoid this, either the agitator is offset from the tank axis or else baffles (vertical finlike strips) are

Figure 4-25 Propeller agitator. (Mixing Equipment Company, Inc. A unit of General Signal © 1982, by permission.)

placed on the inner vessel walls. Although an infinite combination of scale parameters is possible, most turbine-agitated vessels are designed with a ratio of tank diameter to impeller diameter ranging from about 2 to 5. They typically have four vertical baffles about one twelfth as wide as the tank diameter. The impeller is normally located one third to one fourth of a tank diameter above the bottom. As a rule of thumb in both propeller- and turbine-agitated vessels, the liquid height ranges from 0.75 to 1.5 times vessel diameter. If it must be substantially greater, two or more impellers are mounted on the same shaft.

Like most process equipment, final selection of a fluid agitator should not be made without consulting a specialist. For preliminary design, however, it is convenient to know that most commercial applications fit within a rather narrow band of specifications. For example, power consumption typically falls within the

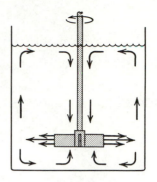

(a)

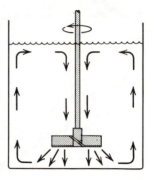

(b)

Figure 4-26   Circulation patterns with (a) radial flow and (b) axial flow impeller on a turbine agitator. [Chemineer-Kenics, Inc. Excerpted by special permission from CHEMICAL ENGINEERING (December 8, 1975). Copyright © 1975, by McGraw-Hill, Inc., New York, N.Y. 10020.]

range of 0.03 to 0.2 kW/m³ for mild, 0.2 to 0.5 kW/m³ for vigorous, and 0.5 to 2 kW/m³ for intense agitation. This is overall power consumption, assuming 90 percent efficiency within the driver speed reducing system. Power consumption is related to agitation intensely in a stirred vessel much as it is related to flow rate in a pipe. At conventional turbulence conditions, friction coefficients are relatively constant with Reynolds number and, thus, independent of viscosity. Hence, the guidelines are valid for viscosities up to about 25 Pa · s. At higher viscosities, up to 500 Pa · s, special impellers such as the anchor or helical types are usually employed. For design in this range, vendors or more specialized literature should be consulted.

For more precise design within the low viscosity range, you should be warned that power is a rather poor index of mixing effectiveness. Torque, the power divided by rotational speed, is theoretically more satisfying and usually is employed by mixing experts. This suggests that one can reduce power consumption by increasing impeller size while decreasing rotational speed. At constant torque, the mixing quality would remain essentially constant. This is, in fact, attempted. However, as rotational speed is decreased, impeller size and gear reduction requirements reach practical limits. This narrows practical power ranges to those cited in Table 4-16. Reference 18, particularly Section 12, can help you narrow the ranges even more. Here, Gates, Dickey, and Hicks have categorized 58 industrial agitation applications into 10 degrees of difficulty. Typical power levels and rotational speeds, can then be found according to vessel size and mixing difficulty in Section 4, 5, and 6 of reference 18.

Although impeller speed affects cost, it can be disregarded in predesign economic estimates. On the other hand, cost is so sensitive to the method of mounting an agitator that cost data for three types are presented in Figure 5-42. Open tanks are the cheapest configuration. Here, the agitator is simply mounted on a frame above the roofless tank. To contain pressures up to 10 barg with noncritical or nonhazardous fluids, a moderately expensive stuffing box shaft seal can be used. For critical or toxic fluids and pressures up to 80 barg, expensive mechanical seals are required. Custom mechanical seals have been used for pressures as high as 350 barg.

For assistance in preliminary design, the following data and characteristics of agitators for particular types of service are provided.

MIXING OF MISCIBLE LIQUIDS AND SOLUTIONS   (*McCabe and Smith 248*)

For this operation, moderate agitation is recommended, with either a propeller or a turbine impeller requiring a specific power consumption of 0.2 to 0.5 kW/m^3. The mixing time for a batch vessel, taken from data of McCabe and Smith, can be calculated by

$$\theta = 12,000 \left(\frac{\mu V}{P} \right)^{1/2} \left(\frac{V}{1.0 \ \text{m}^3} \right)^{1/5} \tag{4-75}$$

where μ is viscosity in pascal-seconds (Pa · s), P is in watts, and V is in cubic meters. For water at room temperature, in a 5 m^3 vessel, agitated with a specific power of consumption of 500 W/m^3, the estimated mixing time is 23 s. For continuous flow vessels, a residence time equal to this batch mixing time is adequate.

DISPERSION OF IMMISCIBLE LIQUIDS   (*Perry 21-6; McCabe and Smith 255, 260; Treybal 521*)

For extraction, where liquids will be separated again by sedimentation, droplets should be between 0.1 and 1 mm in diameter. Such can be dispersed in volume fractions up to 0.6 or 0.7 with turbine axial impellers at moderate to vigorous agitation intensity. For extraction, a residence time of 60 s is usually adequate to provide 90 percent or more of a theoretical stage.

SUSPENSION OF SOLID PARTICLES   (*Perry 19-10; McCabe and Smith 245; Treybal 726*)

Parameters required in this application depend on settling characteristics of the particles. If the settling velocity is less than 0.02 m/s, mild agitation will suffice with either a radial or axial turbine impeller. Up to 0.05 m/s settling velocity, an axial impeller under vigorous agitation conditions is recommended. From 0.05 to 0.1 m/s, intense agitation with a propeller or axial turbine is recommended. (Settling velocities can be estimated with the aid of Figure 4-40.) Suspension of more rapidly settling particles requires even more intense agitation, but particles with settling velocities of up to 10 m/s have been suspended successfully with axial flow turbines. For easily leached or fast-reacting solids, 60 s in a vessel is an adequate residence time. Contacting efficiency will approach 100 percent under these conditions, but the stage efficiency is considerably less than this because exiting sludge contains about 50 percent liquid.

EMULSIFICATION   (*Perry 21-11; Treybal 526; McCabe and Smith 228*)

Emulsions are suspensions of immiscible liquids in which the dispersed droplets, 1 to 1.5 μm in diameter, are too small to coalesce and separate. To create an emulsion, axial agitators are employed with high specific power and high blade tip speeds. Occasionally, a draft tube is employed with propeller agitators. This is a cylindrical, open-ended, submerged shroud around the impeller, which forces all circulating liquid to pass through the high shear mixing zone near the propeller blades.

DISPERSION OF GASES   (*Perry 18-75, 18-80; McCabe and Smith 256; Treybal 153*)

Mechanical agitation is employed in a sparged tank when longer gas residence times are desired. In agitated vessels, volumetric gas holdup can be increased from 10 (volume) percent, mentioned above for the unaided sparged tank, to 30 percent with intense mechanical agitation (approximately 1.0 to 2.0 kW/m^3 based on mixture volume). With gas fluxes mentioned in note *a* of Table 4-16, average superficial bubble rise velocities in mechanically aided tanks are reduced by a factor of up to four, and residence times are lengthened accordingly. Fluxes, with agitation, can be as high as 0.1 m^3/s $\cdot$ m^2. Bubble diameters remain relatively constant, with increasing agitation intensity, at about 1 to 5 mm.

If a vortex is created by intense agitation with an axial impeller, gas above the liquid surface will circulate through the froth continuously as though in a well-stirred vessel. This alleviates the need for a sparger. The average residence time can be calculated accordingly with 30 to 40 percent of the total vessel characterized as gas residence volume. Such mechanically aided surface aerators are popular for oxidation of activated sludges in waste water treatment. Data on the performance of this equipment are included in Perry 18-77 to 18-82. Specially designed units having power consumptions of 0.02 to 0.06 kW/m^3 can supply 3×10^{-4} to 9×10^{-4} kg of oxygen per kilowatt-second (kW $\cdot$ s).

Mass transfer rates for fluid and mechanically aided sparged tanks can be estimated from information in Perry 18-80 and 18-88. According to Perry, 18-77, mechanical agitation is economical in sparged reactors when the overall reaction rate, without agitation, is five or more times larger than the mass transfer coefficient. One significant application of sparged tanks, both fluid and mechanically agitated, is in separation of minerals and ores by way of air flotation (Perry 21-65). Power consumption in such operations is typically in the range of 0.002 to 0.003 kW per kilogram of solid feed. Mechanically aided sparged tanks are employed for numerous aeration, hydrogenation, and fermentation operations. For more details, see reference 18, Parts 6 and 12.

Solids Mixers

(*Perry 19-14, 21-30; McCabe and Smith 895*)

For mixing pastes, doughs, and polymers that have viscosities greater than 200 Pa · s, simple agitation is often not enough. Simultaneous squeezing, dividing, and folding is necessary with increasing power consumption as the paste becomes more viscous. McCabe and Smith categorize the action in the order of least to greatest difficulty as kneading, dispersion, and mastication. The approximate viscosity ranges for these subdivisions are indicated in Figure 4-27. Machines themselves become larger and heavier as demands increase. In fact, heat generated in such service may require water cooling to avoid damage to the process material or machine. An energy balance can be easily applied as a check on this. With dry powders, energy consumption is considerably lower and equipments costs are reduced accordingly.

The numerous types of solids mixer operated strictly on a batch basis are adequately described in both Perry and McCabe and Smith. Since this text focuses

Figure 4-27 Typical power consumption ranges for mixers processing high viscosity pastes, dought, and polymers.

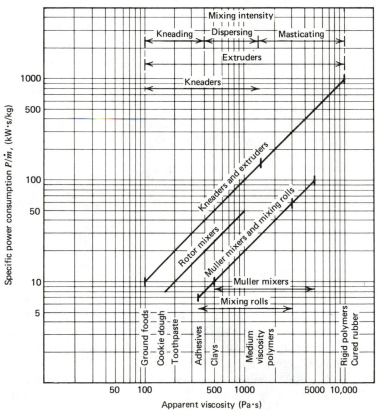

on continuous processing, only machines thus suited are discussed here. The first paste mixer listed in Table 4-16, the so-called *kneader*, is similar to an auger or screw conveyor except that the helix may be segmented, moving in a reciprocal as well as a circular path through stationary teeth attached to the inside of the barrel. These devices accomplish separating, folding, and compressing typical of a kneading operation. They can process up to 1.0 kg/s of heavy, stiff, and gummy materials such as clays, pastes, adhesives, light polymers, and doughs. Power consumption (kW) varies from 25 $\dot{m}$ for low viscosity pastes to 150 $\dot{m}$ for polymers (see Figure 4-27).

Extruders, employed commonly for plastic fabrication (see section on size-enlargement equipment), can also serve as mixers. A screw of variable pitch and

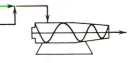

diameter rotating inside a straight or multitapered barrel forms the mixing chamber. Extrusion action is ideal for pastes or nonabrasive semisolids that are reasonably well dispersed but require high shear under pressure to give improved consistency. Extruders are suitable for materials of the viscosity processed by kneaders as well as stiffer polymers and elastomers. Power consumption (kW), with the latter, may be as high as 1000 $\dot{m}$. Both kneaders and extruders require cooling if the process stream is heat sensitive or if power consumption is significantly more than the minimum.

Mixing rolls are similar to roll mills, that is, multiple rotating drums that pinch the process medium as it passes through the gap between them. The rollers, often

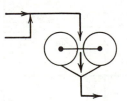

rotate at different speeds to create shear as well as compression. Employed commonly for dispersing additives and pigments into heavy polymer suspensions, they are generally not as versatile for compounding operations as are kneaders or extruders. Of simple mechanical design and efficient, mixing rolls are relatively low in capital cost and usually are selected for the few applications for which their rather limited mixing characteristics are adequate.

Almost identical to pan crushers described earlier, *muller mixers* masticate with a smearing action much like mortar and pestle. The wheels roll over the paste, smashing and rubbing it against the surface of the pan. One unit designed for continuous operation is composed of two pans and muller assemblies designed to provide reasonable residence times and steady flow. Muller mixers are useful for blending and dispersing viscous semisolids. They are efficient for uniformly coating a granular material with liquid, but they do not function well with fluids that are low in viscosity and sticky. Mullers are also useful for mixing of some dry powders. The appropriate range of fluid mixing and power consumption are indicated in Figure 4-27. For powder mixing, capacities are similar to those shown for rolling compression (pan) crushers in Table 4-5. Power consumption (kW) will be in the range of 1 $\dot{m}$ to 5 $\dot{m}$ for this service.

The *rotor mixer* represents another of many variations on the auger conveyor. Rotor mixers themselves can be found with numerous types of rotor, one or two in a single casing. The rotors may be helices as in screw conveyors. In the misnamed pug mill (not really a mill but a mixer), rotors have propellerlike blades that cut and mix as they convey. Some rotors operate at relatively high speeds and have projections on the shaft that impact and disintegrate the process medium by shearing it against the housing or fixed projections mounted thereon.

Rotor mixers are effective with modestly viscous nonsticky pastes and soft lumpy solids or cakes. They can also be used efficiently for mixing of powders. They are easily enclosed for operation with controlled or vacuum atmospheres and are commonly provided with hollow rotors and jacketed housings for removal or addition of heat. Power consumption with pastes falls slightly below the level shown for kneaders in Figure 4-27, and mixing intensity is somewhat less as well.

Another variation of auger conveying is found in the *ribbon mixer* where concentric double-helical counterrotating ribbons lift, disperse, and blend dry

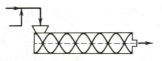

powders. Ribbon blenders are not suitable for sticky pastes or liquids. Because of light construction and mildness of agitation, they are used primarily for free-flowing powders. Specific power consumption (kW) falls, accordingly, within the range of 1 $\dot{m}$ to 6 $\dot{m}$ as shown in Table 4-16.

In addition to machines designed primarily for agitation and mixing of solids, certain *mills*, normally used for size reduction, can be employed. The muller mixer, described earlier, is almost identical with the pan crusher. Drum or vibratory pebble and jet mills can be used for continuous mixing and blending of friable and free-flowing solids, whereas hammer, cage, and attrition mills are commonly employed to blend mixtures of sticky and gummy materials. Capacities and specific power consumption for these units can be assessed from data given under crushing and grinding equipment in Table 4-5. For mixing, power consumption values on the low side and capacities on the high side of ranges shown in that table should be assumed.

PROCESS VESSELS

(Absorbers, Accumulators, Adsorption Towers, Autoclaves, Columns, Cookers, Dehumidifiers, Distillation Towers, Drums, Fluidized Beds, Fractionators, Fume Collectors, Humidifiers, Internals, Ion Exchangers, Kettles, Leaching Equipment, Packed Towers, Plate Columns, Pressure Vessels, Scrubbers, Stills, Stripping Towers, Trays, Vaporizers, Washers)

As the list illustrates, this is an exceptionally popular category of process equipment. The main attribute that ties these numerous items into a single generic type is the characteristic cylindrical, dished-end vessel from which they are formed. Since various reaction, storage, and separation operations are conducted in the same type of vessel, discussion in this section is limited to mass transfer operations such as those itemized above.

Detailed design considerations are legion, constituting most of the chemical engineering college curriculum, and it is out of the question to expound them here. This is done well in the references cited this chapter. On the other hand, abbreviated guidelines exist for preliminary design. These are emphasized here. Capital costs of process vessels can be estimated rapidly and with reasonable accuracy once their dimensions, internal details, construction materials, and maximum temperatures and pressures have been approximated.

Flow Sheet Considerations

Treybal, in his treatise [52],[6] separates the study of mass transfer into six possible combinations of gas–liquid–solid interaction: gas–gas, gas–liquid, gas–solid, liquid–liquid, liquid–solid, and solid–solid. In the first, there are limited commercial applications. *Gas–liquid* phenomena, on the other hand, which include distillation, gas absorption, desorption (stripping), humidification, and dehumidification, are undoubtedly the most prominent in chemical engineering design. *Gas–solids* interactions, except for static bed dryers and adsorbers, generally require agitation. Equipment to promote this has already been discussed under the category of gas–solid contacting equipment. *Liquid–liquid* (e.g., extraction) and *liquid–solid* operations (crystallization, leaching, adsorption) are often conducted in conventional process vessels and are discussed in this section. *Solid–solid* contracting, like gas–gas, is rather limited commercially, reducing the original list of six permutations to four that are commonly encountered in chemical processing. Three of these (gas–liquid, liquid–liquid, and liquid–solid) are usually accomplished in conventional process vessels.

Even though "process vessels" are primarily devoted to the three types of contact just named, they interact with a variety of process streams. At the risk of confusion, I have organized process stream types into 13 categories as shown in Table 4-17.

When phases are distinct, such as liquids or solids entrained in gases, immiscible liquids, and gases or solids entrained in liquids, simple mechanical separation or filtration often is possible. These techniques are discussed in the

---

[6] This reference, of those listed, has been most valuable in the preparation of this section and is highly recommended for detailed as well as preliminary design of process vessels.

TABLE 4-17

POSSIBLE LIQUID–GAS–SOLID COMBINATIONS IN PROCESS STREAMS AND
GENERAL TECHNIQUES EMPLOYED TO SEPARATE THEM

Dispersed Phase	Continuous Phase	Type of Dispersion	Separation Technique	Generic Equipment Reference
Liquid	Gas	Physical	Mechanical	Separators
Liquid	Liquid	Physical	Mechanical	Separators
Liquid	Liquid	Solution	See Figure 4-28	Mixers, process vessels
Liquid	Solid	Various	Drying	Gas–solids contacting equipment
Gas	Gas	Solution	See Figure 4-28	Process vessels
Gas	Liquid	Physical	Mechanical	Mixers, separators
Gas	Liquid	Solution	See Figure 4-28	Process vessels
Gas	Solid	Various	"Drying"	Gas–solids contacting equipment, process vessels
Solid	Gas	Physical	Mechanical	Separators
Solid	Liquid	Physical	Mechanical	Separators
Solid	Liquid	Solution	See Figure 4-28	Mixers, process vessels
Solid	Solid	Physical	See Figure 4-28	Process vessels, crushers, mills, grinders, separators
Solid	Solid	Solution	See Figure 4-28	Process vessels

sections on mixers and separators. For liquids or gases dispersed in solids, separation is conventionally categorized as "drying" and is discussed in the section on gas–solid contacting equipment. When components are more intimately or energetically combined, a choice must be made from among a large number of unit operations conducted, usually, in "process vessels."

Selection from myriad alternatives is aided by the decision guide illustrated in Figure 4-28, which lists the more commonly encountered conventional mass transfer operations. To identify a favorable technique, follow the path that applies to the particular process stream in question. By answering yes or no at each decision point, the path becomes established. These are the same techniques employed in final process design except that decisions will be supported by extensive laboratory work, detailed examination of alternatives, or more experience. A dilemma arises with Figure 4-28 when an answer is "perhaps" or an arrow leads to a box entitled 'Specialized techniques." Two choices remain: you can stop and consult a specialist for help, or you can take the conservative approach, choosing an operation that although more expensive, will accomplish the job. This will permit you to establish a process and define its economics with the assurance that the final process, even though somewhat different, will not be more expensive. For a more extensive list and definition of "specialized processes," refer to King [25, Table 1-1 p. 21] and Perry Sections 16 and 17.

Specific process guidelines, peculiar to the more prominent mass transfer operations, are presented below.

DISTILLATION VERSUS SOLVENT EXTRACTION   (*McCabe and Smith 619, 626; Treybal 478*)

As illustrated in Figure 4-28, alternate separation techniques may, in numerous cases, be employed. For example, methanol could be extracted from aqueous solution using an organic solvent rather than by fractional distillation. The latter is

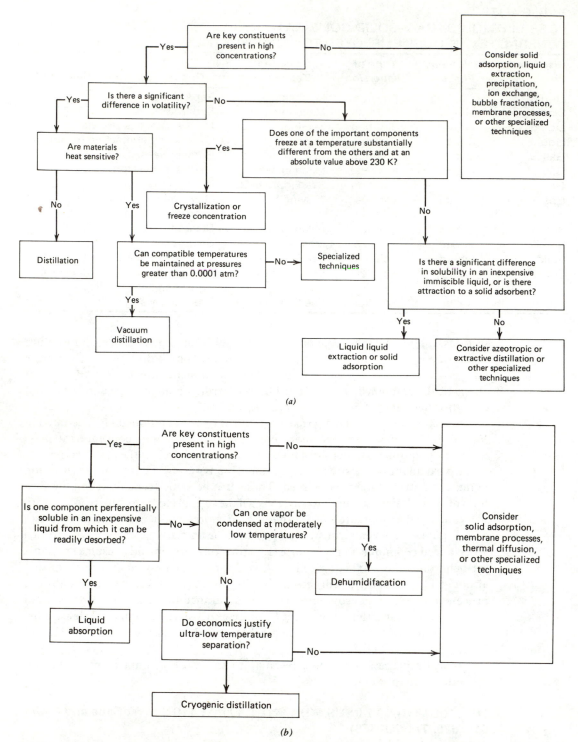

Figure 4-28 Decision guide for selecting among alternate schemes for separation of (*a*) liquid–liquid solutions, (*b*) gaseous mixtures, (*c*) a gas from gas–liquid solutions, (*d*) a solid from solid–liquid solutions, and (*e*) solid–solid mixtures.

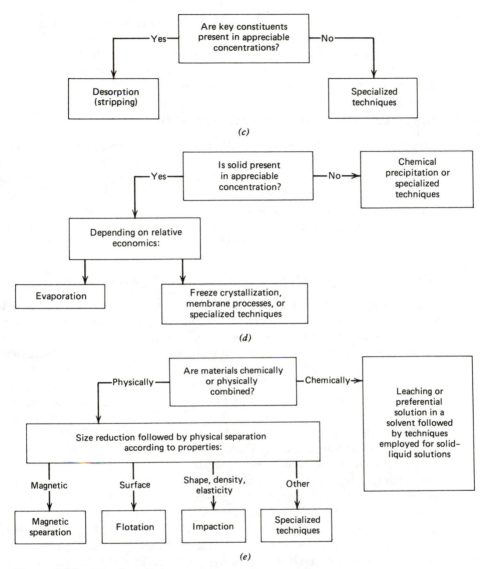

Figure 4-28 (Continued)

generally practiced commercially because in a liquid–liquid extraction, the methanol must subsequently be separated from the solvent, usually by distillation. Thus, as a general rule, distillation is more economical than liquid extraction, which is reserved for heat-sensitive materials, nonvolatile substances, or combinations that do not separate easily on vaporization. These conditions often exist in pharmaceutical manufacture, where liquid–liquid extraction is prominently employed.

DISTILLATION ORDER   (*Treybal 431*)

In a solution containing three liquids, several combinations exist. In one, the split can be made between the least volatile and the intermediate compound in one tower. Then, the two light liquids in the condensate can be split in a second tower. Another approach is to split the lightest compound from the others in the first tower and separate the remaining two by feeding the bottoms to a second tower. In both cases, two towers are required, but the second alternative generally requires less energy and is more economical. [As discussed by Treybal (p. 433), separation by distillation of n species into pure components requires $n - 1$ towers.]

DISTILLATION PRESSURE   (*Treybal 347*)

Since relative volatility decreases with increasing pressure, rectification is usually conducted at the lowest reasonable column pressure. This, in turn, is dictated by the coolant temperature, which controls pressure in the condenser. As a rule of thumb, assuming that cooling water is available at 30° C and a 10° C approach temperature is used in the condenser, the pressure at the top of the tower will often correspond to the saturation pressure of the distillate at 40° C. If this is an inappropriate limitation, use of other coolants usually is required.

ANCILLARY EQUIPMENT   (*Treybal 392, 397, 417*)

It is well known that a distillation column requires a reboiler to supply vapor to the bottom of the tower and a condenser to liquefy the overhead vapors for reflux. In practical terms, the reboiler can be either a kettle type, a thermosyphon, or a forced-convection heat exchanger. The condenser is normally a shell and tube heat exchanger oriented horizontally with condensation occurring on the outside of the tubes. For a conservative preliminary design, it is customary to assume total condensation of overhead vapors and to disregard any fractionation occurring in the reboiler. In practice, of course, use of a kettle reboiler and partial condenser could provide up to two stages in addition to those inside the tower.

To ease cleaning and maintenance, the condenser is usually located at ground level rather than near the top of the tower. To provide controllability and surge dampening, condensate is collected in a reflux accumulator (a drum having approximately 5 min residence time) before being pumped back as reflux to the top of the tower. Distillate product, piped separately from the accumulator, usually passes through another heat exchanger to subcool saturated liquid for storage.

Surge storage for the heavy product is provided by liquid residence volume at the foot of the tower that is free of trays or packing. A volume adequate to allow 5 to 10 min residence for liquid leaving the bottom tray or packing support is normally provided.

For controllability and dampening of process upsets in gas absorption, similar "foot" space is provided in the tower, and a feed storage drum, analogous to the reflux accumulator, having approximately 5 min residence time, is provided for solvent.

NONCONVENTIONAL ARRANGEMENTS

In aqueous systems where water is the bottom product, steam may be added directly to the tower in preference to using an indirectly heated reboiler. For difficult binary separations, a separate solvent or "entrainer" may be added to change relative volatility or to form an azeotrope. For substances, particularly natural products, that are temperature sensitive, vacuum distillation is frequently employed. These schemes are generally used only when necessary for adequate separation or product

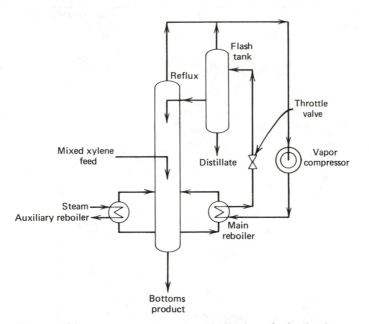

Figure 4-29 Vapor recompression distillation of mixed xylenes (*Hydrocarbon Processing* [48], by permission.)

protection. Useful information on selecting nonconventional distillation techniques can be found in Treybal (pp. 416, 455–462) and Perry 13-36 and 13-43.

An increasingly important variation from traditional distillation involves vapor recompression, in which latent energy contained in the overhead product provides heat in the reboiler. Since reboiler temperatures are higher, the vapor must be compressed to raise its condensing temperature. Capital costs of the added compressor and flash drum are considerably greater than those of the condenser that is eliminated, but this is usually justifiable if the temperature difference between bottoms and distillate is less than about 30°C [24]. A flow sketch for mixed xylene separation is shown in Figure 4-29.

Returning to Figure 4-28, conventional distillation and absorption, because of their prominence and frequency of use, are considered in more detail in this section. Vacuum distillation, desorption, dehumidification, and cryogenic distillation are merely variants that can, of course, be analyzed by similar techniques. Liquid–liquid extraction, leaching, and flotation may be realistically represented on a flow sheet by a series of stirred tanks and settlers. Evaporation is discussed in depth in the section on evaporators and vaporizers.

Equipment Specification

As illustrated in Table 4-18, process vessels can be classified simply as either towers or drums. Even in this regard, the distinction is only one of degree. Towers have the same form and construction as drums except that towers are generally much taller or longer and as the name implies, are oriented vertically. Drums, as a rule, are much shorter and may be oriented with their axes either horizontal or vertical. Both towers and drums are fitted with various "nozzles" or pipe fittings to accommodate flow in and out plus instrument connections and access for cleaning, inspection, or

TABLE 4-18

CRITERIA AND DATA FOR THE PRELIMINARY SELECTION OF PROCESS VESSELS

	Type of Process Vessel			
	Towers			
	Tray		Packed	Bubble (sparged)
	Gas–Liquid	Liquid–Liquid		
Range of Common Equipment Sizes				
Diameter, D (m)	0.5–4	0.5–4	0.3–4	0.3–4
Length or height, L (m)	6–60	6–60	1–60	6–40
Length to diameter ratio	5–30	5–30	5–30	5–10
Normal orientation	Vertical	Vertical	Vertical	Vertical
Type of Operation				
Mass transfer	√	√	√	√
Heat transfer				
Settling				
Storage	√	√	√	√
Suitability				
Distillation				
Moderate to high pressure	A	X	A	D
Vacuum	D	X	B	D
Gas absorption (scrubbing)	A	X	A	D
Gas drying	A		A	D
Humidification (desorption)	A	X	A	B
Leaching	E	C	E	B
Liquid drying	B	B	B	D
Liquid–liquid extraction	X	A	D	X
at small density difference	X	D	E	X
Solids drying	X	X	D	X
Removal of solids from gases	D	X	D	B
Design Characteristics				
Pressure drop				
Per stage or unit (kPa)	0.5–1	0.5–1		1–2
Per meter height (kPa)		0.2–0.6^d		
Liquid flow rate				
Mean mass flux, LM_L/A_t (kg/s · m^2)	<$20D^{-1}$		0.5–100	
Mean velocity, u_L (m/s)	<$0.02D^{-1}$	~0.03	0.005–0.1	
Mass, $\dot{m}_L$ (kg/s)	<$15D^{-1}$			
Gas flow rate				
Mean mass flux, G (kg/s · m^2)			1–4	
Mean velocity, u_g (m/s)	$0.06\,(\rho_l - \rho_g)^{1/2}\rho_g^{-1/2}$		—^{e}	
Mass, $\dot{m}_g$ (kg/s)				
Typical stage dimensions				
Tray separation, H_t (m)	$0.5D^{0.3,f}$			
HETP^g (m)		0.3–3	$0.5D^{0.3}$–$1.0D^{0.3,h}$	L
Residence time, θ (s)				60 (gas)
Stage efficiency (percent)	60–80	15–30		90–100
Diameter of dispersed particle droplet or bubble, d_p (mm)				2–6
Volume fraction of dispersed phase				
Compatibility				
Low gas rates	C		D	D
Low liquid rates	C	B	D	B
Entrained solids	E	E	E	D
High gas rates	B		B	B
High liquid rates	D	E	A	D

KEY

A   excellent or no limitations
B   modest limitations
C   special units available at higher cost to minimize problems
D   limited in this regard
E   severely limited in this regard
X   unacceptable

^{a}Gas inlet duct.

^{b}This, according to Gerunda (19), depends on pressure. Below 18 barg, $L/D = 3$ is recommended; from 18 to 35 barg, $L/D = 4$, and above 35 barg, $L/D = 5$. Flash and knock-out drums may vary from these guidelines as settling velocities and holdup considerations dictate.

^{c}Water-jet scrubbers, because of liquid momentum, are capable of creating drafts of up to 2 kPa per stage.

^{d}See values associated with Figure 4-32 for more precise ranges and lower values for vacuum operation.

^{e}See Figure 4-32.

TABLE 4-18
(Continued)

			Type of Process Vessel			
Towers		**Venturi Scrubber**	**Drums**			
Spray	**Specialized or Proprietary**		**Flash**	**Holdup**	**Mixer**	**Settler**
2–4	0.1–3	0.1–2[a]	0.3–4	0.3–4	0.3–4	0.3–4
6–40	3–10	<4	1–20	1–20	0.5–8	0.5–16
3–10	6–10	1–2	3–5[b]	3–5[b]	1–2	2–4
Vertical	Vertical	Vertical	Vertical	Horizontal	Vertical	Horizontal
√	√	√	√		√	
√					√	
		√	√			√
√		√	√	√		
D	X	X	D			
D	X	X	D			
D	D	D	E			
D	D	D				
B	D	D				
E	C	C			A	A
B	D	E				
B	B	E			A	A
D	A	X			A	C
C	X	D				
B	B	A	E			A
0.2	0.5–1	−2 to +4[c]				
	0.01					
	0.05–30					
<1.2		20				
1–3			$0.06(\rho_\ell - \rho_g)^{1/2}\rho_g^{-1/2}$			
		<60				
L	0.1–1					
.60 (gas)		1–10	600[i]	300–600	60	300–600[i]
90–100	30–90				90–100	100
0.001–0.01		0.0001–0.001			0.1–1.0	
					<0.7	<0.7
B	A	B	D	A	A	
B	A	B	A	A	A	A
D	C	A	B	E	A	C
D		A	A	A	D	
A	C	A	A	A	A	C

[f] For $D > 1$ m. For $D < 1$ m, $H_t = 0.5$ m. In special applications such as cryogenic distillation, values as small as 0.15 m have been used.

[g] Height equivalent to a theoretical plate or stage.

[h] For tower diameters greater than 0.5 m, HETP = $0.5D^{0.3}$ (distillation) and HETP = $1.0D^{0.3}$ (gas absorption). For diameters less than 0.5 m, HETP = D is valid for both distillation and gas absorption. In vacuum distillation, add 0.15 m to the values.

[i] This is for the liquid volume below the entrance point. An additional one meter or one drum diameter (whichever is larger) is provided above the liquid for disengagement. Superficial vapor velocity is calculated from Equation 4-93. See the text for more details.

[j] Perry 21-12. Treybal (p. 528) recommends the equation $D = 8.4\dot{q}^{0.5}$ to calculate settler diameter, where $\dot{q}$ is the volumetric fluid rate of continuous plus dispersed phases (m³/s) and $L/D = 4$.

other purposes. Towers, in most cases, contain trays, packing support plates, baffles, or other "internals" to accomplish specific process functions.

With both towers and drums, the cost engineer's first objective is to define length, diameter, wall thickness, and material of construction. For more refined cost estimates, the internal details of a tower must also be identified.

DRUMS

For drums, specification at the predesign stage is elementary. Volumes are based on average residence times such as those shown in Table 4-18. Unless internal pressures are unusually large, the ratio of length to diameter will be within the range listed. Otherwise, it will be larger because of the ability of small-diameter vessels to withstand higher pressures.

TOWERS

To specify the height of a tower, the number of trays or transfer units must be known plus the height between trays or the height of a transfer unit. To identify tower diameter, the relative internal liquid and gas flows are defined and their appropriate velocities calculated. These objectives, which usually form the bulk of an undergraduate mass transfer course, are almost universally established by computer in the modern process industry. Most design organizations have proprietary programs of their own; others can derive the benefits of sophisticated techniques employed by vendors when purchase is made or bids requested. On the other hand, any engineer who is not suspicious of man, woman, or computer is either inept or inexperienced. To check on computer results or to communicate intelligently with a specialist, preliminary, approximate calculations, such as those presented below, are extremely valuable. Traditionally, packed and tray towers are analyzed by different techniques because the former is a stage contractor and the latter continuous. In practice, neither appearances nor costs of the two are substantially different. For preliminary design, I recommend analyzing both distillation and absorption as stagewise operations to determine the number of theoretical stages or plates. Then, using either the height of packing equivalent to a theoretical plate (HETP) or the recommended distance between plates, one can calculate the height of the tower. (This is despite Treybal's caution, p. 301, against this technique for detailed design.) Establishing liquid flows and vapor flows and determining the number of theoretical stages are discussed separately for distillation and gas absorption because the techniques are different. Once these key parameters have been established, however, sizes of packed and tray towers are determined by the same procedure. This is described later.

Distillation

STEP 1   Equilibrium Relationship

The first step in determining the number of stages required for a given separation is to find the equilibrium relationship. For hydrocarbon systems, distribution coefficients in Perry (Figure 13-6, pp. 13-12 and 13-13) are useful. Other equilibrium data can be found in the traditional reference literature. For a binary system, it is almost automatic to prepare an x–y, vapor–liquid equilibrium diagram so essential for the classical McCabe–Thiele analysis.

STEP 2   Minimum Trays

Considering first a binary distillation, the minimum number of trays is easily determined by traditional methods such as counting or calculating the stages, at total reflux, between the distillate and bottoms compositions.

STEP 3   Minimum Reflux Ratio

Based on feed condition, minimum reflux ratio is evaluated from the operating line equilibrium curve intersection at the pinch point as described in the classical McCabe–Thiele analysis.

STEP 4   Actual Reflux Ratio

Economic considerations dictate that the true reflux ratio be somewhere between 1.05 and 1.5 times the minimum (Table 4-19). This helps define the operating lines. Frank, in an excellent reference for short-cut techniques and practical advice on distillation column design and hardware [16], refines this somewhat for separate cases of refrigerant or water and air-cooled condensers.

STEP 5   Number of Theoretical Stages

The number of perfect trays required for separation is obtained by graphically stepping and counting as in the McCabe–Thiele technique or, if more accurate or convenient, by calculation [25, Chapter 8].

TABLE 4-19
RECOMMENDED REFLUX RATIOS FOR VARIOUS CONDENSING TEMPERATURES IN DISTILLATION COLUMNS (16)

Condensing Utility	Quotient of Actual to Minimum Reflux Ratios, R/R_{min}	Approximate Ratio of Actual to Minimum Number of Theoretical Stages, N/N_{min}
Low temperature refrigerant ($<$ −100°C)	1.05 to 1.1	2.5 to 3.5
Moderate refrigerant (−100 to 0°C)	1.1 to 1.2	2.0 to 3.0
Water- or air-cooled condensers	1.2 to 1.3	1.8 to 2.5

Multicomponent Distillation

Unfortunately, the simple and generally reliable McCabe–Thiele technique is not easily applied to multicomponent distillation. An analogous approach, however (known as the Fenske, Underwood, Gilliland procedure, can be employed to yield an approximate answer. It, like the McCabe–Thiele analysis, is limited to constant or nearly constant molal overflow.

STEP 1   Definition of Keys

Select the so-called light and heavy keys, the lightest component of appreciable concentration in the bottoms product (light key) and the heaviest in the distillate (heavy key). These compositions should be readily available from the material balance table on the flow sheet.

STEP 2   Minimum Stages

Calculate minimum number of theoretical stages from the Fenske equation.

$$N_{min} = \frac{\ln \left[\left(\frac{x_{lk}}{x_{hk}} \right)_D \left(\frac{x_{hk}}{x_{lk}} \right)_B \right]}{\ln \alpha_{lk}} \qquad (4\text{-}76)$$

where x represents mole fraction of heavy or light key. Relative volatility of the light to heavy key, α_{lk}, is assumed to be constant. If it changes substantially from bottom to top, a mean value

$$\overline{\alpha}_{lk} = [(\alpha_{lk})_D \, (\alpha_{lk})_B]^{1/2} \qquad (4\text{-}77)$$

may be employed.

STEP 3   Minimum Reflux Ratio

Calculate this ratio of reflux to distillate from Underwood's equations

$$R_{min} + 1 = \sum_j \frac{\alpha_j x_{j,D}}{\alpha_j - \phi} \qquad (4\text{-}78)$$

and

$$1 - q = \sum_j \frac{\alpha_j z_{j,F}}{\alpha_j - \phi} \qquad (4\text{-}79)$$

The parameter q defines thermal quality (fraction liquid) of the feed (Treybal 407, Frank 114, Perry 13-28), α_j represents the volatility of component j relative to the heavy key, $z_{j,F}$ designates mole fraction in the feed, and ϕ is the Underwood constant, which lies between 1.0 and α_{lk}. When a distributed component (having a relative volatility between the keys) is involved, additional steps are required. These are discussed in the references just cited.

STEP 4   Actual Reflux

Define the actual reflux ratio as 1.05 to 1.3 times the minimum according to guidelines of Frank [16] and elaborated in Table 4-19.

STEP 5   Number of Theoretical Stages

The number of theoretical trays is found from N_{min}, R_{min}, and R, using the Gilliland correlatoin [40], plotted in Figure 4-30, or the Erbar–Maddox correlation (Perry Figure 13-29, p. 13–28).

Note that approximate ratios of actual to minimum theoretical trays are recommended by Frank and listed in Table 4-19. One is tempted to use these directly without employing steps 3, 4, and 5, especially 3, which can be tedious. Unfortunately, the reflux ratio must be known for reboiler and condenser design. Thus, step 3 is essential, regardless, making steps 4 and 5 almost automatic. For assistance, Frank includes charts [16, Figures 2 and 4] that accelerate the process. For a neophyte, detailed calculations outlined in Treybal [52, Illustration 9.13 and

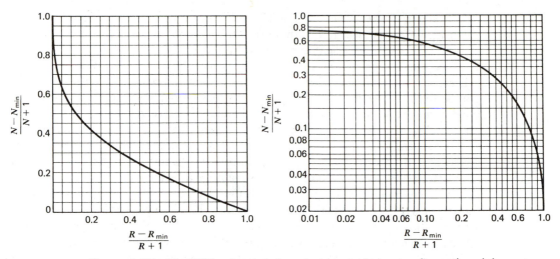

Figure 4-30   The Gilliland correlation relating actual trays to reflux ratio, minimum trays, and minimum reflux ration. Charts are recommended for rapid, approximate analysis of multicomponent distillation towers. [Excerpted by special permission from CHEMICAL ENGINEERING (March 14, 1977). Copyright © 1977, by McGraw-Hill, Inc., New York, N.Y. 10020.]

its continuation through pp. 436–453] should be valuable even though the technique departs near the end from that recommended above. Marinas-Kouris[29] provides a similar illustrative example for comparison. In multicomponent distillation, compositions of nonkey components in bottoms and distillate are not always easily obtained. Yaws et al. [54] suggest a short-cut method that should be useful for this.

This concludes the steps necessary to define liquid and vapor flows and the number of theoretical stages in distillation. (Internal column flows can be easily calculated by traditional material balances, since reflux ratios, feed rate, and product rates are now known.) A similar analysis for gas absorption follows and then, methods for determining tower size are explained.

Gas Absorption

With some noteworthy differences, the liquid and vapor flows and number of theoretical trays in stagewise gas absorption are determined by a series of steps analogous to those applied for distillation. As with distillation, each stage is viewed as an equilibrium contactor with entering streams related by a material balance and leaving streams in equilibrium with one another. Since carrier gas and absorbing liquid usually pass through the tower without appreciable loss or gain, absorption stoichiometry is based on moles of transported species per mole of pure carrier gas Y or absorbing liquid X. This differs somewhat from gas phase y and liquid phase x mole fractions, which are used in distillation analysis.

EQUILIBRIUM RELATIONSHIPS

Equilibrium ratios can be determined from basic thermodynamic phase relationships, given vapor pressures and activity coefficients. Data for specific combinations are found in the references at chapter end. Some common absorber solvents are identified in Perry 14-2 and, with more detail, in Kohl and Riesenfeld [27]. Useful guidelines for solvent selection are given by Treybal 281.

As with distillation, terminal conditions in gas absorption are dictated by flow sheet specifications. If not, a range of tentative inlet and outlet concentrations must be assumed and the corresponding absorbers designed and economically evaluated to determine optimum separation. A potentially serious error arises if one neglects the energy balance. Desirable absorber solvents have a strong affinity for the absorbate gas, causing exothermicity. If a temperature rise shifts the equilibrium curve so that it crosses the proposed operating line, the result is embarrassing and, if not detected, expensive. To remedy this, heat must be removed from the tower or proposed terminal gas compositions must be changed. In the absorption of HCl in water, heat transfer rather than mass transfer is the major constraint on equipment. As an initial approximation, you can assume that the tower is adiabatic and perform an overall energy balance. As a rule of thumb, the leaving gas temperature will be within 2°C of the entering liquid temperature (Treybal 295). Since the entering gas temperature is presumably known, the exiting liquid temperature can be calculated from an enthalpy balance. If there is any vaporization of solvent, its concentration in the leaving gas is assumed to be in equilibrium with entering liquid.

If temperature changes are not significant, the true equilibrium curve can be established. If there is a temperature rise modest enough to obviate the need for external cooling, the equilibrium points at liquid inlet and outlet temperatures can be connected by a straight line to yield a conservative result. This is true of both absorption and desorption.

OPERATING LINE

Based on solute-free molal gas V' and liquid L' values, the operating line is straight with slope equal to L'/V'. It terminates somewhere on the horizontal line that represents entering gas composition. The slope can be any positive value, depending on relative gas and solvent rates. Optimization reveals that economic performance is usually found for absorption when values of L'/mV' lie between 1.25 and 2.0 (Treybal 292), where m is the slope (assumed linear) of the equilibrium curve. Inversely, for stripping, the optimum ratio will fall between 0.5 and 0.8.

THEORETICAL TRAYS

With L' and V' established, an operating line is defined. The number of theoretical stages can be determined simply by stepping off trays graphically (Treybal 289) or by using an algebraic expression such as the Kremser equation (Treybal 128, 291). The number of trays may prove unrealistically low (less than four) or forebodingly large. If so, you may wish to reconsider parameters such as gas exit concentration, adiabatic or nonadiabatic operation, or liquid rate, which are subject to arbitration. How many trays should be used? The answer is, "The most economic number." Unfortunately, establishing this is not trivial. However, if the calculated number falls between 10 and 30 for noncritical separations, further refinement will have little effect on most predesign estimates.

This concludes the determination of flow rates and number of theoretical stages for distillation and gas absorption. The same principles, applied with judgment, can yield similar reasonably accurate parameters for desorption, humidification, dehumidification, liquid and solid absorption, and desorption. Assistance with these operations is available in the references.

Calculation of Tower Size

HEIGHT

Tray Towers

Now that the number of ideal stages is established, vertical tower height H_a can be estimated merely by stacking them one above another if the height per stage is known.

$$H_a = \frac{NH_t}{\epsilon_s} \tag{4-80}$$

For tray towers, theoretical stage height is the actual tray separation distance H_t

divided by overall tray efficiency ϵ_s. Efficiency values for distillation can be estimated from Figure 4-31a. Absorption and desportion efficiences, which are usually lower, can be identified from Figure 4-31b.

The stage separation distance H_t can be assumed to be 0.5 m for diameters up to 1.0 m. In larger diameter towers, Treybal (162) recommends

$$H_t = 0.5 D^{0.3} \tag{4-81}$$

With nonfoaming liquid separations involving many trays and significant heat loss or gain (e.g., cyrogenic distillation of liquid air), tray spacings as small as 0.15 m are used. With a given tray spacing, the active tower height can be determined from Equation 4-80. Total tower height can be determined later by adding inactive volume below the bottom tray for 5 to 10 min liquid surge capacity. This will usually equal 1 to 4 m additional height.

Packed Columns

For packed columns, the height is computed as above except the height equivalent to theoretical plate (HETP) is employed:

$$H_a = N \times \text{HETP} \tag{4-82}$$

After spending endless hours learning how to calculate and combine mass

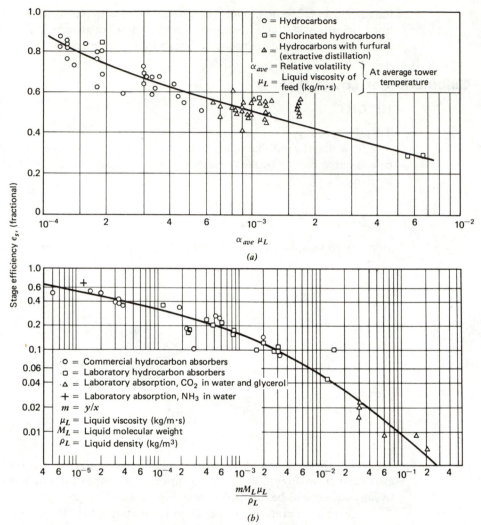

Figure 4-31 Stage efficiencies for tray towers in distillation (*a*) and gas-absorption service (*b*). Bubble-cap trays were used. Results are recommended for conservative enumeration of sieve and proprietary trays. (From *Mass Transfer Operations*, 3rd edition, p. 185, by Treybal, R. E. Copyright © 1980. Used with the permission of McGraw-Hill Book Company.)

transfer coefficients, it is somewhat disquieting to find that HETP values, in commercial towers, fall within a rather narrow range. In distillation, for diameters greater than 0.5 m, Frank [16] recommends HETPs about equal to tray heights given by Equation 4-81 or 4-83.

$$\text{HETP} = 0.5\ D^{0.3} \tag{4-83}$$

For gas absorption, HETPs are approximately twice as large.

$$\text{HETP} = D^{0.3} \tag{4-84}$$

Below 0.5 m diameter,

$$\text{HETP} = D \tag{4-85}$$

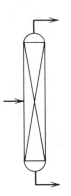

is a good rule of thumb for both distillation and absorption. In vacuum distillation, it is wise to add an extra 0.15 m to the HETP.

DIAMETER

Tray Towers

With modern, high efficiency stages, a new tray tower is usually limited by vapor entrainment or foaming rather than by liquid flow capacity. Thus, for preliminary design, the entrainment limit is recommended as a basis for determining tower cross section. This is the same basic equation employed to calculate cross sections in evaporators, flash drums, and settling tanks. The relationship for tray towers is:

$$u_{s,g} = K_{SB} \left(\frac{\rho_l - \rho_g}{\rho_g} \right)^{1/2} \tag{4-86}$$

where $u_{s,g}$ is the operating superficial vapor flow velocity based on total tower cross section. For most tray towers, the optimum Souders-Brown constant is approximately 0.04 to 0.08 m/s. The lower value applies to gas–liquid systems having large surface tensions or towers with small tray spacing. (For more precise correlations see Perry 18-6 and Treybal 163–167). To calculate tower diameter, employ the overall vapor flow rate in a rearranged form of the continuity equation (assuming a cylindrical tower, of course):

$$D = \left(\frac{4V}{\pi \rho_g u_{s,g}} \right)^{1/2} \tag{4-87}$$

V should be the maximum molar vapor rate in the tower. In distillation, this normally occurs in the stripping section or base, but with partially vaporized feed, it may exist in the rectifier. In vacuum distillation, one should be careful to check the diameter at the top tray, where molar densities may be surprisingly small.

Packed Towers

In contrast to tray units, performance in packed towers is strongly affected by both liquid and vapor rates. Not only is flow area limited, but the gas begins to retard liquid flow at high loadings. Based on extensive research, design velocities for packed towers can be established with confidence. The correct value depends, of course, on shape and size of packing. Nevertheless, for prominently used slotted rings and high efficiency saddles, the relationship, Figure 4-32, originated by Sherwood and amplified by Eckert [10] is recommended, where

$G = VM_g/A_t$, superficial gas mass flux (kg/s · m²)
C_f = packing constant (dimensionless; see insert to Figure 4-32)
μ_l = liquid viscosity (Pa · s)
ρ_g = gas density (kg/m³)
ρ_l = liquid density (kg/m³)
L = liquid flow rate (mol/s)
$A_t = \pi D^2/4$, tower cross-sectional area (m²)
M_l = liquid molecular weight (kg)
M_g = gas molecular weight (kg)
V = gas flow rate (mol/s)

Figure 4-32 Flooding and pressure-drop correlations for packer towers. (From *Mass Transfer Operations*, 3rd edition, p. 195, by Treybal, R. E. Copyright © 1980. Used with the permission by McGraw-Hill Book Company.)

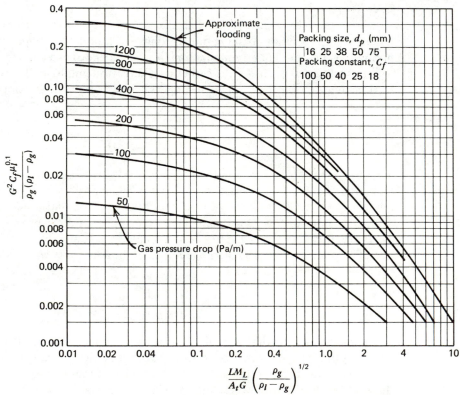

The appropriate curve should be selected using the following pressure drops recommended by Treybal (p. 195).

Recommended Pressure Drop	Pa/m
Absorption and stripping	200–400
Atmospheric and high pressure distillation	400–600
Vacuum distillation	8–40

In establishing packing size, the value selected should be at least eight times smaller than tower diameter.

With superficial gas velocity available from this correlation, tower diameter can be determined from the following.

$$D = \left(\frac{4 V M_g}{\pi G} \right)^{1/2}$$

(4-88)

The final designer may later choose a different packing, affecting this result, but the change will be well within limits of preliminary design accuracy.

TRAY VERSUS PACKED TOWERS

Now that tray and packed towers have been discussed, it is enlightening to consider the merits and disadvantages of each so that, for most cases, selection can be made without the need to evaluate both. General rules of thumb for this choice have been defined and published by numerous authors. Thibodeaux and Murrill [51] assembled a checklist that resembles, to some extent, a qualitative decision chart. It is included in Figure 4-33. Reasons for various selection criteria should be obvious from the foregoing discussion. Others can be discovered by referring to the original paper [51] and to Treybal 210 and Perry 14-13.

For preliminary design, select the alternative with the highest number of checks. If the difference is not decisive, costs will not vary significantly from one tower type to the other.

Once tower size and type have been defined, one needs only identify the construction material before capital costs can be calculated. Another item of importance for flow sheet development is tower pressure drop. Typical values per actual tray or per meter of packing height are listed in Table 4-18, and the table on page 198. Pressure is, of course, higher at the bottom of the tower. Otherwise, gas, which is not aided by gravity, would not flow upward.

OTHER CONSIDERATIONS

Although it has a relatively small impact on total price, the type of tray or packing is important in final design. As a general rule, sieve trays are the choice for most staged contactors. With exceptionally small or large liquid rates or rates that vary appreciably, more flexible proprietary tray designs may be employed. Extensive information on types of trays, their advantages or disadvantages, and detailed design can be found in Perry Section 18, Treybal 158–178, and Frank [16].

Judging from advertising literature, "packing wars" develop occasionally among various suppliers. For analysis here, currently popular slotted ring or high efficiency saddle packings were assumed. This is a conservative design, since we are assured by various manufacturers that their new packing can do better.

Other Types of Liquid–Liquid, Liquid–Solid, and Liquid–Gas Contactors

(Perry 21-21; Treybal 530)

The tray tower, so effective for gas–liquid contacting is also employed for counterflow liquid–liquid processing when liquids are immiscible and their

COLUMN SELECTION GUIDE

	Column Type			Column Type	
	Packed	**Plate**		**Packed**	**Plate**
Factors dependent upon the system			If system requires wide variations in liquid and/or gas rates, check plate; if not, both.	☐	☐
If system has a foaming tendency, check packed; if not, both.	☐	☐	If liquid holdup is undesirable, check packed; if not, both.	☐	☐
If system contains solids or sludges, check plate; if not, both.	☐	☐	If column pressure drop is to be kept low, check packed; if not, both.	☐	☐
If the constituents are corrosive fluids, check packed; if not, both.	☐	☐	*Factors Dependent upon the Physical Nature of the Column*		
If system has heat of solution difficulties check plate; if not, both.	☐	☐	If frequent cleaning is expected, check plate; if not, both.	☐	☐
If operation is intermittent, check plate; if not, both.	☐	☐	If weight is critical, check plate; if not both.	☐	☐
If the scale of the system is small, check packed; if not, both.	☐	☐	If side streams are to be employed, check plate; if not, both.	☐	☐
If system is temperature sensitive, check packed; if not, both.	☐	☐	If diameter of column is less than 2 ft, check packed; if greater, check plate.	☐	☐
If system has close boiling components, check packed; if not, both.	☐	☐	If overhead clearance is critical, check packed; if not, both.	☐	☐
If system is viscous, check packed; if not, both.	☐	☐	If floor space is critical, check plate, if not, both.	☐	☐
Factors Dependent upon the Flow Regime in the Column					
If resistance to mass transfer is controlled by gas phase, check packed; if controlled by liquid phase, check plate; if no phase is controlling, check both.	☐	☐	**Totals** Packed column _____ Plate column _____		

Figure 4-33   Qualitative checklist for selecting between plate and packed columns. [Excerpted by special permission from CHEMICAL ENGINEERING (July 18, 1966). Copyright © 1966, by McGraw-Hill, Inc., New York, N.Y. 10020.]

densities differ by 100 kg/m^3 or more. Design for this service follows the pattern outlined above. Packed towers, because of a more tortuous flow path, are not generally as suitable for liquid–liquid contact and are not recommended.

BUBBLE AND SPRAY TOWERS   (Perry 21-28; Treybal 542)

A relatively cheap counterflow contactor is the simple spray or bubble column, where liquid droplets or bubbles are dispersed into and rise or fall through another

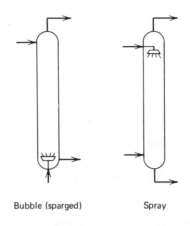

Bubble (sparged)   Spray

continuous phase. The contacting efficiency is poor in these devices, however, and they rarely achieve performance greater than a single stage. For these reasons, they are preferred only when a small number of theoretical stages is required.

SPECIALIZED DEVICES   (Perry 21-24; McCabe and Smith 624; Treybal 544)

Beyond traditional process vessels, there is an astonishing number of specialized or proprietary devices intended to provide more efficient liquid–liquid contacting and

mass transfer. These include at least seven different mechanically agitated tower designs and a centrifugal (Podbielniak) contactor. With most of these, design techniques are not easily accessible to the nonspecialist. For preliminary analysis, I recommend that a more traditional contactor be assumed and designed as the basis of the cost estimate. This is done with the presumption that more advanced devices, if actually used later, will be more economical. This approach is acceptable as long as the conventional, conservative equipment does not represent an inordinate portion of the plant capital. (This seldom is a limitation, since more expensive evaporation or distillation equipment is normally required to purify extract or raffinate.) Otherwise, a specialist should be consulted. In most liquid–liquid or liquid–solid contacting operations, reasonable cost estimates can be based on a series of individual stagewise cascades, each consisting of a mixer and settler, designed within guidelines suggested in Tables 4-16 and 4-18.

VENTURI AND WATER-JET SCRUBBERS   (Perry 20-100; Treybal 186)

Although not strictly process vessels as defined earlier, venturi and water-jet scrubbers are prominent devices employed for gas–liquid mass transfer, especially

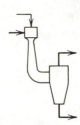

when the gas contains solid particles and a small pressure drop or even a pressure increase is required. These devices are often employed as exhaust scrubbers where large volumes of particle-laden gases are involved. In effect, they behave like a high efficiency spray tower. Venturi scrubbers consist of a water spray injected at high velocity into a confining duct where parallel gas flow is induced. Then, liquid is collected in a separator drum and gas is exhausted. The liquid is usually recirculated by a pump in an external pipe loop. In water-jet units, the water spray is injected into the throat of an aspirator much as steam is employed in an ejector (Figure 4-8). This creates a vacuum that can be used to pump the gas. (Less vacuum can be achieved in practice, however, than is possible with gas-motivated ejectors.)

It should be obvious that venturi scrubbers can provide no more than one stage of contact per unit. In systems that involve chemical reaction as well as absorption, this is often adequate. For more detailed design specifications, refer to Perry (Table 20-42) and McCarthy [31].

Drums

Drums, simple process vessels without complicated internal appendages, are used for storage, settling, separation, and numerous routine process functions. Most

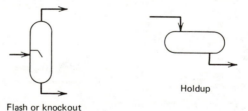

Flash or knockout   Holdup

drums can be specified from information found elsewhere in this chapter. Some useful data are contained in Table 4-18. Since flashing is somewhat like distillation, a few words regarding flash drums are appropriate here.

"Flashing" is a common operation employed to separate and partially enrich a vapor–liquid mixture. It occurs simply and spontaneously when a fluid passes through a pipeline restriction (usually a throttling valve). The pressure drops suddenly, and vaporization occurs. The mixed phase flows into an open drum were liquid and vapor separate, liquid leaving from the bottom of the drum and vapor through a pipeline at the top.

Analysis of flash separation is based on a direct application of chemical engineering fundamentals. Feed temperature, composition, and flow rate are known. The downstream pressure either is known or can be assumed for trial-and-error calculation. The following calculation procedure is recommended.

Step 1   Assume a downstream temperature.

Step 2 For the given temperature and pressure, obtain vapor–liquid equilibrium constants for each of the components from the literature.

Step 3 Assume a value for fraction of original feed vaporized f. A value between 0 and 1.0 should be selected.

Step 4 Calculate liquid composition from

$$Fz_i = F(1 - f)x_i + Ffy_i \qquad (4-89)$$

and

$$y_i = x_i K_i \qquad (4-90)$$

which, combined, becomes

$$x_i = \frac{z_i}{1 + f(K_i - 1)} \qquad (4-91)$$

Step 5 Check

$$\Sigma x_i = 1.0 \qquad (4-92)$$

Return to step 3 and cycle through step 4 until step 5 is satisfied.

Step 6 Calculate vapor composition from Equation 4-90.

Step 7 Use an energy balance to check the assumed temperature. Since the drum is essentially adiabatic, enthalpies of product streams must equal the enthalpy of the feed. If the enthalpy balance is not satisfied, assume a new temperature and return to step 1.

This procedure may seem tedious. For a mixture containing a large number of components, it is. If a substantial number of such calculations is anticipated, it will probably save time to obtain or develop a computer program. For assistance in hand calculation, a plot of f versus temperature and one of h versus temperature for a given mixture will lead to more rapid convergence. For more on this subject and illustrative solutions, see Treybal 360–367 and Perry 13-17.

To determine the size of a flash drum, a liquid holding time of approximately 10 min is employed to define volume below the inlet. Above the inlet, one drum diameter or 1 m, which ever is greater, should be provided [19]. Varpor superficial velocity can be determined according to the Souders-Brown equation

$$u_g = 0.064 \text{ m/s} \left(\frac{\rho_l - \rho_g}{\rho_g} \right)^{1/2} \qquad (4-93)$$

where the constant is based on removal of 100 μm droplets. Drum diameter can be computed from u_g, assuming that it is oriented vertically. If liquid loading is low and L/D as calculated is less than 3, horizontal orientation will yield a more economical result. Here, the same limiting gas velocity and liquid residence time apply, but to a horizontal drum half full of liquid.

PUMPS

(Ejectors, Jets, Syphons: Perry 6-3; McCabe and Smith 179; Foust et al. 580)

The fundamentals and equations employed for pump design are described, as an illustration, at the beginning of this chapter. It remains in this section, therefore, to identify pumps by type and to state guidelines for their selection. In review, using the mechanical energy equation, shaft power for a pump in a given application is calculated from:

$$\dot{w}_s = \frac{\dot{q}\,\Delta p}{\epsilon_i} \qquad (4\text{-}94)$$

This is the same as Equation 4-3 except that the ratio of mass flow rate to density $\dot{m}/\rho$ has been replaced by its equivalent $\dot{q}$, the volumetric flow rate. The intrinsic efficiency ϵ_i compensates for friction and energy losses inside the pump. To calculate power consumption P, shaft power $\dot{w}_s$ must be divided by another efficiency, that of the driver ϵ_d.

$$P = \frac{\dot{q}\,\Delta p}{\epsilon_i \epsilon_d} \qquad (4\text{-}95)$$

Since cost data are related to $\dot{w}_s$ and operating costs are proportional to P, these are the only equations necessary for economic analysis and preliminary design. As mentioned earlier, $\dot{q}$ and Δp can be obtained directly from the process flow sheet and material balance.

To identify the type of pump and its intrinsic efficiency for a given application, Table 4-20 should prove useful. (Driver efficiencies can be obtained from Figure 4-2.) Brief inspection reveals a strong similarity between Table 4-20 and Table 4-9, which pertains to gas movers and compressors. In fact, most types of liquid pump have correspondingly larger, more expensive, and more powerful counterparts, which act as gas movers and compressors. Reasons for these scale distinctions revert to the difference in density between a gas and a liquid and the impact that has on power consumption and equipment size.

Pumps can be categorized according to the physical mode of locomotion, centrifugal force, mechanical displacement, momentum transfer, and volumetric displacement. These are identified more specifically in Table 4-20 and are described in detail below.

Centrifugal Pumps

When an engineer says "pump" unqualified, nine times out of ten a centrifugal pump is implied. More specifically, since Table 4-20 lists three types of centrifugal pumps, the centrifugal radial pump is by far the most prominent and popular pump in the chemical process industry. Axial flow pumps, which act much like ship propellers or propeller agitator blades, are valuable for moving large volumes at low differential

pressures. Conversely, regenerative or "turbine" pumps are more attractive for low flow–high pressure application. Since, however, radial centrifugal pumps can perform the same service at a competitive cost, I focus on that design with the assurance that axial flow or regenerative units, if ultimately chosen in preference, will create a negligible change in process economics. A glance at Table 4-20 reveals why the centrifugal radial (hereafter referred to simply as "centrifugal") pump is so popular. Available off the shelf from numerous suppliers in a myriad of permutations and materials of construction, the centrifugal pump is usually the most economical heavy-duty, reliable mover of low viscosity liquids in large volumes. Its characteristic spiral, snail like shape is a familiar sight to plant operators and maintenance people (Figure 4-34).

To identify the appropriate efficiency in Equation 4-95, choose values near the low end of the range shown in Table 4-20 if service conditions are severe, flow rates are low, or liquid viscosity is greater than 0.5 Pa · s. Choose efficiencies near the maximum for high volume, moderate pressure service with clean liquids having viscosities less than 0.05 Pa · s. For clean, low viscosity fluids, the efficiency versus flow rate relationship can be expressed by [50]:

$$\epsilon_i = 1 - 0.12 \, q^{-0.27} \tag{4-95a}$$

At higher viscosities, see Tan [50] for more precise values.

Because of its prominence, the performance chart for a centrifugal pump, Figure 4-35 (similar to those found in suppliers' catalogs) deserves some attention here. This is a plot of pressure "head" versus volumetric flow rate for a centrifugal pump. The relationship is relatively independent of viscosity up to about 0.05 Pa · s. It does depend, however, on rotational speed as illustrated by the two curves. The use of "head" rather than differential pressure is uniquely adapted to centrifugal pumps because the pressure difference is created by conversion of kinetic to potential energy as the fluid leaving the periphery of the impeller decelerates in the diffuser or tangential exit part. This represents a conversion of momentum to pressure, which can be expressed by the following classical relationship applied to a decelerating fluid.

$$\Delta p = -\dot{m} \, \Delta u \tag{4-96}$$

At a given rotational speed, $\dot{q}$ and u are fixed. Since $\dot{m} = \dot{q}\rho$,

$$\frac{\Delta p}{\rho} = \dot{q} \, \Delta u \tag{4-97}$$

where the term on the right is constant. Thus, the ratio is characteristic of a particular pump at a given rotational speed. From fluid statics, this can also be related to the height of the column of liquid that could be supported at the pump outlet.[7]

$$\frac{\Delta p}{\rho} = g \, \Delta h \tag{4-98}$$

Pump designers have become accustomed to denoting the differential pressure by Δh. This is convenient because it remains constant for a given pump and rotational

---

[7]By an unfortunate set of circumstances, this, in the English system with its conversion constant g_c, has the units ft lb$_f$/lb$_m$. Pump designers have, for the sake of convenience but with sacrifice of rigor, canceled lb$_f$ with lb$_m$ and denoted the outlet pressure merely as "head" in linear dimensions (usually feet). To use Δh rigorously in SI units, one must employ Equation 4-98, where with $g = 9.8$ m/s^2 and Δh expressed in meters, the ratio of pressure differential to liquid density is expressed in Pa · m^3/kg.

TABLE 4-20
CRITERIA AND DATA FOR THE PRELIMINARY SELECTION OF LIQUID PUMPS

	Type of Pump								
	Centrifugal			Rotary (positive displacement)					
							Gear		
	Axial Flow	Centrifugal Radial	Regenerative (turbine)	Cam and Piston	Flexible Liner	Flexible Tube	External	Internal	Lobe
Maximum system pressure (bara)	350	350	50	350	50	10	350	350	350
Temperature range (°C)	−240 to 500	−240 to 500	−30 to 250	−10 to 270	−10 to 200	−10 to 80	−30 to 400	−30 to 400	−30 to 400
Maximum Differential Pressure, Δp (bar)									
Per stage	2	20	35	200	10	10	200	200	17
Overall		200							
Maximum capacity of stock units, $\dot{q}$ (m³/s)	5	10	1.0	0.04	0.01	0.001	0.1	0.1	0.1
Fluid viscosity range, μ (Pa · s)		<0.2	<0.1	0.001–0.1	0.0001–0.1	0.0001–0.1	0.001–400	0.001–400	0.001–1.0
Efficiency range, ϵ_i (percent)	50–85[a]	50–85[a]	20–40[a]	40–85	40–70	40–70	40–85	40–88	40–85
Relative Costs									
Purchase price	Low	Low	Moderate	High	Moderate	Low	Moderate	Moderate	High
Installation	Low	Low	Moderate	Low	Low	Low	Low	Low	Low
Maintenance	Low	Low	Moderate	Moderate	High	Moderate	Low	Low	Low
Utilities	Moderate	Moderate	Moderate	Moderate	Moderate	Moderate	Moderate	Moderate	Moderate
Service Compatibility									
Cavitation conditions	D	E	B	B	B	B	B	B	B
Corrosive liquids	C	C	C	C	A	B	C	C	C
Dry operation	E	E	E	E	A	A	·D	D	D
High flow rates	A	A	E	D	E	E	D	D	D
High pressures	X	C	B	A	D	D	B	B	D
High temperatures	C	C	D	D	E	X	C	C	C
High viscosity liquids	D	E	E	B	B	B	A	A	B
Limited prime or low suction pressures	D	·D	D	A	B	A	B	B	B
Low flow rates	X	D	A	A	A	A	A	A	B
Low viscosity liquids	A	A	A	A	D	B	D	D	D
Non-Newtonian liquids	D	D	D	B	B	D	A	A	B
Particle-laden liquids									
Abrasive	B	C	X	X	D	D	X	X	X
Nonabrasive	B	B	X	X	D	D	D	D	X
Variable capacity service	A	A	B	A	C	C	C	C	C
Variable pressure service	E	D	D	C	C	C	C	C	C
Common Construction Materials									
Carbon steel	✓	✓	✓	✓			✓	✓	✓
Cast iron	✓	✓	✓	✓			✓	✓	✓
Copper alloy	✓	✓	✓	✓			✓	✓	✓
Plastics (conventional)		✓			✓	✓	✓	✓	
Fiberglass		✓					✓		
Fluorocarbon plastics		✓			✓		✓	✓	
Polymer-coated		✓							
Glass-coated		✓							
Stainless steel	✓	✓	✓				✓	✓	✓
Nickel-based alloys		✓					✓	✓	
Titanium		✓					✓	✓	
Carbon		✓					✓	✓	
Ceramics		✓					✓	✓	
Service Problems									
Flow pulsations	A	A	A	B	B	B	B	B	B
Noise	A	A	A	A	A	A ·	A	A	B
Reversibility of flow direction	X	X	X	C	C	A	C	C	C
Leakage of critical fluids	C	C	C	C	C	A	C	C	C
Overpressure protection	A	A	D	C	C	C	C	C	C
Other advantages or disadvantages				_c,d	_c,d	_c,d	_c,d	_c,d	_c,d

KEY

A   excellent or no limitations
B   modest limitations
C   special units available at higher cost to minimize problems
D   limited in this regard
E   severely limited in this regard
F   unacceptable

[a] Independent of viscosity up to 0.05 Pa · s.

[b] Including gas compressor.

[c] Motor gear reducers are often necessary.

[d] Pressure relief protection necessary.

[e] Operated conveniently with stream or compressed air.

TABLE 4-20
(Continued)

	Rotary (positive displacement)			Reciprocating (positive displacement)		Momentum Transfer, Jet (syphon)	Volumetric Displacement	
	Piston (circumferential)	Screw	Sliding-Vane	Piston	Diaphragm		Pressurized Tank	"Air" Lift
Maximum system pressure (bara)	350	350	350	1000	350	350	350	50
Temperature range (°C)	−30 to 370	−30 to 370	−30 to 270	−30 to 370	−30 to 270	−240 to 500	−240 to 500	−240 to 500
Maximum Differential Pressure, Δp (bar) Per stage Overall	17	20	150	1500	70	1.0	350	50
Maximum capacity of stock units, $\dot{q}$ (m³/s)	0.04	0.1	0.1	0.03	0.006	1.0		
Fluid viscosity range, μ (Pa · s)	0.0001–400	0.001–1000	0.001–100	0.001–400	0.001–100		<400	<0.1
Efficiency range, ϵ_j (percent)	40–85	40–70	40–85	60–90	40–70	5–20	40–60	40–60
Relative Costs								
Purchase price	Moderate	High	Moderate	High	Moderate	Low	High[b]	High[b]
Installation	Low	Low	Low	High	Moderate	Low	Low	Low
Maintenance	Low	Moderate	Low	High	Moderate	Low	Low	Low
Utilities	Moderate	Moderate	Moderate	High	Low	Low	Moderate	Moderate
Service Compatibility								
Cavitation conditions	B	B	B	B	B	A	A	A
Corrosive liquids	C	C	C	C	C	A	A	A
Dry operation	A	D	E	E	B	A	A	X
High flow rates	D	D	D	D	E	E	A	B
High pressures	B	B	B	A	B	X	C	B
High temperatures	C	C	C	C	C	A	C	D
High viscosity liquids	A	A	A	A	D	D	B	D
Limited prime or low suction pressures	B	B	A	B	B	A	A	A
Low flow rates	A	A	A	B	A	B	A	B
Low viscosity liquids	D	D	B	B	A	A	A	A
Non-Newtonian liquids	A	A	A	A	D	D	B	D
Particle-laden liquids								
Abrasive	X	E	X	C	B	A	C	A
Nonabrasive	X	A	E	C	B	A	C	A
Variable capacity service	C	C	C	C	C	A	A	B
Variable pressure service	C	C	C	C	C	D	D	B
Common Construction Materials								
Carbon steel	√	√	√	√	√	√	√	√
Cast iron	√		√	√	√	√	√	
Copper alloy	√	√	√		√	√	√	√
Plastics (conventional)		√				√	√	√
Fiberglass		√				√	√	√
Fluorocarbon plastics		√			√	√	√	√
Polymer-coated							√	√
Glass-coated							√	√
Stainless steel	√	√	√	√	√	√	√	√
Nickel-based alloys	√		√	√		√	√	√
Titanium	√					√	√	√
Carbon						√		
Ceramics						√		√
Service Problems								
Flow pulsations	B	A	B	C,D	C	A	A	B
Noise	A	B	B	D	B	B	D[b]	D[b]
Reversibility of flow direction	C	X	X	C	C	X	X	X
Leakage of critical fluids	C	C	C	D	A	A	A	X
Overpressure protection	C	C	C	C	C	A	C	A
Other advantages or disadvantages	_c,d	_c,d	_c,d	_c–e	_c–f	_g	_h–l	_h–m

[f] Diaphragm failure should be anticipated.

[g] Process fluid may be contaminated by motive fluid.

[h] Significant space required.

[i] Limited equipment flexibility.

[j] Pressurized gas or liquid source required.

[k] Gas and liquid must be compatible.

[l] Geometric constraints limit process use.

[m] Restricted to vertical lifting.

speed, whereas the actual differential pressure depends on fluid density. Note in Figure 4-35 that there is a characteristic head created at zero flow that declines slightly as flow increases. Efficiency increases also with flow, from zero to its peak value and then back to zero again. This illustrates how a pump operating at a steady flow rate and pressure differential can be designed for peak efficiency, whereas one requiring flexibility must operate occasionally at lower efficiencies. The curve labeled NPSH in Figure 4-35 denotes the net positive suction head. This is the absolute pressure minus the vapor pressure of the fluid (as mentioned at the beginning of this chapter) required at the pump inlet to prevent cavitation.

The theoretical basis for Figure 4-35 and various factors that affect centrifugal pump performance are cogently described in McCabe and Smith (pp. 187–194). An excellent review of modern centrifugal pump technology is provided by I. J. Karassik, ["Centrifugal Pumps and System Hydraulics," *Chem. Eng.*, pp. 84–106 (Oct. 4, 1982). Also available as *Chem. Eng.* Reprint No. 083.]

Positive Displacement Pumps

When liquid viscosities are large and flow rates are small, or carefully metered liquid rates are desired, positive displacement pumps are attractive. As the name implies, fluid in these devices is pushed, carried, or squeezed by a moving surface. The simplest type to understand is probably a basic reciprocating piston and cylinder

arrangement with check values that prevent backward motion of fluid [21]. The diaphragm pump is merely a variation on this where a diaphragm deflects back and forth to give pistonlike motion without the complexities of a sliding surface that must be sealed.

Far more common, in chemical process applications, are rotary positive displacement pumps. Various ingenious arrangements of gears, lobes, screws, and vanes, on rotating shafts create positive, smooth flow. Piston-type action can be created in an inclined rotary plane by the cam and piston arrangement shown in Figure 4-36a. Squeezing of a flexible liner or tube is effective at moderately low pressure and in small volumes by the action illustrated in Figures 4-36b and 4-36c. High pressure, steady, reliable flow is delivered in gear pumps by trapping liquid between the gear teeth as illustrated in frames d and e for both external and internal gear arrangements. Similar motivation can be created by intremeshing lobes, rotary pistons, screws or sliding vanes as illustrated in Figure 4-36 and in Figure 4-37.

If extremely high pressure, moderate capacity flows are required, a more expensive reciprocating piston pump, like its large compressor relative, will often be required. Otherwise, one of the other positive displacement units is more economical. Each has its unique advantages and disadvantages (see Table 4-20). For

Figure 4-34 Centrifugal pump. (A. R. Wilfley & Sons, Inc., by permission).

Figure 4-35 Representative performance chart for a centrifugal (radial) pump.

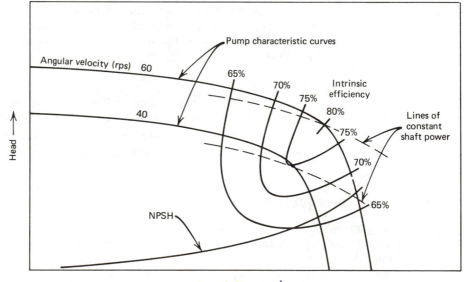

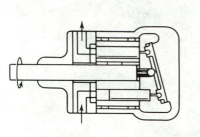

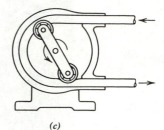

(a) *(b)* *(c)*

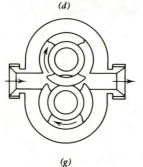
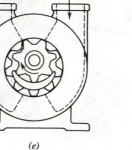
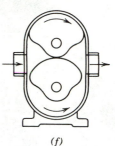

(d) *(e)* *(f)*

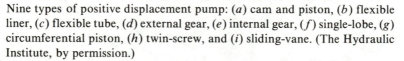

(g) *(h)* *(i)*

Figure 4-36 Nine types of positive displacement pump: (*a*) cam and piston, (*b*) flexible liner, (*c*) flexible tube, (*d*) external gear, (*e*) internal gear, (*f*) single-lobe, (*g*) circumferential piston, (*h*) twin-screw, and (*i*) sliding-vane. (The Hydraulic Institute, by permission.)

more information, the material already cited in Perry, McCabe and Smith, and Foust is useful. Some excellent photographs and sketches are also included in these references. Useful interpretive treatment of rotary positive displacement pumps is found in the review by Neerken [35] and of piston pumps in that by Henshaw [21].

Momentum Transfer (Jet) Pumps

Jet pumps, also known as syphons, are similar to the ejectors described earlier for gas moving and vacuum pumping. For theoretical reasons, pressure differentials are much less when liquid is the motive fluid. Otherwise, liquid jet pumps resemble their more prevalent cousins in appearance and performance.

Volumetric Displacement

With some fluids, direct contact pumping is almost out of the question. In other instances, particularly laboratory or temporary situations, fluid displacement pumping is merely more convenient. In pressurized tank methods (also known as "acid egg" techniques), liquid is moved under the influence of a blanket of pressurized gas. Such equipment is trouble free and, excluding the compressor, inexpensive, although it is not practical for most steady, rugged process applications.

Another type of volumetric displacement occurs with an "air" lift where gas is introduced at the bottom of a liquid-filled tube to create buoyancy and cause flow. A useful application of this technique occurs in the boring of artesian wells when compressed air is admitted at the base of the well to purge the water and wash out drilling sediment. Since jets, pressurized tanks, and air lifts involve conventional equipment of other types, such as compressors, pumps, and vessels, their capital costs can be synthesized from basic component costs. Prices of other pumps listed in Table 4-20 can be estimated with the aid of Figures 5-49, 5-50, and 5-51.

REACTORS

(*Autoclaves, Fluidized Beds*)

In most cases, reactors are process vessels, furnaces, mixers, gas–solid contactors, heat exchangers, or other conventional chemical process equipment adapted or modified for a specific reaction. As a result, reactor capital costs can be determined

Figure 4-37   Twin-screw, rotary, positive displacement pump. (Warren Pumps Div., Houdaille Industries, by permission.)

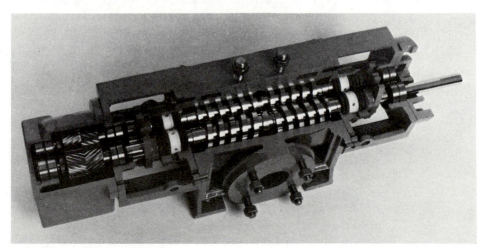

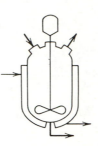

from data on these other equipment types once the size, pressure, construction materials, orientation, and other factors have been determined. The reactors themselves are usually considerably less expensive than the separation modules that process their effluents. Conversion in a reactor, however, exerts a profound influence on the cost of other process equipment and its operation. For this reason, reactor performance is often the key element in optimizing a chemical manufacturing scheme.

Assuming that the reader has taken a course in chemical engineering reactor design, I do not attempt a comprehensive treatment of the subject. Nonetheless, a brief discussion of reactor design steps and potential pitfalls is presented as a review.

Advantages of solid catalyzed reactions have led to extensive use of both fluidized and packed beds as chemical reactors. Although they are employed for other types of fluid–solids contacting as well, their prominence as reactors warrants amplification here.

Reactor Design and Specification

STEP 1   Definition of Terminal Conditions

As listed in Table 4-2, inlet temperature, pressure, and concentration are necessary for specification of a reactor. These are usually apparent from flow sheet storage conditions, and other characteristics of process modules. Sometimes pumps, heat exchangers, and other equipment are added to condition the feed and promote efficient reaction. On the other hand, reactor outlet concentration, temperature, and pressure are uniquely determined by the design and are under your control. Usually, outlet concentration is the most important of these three variables and the other two are allowed to adjust accordingly. If the desired outlet concentration is not obvious, a trial-and-error economic analysis is required to establish the optimum value. In this event, an assumed outlet conversion is employed for trial calculations.

STEP 2   Definition of Kinetic Data

Unfortunately, in contrast to physical equilibrium data, there are no comprehensive, exhaustive sources of rate data. Countless possible combinations of reagents, catalysts, pressures, temperatures, and concentrations make generalized tabulations and correlations impractical if not impossible. For primitive estimates, Perry Tables 4-16, 4-17, and 4-18 should prove useful. Literature searching techniques described in Chapter Two may also lead to tentative, approximate rate relationships. Laboratory data for the specific reaction in question are much more desirable. Additional pilot plant results provide even more security. Since those doing the laboratory work are often unfamiliar with important parameters and constraints in commercial reactor design, you should apply all your analytical skills to checking the testing the reported data. Hill [23, p. 36] goes a step further and suggests independent analysis by a colleague before laboratory or pilot plant rate data are used for design. If there are uncertainties, these should be identified and discussed with your supervisor at the outset. There is nothing more disappointing and nonproductive than an elegant reactor design that fails because of faulty kinetic data. Often tentative or approximate data must be used, but when this is necessary, the risks should be clearly identified and appropriately shared.

STEP 3   Equilibrium Analysis

In contrast to kinetic information, thermodynamic data are readily available and accurate. Equilibrium constants should be calculated and employed to confirm that an impossible result is not being sought and to define the limits of conversion that can be realized. Though trivial and fundamental, it is surprising how much effort is wasted by neglect of this step.

STEP 4   Reactor Selection

Traditionally, reactors are classified according to flow characteristics. Most are perfectly stirred, plug flow, batch, continuous, or somewhere in between. Because of their commercial prominence, only continuous reactors are discussed in this limited treatment. In reality, no reactor behaves in the ideal, plug flow or perfectly stirred mode, but most commercial units are similar enough to these models to validate idealized design equations.

In many reactor designs, other factors such as heat transfer and mixing are more significant than chemical kinetics. This is true, for example, of coal-burning furnaces, cement kilns, and hydrocarbon alkylation reactors. As indicated at the beginning of this section, reactors are usually merely modifications of other types of process equipment. This is illustrated in Table 4-21, which has been assembled from other sections of this chapter. It includes types of conventional process equipment that have been and are now used for promoting chemical transformations on an important commercial scale. Note that only six of the reactions in the lower section of the table employ what could be termed custom reactors. Two of those are electrochemical cells. For carbon black and other flame synthesis processes, reactors, although custom designed, are seldom very expensive. The remaining custom reactors, for ammonia and catalytic cracking, are merely extensions of more conventional designs. Thus, for preliminary design, selection of a suitable reactor or reactor type should be possible from your knowledge of the process and the guidelines in Table 4-21.

In making a selection, some basic rules of thumb should be kept in mind.

1   In isothermal reactors or those in which temperature drops with conversion, the volume required for a given conversion is larger in a well-stirred compared with a plug flow operation.

2   In adiabatic exothermic reactions, well-stirred reactors require the least volume except near equilibrium conversions.

3   Isothermal performance and general temperature and concentration control are accomplished more easily in well-stirred reactors.

TABLE 4-21

TYPES AND CHARACTERISTICS OF PROCESS EQUIPMENT PROMINENTLY EMPLOYED AS INDUSTRIAL CHEMICAL REACTORS

| | Conveyors | | Furnaces | | | Gas–Solids Contactors | | | | |
	Auger	Pneumatic (transport reactors)	Combustion Side	Tube-Side	Incinerators	Rotary Kilns	Gravity Shafts	Fluid Beds L/D < 4	Fluid Beds L/D > 4	Spray Towers
Flow Characterization										
Plug flow	√	√	√	√	√	√	√		√	
Well-stirred			√		√			√		√
Other										
Design-Controlling Process										
Chemical kinetics	√	√							√	√
Heat transfer	√	√	√	√		√	√			√
Diffusion	√				√	√	√			√
Mixing			√		√					
Reaction Type										
Catalytic		√		√					√	√
Noncatalytic	√	√	√	√	√	√	√	√		√
Single-phase		√	√	√						
Multiple-phase	√	√	√		√	√	√	√		
Solid–Fluid	√	√	√		√	√	√	√		
Thermal Behavior										
Adiabatic	√	√	√		√			√		√
Isothermal								√		
Intermediate	√	√	√	√	√	√	√			√
Endothermic	√	√		√		√	√			√
Exothermic	√	√	√		√			√		
Typical Process Stream										
Gas		√	√	√	√			√		
Gas–liquid mixture					√					√
Liquid–liquid mixture					√					√
Viscous liquid	√				√					√
Liquid–solid mixture	√				√	√				
Gas–solid mixture		√	√					√		
Solid	√				√	√	√			
Typical Stream Conditions										
Pressure										
100–500 bara										
10–100 bara		√		√					√	√
<10 bara	√	√		√					√	√
<2 bara	√	√	√		√	√	√	√		√
Temperature										
1000–2000°C			√		√					
500–1000°C		√		√	√	√	√			
500°C	√	√			√	√	√	√		√

Typical Commercial Reactions

Product	Ref.[b]	Auger	Pneumatic	Combustion Side	Tube-Side	Incinerators	Rotary Kilns	Gravity Shafts	L/D < 4	L/D > 4	Spray Towers
Ammonia	276										
Aluminum	224										
Butadiene	631									√	
Carbon black	121			√							
Chlorine	214										
Coal, coking								√			
Coal, combustion				√							
Esterification											
Ethylene glycol											
Ethylene oxide											
Flame synthesis											
Hydrocarbons											
Alkylation	683, 21-7										
Catalytic cracking	678									√	
Desulfurization										√	
Hydrogenation	683									√	
Isomerization	684										
Partial oxidation	103, 115									√	
Steam reforming	99				√						
Thermal cracking					√						
Lime calcination	165						√	√			
Monomers	604										
Nitric acid	288										
Oil combustion				√							
Pharmaceuticals	753										
Phosphates	244	√									
Polymers	580										
Portland cement	156						√				
Phthallic anhydride	599										
Pyrolysis products					√						
Reforming										√	
Saponification											
Steel smelting								√			
Sulfuric acid	296										
Synthesis gas	105				√					√	
Water gas	100									√	

[a] Agitated tanks in series vary from well stirred for one tank to plug flow for a large number of separate units. See Smith 46, p. 80] or Hill [23, p. 279] for amplification.

[b] References are to Shreve and Brink [42] except for hyphenated numbers, which denote pages in Perry [37].

[c] Special electrolytic cells.

214

TABLE 4-21
(Continued)

	Heat Exchangers			Mixers				Process Vessels		
	Shell-Side (agitated)	Tube-side	Vessels (jacketed or having internal coils)	Fluid Jet Mixed Vessels	Gas-Sparged Vessels	Agitated Tanks	Agitated Tanks in Series	Packed Columns	Open Vessels	Custom Reactors
Flow Characterization										
Plug flow		√					√	√	√	√
Well-stirred	√		√	√	√	√	√		√	√
Other							—[a]			
Design-Controlling Process										
Chemical kinetics	√	√	√			√	√	√	√	√
Heat transfer	√	√	√		√			√	√	√
Diffusion				√	√	√	√	√		
Mixing										√
Reaction Type										
Catalytic	√	√	√		√	√	√	√		√
Noncatalytic		√	√	√	√	√	√		√	√
Single-phase	√	√	√	√		√	√	√		√
Multiple-phase	√	√	√	√	√	√	√	√		√
Solid–Fluid			√	√	√	√	√	√	√	√
Thermal Behavior										
Adiabatic				√	√	√	√	√		√
Isothermal	√		√	√	√	√	√	√		√
Intermediate	√	√		√						√
Endothermic	√	√	√		√	√	√	√		√
Exothermic	√	√	√	√	√	√	√	√	√	√
Typical Process Stream										
Gas		√		√	√			√	√	√
Gas–liquid mixture		√	√	√	√	√	√	√	√	√
Liquid–liquid mixture	√	√	√		√	√	√	√	√	√
Viscous liquid		√	√			√	√			√
Liquid–solid mixture			√		√	√	√		√	√
Gas–solid mixture				√						√
Solid										√
Typical Stream Conditions										
Pressure										
100–500 bara		√		√				√	√	√
10–100 bara	√	√	√	√	√	√	√	√	√	√
<10 bara	√	√	√	√	√	√	√	√	√	√
<2 bara	√	√	√	√	√	√	√	√	√	√
Temperature										
1000–2000°C				√						√
500–1000°C		√		√				√	√	√
500°C	√	√	√	√	√	√	√	√	√	√

Typical Commercial Reactions											
Product	Ref.[b]										
Ammonia	276		√						√		√[c]
Aluminum	224										—[c]
Butadiene	631								√		
Carbon black	121				√					√	√[c]
Chlorine	214										—[c]
Coal, coking											
Coal, combustion				√	√						
Esterification				√			√				
Ethylene glycol											
Ethylene oxide											
Flame synthesis				√	√					√	√
Hydrocarbons											
Alkylation	683, 21-7	√									
Catalytic cracking	678										√
Desulfurization									√		
Hydrogenation	683			√			√	√	√		
Isomerization	684	√									
Partial oxidation	103, 115				√					√	
Steam reforming	99										
Thermal cracking											
Lime calcination	165										
Monomers	604			√			√	√			
Nitric acid	288		√						√		
Oil combustion				√	√						
Pharmaceuticals	753			√			√	√			
Phosphates	244										
Polymers	580			√			√				
Portland cement	156										
Phthalic anhydride	599		√						√		
Pyrolysis products									√		
Reforming					√				√	√	
Saponification				√			√				
Steel smelting											
Sulfuric acid	296				√				√	√	
Synthesis gas	105				√					√	
Water gas	100								√		

4   Some well-stirred exothermic reactors have multiple stable operating modes.

5   Plug flow behavior can be approached in well-stirred reactors by placing multiple reactors in series.

STEP 5   Energy Balance

Although most laboratory reactors, for ease of interpretation, are operated isothermally, this is not always the case commercially. Since reaction rates are so sensitive to temperature, thermal behavior of a reactor must be characterized for the design if it is to have any meaning. Conventional techniques involving heat transfer, thermodynamic, and chemical kinetic principles are employed. There should be no excuse for incomplete or inaccurate thermal analysis.

STEP 6   Determination of Reactor Size

This step, involving the traditional material and energy balance equations of reactor design, defines the reactor volume necessary to achieve the desired yield.

STEP 7   Pressure-Drop Calculation

Principles of fluid mechanics are employed to characterize pressure drop through the reactor and to reconcile this with the flow sheet. If pressure drops are large, as through a long plug flow or packed-bed reactor, it may be necessary to repeat preceding steps using incremental reactor volumes.

STEP 8   Reevaluation

Examine the results from a process performance standpoint. If any conditions or results are illogical, some of the bases and assumptions made earlier may require modification.

STEP 9   Economic Evaluation

Invariably, there will be a tradeoff between conversion and costs. With reactor size, configuration, temperature, and pressure specified, a capital cost can be assembled readily from data provided in Chapter Five. Operating costs can be estimated as described in Chapter Six. Considering these costs as a function of conversion, an optimum operating design can be established. Complete optimization may require repetition of the design sequence for a number of conversions and, perhaps, even a number of alternate reactor types.

Fluid and Fixed-Bed Reactors

(Perry 11-43, 20-64; McCabe and Smith 159, 146; Treybal 608, 697; Foust 637, 642)

Fluid and fixed beds are comprised basically of process vessels filled with particulate solids and having an appropriate support plate and flow distribution fittings. Fluid beds are particularly valuable as reactors because of their ability to transform a difficult solid into a controllable, transferable fluid state. Intimate fluid–solid contact, efficient heat transfer, and temperature uniformity are additional valuable features of fluid bed reactors. (Temperature differences from point to point in a fluid bed seldom exceed 5°C.) Packed beds, on the other hand, because of inefficient gas–solid contact and poor heat transfer, can easily develop "hot spots" with exothermic reactions. Fixed beds normally create a less severe pressure drop, and they are not as sensitive to solid particle size as fluid beds.

Liquid or gas must flow up to fluidize a system, whereas it normally flows down in packed beds to avoid fluidization. When greater fluid mixing is desired in fluidized beds, ratios of bed length to diameter are generally smaller, on the oder of 0.5 to 2. When plug flow of fluid is desired, length to diameter ratios should be greater than 4. In either event, a fluid bed is usually well mixed with regard to the solid. In a packed bed, the solid is stationary, and fluid motion is basically in plug flow. Approximate parameters, useful for preliminary design, are listed in Table

TABLE 4-22
CHARACTERISTICS OF FLUIDIZED AND FIXED BEDS

	Type of Bed			
	Fluidized		Fixed	
	Liquid	Gas at Standard Conditions	Liquid	Gas at Standard Conditions
Range of Common Equipment Sizes				
Bed diameter, D (m)	1–10	1–10	0.3–4	0.3–4
Bed height, L (m)	0.3–15	0.3–15	0.3–30	0.3–30
Freeboard (space above bed), (m)	0.4D	0.4D	—[a]	—[a]
Fraction void volume (porosity), ϵ	0.6–0.8	0.6–0.8	0.35–0.70	0.35–0.70
Particle size range D_p (m)	1×10^{-4} to 1×10^{-1}	1×10^{-5} to 1×10^{-2}	<0.1D	<0.1D
Fluid superficial (empty tower) velocity,[b] u_g (m/s)	0.1–0.5[c]	0.1–5[c]	0.0005–0.1	0.005–1
Pressure drop (kPa/m)	3–10[d]	5–15[d]	0.001–10[e]	0.001–1[e]
Maximum Vessel Temperatures (°C)				
Carbon steel		450		450
Stainless steel		750		750
Nickel-based alloy		1200		1200
Brick-lined		1500		1500
Mean overall heat transfer coefficient (J/s · m² · K)	400–2000	400–800	400–2000	20–80

[a]Space within dished end is adequate.

[b]Actual average fluid residence time, because of higher velocities in the interstices of the bed, is affected by porosity (i.e., $\theta = V\epsilon / \rho_g u_g$).

[c]This is only an approximation. For more accurate values, refer to McCabe and Smith (Equations 7-69 and 7-70, p. 164). Actual fluid velocities are approximately five time the minimum fluidization values derived from these equations.

[d]Approximate values. Actual values are easily estimated as the bed weight per unit area less buoyancy $[\Delta p / L = (\rho_s - \rho_f)(1 - \epsilon)g]$. For more information, see Perry 5-54.

[e]Approximate values. For more accurate values, refer to the Ergun equation (Equation 7-20, McCabe and Smith, p. 149; Equation 6.66, Treybal, p. 200; or Equation 22.86, Foust, p. 640). See also Perry 5-52.

4-22. For more detailed and accurate design, the comprehensive treatment of McCabe and Smith is recommended.

SEPARATORS

(Bag filters, Centrifuges, Clarifiers, Classifiers, Coalescers, Collectors, Concentrators, Cyclones, Decanters, Dewaterers, Digesters, Drums, Dust Collectors, Electrostatic Precipitators, Extractors, Filters, Flocculators, Grit Separators, Presses, Purifiers, Screens, Settlers, Sifters, Strainers, Thickeners)

As illustrated by the length of the list (taken from Table 4-1), this family of process equipment is prolific and diverse. Much of chemical engineering is devoted to separation of various mixtures. Possible equipment combinations and permutations to accomplish such, are listed in Table 4-17. As noted, process vessels are indicated for separation of solutions and other intimately bound mixtures. Equipment designated generically here as "separators" is devoted to refinement of physically dispersed mixtures.

Phase separations considered in this section are illustrated in Figure 4-38, which gives combinations of possible physically dispersed mixtures and types of equipment employed to refine them. Note, in particular, the large variety and number of alternatives. This arises from the limitless range of properties and characteristics possible in physically dispersed mixtures.

In general, since phase mixtures are not chemically combined, an engineer can employ mechanical or other forces to accomplish physical separation. One could, of course, use thermal or other types of energy as is done in distillation, absorption,

Figure 4-38   Possible combinations of physically dispersed mixtures and equipment commonly employed for their separation. (Adapted from Fitch [13], *Chemical Engineering Progress*, by permission.)

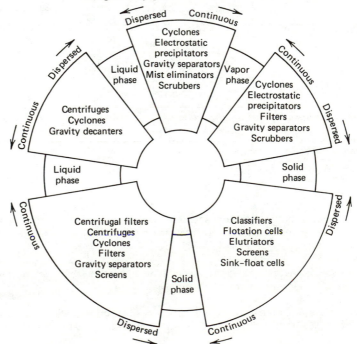

and evaporation, but such energy costs are much higher than those typical of mechanical separations. In addition to mechanical forces, which are employed in filtration or screening, the equipment listed in Figure 4-38 also includes devices that depend on gravitatonal, centrifugal, and electrostatic force fields as well as differences in momentum and surface properties. Various equipment alternatives are listed in Tables 4-23 according to mechanism of separation. This material includes data useful for identifying and tentatively characterizing alternatives for a perplexing flow sheet junction. There is probably no generic equipment category where the services of specialists (including laboratory and even pilot plant trials) are more necessary for accurate final design. In preliminary analysis, it will be frustrating to find a number of alternatives in Tables 4-23 that seem to be logical for a given separation. Proficiency in selection will improve, of course, with judgment. It should be reassuring to the beginner, nevertheless, to realize that any one of the promising alternatives will yield a reasonable flow sheet and cost estimate for preliminary evaluation. In making your final selection, the decision chart discussed in Chapter Two should be useful.

It has been said that the most important step in separator selection is taken during flow sheet synthesis. Often separators are combined at this time. With creative directing or recycle streams, some separations can be avoided. In other situations, such as classification of slurry from a mill, the oversize fraction can be returned directly to upstream equipment.

Not included in Tables 4-23 are three devices for solid–solid separations: flotation cells, elutriation separators, and sink–float cells. These are commonly employed for ore separations, which can be accomplished rather directly in process vessels. However, since each such separation is unique, it is difficult to define general rules. When faced with the promising use of these alternatives, one should consult the reference literature or a specialist. Other separators of Tables 4-23 are discussed briefly in the section that follows.

Centrifugal Separators

(Perry 18-83, 19-89, 19-94, 20-81, 21-46; McCabe and Smith 948, 963, 966; Foust 620, 622, 674)

Many physical mixtures can be separated because of differences in density or settling velocity. These differences are employed quite conveniently in common settling devices that depend on gravity. The force of gravity is, of course, limited. Thus, for mixtures that settle slowly or where cleaner separations are desired, the process can be accelerated by application of a centrifugal force field. Centrifugation must be conducted in a device of circular geometry rotated at high speed. Commercial units of various designs are, through this means, able to produce forces greater than 10,000 times that of earth's gravity. Because of their relatively small size, centrifuges are especially useful for high pressure separation or with critical, volatile, or hazardous materials that must be confined. They function with little operator attention. High capital and power costs, on the other hand, counteract these advantages.

TABLE 4-23

CRITERIA AND DATA FOR THE SELECTION AND PRELIMINARY DESIGN OF SEPARATORS

TABLE 4-23a Centrifugal, Gravity, and Impingement Separators

	Type of Separator						
	Centrifugal						
	Centrifuge		**Centrifugal filter**		**Cyclone**		
	Sedimentation (disk, bowl)	Helical Conveyor (solid bowl)	Continuous Conveyor	Auto Short-Cycle	Gas	Liquid	Electrostatic Precipitator
Continuous phase	Liquid	Liquid	Liquid	Liquid	Gas	Liquid	Gas
Dispersed phase	Liquid or solid	Solid	Solid	Solid	Liquid or solid	Liquid or solid	Liquid or solid
Range of Equipment Sizes							
Length, L (m)	0.5–2.0	0.5–10	0.5–2.0	0.5–1.0	0.3–8	0.3–3.0	<20
Diameter or width, D (m)	0.2–1.0	0.1–2.0	0.3–1.4	0.5–1.0	0.05–2	0.01–1.0	<50
Nominal area, A (m²)					$0.1D^{2,b}$	$0.1D^{2,b}$	$0.1\dot{q}$
Pressure range, p (bara)	0.1–10	0.1–10	0.1–10	0.1–10	1–500	0.1–500	Near 1.0
Pressure drop, Δp (kPa)					$1p-2p^c$	30–400	0.06–0.3
Centrifugal force	<10,000g	300g–3000g	100g–600g	<4000g	<2500g	<250g	
Temperature range, T (°C)	−90 to 315	−90 to 315	−90 to 260	−90 to 260	−250 to 1000		0 to 800
Particle size of separated solids, D_p (μm)^d	>0.1	>0.1	>150	>100	>5 (80% efficiency)	>50	0–100 (95% + efficiency)
Flow Rates							
Total feed stream							
$\dot{m}$ (kg/s)							
$\dot{q}$ (m³/s)	0.0005–0.03				$10A-20A^b$	$3A-6A^b$	
u (m/s)					10–20	3–6	1–3
Residence time, θ (s)							1–15
Dry solids feed, $\dot{m}_s$ (kg/s)		0.1–15	0.5–50	0.4–4			
Liquids, $\dot{q}_l$ (m³/s)		0.001–0.05					
Wash liquid							
$\dot{m}_w$ (kg/s)		0.25$\dot{m}_s$	0.1$\dot{m}_s$	0.1m_s			
Concentration of Dispersed Phase in Feed							
w (weight percent)	<1	0–60	30–60	<10			
ϕ (volume percent)						<20	
(g/std m³)							<500
Performance Characteristics							
Cake dryness	E	B	A	A		D	
(weight percent solids)	(20–40)	(50–80)	(80–95)	(80–95)		(20–70)	
Cake fragility	B	D	B	D	B	B	A
Classification of solids	C	C	C	X	C	A	X
Compressible cakes	A	B	B	B	A	A	A
Contamination of cake	A	A	A	A	A	A	A
Contamination of filtrate	A	A	A	A	A	A	A
Critical materials, toxic or hazardous	A	A	A	A	A	A	A
Ease of cake discharge	D	B	B	B	C	A	A
Filtrate clarity	A	E	E	D	D	E	A (<0.1 g/m³)
Filtrate loss	A	B	A	A	A	A	A
Volatile filtrates	A	A	A	A		A	A
Floor space	A	A	A	A	A	A	D
Large capacities	A	A	A	A	A	A	A
Odor control	A	A	A	A	A	A	A
Thick cakes	X	A	B	A	A	A	B
Thin cakes	A	B	B	D	A	A	B
Viscous liquids	A	A	D	E		E	
Washability of cake	X	C, D	C	A	E	E	C
Suitability							
Gas–solid mixtures	X	X	X	X	A		A
Gas–liquid (dilute)	X	X	X	X	A		A
Gas–liquid (concentrated)	X	X	X	X	A		D
Liquid–solid mixtures							
Dilute slimes	A	E	X	X		X	X
Concentrated slimes or clays	X	D	X	E		X	X
Concentrated ores or cement dispersions	X	B	E	D		D	X
Granular, sandy, or crystalline solids	X	A	A	A		A	X
Free-filtering fibers of pulps	X	B	B	D		D	X
Viscous syrups or oils	X	B	B	D		D	X
Multiple-particle-size classification	X	C	D	X		A	X
Solid–solid mixtures							
Coarse (D_p > 1 mm)	X	X	X	X			
Fine (1 mm > D_p > 1 um)	X	X	X	X			
Ultra-fine (D_p < 1 um)	B	B	B	X			
Multiple particle sizes	X	C	D	X	C	C	
Liquid–liquid mixtures	A	C	C	X		D	X
Power consumption (including liquid and vacuum pumps), P (kW)	1000$\dot{q}_l$–10,000$\dot{q}_l$	3000$\dot{q}_l$–15,000$\dot{q}_l$ $3\dot{m}_s$–8$\dot{m}_s$	3$\dot{m}_s$–30$\dot{m}_s$	10$\dot{m}_s$–40m_s	1$\dot{q}_g$–3$\dot{q}_g$	100$\dot{q}_l$–300$\dot{q}_l$	0.4$\dot{q}$–1.2$\dot{q}$
Relative Costs							
Capital	High	High	High	High	Low	Low	Low
Labor	Low	Low	Low	Low	Low	Low	Low
Maintenance	Moderate	Moderate	Moderate	Moderate	Low	Low	Moderate
Other operating costs	High	High	High	High	Moderate	Moderate	High

KEY

A excellent or no limitations
B modest limitations
C special units available at high cost to minimize problems
D limited in this regard
E severely limited in this regard
X unacceptable

TABLE 4-23

TABLE 4-23a (Continued)

	Type of Separator					
	Gravity				Impingement	
	Classifier	Clarifier–Thickener	Decanter	Settling Chamber (settler, sedimentation Drum)	Mist Eliminator	Scrubber[a]
Continuous phase	Liquid	Liquid	Liquid	Gas or liquid	Gas	Gas
Dispersed phase	Solid	Solid	Liquid	Liquid or solid	Liquid	Liquid or solid
Range of Equipment Sizes						
Length, L (m)	0.6–12		3–20	3–20	0.1–0.2	<4
Diameter or width, D (m)	0.3–8	3–180	1–4	0.3–4	0.2–4	0.1–2
Nominal area, A (m^2)						
Pressure range, p (bara)	Near 1.0	Near 1.0	0–400	0–400	No limit	0–400
Pressure drop, Δp (kPa)			5	5	<1	<4 (gases)
Centrifugal force						
Temperature range, T (°C)	Ambient	Ambient			−250 to 600	10–90
Particle size of separated solids, D_p (μm)[d]	>800	>0.1	>100	>100	>10	>1 (90% efficiency)
Flow Rates						
Total feed stream						
$\dot{m}$ (kg/s)	$6D^{1.5}$			See Table 4-25		$16D^2$
$\dot{q}$ (m^3/s)		$1 \times 10^{-4}D^2$ to $4 \times 10^{-4}D^2$			$4D^2$	
u (m/s)			<0.03	<3.0 (gases) <0.03 (liqs.)	3–5	
Residence time, θ (s)						
Dry solids feed, $\dot{m}_s$ (kg/s)						
Liquids, $\dot{q}_l$ (m^3/s)						
Wash liquid						
$\dot{m}_w$ (kg/s)						$1m$–$10\dot{m}$
Concentration of Dispersed Phase in Feed						
w (weight percent)	5–40	0–20				
ϕ (volume percent)			0–60	0–30		
(g/std m^3)]						<500
Performance Characteristics						
Cake dryness	D	D		E		E
(weight percent solids)	(70–80)	(20–60)		(20–60)		
Cake fragility	D	A		A		B
Classification of solids	A	E	X	C	X	X
Compressible cakes	D	A		A		A
Contamination of cake	A	A		A		D
Contamination of filtrate	A	A	A	A	A	B
Critical materials, toxic or hazardous	E	E	A	A	A	A
Ease of cake discharge	A	A		C		A
Filtrate clarity	E	A	D	D	B	B(<0.2 g/m^3)
Filtrate loss	B	D	D	D	A	B
Volatile filtrates	D	B	A	A	A	A
Floor space	D	E	A	A	A	D
Large capacities	B	A	A	A	A	A
Odor control	E	E	A	A	A	A
Thick cakes	A	A		A		B
Thin cakes	D	A		A		A
Viscous liquids	D	D	B	B	D	
Washability of cake	E	D	D	D		D
Suitability						
Gas–solid mixtures	X	X		A	X	A
Gas–liquid (dilute)	X	X		A	A	A
Gas–liquid (concentrated)				A	A	A
Liquid–solid mixtures						
Dilute slimes	X	A		E	X	X
Concentrated slimes or clays	X	A		E	X	X
Concentrated ores or cement dispersions	D	A		D	X	X
Granular, sandy, or crystalline solids	A	D		A	X	X
Free-filtering fibers of pulps	B	A		D	X	X
Viscous syrups or oils	E	E	D	D	X	X
Multiple-particle-size classification	A	E		X	X	X
Solid–solid mixtures						
Coarse (D_p > 1 mm)						
Fine (1 mm > D_p > 1 um)						
Ultra-fine (D_p < 1 um)						
Multiple particle sizes						
Liquid–liquid mixtures	X	X	A		X	X
Power consumption (including liquid and vacuum pumps), P (kW)	$0.25\dot{m}$	$0.003D^2$–$0.006D^2$	—[e]	—[e]	—[e]	—[e]
Relative Costs						
Capital	Moderate	Moderate to high	Low	Low	Low	High
Labor	Low	Low	Low	Low	Low	Low
Maintenance	Moderate	Low	Low	Low	Low	Moderate
Other operating costs	Moderate	Low	Low	Low	Low	Moderate

[a]For more details, see section on process vessels in this book and page 20-94 in Perry.

[b]Flow cross section of cyclone entrance.

[c]Pressure differential ΔP is given in kPa with pressure p in bara, consistent with units shown. Near atomspheric pressure and temperature, Δp is usually in the range of 1–2 kPa. At other temperatures, multiply by $(T + 273)/273$ to obtain the corrected Δp.

[d]See Buonicore [6] for detailed particle-removal efficiencies for gas-cleaning devices and Bergmann [3] for information on bag filter fabrics.

[e]Compute from power requirements of pumps and blowers necessary to move liquids and gases.

TABLE 4-23
CRITERIA AND DATA FOR THE SELECTION AND PRELIMINARY DESIGN OF SEPARATORS

TABLE 4-23b Mechanical Separators

	Filters			Liquid–Solid Process Filters[a] — Horizontal			Rotary	Rotary Drum Liquid–Solid Process Filters[a] — Precoat		Multicompartment		Single Compartment	
	Bag	Cartridge	Sand	Belt	Table	Tilting Pan	Disk	Pressure	Vacuum	Pressure	Vacuum	Pressure	Vacuum
Continuous phase	Gas	Gas or liquid	Gas or liquid	Liquid	Liquid	Liquid	Liquid	Liquid	Liquid	Liquid	Liquid	Liquid	Liquid
Dispersed phase	Solid	Liquid or solid	Solid	Solid	Solid	Solid	Solid	Solid	Solid	Solid	Solid	Solid	Solid
Range of Equipment Sizes													
Length, L (m)	<20	1	<4	0.5–15			1–4	1.5–5	1.5–5	1.5–7	1.5–8	1.5–4	1.5–4
Diameter or width, D (m)	<50	1	<12	0.2–3	2–8	6–25	2–6	0.6–2.5	0.6–2.5	0.6–3.5	0.6–4	0.6–2	0.6–2
Nominal Area, A (m²)	$<3\times10^5$	<50	$\pi D^2/4$	0.2–100	1–60	6–200	10–600	0.5–70	0.5–120	0.5–70	0.25–120	0.5–15	0.5–15
Pressure range, p (bara)	Near 1 atm	0–400	0–400	Near 1 atm	Near 1 atm	Near 1 atm	Near 1 atm	2–7	Near 1 atm	2–30	Near 1 atm	2–7	Near 1 atm
Pressure drop, Δp (kPa)	0.5–2.0	10–100	<5	<100	<100	<100	10–100	50–300	10–100	50–300	10–100	50–300	10–100
Temperature range, T (°C)	0–250	–250 to 500	–250 to 800	10–100	10–100	10–100	10–100	10–100	10–50	10–400	10–50	10–100	10–50
Particle size of separated solids.	all (>99% efficiency)[b]	all (>99% efficiency)											
Flow Rates													
Liquids													
$\dot{q}_l$ (m³/s)		0.001A	0.001A										
u_l (m/s)		0.001	0.001			*Flow Rates* *See Table 4-23c*							
Gases													
$\dot{m}_g$ (kg/s)													
$\dot{q}_g$ (std m³/s)	0.005A–0.5A	0.1A	0.1A										
u_g (std m/s)	0.005–0.5	0.1	0.1										
Wash liquid, $\dot{m}_w$ (kg/s)	—[c]			$10\dot{m}_s$–$20\dot{m}_s$	$1\dot{m}_s$–$2\dot{m}_s$	$0.5\dot{m}_s$	$\dot{m}_s$	$2\dot{m}_s$		$2\dot{m}_s$	$2\dot{m}_s$	$2\dot{m}_s$	$2\dot{m}_s$
Solids batch capacity (m³)		$0.01V^d$											
Concentration of Dispersed Phase in Feed													
w (weight percent)		<0.1		0.1–5	5–30	10–30	5–20	0.1–1	0.1–1	1–30	1–30	10–30	10–30
(g/std m³)	<500												
Performance Characteristics													
Cake dryness	A	E	E	B	B	B	D	B	B	B	B	B	B
(weight percent solids)		10–20	50–70	50–70	50–70	50–70	40–60	50–70	60–80	50–70	60–80	50–70	60–80
Cake fragility	A	A	B	B	D	B	B	B	B	B	B	B	B
Classification of solids	X	X	E	X	X	X	D	X	X	X	X	X	X
Compressible cakes	B	B	B	D	E	D	B	A	B	B	D	B	D
Contamination of cake	A	A	A	A	A	A	A	E	E	A	A	A	A
Contamination of filtrate	A	A	A	A	A	A	A	A	A	A	A	A	A
Critical materials, toxic or hazardous	A	C	B	E	E	E	E	A	E	A	E	A	E
Ease of cake discharge	A	D	D	C	D	B	B	B	A	D	C	D	C
Filtrate clarity	A (<0.05 g/m³)	A	A	D	D	D	D	A	A	D	D	D	D
Filtrate loss	A	A	B	B	B	B	B	B	B	B	B	B	B
Volatile filtrates	A	A	A	D	D	D	E	A	E	A	E	A	E
Floor space	D	A	B	E	E	E	A	D	D	D	D	D	D
Large capacities	A	C	A	B	A	A	A	D	A	D	A	D	A
Odor control	A	A	A	D	D	D	D	A	D	A	D	A	D
Thick cakes	A	X	D	A	A	A	A	E	E	A	A	A	A
Thin cakes	A	A	A	C	D	D	D	A	A	C	C	B	B
Viscous liquids	A	D	D	D	D	D	D	B	D	B	D	B	D
Washability of cake	X	E	C	A	A	A	C	B	B	B	B	A	A
Suitability for													
Gas–solid mixtures	A	A	C	X	X	X	X	X	X	X	X	X	X
Gas–liquid (dilute)	X	A	D	X	X	X	X	X	X	X	X	X	X
Gas–liquid (concentrated)	X	X	X	X	X	X	X	X	X	X	X	X	X
Liquid–solid mixtures	X												
Dilute slimes	X	A	X	X	X	X	D	A	A	D	D	E	E
Concentrated slimes or clays	X	X	X	C	D	D	C	D	D	C	C	D	D
Concentrated ores or cement dispersions	X	X	E	A	B	A	A	X	X	A	A	D	D
Granular, sandy, or crystalline solids	X	C	B	A	A	A	B	X	X	B	B	B	B
Free-filtering fibers or pulps	X	B	A	B	B	B	A	D	D	A	A	B	B
Viscous syrups or oils	X	D	D	D	D	D	D	X	X	C	D	C	D
Multiple-particle-size classification	X	X	X	X	X	X	E	X	X	X	X	X	X
Solid–solid mixtures	X	X	D	X	X	X	X	X	X	X	X	X	X
Coarse ($D_p > 1$ mm)	X	X	D	X	X	X	X	X	X	X	X	X	X
Fine (1 mm > D_p > 1 μm)	X	X	D	X	X	X	X	X	X	X	X	X	X
Ultra-fine ($D_p < 1$ μm)	X	X	E	X	X	X	X	X	X	X	X	X	X
Multiple particle sizes	X	X	D	X	X	X	X	X	X	X	X	X	X
Liquid–liquid mixtures	X	D	X	X	X	X	X	X	X	X	X	X	X
Power consumption (including liquid and vacuum pumps), P (kW)	—[e]	—[e]	—[e]	$A^{0.75}$	$A^{0.75}$	$A^{0.75}$	$0.5A^{0.75}$–$A^{0.75}$	$0.5A^{0.75}$	$A^{0.75}$–$2A^{0.75}$	$0.5A^{0.75}$	$A^{0.75}$–$2A^{0.75}$	$0.5A^{0.75}$	$A^{0.75}$–$2A^{0.75}$
Relative Costs													
Capital	High	Low	Low	High	Moderate	High	Moderate	Extra high	Moderate	Extra high	Moderate	Extra high	High
Labor	Low	High	Moderate	Low	Low	Low	Low	Low	Low	Low	Low	Low	Low
Maintenance	Moderate	Low	Low	Moderate	Moderate	Low	Low	High	Moderate	High	Moderate	High	Moderate
Other operating costs	Moderate	Moderate	Moderate	Moderate	Moderate	Moderate	Moderate	Moderate	Moderate	Moderate	Moderate	Moderate	Moderate

KEY

A   excellent or no limitations
B   modest limitations
C   special units available at higher cost to minimize problems
D   limited in this regard
E   severely limited in this regard
X   unacceptable

TABLE 4-23

TABLE 4-23*b* (Continued)

	Plate and Frame (filter press)		Shell and Leaf Filter	Presses		Screens	
Liquid-Solid Process Filters[a]	Filter	Thickener		Screw	Roll	Grizzly	Vibratory
Continuous phase	Liquid	Liquid	Liquid	Liquid	Liquid	Gas, liquid, or solid	
Dispersed Phase	Solid	Solid	Solid	Solid	Solid	Solid	
Range of Equipment Sizes							
Length, L (m)	0.5–20		2–5				2–5
Diameter or width, D (m)	0.1–2		0.5–2				0.5–1.5
Nominal area, A (m^2)	1–1000		0.1–300				1–7.5
Pressure range, p (bara)	3–70		3–15	Near 1 atm	Near 1 atm	Near 1 atm	Near 1 atm
Pressure drop, Δp (kPa)	10–1000		10–600				
Temperature range, T (°C)	0–250		0–250	0–250	0–250	−250 to 500	−50 to 350
Particle size of separated solids, D_p (μm)	>0.1		>0.1			>50,000	50–50,000
Cake Thickness, t (mm)	<200		<50				
Flow Rates	See Table 4-23c						
Dry solids feed, m_s (kg/s)						$6 \times 10^{-6} AD_p$ to $2.5 \times 10^{-5} AD_p$	$2.5 \times 10^{-5} AD_p$ to $1.0 \times 10^{-4} AD_p$
Wash liquid, $\dot{m}_w$ (kg/s)	$2\dot{m}_s$		$2\dot{m}_s$				
Solids batch capacity (m^3)	0.02A–0.1A		0.01A–0.03A				
Concentration of dispersed phase in feed, w (weight percent)	0.1–20	0.1–20	0.1–20				
Performance Characteristics							
Cake dryness	A	D	B	A	A		
(weight percent solids)	60–80	20–40	50–70	80–90	80–98		
Cake fragility	B	A	B	X	X	D	B
Classification of solids	X	X	X	X	X	A	A
Compressible cakes	A	A	A	A	A	D	C
Contamination of cake	A	A	A	A	A	A	A
Contamination of filtrate	A	A	A	A	A		
Critical materials, toxic or hazardous	C	C	B	C	C	C	C
Ease of cake discharge	B	A	B	B	B	D	A
Filtrate clarity	B	X	B	D	D		
Filtrate loss	B	A	D	A	A		
Volatile filtrates	C	A	B	C	C		
Floor space	B	B	B	B	B	A	A
Large capacities	C	A	B	A	A	A	A
Odor control	C	C	A	C	C	C	A
Thick cakes	B	A	D	A	A		
Thin cakes	A	A	B				
Viscous liquids	B	B	D	A	B	X	
Washability of cake	C	C	B	B	B	A	
Suitability for							
Gas–solid mixtures	X	X	X	B	B	A	A
Gas–liquid (dilute)	X	X	X	X	X	X	X
Gas–liquid (concentrated)	X	X	X	B	B	X	X
Liquid–solid mixtures							
dilute slimes	B	A	B	X	X	X	X
concentrated slimes or clays	A	A	B	B	D	X	X
concentrated ores or cement dispersions	B	B	B	E	B	X	X
granular, sandy, or crystalline solids	B	B	B	X	D	E	A
free-filtering fibers or pulps	A	A	A	A	A	E	D
viscous syrups or oils	B	B	D	A	A	A	C
multiple-particle-size classification	X	X	X	X	X	A	A
Solid–solid	X	X	X	X	X	A	A
Coarse ($D_p > 1$ mm)	X	X	X	X	X	A	D
Fine (1 mm > D_p > 1 μm)	X	X	X	X	X	X	A
Ultra-fine ($D_p < 1\ \mu$m)	X	X	X	X	X	X	D
Multiple particle sizes	X	X	X	X	X	X	X
Liquid–liquid	E	E	E	X	X	X	X
Power consumption, P (kW)	—[e]	—[e]	—[e]	$25\,\dot{m}^{0.5}$	$25\,\dot{m}^{0.5}$		$1600\,\dot{m}_s/D_p^f$
Relative Costs							
Capital	Low	Low	Low	High	High	Low	Moderate
Labor	High	Low	High	Low	Low	Moderate	Low
Maintenance	Low	Low	Low	Moderate	Moderate	Low	Moderate
Other operating costs	Moderate	Moderate	Moderate	Moderate	Moderate	Low	Moderate

[a]For flow rates through Liquid–solid process filters, including rotary drums, see Table 14-23c.

[b]See Buonicore (6) for detailed particle-removal efficiencies for gas-cleaning devices and Bergmann (3) for information on bag filter fabrics.

[c]In pulse-cleaned bag filters, cleaning air volume is approximately 1% of process gas volume.

[d]Based on volume of bed.

[e]Compute from power requirements of pumps and blowers necessary to move liquids and gases.

[f]D_p is mesh size of screen.

TABLE 4-23

CRITERIA AND DATA FOR THE SELECTION AND PRELIMINARY DESIGN OF SEPARATORS

TABLE 4-23c   Flow Rates for Liquid–Solid Process Filters, Including Rotary Drum

Suspensions	Dry Solids Feed Rate m_s (kg/s)	Liquid Flow Rate q_l $(m^3/s)^a$
Dilute slimes (<1% solids)	0.002A–0.02A	$7 \times 10^{-6}A - 1 \times 10^{-3}A$
Concentrated slimes or clays (>5% solids)	0.02A–0.1A	$1 \times 10^{-4}A - 3 \times 10^{-3}A$
Concentrated ores or cements	0.02A–0.1A	$1 \times 10^{-4}A - 3 \times 10^{-3}A$
Sands	0.2A–1.0A	$>3 \times 10^{-3}A$
Fibers or pulps	0.1A–0.7A	$0.001A–0.015A$
Thin syrups (<0.1 Pa·s)		$1 \times 10^{-4}A - 4 \times 10^{-4}A$
Viscous syrups (>0.1 Pa·s)		$1 \times 10^{-5}A - 2 \times 10^{-4}A$

Source: Based on Perry, Tables 19-18, 19-25, and 19-28.

[a] Flow rates are based on nominal filter area (i.e., $q_l = u_l A$). For batch devices, these are averages, including cleaning and washing times.

For preliminary design and cost estimation purposes, centrifugal separators are organized into the six categories listed in Table 4-23a. The first includes *sedimentation (disk or bowl) centrifuges*. These are designed for removal of liquid from two or more radial positions. Thus, mixtures of immiscible liquids can be easily separated. This centrifuge is employed primarily for "breaking" different emulsions where simple gravity separators are impractical. It is among the most ancient items of process equipment, having been developed originally for separation of cream from raw milk.

Because of limited capacity for solids, sedimentation centrifuges cannot process concentrated solid–liquid slurries. They are ideal, however, for clarification of dilute liquid–solid systems where large quantities of liquid can be treated with only an occasional shutdown for cleaning of the bowl. Some disk units have been devised for discharge of concentrated slurries from the periphery of the bowl. In these cases, the residue is quite wet and must generally be treated further for removal of solvent. Often, as with oil dewatering, liquid–liquid separation is accomplished simultaneously with sedimentation of solid contaminants.

For streams that contain higher concentrations of solids, an ingenious adaptation, known as the *helical conveyor* centrifuge, was developed (Figure 4-39). Its solid bowl contains an outer helix that rotates at a slightly different speed from the drum itself. Solids that are thrown to the wall are gradually conveyed to the discharge end of the unit. The drum is designed, usually, with a decreasing diameter through the solids transport path, creating a "beach" where solids drain more completely and can be washed by a continuous spray. Helical conveyor units are the most promising centrifuges for producing relatively dry cakes from large flows of concentrated slurries containing moderately fine clays or ores. They are limited, on the other hand, in cake washability and filtrate clarity.

Another useful variation employs a perforated wall to allow passage of liquid while retaining solids on the screen or filter. Operating much like the spin cycle of an automatic clothes washer, centrifugal filters are capable of producing especially dry cakes. *Continuous conveyor centrifugal* filters are devised with a vibrating pusher or other arrangement that promotes continuous flow of solid. The bowl can be designed to provide considerable washability, plus control and separate collection of filtrate and wash streams. The major limitation of continuous conveyor centrifugal filters stems from screen size, which prevents separation or clarification of fine-particle suspensions. Because of their ability to produce low residual liquid concentrations, they are used in numerous applications involving sandlike particles or crystals having diameters larger than 150 μm.

An adaptation of batch technology is represented by the *auto short-cycle centrifugal filter*. This, like the continuous conveyor unit, contains a rotating perforated bowl. Cake, however, is not conveyed but accumulates instead inside the bowl. The system passes through a programmed, automatically controlled and executed cycle of filtration, washing, and cake removal. Discharge is caused by a knife, blade, or plow, which peels cake from the wall. Each cycle is so short that for all practical purposes, this equipment can be considered to be continuous rather than batch type. Because of intense shear experienced by the cake, fragile solids are degraded in this centrifuge. Solids that deform, smear, or block the filter are also not compatible with it. Otherwise washability and filtrate clarity are somewhat better than for centrifuges as a whole.

A common and inexpensive application of centrifugal force occurs in *cyclone separators*, which separate solids from liquid streams and solids or liquids from gas

streams. The cyclone, a nonmechanical scroll-like cylinder, is a familiar appendage to the roof of most woodworking or other sawdust-generating plants. Such large-diameter units, however, are little more effective than gravity settling chambers. Process cyclones, to generate high centrifugal fields, are usually less than 0.5 m in diameter for gas streams and even smaller for liquids. If stream flow exceeds capacity, they are grouped in parallel, forming so-called multicyclone collectors. Cyclones are designed on the basis of fluid entering velocity, which is near 10 to 20 std m/s for gases and about one third this for liquids. Higher velocities cause reentrainment and excessive pressure losses. Because of these limits, gas cyclones are relatively ineffective for particles less than 5 μm and liquid cyclones for particles less than 10 μm [32]. Power consumption is based on that needed to pump the fluid. Thus, power ranges shown in Table 4-23a are based on the pressure differences and flow rates listed.

Because of their simplicity and mechanical integrity, cyclones are superb for removing solid or liquid particles from gases if entrainment or smaller particles can be tolerated. For environmental applications, cyclones must generally be followed by a device which removes micrometer- and submicrometer-sized particles more efficiently. Gas cyclones are limited at subatmospheric pressures because of leakage through the solids discharge port. This is often unavoidable and causes reentrainment. Liquid cyclones, as liquid–solid separators, are competitive with but not significantly superior to gravity settlers. As solids classifiers, on the other hand, liquid cyclones are uniquely effective. They often are found in the effluent stream of a wet mill or grinder, classifying the solids and returning oversize material to the feed point. In dry grinding, on the other hand, screens, aided by gravity, are often less expensive than a gas cyclone with its fan and auxiliary equipment.

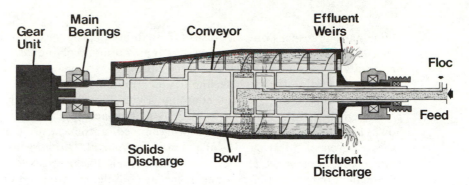

Figure 4-39 Helical conveyor solid bowl centrifuge. (Bird Machine Company, Inc., South Walpole, Mass., by permission.)

Electrostatic Precipitators

(Perry 20-103; Foust 680)

Founded on the principle that charged particles migrate in an electric field, electrostatic precipitators act through high voltage applied between a central axial

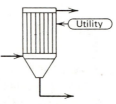

electrode wire and a tube. Particles passing through the field pick up charge and migrate to the tube surface, where they become neutralized. Periodic rapping of the tube causes the precipitated cake to slough off and fall into a hopper below the tube. The range of potential construction materials is broad, pressure drop is low, and dust collection efficiency is high. For these reasons, electrostatic precipitators were originally developed for collection of acid mists. In more recent years, they have been used extensively for fly-ash collection from electric power plants. Now through the development of temperature-resistant fabrics, bag filters, which collect at higher efficiency and are not sensitive to ash resistivity, seem to have a competitive edge in this application. Electrostatic precipitators tend to be more expensive than competitive dust collectors except in the highest capacity ranges. In terms of operating costs, electricity for particle charging compensates for savings in blower costs resulting from the exceptionally low pressure drop in a precipitator. In severe corrosive or especially high temperature surroundings, electrostatic precipitators are unchallenged for high efficiency collection of ultra-fine particles.

Gravity and Impingement Separators

In many systems (liquid droplets dispersed in a second immiscible liquid continuum, liquid droplets in gases, and solids in gases and liquids), natural separation occurs by settling due to gravity, that is, sedimentation. Several types of equipment are available to take advantage of this phenomenon. If particles or droplets are large and density differences great, separation is simple. Equipment is small and inexpensive. As particle diameters and density differences decrease, equipment becomes larger (up to 180 m in diameter for some thickeners and clarifiers). If sedimentation is very slow, separation rates must be increased by external means such as centrifugal force, and gravity separators will not suffice.

SETTLING CHAMBERS AND DECANTERS   *(Perry 18-82, 21-11; McCabe and Smith 954; Treybal 527; Happel and Jordan [20], 482)*

In its simplest form, a gravity decanter or sedimentation drum consists of a horizontal vessel. Length is sufficient for the dispersed solid or liquid to separate

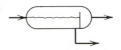

from the continuous phase as it flows smoothly from one end to the other. If the particle or droplet terminal velocity u_t is known, design is straightforward. In the time of residence, computed by the following,

$$\theta = \frac{V}{\dot{q}_C + \dot{q}_D} = \frac{L}{u_{ave}} \tag{4-99}$$

the particles, as a rough approximation, should be given time to migrate across half the depth or diameter of the chamber. Thus, the residence time is also given by

$$\theta \geq \frac{D}{2u_t} \tag{4-100}$$

combination of Equations 4-99 and 4-100 yields the following expression for drum length to diameter ratio.

$$\frac{L}{D} \geq \frac{0.5u_{ave}}{u_t} \tag{4-101}$$

For economic reasons, the L/D ratio for settling drums, like other cylindrical process vessels, is dictated according to the following approximate guidelines [19], summarized in Table 4-24. If, for example, the operating pressure is 30 barg, $L/D = 4$; and

$$u_{ave} \leq 8u_t \tag{4-102}$$

TABLE 4-24

LENGTH TO DIAMETER RATIOS AS A FUNCTION OF PRESSURE FOR DRUM SETTLERS

Internal Pressure (bara)	Length to Diameter Ratio, L/D
0–20	3.0
20–35	4.0
35 and higher	5.0 or greater

For solids or liquids in a gas stream, the terminal velocity can be estimated from the set of equations in McCabe and Smith, p. 155. In the Stokes law range ($0 < N_{Re} < 2$ or $0 < K < 3.3$), we have

$$u_t = g \frac{D_p^2}{18\mu} (\rho_p - \rho) \qquad (4\text{-}103)$$

in the intermediate range ($2 < N_{Re} < 500$ or $3.3 < K < 43.6$):

$$u_t = \frac{0.153 \, g^{0.71} \, D_p^{1.14} (\rho_p - \rho)^{0.71}}{\rho^{0.29} \mu^{0.43}} \qquad (4\text{-}104)$$

and in the Newton's law range ($N_{Re} > 500$ or $K > 43.6$):

$$u_t = 1.74 \left[\frac{g D_p (\rho_p - \rho)}{\rho} \right]^{1/2} \qquad (4\text{-}105)$$

Indices are given by

$$N_{Re} = \frac{\rho u_t D_p}{\mu} \qquad (4\text{-}106)$$

and

$$K = D_p \left[\frac{g \rho (\rho_p - \rho)}{\mu^2} \right]^{1/3} \qquad (4\text{-}107)$$

(K is a more convenient index to use than Reynolds number because it does not contain u_t and, therefore, does not require trial-and-error calculation.)

For solid particles or liquid droplets in a liquid medium, motion of one particle may be hindered by that of its neighbors. In this case, the nomographs developed by Zanker [55] are recommended. These are shown in Figure 4-40 for two ranges of particle sizes. They yield terminal velocities identical to those from Equations 4-103 to 4-105 when the volume percent of particles approaches zero. Thus, they are useful for both hindered and unhindered settling. Liquid droplets, because of internal circulation, settle at velocities higher than those determined from the equations or nomographs which, therefore, yield conservative results.

As a basis for design, a particle cutoff diameter of 0.1 mm has, by experience, been found appropriate for settling drums. With D_p thus defined and fluid properties known, u_t can be calculated from the appropriate equation above or from the nomograph.[8] The u_{ave} for horizontal drums can be specified according to Equation 4-101. To prevent turbulence and reentrainment, u_{ave} should be no greater than 3 m/s for gases and 0.003 m/s for liquids. This defines limits on the values yielded by Equation 4-102. With u_{ave} and L/D known, drum dimensions can be determined by application of the continuity equation:

$$u_{ave} = \frac{\dot{m}}{\rho A} = \frac{4(\dot{q}_C + \dot{q}_D)}{\pi D^2} \qquad (4\text{-}108)$$

which, rearranged, becomes

---

[8] With D_p equal to 100 μm, particle motion is almost invariably within the Stokes law range when the continuous phase is liquid and within the intermediate range when it is a gas.

$$D = \left[\frac{4(\dot{q}_C + \dot{q}_D)}{\pi u_{\text{ave}}} \right]^{1/2} \cong \left[\frac{\dot{q}_c + \dot{q}_D}{2\pi \, u_i} \right]^{1/2} \qquad (4\text{-}109)$$

To permit adequate space for nozzles and other hardware, the diameter should be at least 1 m or greater for horizontal vessels. Guidelines and rules of thumb for settling drums are summarized in Tables 4-23a and 4-25. In liquid–liquid settlers or decanters, the entering droplet size is often enlarged by flow-through mats or pads of fibers (coalescers). In gas–liquid separators, a similar device, known as a *mist eliminator*, is almost always used in modern practice (Perry 18-84). Otherwise, the design diameter of the drum, calculated above, must be increased by a factor of 2.5 [19].

In gas–liquid and liquid–liquid systems, separated fluid streams exit simply through outlets in the top and bottom of the vessel (see Figure 4-41). When the sediment is a solid, auxiliary equipment such as an auger conveyor or sludge pump is often required.

In gas–liquid separators such as flash and knockout drums, vessels are generally oriented vertically to provide more flexibility in liquid surge capacity. In this condition, diameter is dictated by entrainment limits within the gas. Even though sedimentation for the 100 μm cutoff droplet generally occurs in the intermediate range. Newton's law, Equation 4-105, is a convenient approximation that is employed to compute a pseudo-settling velocity u_t'. With appropriate substitution, it becomes

$$u_t' = 0.06 \left(\frac{\rho_p - \rho}{\rho} \right)^{1/2} \qquad (4\text{-}110)$$

This should be familiar to the reader by now as the Souders-Brown equation, which is used to define velocities in tray towers, vaporizers, and other vertical drums where entrainment may occur.

Vertically oriented, drum diameter is independent of length and can be computed from

$$D = \left(\frac{4\dot{q}_C}{\pi u_t'} \right)^{1/2} \qquad (4\text{-}111)$$

In comparison with Equation 4-109, this may seem to yield a much larger diameter than for a horizontal drum. Actually, u_t' is about four times greater than the true settling velocity u_t. Hence, the calculated diameter for a vertical drum is only about 40 percent larger than that of a horizontal one for the same service.

In practice, if the settling velocity is less than 0.1 m/s and a pressure differential of 10 kPa is available or reasonably obtainable, a cyclone separator will generally be less expensive than a settling drum for removing solid or liquid particles from a gas stream.

In liquid settling drums, sedimentation velocities should be 0.001 m/s or greater. Residence times are seldom greater than 500 s in practical systems. If settling velocities are smaller or residence times greater, centrifugal separators or, in the case of nonvolatile, nonhazardous liquids, radial flow thickeners or clarifiers should be considered.

Capital costs for drum settlers can be assembled from data for process vessels and other components provided in Chapter Five.

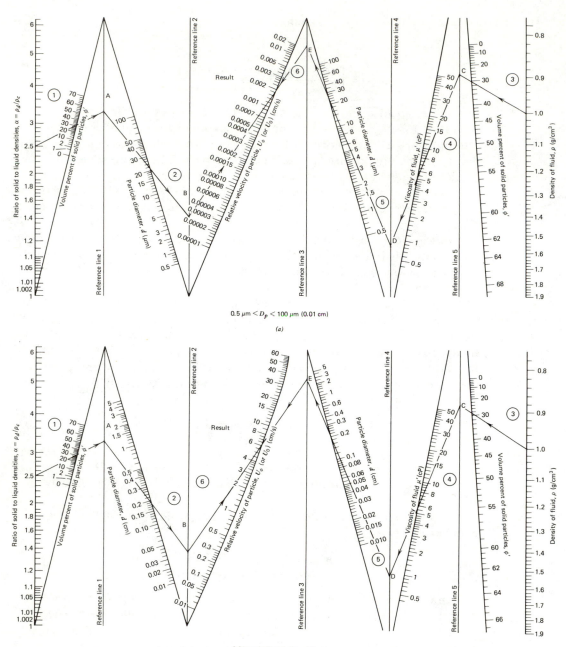

Figure 4-40 Settling velocities for solid particles dispersed in a liquid. Settling velocity is found from solid and liquid densities, concentration, viscosity, and particle diameter. Follow the arrow-marked lines going to reference points A, B, C, D, E, each step in sequence, as numbered. The result is found by connecting reference points B and E. (*a*) Nomograph for smaller particles. (*b*) Nomograph for larger particles. Examples shown are for solid particles of density 2.5 g/cm³ in a fluid of 1.0 g/cm³ density and 5 cP viscosity. Solids loading is 30 percent by volume. For a 20 μm particle, the settling velocity is found to be 0.002 cm/s from nomograph (a). For a 0.2 cm particle, the settling velocity, from nomograph (b), is 6 cm/s. [Excerpted by special permission from CHEMICAL ENGINEERING (May 19, 1980). Copyright © 1980, by McGraw-Hill, Inc., New York, N.Y. 10020.]

TABLE 4-25
GUIDELINES FOR THE SPECIFICATION OF GRAVITY SETTLING DRUMS

	Drum Settlers			
	Flash and Knockout Drums		Decanters	Sedimentation Tanks
Continuous phase	Gas		Liquid	Gas or liquid
Dispersed phase	Liquid		Liquid	Solid
Normal orientation	Vertical	Horizontal	Horizontal	Horizontal
Range of Common Equipment Sizes				
Diameter, D (m)	0.3–4	1–4	1–4	1–4
Length, L (m)	1–20	3–20	3–20	3–20
Length to diameter ratio	Various; usually between 3 and 5		Operating Pressure (bara) / L/D: 0–20 → 3; 20–35 → 4; >35 → 5	
Height above settled phase (m)	1.0 or D (whichever is larger); allow 0.3 m above mist eliminator	0.5D	0.5D	0.5D
Dispersed phase particle cut off diameter, D_p (mm)	0.1	0.1	0.1	0.1
Terminal velocity, u_t (m/s)	<0.4	<0.4	<0.004	<0.004
Storage residence time, θ (s)	<300[a]	<600[a]		
Continuous phase superficial velocity, u_{ave} (m/s) (approximately $4u_t$)	$u_{ave}=u_t'=0.06\left(\dfrac{\rho_p-\rho}{\rho}\right)^{1/2}$			
Combined stream superficial velocity, u_{ave} (m/s)		<8u_t	<8u_t	<8u_t
Design diameter, D (m)	$[4\dot{q}_c/\pi u_{ave}^{1/2}]$		$[4(\dot{q}_c+\dot{q}_D)/\pi u_{ave}]^{1/2}$	
Pressure drop, Δp (bar)	0.001	0.001	0.05	0.05

[a]Liquid storage capacity for knockout drums, upstream of compressors, is normally equivalent to a residence time of 600 to 1200 s.

CLARIFIERS–THICKENERS *(Perry 19-44; McCabe and Smith 956; Treybal 733; Foust 629)*

For gravity separation of very fine solid particles from a liquid, long residence time in a quasi-quiescent state is required. Settlers devised for this purpose, known as clarifiers and thickeners, are generally cylindrical pools, up to 180 m in diameter and several meters deep (see Figure 4-42). They are typically found in water pollution

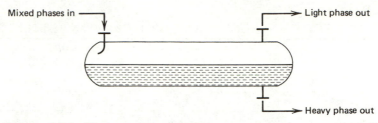

Figure 4-41   Horizontal drum settler.

control or treatment plants and, because of their size, are obvious even to the casual visitor.

Clarifiers are almost identical to appearance and cost to thickeners. The difference depends on emphasis—either to clarify the liquid or to concentrate the solid. Both accomplish these two purposes, but thickeners provide more solid residence capacity because of the higher concentrations encountered. In some cases, flocculating agents are added with the entering fluid at the axis to promote particle growth and more rapid sedimentation as the slurry flows to the discharge at its periphery. Such units are referred to as flocculators or digesters. The bottom of a clarifier or thickener is sloped gradually toward the center. Radial spokelike rakes rotate through the pool at peripheral velocities of 0.1 to 0.3 m/s and gently agitate the residue, moving it toward the center of the cone-shaped base while allowing it to further densify. Concentrated sediment is removed from the tank by a sludge pump.

Because of size, clarifiers and thickeners are seldom enclosed. Thus, their use is generally restricted to aqueous-based or other nontoxic, nonhazardous systems. For more critical applications that must function enclosed or under pressure, smaller filters or centrifugal-type separators are employed. In most cases, the sludge from a clarifier or thickener is further concentrated in a centrifuge or filter.

For rapid design of aqueous-based systems, criteria and operating conditions of Table 4-23a may be employed. Equipment size and power consumption are determined by obvious application of these criteria. Note that power consumption is small (less than 100 kW for the largest units) becuase of low raking rates and fluid velocities.

Capital costs are illustrated in Figure 5-52. Additional useful information can be found in the article by Raynor and Porter [39].

CLASSIFIERS   (*Perry 21-45; McCabe and Smith 955; Foust 619*)

Several separators listed under other categories such as cyclones, settlers, and centrifuges can be called classifiers; that is, they separate solids into sized fractions. Two prominent mechanical devices designed exclusively for that purpose are *rake* and *spiral* classifers. Each consists of an inclined tank or trough having a gyrating rake or an auger conveyor which withdraws larger particles that settle quickly to the bottom of a slurry. They are effective in separating coarse solids or sands from the overflow of a mill. For wet classification of finer particles, cyclones are more effective. Data pertaining to the preliminary selection and design of rake and spiral classifiers are listed in Table 4-23a. More details on these and other types of classifiers can be found in Perry 21-45.

Figure 4-42 Thickeners employed to concentrate slurry for feeding kilns in a cement plant. (Ideal Cement Company, Devil's Slide, Utah, by permission.)

Filters

This category of mechanical separators represents a large and diverse collection of equipment based on the interception of dispersed particles by a porous medium. In this section, the first three rather specialized filters in Table 4-23b (bag, cartridge, sand) are discussed briefly. Then the remaining filters, which can be considered to be competitive—large liquid–solid process devices—are discussed as a group and individually.

BAG FILTERS *(Perry 20-89)*

Based on the same principle as the domestic vacuum cleaner, bag filters collect dust from a gas by interception on a fabric surface. Depending on the type of fabric, pore

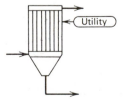

sizes can be made so small that essentially all particle sizes can be efficiently removed, either by direct interception or by Brownian diffusion. As with most filters, collection occurs primarily on a cake of predeposited solids. Fabric area is the variable that determines filter capacity. To provide extended area within a minimum volume, bags are suspended from a plenum flange and supported internally with wire cages or by tension to prevent collapse. Cleaning is accomplished by mechanical shaking or by backflushing with compressed gas. Solids drop to a hopper, from which they are conveyed for disposal or storage. As indicated in Table 4-23*b*, superficial gas flow rates range from 0.005 to 0.5 m/s at the fabric surface.

Because of vessel size, bag filters are seldom designed for high pressures, except in units of small capacity. Temperatures are limited by the fabric. Values as high as 250° C are possible with glass fiber, fluorocarbon- and nylon-based bags. In unusual custom designs, sintered metal cartridges can be employed to function at higher temperatures.

Compared with electrostatic precipitators, bag filters cannot collect liquids effectively nor separate with as small a pressure drop or at as high a temperature. On the other hand, fabric filters tend to be less expensive, particularly in smaller sizes, and their superior efficiency for submicrometer particles favors their selection for air pollution control [6]. Scrubbers, because of high maintenance, even poorer efficiency, and technological immaturity are even less competitive. The primary maintenance expense with fabric filters is bag replacement, which is normally performed about once a year.

CARTRIDGE FILTERS   *(Perry 19-83; McCabe and Smith 925)*

Use primarily for final cleaning or "polishing" of an effluent stream, cartridge filters are vessels containing replaceable or manually cleanable filter elements. They are

employed only when residue concentrations are very low, such that element replacement is no more frequent than once or twice a week. When employed in a continuous process, duplicate units are often installed to prevent interruption during cleaning. Cartridge elements are fabricated from a variety of media to promote high efficiency and large fluid capacities. For high temperatures or corrosive fluids, sintered metal elements are readily available.

SAND FILTERS   *(Perry 19-65; Foust 656)*

Sand or "media" filters are much like fixed-bed reactors except that the bed, comprised of carefully sized particles, entrains the residue. Used primarily for

clarifying culinary water in treatment plants, sand filters function semicontinuously until the bed becomes saturated and must be flushed by a counterflow of liquid. As with cartridge filters, duplicate units or adequate in-process storage must be provided to prevent interruption of flow during cleaning. Sand filters can be designed on the basis of criteria given in Table 4-23*b* and information on packed and fluid beds in Table 4-22. Capital costs can be estimated from data for process vessels.

Liquid–Solid Process Filters

Because of the numerous alternatives and complexities that exist in liquid–solid filtration, ultimate selection of a process filter from the 13 listed in Table 4-23*b* is

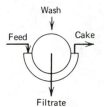

done by a specialist. Laboratory and pilot plant trials in process or at a vendor's test site would, undoubtedly, be a step in the selection process. The information listed in Table 4-23*b* should, nevertheless, aid in identifying a reasonably competitive, if not optimum, device. Before discussing each filter in detail, some general guidelines are useful for limiting the possibilities. More detailed assistance can be found in the work of Fitch [13] and Emmett and Silverblatt [11]. The article by Flood et al. [14], though somewhat dated, is also excellent. Much of the quantitative data found there is, however, available in Section 19 of Perry's *Handbook*. In preliminary filter selection, Fitch [13] suggests that with modest literature research or logical analysis, one can identify equipment that is currently used for the same separation or one that is analogous.

In violation of my general policy in this text, I must consider batch devices in this section. Plate and frame presses and shell and leaf filters, until recently, could not be replaced by any single type of continuous device. Thus, despite high labor costs, they are still numerous among the rare items of batch equipment employed in modern chemical processing. Criteria listed in Table 4-23*b* will help you to identify situations in which these batch units should be considered. If used, at least two filters are recommended to allow for cleaning and washing. The ultimate decision between continuous and batch filters is an economic one. As a tentative guideline (Perry 19-85), continuous filters become less competitive at solids rates below about 0.1 kg/s, and batch devices are questionable at solids rates much above 2 kg/s.

HORIZONTAL FILTERS   (*Belt, Table, Tilting Pan: Perry 19-80; McCabe and Smith 930; Foust 663*)

One common horizontal filter consists of a porous continuous *belt* that conveys a slurry over a vacuum trough or drain channel. *Table filters* convey the slurry in segmented pie-shaped segments that form a large horizontal filter disk. Each pan passes through various filling, extraction, washing, drying, and discharge steps while rotating full circle. Discharge is affected by an auger or rake. Because of this, table filters are also known in the industry as "scroll-discharge" filters. This discharge mechanism leaves a residual cake, which can be a disadvantage in some instances. The *tilting pan* filter, (Figure 4-43) at the expense of complexity and cost, overcomes this disadvantage by tipping each segment at the discharge point for efficient removal. For separating conventional slurries, horizontal filters are

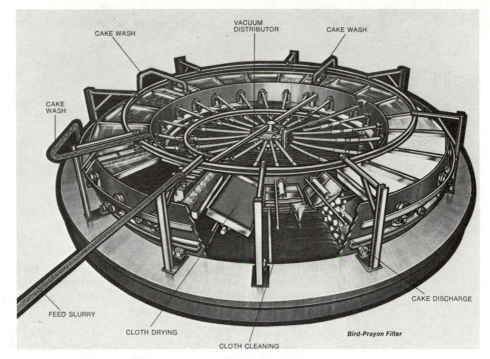

Figure 4-43 Tilting pan filter. (Bird Machine Company, Inc., South Walpole, Mass., by permission.)

adequate and cake washability and flexibility are excellent. They are among the few continuous devices that can take advantage of gravity. On the other hand, floor space, capital costs, and maintenance expenses are relatively high. Hence, other factors being equal, a rotary drum filter often is more attractive.

ROTARY DISK FILTERS   (*Perry 19-79; Foust 663*)

Similar to table filters in geometry, disk filters employ a series of circular disks rotating about a *horizontal* axis. This allows submersion of up to half the disk continuously in a slurry. Vacuum is drawn inside each hollow disk to promote flow. Disk filters are relatively inexpensive to purchase and operate, and they are among the most efficient in use of floor space. For relatively easy noncritical separations where flexibility and efficient cake washing are not essential, disk filters are an excellent alternative.

ROTARY DRUM FILTERS   (*Precoat, Multicompartment, Single-Compartment: Perry 19-25, 19-75; McCabe and Smith 928; Foust 665*)

This family, with its populous and diverse children, performs most of the world's liquid–solid chemical process filtration. The basic element is a cylindrical drum that rotates about a horizontal axis. Filter cloth or mesh forms a continuous band around the outside surface of the drum. Rotating through a pool of slurry, filtrate is forced into the center of the drum, which is at lower pressure. Solids accumulate on the mesh and are washed as they rotate above the pool. Cake is removed during downward travel at the end of the cycle (Figure 4-44). Several ingenious techniques for cake removal deserve mention. In one, a series of parallel strings is wound about the drum and a smaller satellite roller. During downtravel, strings separate from the drum, flaking the cake and aiding discharge. In another variant, the filter medium itself leaves the drum and discharges the cake during an abrupt change of direction cause by rotation about the satellite cylinder. Some filtrations use coillike springs that expand at the satellite roll because of tension, aiding discharge.

Although mechanically somewhat complex, drum filters, when open to the atmosphere, are only moderately expensive. This is an advantage in operations involving water and nontoxic or odor-free solids. The pressure differential, under these circumstances, is created by a vacuum pumps connected to the drum interior. With toxic materials, liquids of high vapor pressure, or where higher external pressures become necessary, the total units must be enclosed at much expense and loss of convenience. This is a situation that makes filter presses or shell and leaf units (batch filters) more attractive.

The *precoat rotary drum* filter is rather unique in its use of a heavy layer of solids to form a filtration cake, several centimeters thick, on the outside of the drum. This cake is extremely effective in removing cloudiness from dilute fine particle slimes. Residue is removed by a slowly advancing knife. In principle, this is a batch device because filtration must be interrupted to recoat the filter. In practice, since it is employed only for dilute mixtures and because of the precision of the knife advancement mechanism, recoating is necessary at a frequency of only one to ten days.

The so-called *multicompartment* unit is the most common and least expensive type of rotary drum filter. In fact, the precoat filters operate on this principle. Here, the drum arc is divided into segments, each attached to a pipe manifold and programmed for the appropriate filtration, washing, and discharge steps. In *single-compartment* devices, the total drum is open and special stationary plena are designed for cake washing and drying. This allows much higher rotation speeds and

Figure 4-44   Rotary drum filter. (Bird Machine Company, Inc., South Walpole, Mass., by permission.)

filtration rates with certain free-filtering solids, but the increased precision necessarily results in a more expensive filter.

Rotary drum precoat units are the only continuous filters capable of ultimate crystal-clear cleaning of effluent streams. Multicompartment media-covered drums, which are flexible and low in cost, are the most popular variety in this family. Single-compartment filters are more efficient in certain separations, allowing a smaller drum size, which compensates for the relatively large capital cost.

BATCH FILTERS   *(Plate and Frame Press, Shell and Leaf: Perry 19-65, 19-69, 19-80; McCabe and Smith 924; Foust 658, 661; Moir [33]; Brown [5])*

The *plate and frame filter press* is comprised of rectangular rings pressed between two plates covered with cloth. Frames, cloths, and plates are alternated in

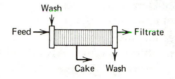

horizontal stacks of up to several meters long and compressed by a hydraulic mechanism, forming a series of cavities having porous walls. Slurry is pumped into these cavities, and filtrate passes through the cloth, leaving solids inside the cavities. When the frames are saturated with cake, a programmed wash–dry cycle is executed and the press automatically opens, dropping cake by gravity into a trough. Since pressures are confined to the inside cavities, high values (up to 70 bara) can be readily accommodated in standard presses. Flexibility of design, media, and operation keep this rather archaic device competitive in a number of applications. On the other hand, inevitable leakage and dripping plus the large amount of operator attention required are disadvantages that may tip the balance in favor of continuous equipment.

An important relative to this filter is the plate and frame thickener. Construction is similar except that frames are designed differently and concentrated slurry is allowed to leave. This, then, is not a true filter, but a densifier to prepare a concentrated slurry for subsequent processing.

The *shell and leaf filter* consists of a pressure drum fitted with a series of bag-covered, porous, hollow cavities or leaves, all manifolded together. Pressure is exerted on slurry in the drum, causing filtrate to exit through the manifold. Cake is discarded by withdrawing the "branch" and attached leaves from the drum and manually scraping or sluicing the cake into a trough below. Because of design, shell and leaf filters are somewhat more tidy than filter presses. Thus, they can be used more conveniently for toxic, odorous, or hazardous materials. However, pressure limits are more severe and steps such as washing and discharge are not as convenient. Capital costs are similar.

Presses

(Perry 19-101)

If a wet solid is to be dried by heating, mechanical dewatering through squeezing or "expression" may be economically attractive. This is also an important technique for recovering oils or syrups from vegetable or food pulps. Devices designed to accomplish this are similar to extruders and other high pressure solids compaction equipment. In *screw presses*, wet solids enter a chamber having a rotating auger with decreasing pitch. This compresses the solid and expels liquid through rivulets or a porous wall. Screw presses are used extensively to express vegetable oils and to dewater paper pulp, plastics, and rubber. They are suitable for a variety of sludges and residues except those that contain coarse and abrasive solids.

Roller presses, similar to roller crushers, compress feed within the gap between two or three rotating smooth or corrugated cylinders. Three-roll presses are prominent in the expression of syrup from sugarcane. Two-roll presses are widely used for dewatering paper and fabrics. Roller presses are more suitable than screw presses for abrasive materials, but inferior for sticky residues where discharge may be a problem.

Screens

(Perry 21-39; McCabe and Smith 914; Beddow [2])

Screens are a familiar means of separating solids according to particle size. In the laboratory, stacks of screens, decreasing in mesh size from top to bottom, are

shaken to yield fractions that fall between intermediate decks. As with distillation, it requires $n - 1$ separators to yield n fractions. In commercial applications, it is usually more efficient to employ a series of single-deck screens rather than multiple-deck units. If particles are large (greater than 5 cm) and will separate effectively without agitation, stationary screens are usually adequate. These "grizzlies" are usually inclined slightly and are made of heavy rods or bars rather than wire mesh. For separation of smaller particles, screens are vibrated and, occasionally, wet screening is employed. Costs for grizzlies are insignificant in most mills and are not included in Chapter Five. Vibrating screens, on the other hand, are more sophisticated and expensive. Capital costs for these are shown in Figure 5-59. General guidelines for screen sizing are contained in Table 4-23*b*. For more breadth and accuracy, the excellent review of all "dry separation techniques" by Beddow [2] is recommended.

SIZE-ENLARGEMENT EQUIPMENT

(Perry, 8-57)

In a constant fight against entropy, it seems ironic that we sometimes need to enlarge particle size after having invested substantial energy to crush, grind, or pulverize a solid. When necessary, size-enlargement satisfies one of three general purposes. The first is enlargement for handling and marketing convenience. Carbon black, for example, which is generated as a smoke, is pelletized to minimize dust during shipping and handling. The second purpose is to provide convenience to the consumer. Most people, for instance, do not like to swallow powders. Thus, most solid forms of medicine are converted to tablets or pills. End-use convenience justifies size enlargement for a number of consumer products ranging from foods to detergents. The third reason for increasing size is to fabricate special shapes. Conversion of plastic pellets to rods, tubes, or sheets is a good example.

Preliminary design of size-enlargement equipment is relatively simple. Selection of equipment type can be eased using data in Table 4-26 plus other information here. Once equipment type has been defined, power consumption can be calculated directly from solids flow rate and specific power consumption listed in the table. Some enlargement processes require addition of a liquid binder or lubricant. Guidance to binder types and quantities is also provided below.

Many special techniques for size enlargement exist in unique and unusual situations. Some of these are described in Perry. Only conventional, large-volume techniques are discussed here.

Size enlargement requires either frequent or intense contact of particles under conditions that will promote fusion or sticking. Machines enumerated in Table 4-26 fall into these categories according to the type of physical contact. Tableting presses, roll presses, pellet mills, and extruders, which produce hard, resilient briquets, pellets, or other shapes, depend on compression contact. Pan or drum agglomerators yield light, fragile pellets by multiple, gentle, tumbling contact. Prilling creates large particles by melting a powder and spraying the liquid into a cooling chamber.

Tableting Press

(Perry 8-59)

This high speed mechanical press has a large number of cavities that are automatically charged with precise quantities of powder. Each charge is compressed

by a plunger and discharged from the press in tablet form. As the name suggests, these machines are employed extensively in pharmaceutical manufacture. They are also used to prepare catalyst pellets, ceramics, porous metals, and other industrial chemical products. Liquid binders or lubricants are often employed to create a uniform rugged product. Liquid mass fractions normally fall between 1 and 5 percent. For types and amounts of binders used for various products, see Perry (Table 8-57, p. 8-60). As Table 4-26 indicates, tableted products are extremely uniform in size and composition and have excellent mechanical strength. Equipment operating and capital costs are high, however, and capacity is limited. These machines are obviously limited in their ability to process sticky or gummy materials.

Roll Press

(Perry 8-59)

Similar to roll crushers or twin-roll dryers, this rugged, high speed machine produces egg- or pillow-shaped briquets by compression of raw material between

TABLE 4-26

CRITERIA AND DATA FOR THE PRELIMINARY DESIGN OF SIZE-ENLARGEMENT EQUIPMENT

| | Equipment Type | | | | | | | |
| | Pressure Compaction | | | | | Tumbling Compaction | | Other |
	Tableting Press	Roll-Type Press	Pellet Mill	Screw Extruder	Pug Mill Extruder	Disk (pan)	Drum	Prilling Tower
Maximum capacity, $\dot{m}_s$ (kg/s)	1.0	15	5	0.1	6	15	30	50
Maximum pellet, tablet, or briquet dimension (cm)	5	10	1	12	12	3	0.5	0.3
Quantity of liquid lubricant or adhesive normally employed (wt %)	1–5	1–5	1–10	0–10	1–10	10–30	10–30	—[a]
Compatibility or Suitability								
Sticky or gummy feeds	E	C	A	A	A	E	E	A
Uniform product dimensions and properties	A	D	B	A	A	B	D	B
Mechanically strong product	A	B	D	B	B	D	D	A
Flexibility of product shape	D	B	D	A	B	D	D	E
Low dust product	A	A	A	A	A	A	A	A
Atmospheric control	D	D	C	C	C	D	C	A
Capital cost	High	Modest	Low	High	Modest	Very low	Low	Modest
Specific power consumption (kW · s/kg)	30–50	8–16 (Moh 1–2) 16–32 (Moh 3–4) 36–64 (Moh 5+)	30	1500–5000	500–2000	3–4	6–8	—[b]

KEY

A   excellent or no limitations
B   modest limitations
C   special units availabe at higher cost to minimize problems
D   limited in this regard
E   severely limited in this regard
F   unacceptable

[a]Feed material must be molten.

[b]Utilities must be estimated for specific application.

two synchronized rollers having indentations designed for the desired shape. Smooth rolls are employed in some instances to produce sheets, which are diced by a rotary knife cutter (see the section on size-reduction equipment). Roll presses operate at high capacity and low cost, but the product is less uniform than that from tableting presses. Rolls can handle some solids that will not flow into or easily discharge from tablet presses, but most sticky and gummy solids are better enlarged in pellet mills.

Roll presses operate with or without liquid binders depending on the solid. Types of materials and conditions employed are listed in Perry (Table 8-59, p. 8-61).

Pellet Mill

(Perry 8-60)

An ideal compromise for large-scale production of pellets from sticky solids, the pellet mill ("mill" being an inaccurate designation) employs a rotating perforated

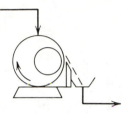

drum that is fed from the center. An internal roller squeezes paste through the holes or dyes. These resulting cylindrical projections are sheared into pellets by external fixed cutting blades.

Force due to extrusion through the dye is the only means of compaction. Thus, pellet strength is limited. Uniformity, however, is quite good. The feed must, of course, be capable of plastic flow. Pellet mills are excellent for pastes that gum and clog other presses.

Screw Extruder

(Perry 8-61)

Employed for polymers and other very viscous plastic materials, screw extruders are much like screw or auger conveyors except that conveying volume decreases with progression through the machine. This produces high compression forces and

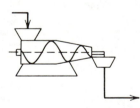

generates substantial heat, which, in polymer production, melts the granular or pelletized feed. The resulting viscous liquid is forced through a die to yield a rod, tube, sheet, or other shape as desired. When coupled with a cutter, pellets, hollow cylinders, or other shapes can be produced. A glance at Table 4-26 reveals that these machines are highly versatile but also extremely expensive in both capital and operating costs as well as limited in capacity. Screw extruders are employed extensively where heat is required and shape of the final product is a major consideration.

Pug Mill Extruder

Capable of mixing, densifying, and extruding in one machine, pug mill extruders are employed for large-capacity production of ceramic shapes, catalysts, fertilizers, and animal feeds. The name comes from the propellerlike auger conveyors discussed earlier. Pug mills are more expensive than pellet mills, and their power consumption is greater. Quality, strength, and versatility of the product are also greater. These extruders are used typically to fabricate green bodies for subsequent firing into ceramic products. They are superior to screw extruders in their mixing function, thus making them ideal for situations where mixing is desired but heat addition is not.

Disk or Pan Agglomerator

(Perry 8-61)

Designed primarily to convert dusty, difficult-to-handle powders into spherical balls, these contactors consist of a rotating inclined disk or pan with a raised rim.

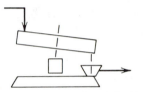

Powder is fed at the center with a significant quantity of liquid, usually water. As the powder migrates toward the rim, agglomerates form and grow like snowballs into compact, rather fragile spheres. These units are relatively inexpensive and have high capacity and low power consumption. Table 4-26 indicates why tumbling compaction is chosen over pressure compaction when the primary goal is ease of handling rather than final product strength and versatility.

Drum Agglomerator

(Perry 8-62)

Similar to drum dryers, rotary kilns, and media mills, a drum agglomerator consists of a rotating inclined cylinder where powder, fed at one end, becomes agglomerated

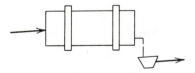

as it migrates to the other. From data contained in Table 4-26, the choice between disk or drum agglomerator is not obvious because the capital costs and energy consumptions are similar. Disk units generally produce a more uniform product because of classification created by the centrifugal path and in the rim of the pan. Capital and operating costs are lower for disk machines, and they can produce a larger pellet.

Drum agglomerator are capable of higher capacities and can be more easily enclosed where dust and fumes create hazardous problems. Longer retention times are possible in drum machines for materials that are difficult to agglomerate.

Prilling Tower

(Perry 8-64)

Prilling towers or chambers resemble spray dryers and contain atomizers at the top. In contrast to dryers, liquid droplets are formed from a melt rather than solution

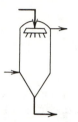

and cool or freeze rather than dry as they fall through the chamber. Prilling normally requires a feed material such as ammonium nitrate, which forms a low viscosity liquid at relatively low temperatures, although it is also used to form powders from liquid metals. Utility consumption and size of a prilling operation can be estimated from the falling velocities and heat transfer characteristics of spherical droplets. For design, falling solids fluxes range from 0.1 to 0.3 kg/m^2 · s for 0.1 to 0.3 cm droplets (based on total tower cross section). Typical heat transfer coefficients range from 300 to 500 J/m^2 · s · K based on droplet surface area. The quantity of cooling gas required can be computed from the latent heat of fusion and sensible heat change of the product.

Prilling is excellent for sticky, gummy materials if they can be melted to a free-flowing liquid. The equipment is moderate in cost and is easily adapted for materials that for safety or quality reasons, must be processed in a controlled atmosphere. Energy costs are significant because of the need to supply and remove heat. But heats of fusion are much smaller than latent heats, so the penalty is not as great as in drying. Once the size of a prilling tower has been defined, its capital cost can be assembled from data on process vessels.

STORAGE VESSELS

(Accumulators, Bins, Drums, Hoppers, Silos, Tanks: Perry 6-85 to 6-105, 7-22 to 7-50; Treybal, 397)

Tanks are receptacles employed to hold, transport, or store liquids or gases. Bins perform the same functions for solids. Hopper refers to the funnel-shaped outlet typical of most bins. Hence, the entire vessel is often known as a hopper. If a bin is cylindrical, it can be termed a silo.

Accumulators are vessels auxiliary to major equipment items. They provide intermediate storage to compensate for minor process upsets and to allow surge capacity or time for control action to be taken. A familiar example is the accumulator provided to hold reflux in distillation operations.

Drums, named after and shaped like musical percussion instruments, are common storage receptacles employed when pressures do not differ more than a few kilopascals from ambient. Because of the tendency for stress to focus at the joints between disk-shaped ends (or heads) and the cylindrical body, most process tanks have hemispherical or ellipsoidal "dished" or "shaped" ends. Because of their appearance, such vessels are sometimes called bullet tanks. Despite their shape, however, they are still, by tradition, called rums. The vulnerability of flat disk ends is revealed by occasional examples where flat-ended drums have inadvertently developed dished-ends through the application of excessive internal pressure.

Many process vessels are basically tanks. A distillation tower, for example, is a long vertical tank having shaped ends and containing internal plates, packing, or other modifications to accomplish its function. This discussion is limited to storage vessels. Process vessels are treated separately.

Preliminary design of a storage receptacle is direct and elementary. One merely needs to know its size and the temperature, pressure, and exposure conditions anticipated. All but the first will be evident from the flow sheet. Rules for determining size are based on common sense and experience. For the beginner who has a scarcity of both, some criteria are formalized in Table 4-27.

Large tanks for storage of raw materials and final products are not usually included explicitly in preliminary economic evaluations. Instead, they comprise a category of *service* or *yard* facilities (see Table 4-3). Their cost is included as an added fraction of direct process equipment. Bulk storage tanks are described here, nevertheless, for the sake of completeness and breadth.

Atmospheric Pressure Storage

Large atmospheric pressure storage tanks are employed universally to hold raw materials for processing or products awaiting shipment. In most modern processes, vapors cannot be vented to the atmosphere. Thus, to compensate for changes in ambient temperature and pressure, or accumulation and depletion of contents, either an internal vent system or a floating roof is employed. These tanks are usually

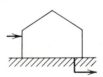

like large disks, up to 15 m high and more than 30 m in diameter with cone-shaped roofs. They are the most obvious visual components of oil refineries and other operations that process large volumes of liquids.

To assure uninterrupted operation, the designer usually provides about 30 days storage capacity for raw materials and manufactured products. In some cases, this is exceeded, since storage tanks should be at least 1.5 times the size of the vessel (i.e., rail car, truck, or ship) intended for transport. Representative capacities of transport vessels are shown in *Note a* of Table 4-27.

Because of their large surfaces, storage tanks cannot sustain appreciable pressure differences. Pressure is relieved automatically in tanks having *floating roofs*, which are free to move. Because of size and cost, steel or concrete is normally used for construction. If special alloys or plastics are needed for corrosion resistance, they are usually clad or bonded to the steel or concrete wall that provides structural support.

Gas holders are storage tanks having a double wall filled with liquid, usually water. A second cylindrical shell, closed with a roof, mates with the wall, creating a

seal. Variable volume is accommodated by up and down motion of the movable canister.

Bins are vertically oriented vessels, commonly square in cross section and having a funnel-shaped base for gravity discharge of solids. If pressurized, a circular cross section is employed, as with liquid storage tanks, and a dished cover is placed

on top. Since many solids can be exposed to the atmosphere, most large-scale storage (e.g., coal) is in piles in *open yards*. If moisture from precipitation is a

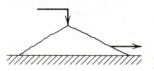

problem, the piles can be protected with an inexpensive open-sided building. The base of a circular pile of given volume depends on the angle of repose of the solid. The volume is given by

$$V = \frac{\pi L D^2}{12} \tag{4-112}$$

For an angle of repose equal to a typical value of about 40 degrees, $L = 0.4D$ and $V = 0.11D^3$.

Bins, which are more costly than yard storage, are most frequently employed for in-process use, between grinders, conveyors, calciners, and other such equipment, to provide surge capacity in case of process malfunctions. Capacity equal to 8 h throughput is recommended upstream of equipment that is difficult to shut down or would be harmed if operated empty.

Pressure Storage

In many situations, low pressure storage is impractical or undesirable. Higher pressures are accommodated in *spherical or bullet-shaped tanks* having thicker walls. Practical and economic considerations limit the sizes of these units well below

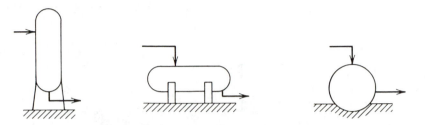

those of atmospheric pressure tanks. With gases, the density increases to provide added storage capacity at elevated pressures. Even so, large-scale storage of gases is difficult. One common solution is to liquefy the gas and store it at cryogenic temperatures and relatively modest pressures. Pressure vessels, either spherical or cylindrical are ideal for this. Cylindrical tanks with formed ends are less expensive. They are generally oriented, for convenience of access, with the axis horizontal. The volume of a partially filled vessel is not a linear function of height in this situation as it is with vertical orientation. Evaluation of the height–volume relationship is a challenging review of geometry and trigonometry. For those in a hurry, Perry Tables 6-52 and 6-54 (pp. 6-87 and 6-88) will be useful.

For cryogenic storage, tanks are insulated or both insulated and buried. Burial is particularly advantageous for volatile, combustible, or hazardous liquids, since subsoil temperatures are not only moderate but relatively constant. Also, accidental damage is unlikely. With metal tanks advantages of burial are offset by the potential for corrosion on the outside. Fiberglass-reinforced plastic tanks, on the other hand, are free of this defect. Such vessels, up to 200 m^3 in volume, have been used successfully to store gasoline, fuels, and volatile corrosive chemicals [28].

In-Process Storage

Day tanks, like many bins, are installed between process modules in a complex process for storage of intermediate liquids and gases. This provides surge capacity and a margin of time to repair defective equipment in a single module without shutting down the entire plant. Eight hours is generally considered an optimum time, and day tanks are sized accordingly. Otherwise, they are similar to pressure storage vessels and are designed according to data in the preceding section. A length to diameter ratio between 3:1 and 5:1 is most common. It is based on pressure with more precise values given in Table 4-24. *Feed tanks* are provided ahead of individual equipment items such as furnaces, which would be harmed if operated empty. Typical residence times in feed tanks are about 1800 s.

Accumulators and *knockout "drums"* are smaller vessels (almost always cylindrical with formed ends) designed like pressure storage and in-process storage tanks. Accumulators are usually oriented horizontally to provide a large liquid volume within a small vertical level range (for control flexibility). A time between control levels of 300 s is typical. This requires a total tank residence volume equivalent to about 600 s. More information on accumulators is provided in the

TABLE 4-27

CRITERIA AND DATA FOR THE PRELIMINARY SPECIFICATION OF STORAGE VESSELS

	Type of Receptacle				
	Atmospheric Pressure Storage				
	Fixed (conical) Roof	Floating Roof	Gas Holder	Bin	Open Yard (pile)
Typical Maximum Size					
Volume, V (m^3)	100,000	100,000	20,000	4,000	200,000
Height or length, L (m)	15	15	30	50	50
Diameter of width, D (m)	90	90	30	10	120
Length to diameter ratio	<2	<2	1–2	2–5	0.4
Stored Medium					
Solid				√	√
Liquid	√	√			
Gas			√		
Orientation					
Axis vertical	√	√	√	√	√
Axis horizontal					
Modifications Frequently Used					
Burial					
Cryogenic service					
Insulation					
Discharge Mechanism					
Pump, solids conveyor, or blower	√	√		√	√
Intrinsic pressure			√		
Gravity				√	
Residence or storage time[a]	30 days	30 days	30 days	8 h	30 days
Maximum pressure (barg)	0.2	0.2	0.2	—[c]	0
Temperature range (°C)	−20 to 40	−20 to 40	−20 to 40	−20 to 40	−20 to 40
Common Materials of construction					
Carbon steel	√	√	√	√	
Concrete	√	√		√	√
Plastics or fiberglass				√	
Alloys and coated or clad steel	√	√	√		

[a]Storage times are typical values. If the plant capacity is small or the raw materials delivery vessel is large, storage vessels will be larger. They should be at least 1.5 times the size of the delivery vessel. For trucks, this is typically 25 m^3 or 20,000 kg (whichever is smaller). For rail cars, capacitites are 25, 60 or 130 m^3 (20,000, 45,000, or 100,000 kg), and for barges or ships, 1500 m^3 (1,000,000 kg) are larger.

[b]See Table 4-18, Table 4-25 and the discussion of process vessels and of separators.

[c]Limits on pressure in these vessels are economic. See section on process vessels.

[d]For steel, the range is −20 to 600°C; for aluminum, −250 to 200°C; for stainless steel, −250 to 800°C; for nickel-based alloys, −200 to 700°C.

TABLE 4-27
(Continued)

Type of Receptacle					
Pressure Storage		In-Process Storage			
Cylindrical (bullet) Tank	Spherical Tank	Day Tank	Feed Tank	Accumulator	Knockout Drum
1600	15,000	1600	Various	Various	Various
20	30	20	Various	Various	Various
10	30	10	Various	Various	Various
2–5	1	3–5	3–5	3–5	3–5
√	√	√	√	√	√
√	√	√			√
√		√			√
√		√	√	√	
√	√	√			
√	√				
√	√	√	√	√	√
		√	√	√	
√	√	√	√	√	√
30 days	30 days	8 h	1800 s	600 s	—[b]
17	14	—[c]	—[c]	—[c]	—[c]
—[d]	—[d]	—[d]	—[d]	—[d]	—[d]
√	√	√	√	√	√
√			√		
√	√	√	√	√	√

discussion of process vessels. Knockout drums, employed for rough separation of vapor–liquid mixtures, are oriented vertically so that vapor velocity, hence, entrainment, are independent of liquid depth. Accumulators and knockout drums are described in more detail in sections on process vessels and separators.

Capital costs for small in-process storage tanks can be derived directly from the charts pertaining to process vessels in Chapter Five. Separate charts for larger atmospheric and pressure storage vessels and bins are also presented in Chapter Five.

MATERIALS OF CONSTRUCTION
Properties of Materials

STRENGTH

For conventional steel construction, temperatures above 600° C are impractical because they result in poor strength and poor oxidation resistance. Temperatures up to 800° C are possible with high alloy or stainless steels, but strength at this temperature is 10 times less than that at room temperature. The maximum pressure of a vessel is directly proportional to metal thickness. An algebraic relationship can be derived easily from a force balance on a shell cross section. For a cylindrical shell subject to internal pressure, the force balance, appropriately corrected by safety factors and joint efficiencies, becomes

$$t = \frac{pR}{0.9S - 0.6p} \tag{4-113}$$

or

$$p = \frac{0.9St}{R + 0.6t} \tag{4-114}$$

For a spherical shell, the equations are

$$t = \frac{pR}{1.8S - 0.2p} \tag{4-115}$$

or

$$p = \frac{1.8St}{R + 0.2t} \tag{4-116}$$

where t is shell thickness, p the pressure, R the inside vessel radius, and S the allowable tensile stress. These equations are dimensionally consistant, since stress and pressure have the same units as do t and R. An additional allowance, typically 3 mm, is added to the thickness to compensate for corrosion of vessel walls. (For vessels subject to external pressure, the discussion in Perry, p. 6-92, and the corresponding chart in Fig. 6-133 can be employed profitably.) Approximate values of S for carbon steel, aluminum, and stainless steels are given as a function of temperature in Perry Tables 6-57, 6-58, and 6-59, (pp. 6-96 and 6-97). Additional stress data for pipe including a more extensive list of materials are found in Perry Tables 6-3 and 6-4. Some of these data are summarized for quick reference in Figure 4-45.

It is obvious from Figure 4-45 that process temperatures above 800° C cannot be tolerated by conventional metals and alloys. Above this level, refractory metals, graphite, or ceramics are necessary. Most refractory metals and graphite burn readily in air and, if used, must be protected from the atmosphere or any other oxygen-containing gases. Numerous inexpensive ceramics are strong up to 1800° C, but they are fragile and brittle. Kilns, heaters, and several other types of high temperature equipment have composite construction. Mechanical and atmospheric integrity is provided by an outer metal skin or shell, which is protected from high process temperatures by refractory batting or brick lining.

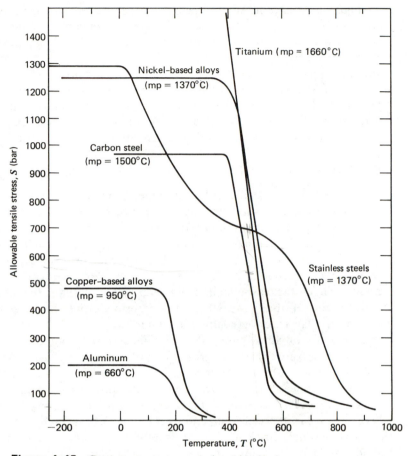

Figure 4-45 Stress-temperature relationships for important process metals and alloys.

CORROSION

Beyond temperature considerations, chemical processes are noted for their antagonistic, corrosive chemical environment. Perry, Section 23, discusses corrosion and its control in excellent and complete detail. Characteristics of practical metallic and nonmetallic construction materials are also discussed at length. The comprehensive review by Kirby [26] is highly recommended for experts and beginners alike. In the brief treatment presented here, I cannot approach the thoroughness of these references but attempt to present guidelines that will permit you to make a reasonably intelligent selection from common construction materials.

The need for experience and expert guidance becomes evident very quickly to chemical engineers, who learn through experience the pitfalls that exist. A neophyte, for example, might think that the blower downstream of an acid scrubber could be made of carbon steel because acid vapors will have been removed by the time gases reach the blower. The expert (often because of sad experience) will be quick to point out that this environment is, in fact, one of the most corrosive because there are traces of acid gases in a moist airstream (not to mention the devastating effects of process upsets). To be safe, one would specify blower materials suitable for the worst possible situation. At the other extreme, given a high temperature gas stream containing air, hydrogen chloride, and water, an engineer would automatically consider expensive nickel-based alloys or ceramics, yet I know of one such exposure where inexpensive aluminum metal is used. As long as gases are below the maximum service temperature of the metal and above the acid dew point, aluminum is superb—significantly better than stainless steel in this environment. Surprises such as this are welcomed by project managers. Those of a reverse nature are not. In general, we should select a material that either has been proved by past experience or is promising according to general performance criteria.

SPECIFIC MATERIALS

Cost data are readily available for equipment constructed of carbon steel, alloy steel, conventional plastics, fiberglass, fluorocarbon plastics, aluminum and its alloys, copper and its alloys, stainless steels, nickel-based alloys, and occasionally, more exotic metals or alloys such as titanium, tantalum, tungsten or molybdenum, zirconium, or the "stellites." A few words about each group are appropriate.

Based on strength, cost, and temperature resistance, carbon steel is outstanding. That is the reason for its overwhelming popularity in the chemical process industries. Susceptibility to atmospheric corrosion is often corrected by applying a paint or similar exterior coating. Its limitations in the presence of aqueous electrolytes are obvious, however, to anyone who owns an automobile near the seacoast or in severe climates where salt is applied to the highways for snow and ice control. For more substantial internal protection, carbon steel is frequently lined or "clad" with rubber, glass, stainless steel, nickel alloy, or titanium. This provides corrosion resistance typical of the cladding with the strength and economic advantages of carbon steel. Temperature is limited, of course, by the partner having the least tolerance. Brick linings are usually much thicker than cladding. These provide insulation also, lowering the thermal exposure of the shell.

Conventional inexpensive plastics such as polyethylene, polypropylene, and ABS, because of their inherent resistance to ionic solutions, are becoming more popular as their mechanical characteristics improve. Because of low strength and low decomposition temperatures, however, process applications are limited. Fiberglass composites, having greater strength and reasonably good temperature tolerance, are employed for some applications, such as atmospheric storage tanks, in competition with carbon steel and other traditional materials. Fluorocarbon plastics, as mentioned at the beginning of this book, exhibit superior corrosion resistance and (for polymers) temperature tolerance. This comes at a high purchase price that, combined with low strength, restricts their use considerably.

Aluminum is a moderately priced alternative to carbon steel or copper-based metals, especially in heat transfer applications where high thermal conductivity is an asset. Aluminum is also superior at low temperatures (-250 to $-20°$ C) where carbon steel becomes brittle and inadequate. Its light weight compensates somewhat for relatively lower strength (see Figure 4-45). The strength of aluminum declines seriously at about $200°$ C. It is generally corroded badly by aqueous electrolytic solutions.

Copper and copper-based alloys, because of good corrosion resistance and superior thermal conductivity, have been employed traditionally in heat transfer equipment. In many applications, however, metal thermal resistance is insignificant compared with film resistance, and copper has been supplanted to some extent, especially in the food-processing industry, by stainless steel. The advantages of copper and its alloys in salt solutions are witnessed by the prominence of brass trim on ships. In fact, the name "admiralty" has been employed to designate a copper alloy having superior resistance to marine environments.

The advent of stainless steel and its modern availability at a reasonable price have made it second only to carbon steel in chemical process use. It is routinely, almost automatically, specified for tower trays, mist-eliminator mesh, and other items when the availability of a single versatile material often is more economical than providing a spectrum of less flexible alternatives. Stainless steel is limited, as mentioned earlier, in some environments, especially those containing chlorides. Also, as with all electrically conductive metals, stainless steel should not be in direct contact with a dissimilar metal in the presence of a liquid electrolyte. This creates the equivalent of a short-circuited electrochemical cell, leading to rapid corrosion in what would otherwise be an innocuous application.

In Perry Table 23-5 approximately 50 varieties of stainless steel are identified. This represents only a fraction of those commercially available. The choice in a particular situation usually depends on a combination of factors such as machinability, wear resistance, acid tolerance, and high temperature performance. In this text, "stainless steel" is employed generically because most metals so designated have similar corrosion resistance and prices. In the cost charts of Chapter Five, the most expensive stainless steel for a typical service is assumed. This assures flexibility for final alloy selection without invalidating economic projections. The same is true of copper-based and nickel-based alloys that as generic groups have similar corrosion, temperature, and price characteristics.

Nickel and its alloys, known by such trade names as Hastelloy, Inconel, Incoloy, Monel, and Nimonic, have been developed in a range of compositions approaching those of stainless steel. Stainless steel itself contains a large amount of nickel as well as chromium and iron. The so-called nickel-based alloys contain nickel as the major ingredient rather than iron. As illustrated in Figure 4-45, the strength and thermal properties of the two alloy types are similar. Nickel and its alloys are superior in resistance to corrosion, especially by chlorides that attack stainless steel. Applications in such environments, which are quite common, justify the higher price (about double that of stainless steel) of nickel-based alloys.

Titanium and other premium metals or alloys are reserved for applications in which servere, corrosive, abrasive, or high temperature exposure precludes the foregoing alternatives. Heat exchangers, pumps, and process vessels are on the list of equipment that can be readily obtained in titanium and other less common metals.

Selection Guide

As a quick guide to materials selection, Table 4-28 has been prepared from data in handbooks and miscellaneous suppliers' catalogs. It is for tentative selection only. (Don't blame me if you use it for formal design and the equipment dissolves.) Ultimate selection usually is made by a materials specialist in consultation with vendors. This might even include extensive corrosion testing in a pilot plant or laboratory. The guidelines presented here, however, should be adequate for

TABLE 4-28
CORROSION GUIDE FOR COMMON MATERIALS USED IN CHEMICAL PROCESS CONSTRUCTION

Exposure	Carbon Steel (cs) and Alloy Steel (as)	Stainless Steels (ss)	Aluminum-Based (Al)	Copper-Based (Cu)	Nickel-Based (Ni)
Aqueous Solutions					
Acetates	XXXX	AAAA	AAAA	AAAA	BBBB
Ammonium salts	BDDE	AAAA	BBCC	DEXX	AABC
Carbonates	BBBB	AAAA	DDEE	AABC	BBCC
Chlorides	DDEX	CCCD	DDEE	AACC	CCCC
Nitrates and nitrites	AABB	AAAA	BBDD	BBCD	AABC
Sulfates and Sulfites	BBBB	AAAA	DDDD	BBCC	AAAA
Acid Solutions and Wet Acid Vapors					
HBr	XXXX	XXXX	XXXX	XXXX	CCCC
HCl	XXXX	XXXX	XXXX	CCXX	CCCC
HF	XXXX	XXXX	XXXX	CCCC	CCCC
HNO$_3$	XXXX	AAAA	XXXX	XXXX	AAAA
H$_2$SO$_4$	XXXX	BBDX	DDXX	XXXX	ABCX
H$_2$PO$_4$	XXXX	AABC	XXXX	CCXX	AABC
Organic acids	XXXX	AABC	AABX	AABX	AAAA
Basic Solutions and Wet Vapors					
Ca (OH)$_2$	BBDD	AAAA	XXXX	AAAA	AAAA
NaOH	BBBB	AAAA	XXXX	XXXX	AAAA
NH$_3$OH	XXXX	AAAA	BBDD	XXXX	AABC
Food Intermediates					
Dairy products	XXXX	AAAA	DDXX	BBDD	AAAA
Fruit juices	XXXX	AAAA	ABDX	BBBB	BBBB
Sugar syrups	BBBB	AAAA	AAAA	BBBB	AAAA
Vegetable oils	BBBBBBBB	AAAAAAAA	AAAAA	BBDDDD	AAAAAAAAA
Vinegar	XXXX	AAAA	BBBB	BBBB	AAAA
Gases (moist) and Cryogenic Liquids					
Air	AAAAAAABBDDDD	AAAAAAAAAAAAAABBBBBBDDE	AAAAAAAAAD	AAAAAAABDD	AAAAAAAAAAAAAABBD
Br$_2$	XXXXXXXXXXX	XXXXXXXXXXXXXXXX	XXXXX	CCCCCC	AAAAAAAAAAAA
CO$_2$	XXXXBBBBBBBBBB	DDDDAAAAAAAAAAAA	BBBBAA	AABEEBBB	AAAAAAAAAAAAAA
Cl$_2$	XXXXXXXXXXXXX	ABXXXXXXXXXXXXXXX	XXXXXXX	XXXXXXXX	AAAAAAAAAAAAAA
F$_2$	XXXXXXXXXXX	AAABBXXXXXXXXXXXXXX	XXXXXXXXXX	XXXXXXXXX	AAAAAAAAAADDDDD
Flue gases	XXXBBBBBBBBB	XXXBBBBBBBBBBBBBB	XXXBB	CCCCCC	AAAAAAAAAAAA
HBr	XXXXXXXXXXXXX	BBDXXXXXXXXXXXXXXX	XXXXXAA	XXXXXXXX	AABBBBBBBBBBBBBB
HCl	XXXXXXXXXXXXX	BBDXXXXXXXXXXXXXX	XXXXXAA	XXXXXXXX	AABBBBBBBBBBBBBB
HF	XXXXXXXXXXXXX	BBDXXXXXXXXXXXXXX	XXXXXBB	XXXXXXXX	AABBBBCCCCEEEEE
H$_2$	DDDXXXXXXXXXX	AAAAAAAAAAAAAAAAAAAA	AAAAAAAAAA	AAAAAAAAA	AAAAAAAAAAAAAAAA
H$_2$S	ABDDBBBBBBBBB	AAAAAAAAAAAAAAAAAA	AABDDBB	XXXXXXXX	AAABBBBBBBBBBBBB
Halogenated hydrocarbons	BBBDEXXXXXXXX	BBBBBBBBBBBBBBBBBB	XXXXXX	CCCCCCC	AAAAAAAAAAAA
Hydrocarbons	BBBBBBBBBBBBB	BBBBBBBBBBBBBBBBBBBBB	AAAAAAAAAA	CCCCCCCCCC	AAAAAAAAAAAAAA
NH$_3$	BBBBBBBBBBBBB	BBBBBBBBBBBBBBBBBB	BBBBBBB	XXXXXXXX	AAAAAAAAAAAAAA
N$_2$	AAAAAAAAAAAAA	AAAAAAAAAAAAAAAAAA	AAAAAAAAAA	AAAAAAAAA	AAAAAAAAAAAAAA
O$_2$	AAADXXXXXXXXX	AAAAAAAAAAADDDDEEXXXX	AAAAAAAAAA	AAAAAAABDD	AAAAAAAAAAAAAABBB
SO$_2$	BBXXBBBBBBBBB	BBBXXBBBBBBBBBBBB	BBXXXBB	CCCCCCCC	AAAAAAAAAAAA
SO$_3$	BXXBBBBBBBBB	BXXBBBBBBBBBBBBB	XXXBB	CCCCC	AAAAAAAAAAAA
Steam	AAAAAAAAAAAA	AAAAAAAAAAAAAAAA	BBBBB	BBBCCC	AAAAAAAAAAAA
Liquids and Solvents					
Acetone	AAAA	AAAA	AAAA	AAAA	AAAA
Alcohols	BBBB	AAAA	AAAA	AAAA	AAAA
Dowtherm	AAAABBBBBB	AAAAAAAAAA	AAABX	AAABBXX	AAAAAAAAAA
Ethers	BBDD	AAAA	AAAA	AAAA	AAAA
Freon	XXXX	AAAAA	AAAAA	AAAAA	AAAAA
Glycols	AAAA	AAAA	AAAA	AAAA	AAAA
Halogenated hydrocarbons	BBBD	BBBB	XXXX	CCDE	AAAA
Hydrocarbons	BBBB	BBBB	BBBB	CCCC	AAAA
Mercury	AAAAAAAAAAAA	CCCCCCCCCCCCCCCC	XXXXXXX	XXXXXXX	CCCDDDDDDDDEEEX
Molten alkali metals	AAAAAAAAAA	AAAAAAAAAAAAAAA	XXXXX	AAAAA	AAAAAAAAAAAA
Molten salts					
Halides	AAA	AAAAAAA	X	X	AAAAA
Nitrates	AAAAAAAA	AAAAAAAA	XX	XX	AAAAAAAA
Sulfates	AAAAAAAAAA	AAAAAAAAAAAAAA	XXXX	XXXX	AAAAAAAAAA
Water					
Boiler feed	AAA	AAA	CCC	AAA	AAA
Brackish	XXX	BBC	XXX	ABD	AAA
Cooling tower	AAA	AAA	CCC	AAA	AAA
Fresh	XXX	AAA	XXX	ABD	AAA
Sea	XXX	CCD	XXX	CCC	AAA

KEY

A excellent or no limitations; B modest limitations; C special materials available at higher cost to minimize problems; D limited in this regard; E severely limited in this regard; X unacceptable

TABLE 4-28 (Continued)

	Construction Material					
	Plastics		Linings		Refractories	
Titanium-Based (Ti)	Conventional (cp) and Fiberglass-Reinforced (frp)	Fluorocarbon Plastics (fp)	Rubber-Lined (rl)	Glass-Lined (gl)	Ceramics (c)	Graphite (g)
−200 0 200 400 600	0 200	−200 0 200	−200 0 200	0 300	0 1000 2000	0 1000 2000
AAAA	AABCC	AAAAAAD	AABCCCD	AAA	A	A
AAAA	AAACC	AAAAAAD	AAACCCD	AAA	A	A
AAAA	AAACC	AAAAAAD	AAACCCD	AAA	A	A
AAAA	AAACC	AAAAAAD	AAACCCD	AAA	A	A
AAAA	AAACC	AAAAAAD	AAACCCD	AAA	A	A
AAAA	AAACC	AAAAAAD	AAACCCD	AAA	A	A
BBBB	AABCC	AAAAAAD	AABCCCD	AAA	A	A
CCCC	AABCC	AAAAAAD	AABCCCD	AAA	A	A
XXXX	ABCCC	AAAAAAD	ABCCCCD	XXX	C	X
BBBB	CCCEE	AAAAAAD	CCCCCCD	AAA	A	X
CCCX	CCCEE	AAAAAAD	CCCCCCD	AAA	A	X
AAAB	AABCC	AAAAAAD	AABCCCD	AAA	A	E
AAAA	AABCC	AAAAAAD	AABCCCD	AAA	A	B
AAAA	AAACC	AAAAAAD	AAACCCD	AAD	A	A
DDDD	AAACC	AAAAAAD	AAACCCD	AAD	C	A
AAAA	AAACC	AAAAAAD	AAACCCD	AAD	A	A
AAAA	AAACC	AAAAAAD	AAACCCD	AAA	A	A
AAAA	AAACC	AAAAAAD	AAACCCD	AAA	A	A
AAAA	AAACC	AAAAAAD	AAACCCD	AAA	A	A
AAAAAAAA	ABCCC	AAAAAAD	ABCCCCD	AAAA	A	A
AAAA	AABCC	AAAAAAD	AABCCCD	AAA	A	A
AAAAAAAAAAAABBBBD	AAACC	AAAAAAAAAD	CCCCAAACCCD	AAAA	AAAAAAAAAA	AADXXXXXXX
XXXXXXXXXXXXX	CCCCC	AAAAAD	CCCCCD	AAAA	AAAAAAAAAA	BDXXXXXXXX
AAAAAAAAAAAAAAAA	AAACC	AAAAAAD	CCAAACCD	AAAA	AAAAAAAAAA	AADXXXXXXX
XXXXXXXXXXXXXXXX	CCCCC	AAAAAAD	CCCCCCD	AAAA	AAAAAAAAAA	AAAAAAAAAA
XXXXXXXXXXXXXXXX	DDDEE	DDDDDDDEE	DDDDDDDDEEE	DDDD	DDDDDCCCCC	DXXXXXXXXX
BXXBBBBBBBBBBB	AAACC	AAAAAD	AAACCD	AAA	AAAAAAAAAA	AADXXXXXXX
BBBBBBBBBBBBBBB	AABCC	AAAAAAD	CCAABCCD	AAAA	AAAAAAAAAA	AAAAAAAAAA
XXXXXXXXXXXXXXXX	AABCC	AAAAAAD	CCAABCCD	AAAA	AAAAAAAAAA	AAAAAAAAAA
XXXXXXXXXXXXXXXX	DDDEE	DDDDDDDD	DDDDDDDD	XXXX	CCCCCCCCCC	DXXXXXXXXX
AAAAAAAAAAAAAAAA	AABCC	AAAAAAD	CCAABCCD	AAAA	AAAAAAAAAA	AAAAAAAAAA
AAAAAAAAAAAAAA	AABCC	AAAAAAAAD	CCCCCAABCCD	AAAA	AAAAAAAAAA	AAAAAAAAAA
AAAAAAAAAAAAAA	DDDDD	DDDDDDD	DDDDDDD	AAAA	AAAAAAAAAA	AAABBDEXXX
AAAAAAAAAAAAAAAA	CCCCC	AAAAAAAAAD	CCCCCCCCCD	AAAA	AAAAAAAAAA	AAABBDEAAA
BBBBXXXXXXXXXXXXXX	AACCC	AAAAAAD	CCAACCCD	AAAA	AAAAAAAAAA	AAAAAAAAAA
AAAAAAAAAAAAAAAA	AAACC	AAAAAAAAAD	CCCCCAAACCD	AAAA	AAAAAAAAAA	AAAAAAAAAA
AAAAAAAAAAABBBDD	AAACC	AAAAAAAAAD	CCCCCAAACCD	AAAA	AAAAAAAAAA	AADXXXXXXX
BBBXXBBBBBBBBBBB	ABCCC	AAAAAAD	CCABCCCD	AAAA	AAAAAAAAAA	AADXXXXXXX
BXXBBBBBBBBBB	CCCEE	AAAAAD	CCCCCD	AAA	AAAAAAAAAA	AADXXXXXXX
AAAAAAAAAAAAAA	AAACC	AAAAAD	AAACCD	AAA	AAAAAAAAAA	AADXXXXXXX
AAAA	AACCC	AAAAAAD	AACCCCD	AAA	A	A
AAAAAAAA	AABCC	AAAAAAD	AABCCCD	AAA	A	A
AAAAAAAA	AAACC	AAAAAAD	AAACCCD	AAAA	AAA	AAA
AAAA	CCCEE	AAAAAAD	CCCCCCD	AAA	A	A
AAAAA	AAAACC	AAAAAAAD	AAAACCCD	AAA	A	A
AAAA	AAACC	AAAAAAD	AAACCCD	AAA	A	A
AAAA	DDDDD	DDDDDDD	DDDDDDD	AAA	A	A
AAAA	CCCCC	AAAAAAD	CCCCCCD	AAA	A	A
EEXXXXXXXXXXXXX	AAACC	AAAAAAD	AAACCCD	AAAA	AACCCCCCCC	AAAAAAAAAA
AAAAAAABDEXX	DDD	DDDEE	DDDEE	AA	AACCCCCCCC	AAAAADDDDD
XXXX					AAAAAAAA	AAAAAAAAAA
BBBBBBBB		AD	CD	A	AAAAAAAAAA	AAAAAAAAAA
BBBBBBBBBB	XX	AAAD	CCCD	AA	AAAAAAAAAA	ABDXXXXXXX
AAA	AACC	AAAAAD	AACCCD	AAA	A	A
AAA	AACC	AAAAAD	AACCCD	AAA	A	A
AAA	AACC	AAAAAD	AACCCD	AAA	A	A
AAA	AACC	AAAAAD	AACCCD	AAA	A	A
AAA	AACC	AAAAAD	AACCCD	AAA	A	A
−200 0 200 400 600	0 200	−200 0 200	−200 0 200	0 300	0 1000 2000	0 1000 2000

defining a likely construction material and one on which an adequate predesign cost estimate can be based.

Table 4-28 includes materials discussed above: carbon steel (cs), alloy steel (as), conventional plastics (cp), fiberglass-reinforced plastic (frp), fluorocarbon plastics (fp), aluminum and its alloys (Al), copper and its alloys (Cu), stainless steel (ss), nickel-based alloys (Ni), and titanium (Ti) plus ceramics, graphite, and some common linings. The abbreviations in parentheses are used to designate these generic materials generally and in the cost charts of Chapter Five. Detailed physical information and thermal data on individual metals, alloys, plastics and ceramics are found in Perry Tables 23-5, 23-10, and 23-21.

REFERENCES

1   Babcock and Wilcox, *Steam*, Babcock and Wilcox, New York (1978).

2   Beddow, J.K., "Dry Separation Techniques," *Chem. Eng.,* **6,** pp. 70–84, (Aug. 10, 1981). Also available as *Chem. Eng.* Reprint No. 059.

3   Bergmann, L., "Baghouse Filter Fabrics," *Chem. Eng.,* pp. 177–178 (Oct. 19, 1981).

4   Brown, R., "Design of Air-Cooled Exchangers," *Chem. Eng.,* pp. 108–111 (March 27, 1978).

5   Brown, T.R., "Designing Batch Pressure Filters," *Chem. Eng.,* pp. 58–63 (July 26, 1982).

6   Buonicore, A.J., "Air Pollution Control," *Chem. Eng.,* pp. 81–101 (June 30, 1980).

7   Buse, F., "Using Centrifugal Pumps as Hydraulic Turbines," *Chem. Eng.,* pp. 113–117 (Jan. 26, 1981).

8   DeVore, A., G. Vago, and G. Picozzi, "Specifying and Selecting Heat Exchangers," *Chem. Eng.,* pp. 113–148 (Oct. 6, 1980).

9   Doll, T.R., "Making the Proper Choice of Adjustable-Speed Drives," *Chem. Eng.,* pp. 46–60 (Aug. 9, 1982).

10   Eckert, J.S., *Chem. Eng.,* p. 70 (Apr. 14, 1975).

11   Emmett, R.C., and C.E. Silverblatt, "When to Use Continuous Filtration," *Chem. Eng. Prog.,* **70,** pp. 38–42 (December 1974).

12   Finn, D.P., "Select Equipment Drives to Cut Operating Energy Costs," *Chem. Eng.,* pp. 121–124 (March 24, 1980).

13   Fitch, B., "Choosing a Separation Technique," *Chem. Eng. Prog.,* **70,** pp. 33–38 (December 1974).

14   Flood, J.E., H.F. Porter, and F.W., Rennie, "Filtration Practice Today," *Chem. Eng.,* pp. 163–181 (June 20, 1966).

15   Foust, A.S., L.A. Wenzel, C.W. Clump, L. Mais, and L.B. Anderson, *Principles of Unit Operations*, 2nd edition, Wiley, New York (1980).

16   Frank, O., "Shortcuts for Distillation Design," *Chem. Eng.,* pp. 111–128 (March 14, 1977). Also available as *Chem. Eng.* Reprint, No. 276.

17   Ganapathy, V., *Chem. Eng.,* pp. 112–119 (March 27, 1978).

18   Gates, L.E., T.L. Henley, J.G. Fenic, D.S. Dickey, R.W. Hicks, J.R. Morton,

P.L. Fondy, R.S. Hill, D.L. Kime, W.D. Ramsey, G.C. Zoller, W.S. Mayer, R.R. Rautzen, and R.R. Corpstein, "Liquid Agitation," *Chem. Eng.* Reprint, No. 261 (1976). A list of the individual articles comprising this collection is published in a summary article, *Chem. Eng.,* p. 84 (Dec. 6, 1976).

19 Gerunda, A., "How to Size Liquid–Vapor Separators," *Chem. Eng.,* pp. 81–84 (May 4, 1981).

20 Happel, J., and D.G. Jordan, *Chemical Process Economics*, 2nd edition, pp. 482–485, Dekker, New York (1975).

21 Henshaw, T.L., "Reciprocating Pumps," *Chem. Eng.,* pp. 105–123 (Sept. 21, 1981).

22 Hickok, H.N., "Save Electrical Energy. Part I," *Hydrocarbon Process*, pp. 131–141 (July 1978).

23 Hill, C.G., *Chemical Engineering Kinetics and Reactor Design*, Wiley, New York (1977).

24 Kenney, W.F., *Chem. Eng. Prog., 75,* pp. 68–71 (March 1979).

25 King, C.J., *Separation Processes*, 2nd edition, McGraw-Hill, New York (1980).

26 Kirby, G.N., "How to Select Materials," *Chem. Eng.*, pp. 86–131 (Nov. 3, 1980); also available as *Chem. Eng.* Reprint No. 046.

27 Kohl, A.L., and F.C. Riesenfeld, *Gas Purification,* 2nd edition, Gulf, Houston (1974).

28 Kraus, N.J., "FRP Underground Horizontal Tanks for Corrosive Chemicals," *Chem. Eng.,* pp. 125–128 (Feb. 11, 1980).

29 Marinas-Kouris, D.S., "A Shortcut Method for Multicomponent Distillation," *Chem. Eng.,* pp. 83–86 (March 9, 1981).

30 McCabe, W.L., and J.C. Smith, *Unit Operations of Chemical Engineering*, 3rd edition, McGraw-Hill, New York (1976).

31 McCarthy, J.E., "Scrubber Types and Selection Criteria," *Chem. Eng. Prog., 76,* p. 58 (May 1980).

32 Merrill, F.H., "Program Calculates Hydrocyclone Efficiency," *Chem. Eng.,* pp. 71–78 (Nov. 2, 1981).

33 Moir, D.N., "Selecting Batch Pressure Filters," *Chem. Eng.* pp. 47–57 (July 26, 1982).

34 Neerken, R.F., "Use Steam Turbines as Process Drivers," *Chem. Eng.,* p. 63 (Aug. 25, 1980).

35 Neerken, R.F., "How to Select and Apply Positive-Displacement Rotary Pumps," *Chem. Eng.*, pp. 76–87 (Apr. 7, 1980); also available as *Chem. Eng.* Reprint No. 035.

36 Oldshue, J. Y., "Fluid Mixing Technology and Practice," *chem. Eng.*, pp. 82–108 (June 13, 1983).

37 Perry, R.H., and C.H. Chilton, *Chemical Engineers' Handbook*, 5th edition, McGraw-Hill, New York (1973).

38 Peters, M.S., and K.D. Timmerhaus, *Plant Design and Economics for Chemical Engineers*, 3rd edition, McGraw-Hill, New York (1980).

39   Raynor, R.C., and E.F. Porter, "Thickeners and Clarifiers," *Chem. Eng.,* pp. 198–202 (June 20, 1966).

40   Robinson, C.S., and E.R. Gilliland, *Elements of Fractional Distillation*, 4th edition, McGraw-Hill, New York (1950), pp. 348–349.

41   Ryans, J.L., and S. Croll, "Selecting Vacuum Systems," *Chem. Eng.,* pp. 72–90 (Dec. 14, 1981); also available as *Chem. Eng.* Reprint No. 067.

42   Shreve, R.N., and J.A. Brink, *Chemical Process Industries*, 4th edition, McGraw-Hill, New York (1977).

43   Shultz, J.M., "The Polytropic Analysis of Centrifugal Compressors," *J. Eng. Power, Trans. Am. Soc. Mech. Eng.,* pp. 69–82 (January 1962).

44   Singer, J.G. (ed.), *Combustion*, Combustion Engineering, Windsor, Conn. (1981).

45   Singh, J., "Selecting Heat-Transfer Fluids for High-Temperature Service," *Chem. Eng.,* pp. 53–58 (June 1, 1981).

46   Smith, J.M., *Chemical Engineering Kinetics*, 3rd edition, McGraw-Hill, New York (1981).

47   Smith, J.V., "Improving the Performance of Vertical Thermosyphon Reboilers," *Chem. Eng. Prog.,* **70,** pp. 68–70 (July 1974).

48   Stringle, R.F., and D.A. Perry, *Hydrocarbon Process.,* pp. 103–107 (February 1981).

49   Summerell, H.M., "Consider Axial-Flow Fans When Choosing a Gas Mover," *Chem. Eng.,* pp. 59–62 (June 1, 1981).

50   Tan, S.H., "Chart Gives Centrifugal-Pump Power Needs," *Oil Gas J.,* pp. 140–142 (Nov. 2, 1981).

51   Thibodeaux, L.J., and Murrill, P.W., *Chem. Eng.,* p. 155 (July 18, 1966).

52   Treybal, R.E., *Mass-Transfer Operations*, 3rd ed., McGraw-Hill, New York (1980).

53   Wimpress, R.N., "Handy Rating Method Predicts Fired Heater Operation," *Oil Gas J.,* pp. 144–154 (Nov. 21, 1977).

54   Yaws, C.L., D.M. Patel, F.H. Pitts, and C.S. Fung, "Estimate Multicomponent Recovery," *Hydrocarbon Process.,* pp. 99–103 (February 1979).

55   Zanker, A., *Chem. Eng.,* pp. 147–150 (May 19, 1980).

PROBLEMS

4-1   Rules of Thumb

Read the beginning pages of Chapter Four plus sections on auxiliary facilities, heat exchangers, process vessels, materials of construction, and two other sections as assigned. Formulate a set of rules of thumb for flow sheet preparation and equipment design for each section. For guidance, read rules of thumb in Appendix B on sections that you have not been assigned.

(Additional exercise in the use of this chapter will occur with the problems at the ends of Chapters Five and Six plus the case studies of Chapters Seven and Eight.)

Section 2
ECONOMIC ANALYSIS

ECONOMIC ANALYSIS

Some pure scientists, operating near the limits of knowledge, find little immediate economic impetus for their work, but it is rare for chemical engineers, even those in fundamental research, to work on projects that have no underlying economic justification. Development of the atomic bomb, synthetic rubber, and penicillin during World War II were, perhaps, among the exceptions.

Considering the pervasiveness of economics in our discipline, it is vital that we understand fundamental cost accounting practices and techniques. The broadly educated chemical engineer should be able to execute an economic evaluation of any project, existing or proposed. This is as true for researchers who must evaluate the relevance of their work as for plant managers who compete in the marketplace. A design engineer must have not only the skills to analyze a project economically, but the ability to make wise decisions during its conception phase.

In modern industrial society, so strongly influenced by national and multi-national corporations, a minuscule improvement in operating cost or product quality can mean millions of dollars in profit. For this reason, technology-based organizations can gamble in supporting scientists and engineers to perform work that in a large fraction of cases, will never be commercialized. However, as described in Chapter One and illustrated in Figure 1-1, the longer a project remains active, the more costly it becomes because of the human labor exerted on it. Thus, nonviable projects must be identified early before unnecessary resources are dissipated in their pursuit.

The four chapters that follow emphasize fundamental chemical engineering economics. Chapter Five presents techniques and tools for determining capital costs of process plants. Chapter Six discusses procedures for evaluating plant operating costs, including means for translating capital and operating costs into an annual expense on the balance sheet. Chapter Seven gives techniques for determining the optimum design from a group of alternatives. Chapter Eight contains a discussion of how investors judge the viability of a project, the economic effects of the passage of time, and cash flow techniques commonly employed for profitability analysis.

As an introduction to economic evaluation, cash flow is a useful concept. Thinking of money as a fluid, one can visualize flow into or out of a project. The system can be viewed as a pipeline network with streams diverging through multiple tubes or channels to and from sources and sinks. Think of a plant that manufactures a commodity for sale (polyethylene, gasoline, electricity, shoes, automobiles, etc.). The annual sales income A_S is expressed[1] by

$$A_S = RC_S$$

where R = annual rate of production
C_S = selling price

For example, if an electric power plant produces 2 billion kWh annually at 10 cents/kWh, $R = 2 \times 10^9$, $C_S = \$0.10$, and the annual sales income A_S is $200 million.

Obviously, not all sales income is available to compensate owners. The annual cash income A_{CI} is that available from sales minus the annual expenses required to produce and market the product, A_{TE}.

$$A_{CI} = A_S - A_{TE}$$

As entrepreneurs discover all too soon, this cash income does not find its way unchallenged into their coffers. Instead, approximately half is paid to the federal government as a corporate income tax. The balance, A_{NNP}, known inappropriately at times as profit, is part of the cash flow to (or from) the investor. Cash flow schemes for two ventures, one making a profit and the other operating at a loss, are illustrated in Figure S2-1. (These schemes are based on a simple balance sheet approach where depreciation is included as part of the annual total expense. For investment analysis, depreciation is more meaningfully treated as a separate cash flow. This is discussed in Chapter Eight.)

Numerous cost elements must be identified and calculated to assess the fiscal health of a proposed or existing manufacturing process. Many of these expenses are associated with the capital cost or "purchase price" of the plant. This is the element most influenced by the chemical engineer, since the plant must be engineered to some extent before its cost can be assessed. A second basic expense is that associated with operating the plant. This can be estimated by applying judgment and rules of thumb—using raw materials, utilities, and labor requirements based directly on the process flow sheet. As illustrated in Chapter Six, other manufacturing and general expenses can be derived from these data using appropriate scaling factors.

The fundamental concepts and mathematical manipulations required for basic economic analysis are elementary to one trained in engineering sciences. The jargon and technical–statistical manipulations of higher order accounting, however, are not. For example, perishability or time-enhanced value of money (due to inflation

---

[1]A table of economic nomenclature begins on page 262. It corresponds, by and large, with that of Holland, Watson, and Wilkinson [1].

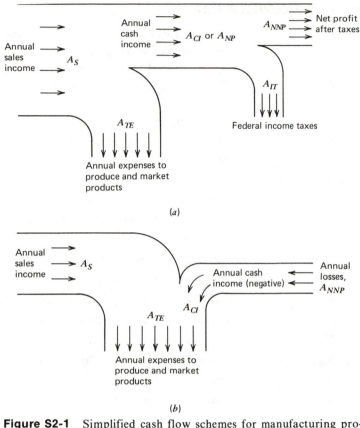

Figure S2-1 Simplified cash flow schemes for manufacturing processes: (1) a profitable venture and (*b*) a venture operating at a loss.

and interest) is a concept foreign to one trained in the inviolate laws of mass and energy conservation. Engineers who are interested in such concepts can, with modest additional training, become versatile in their application. The accountant. on the other hand, cannot easily learn process design. Thus, the engineer must provide basic design and cost data. It is also essential that he or she translate these into a format that can be interpreted by and discussed intelligently with a financial specialist. The material contained in this section is adequate for these purposes. Beyond that, the reader is referred to the more detailed treatment in reference 1.

REFERENCE

1 Holland, F.A., Watson, F.A. and J.K. Wilkinson, "Engineering Economics for Chemical Engineers," a 20-part series in *Chemical Engineering*, beginning June 25, 1973, p. 103, and ending Oct. 28, 1974. The series is available in total as *Chem. Eng.* Reprint No. 215 (1975).

ECONOMIC NOMENCLATURE

a	size or capacity exponent (see Equation 5-1; dimensionless); also, as a superscript on price data and factors, denotes alloy or special material of construction
A_A	allowances credited to an operation through special tax laws ($/yr)
A_{BD}	depreciation ($/yr)
A_C	constant annual expenses; those that are independent of the optimization variable ($/yr)
A_{CF}	net annual cash flow ($/yr)
A_{CI}	annual cash income ($/yr)
A_{cw}	annual costs for cooling water ($/yr)
A_{DCF}	net annual discounted cash flow ($/yr)
A_{DME}	annual direct manufacturing expense ($/yr)
A_e	annual costs for electricity ($/yr)
A_{FC}	annual expenses due to fixed capital ($/yr)
A_{GE}	annual total general expense ($/yr)
A_I	annual cash investment ($/yr)
A_{IME}	annual indirect manufacturing expense ($/yr)
A_{IT}	annual amount of income tax ($/yr)
A_L	annual costs for labor ($/yr)
A_{ME}	total annual manufacturing expense ($/yr)
A_{NCI}	annual net (aftertax) cash income ($/yr)
A_{NNP}	net annual profit after tax ($/yr)
A_{NP}	net annual profit ($/yr)
A_{OE}	annual operating expense ($/yr)
A_R	annual income recovered through investment of additional capital ($/yr)
A_{RM}	annual raw materials costs ($/yr)
A_s	annual costs for steam ($/yr)
A_S	annual revenue from sales ($/yr)
A_{Su}	annual costs for plant supervision ($/yr)
A_{TE}	total annual expense ($/yr)
A_u	annual cost of a utility ($/yr)
A_V	annual variable expense ($/yr)
(BEP)	break-even period (yr)
C	capital ($)
C_{AB}	capital cost of auxiliary buildings ($)
C_{BM}	bare module capital cost ($; see Table 5-2)
C_C	contingency expense ($)
C_{CFC}	constant fixed capital in Equation 7-2 ($)

C_E	contractor engineering expenses ($)
C_F	contractor's fee ($)
C_{FC}	fixed capital cost ($)
C_{FIT}	freight, insurance, and taxes to procure and install equipment ($)
C_{GR}	total grass-roots capital ($)
C_L	labor cost associated with installation of process equipment ($)
C_I	cost of land and other nondepreciable items ($)
C_M	cost of materials required to install an item of process equipment ($)
C_O	construction overhead expense ($)
C_{OS}	capital cost of offsite facilities ($)
$C_{P,v,r}$	purchase cost of process equipment having size or capacity v and in the year r ($) (A superscript, when employed, indicates material of construction. Abbreviations are included in the headings of Table 4-28; cs = carbon steel, as = high alloy steel, ss = stainless steel, etc.)
C_S	selling price per unit of product ($/unit)
C_{SD}	site development expense ($)
C_t^a	capital cost of piping materials ($)
C_{TBM}^a	total bare module cost $= \sum_i C_{BM,i}^a$ ($) (The superscript refers to construction material. When it is absent, costs are for base material, usually carbon steel, construction; see Table 5-5.)
C_{TC}	total capital ($)
C_{TM}	total capital of a process module ($)
C_{VFC}	variable fixed capital in Equation 7-2 ($)
C_{WC}	working capital ($)
ΔC_{BM}^a	increase in bare module capital above that of carbon steel due to nonbase materials of construction ($)
(DBEP)	discounted breakeven period (yr)
(DCFRR)	discounted cash flow rate of return (percent)
f_a	annuity factor employed to convert capital cost to annual expense (yr^{-1}; see equation 7-3)
f_d	discount factor based on annual compound interest (dimensionless; see Equation 8-7)
f_d'	discount factor based on continuously compounded interest (dimensionless; see Equation 8-8)
f_o	operating factor (dimensionless); the fraction of elapsed time that a plant is in equivalent full-scale production
f_t	ratio of costs for piping materials relative to those of purchased equipment (dimensionless)
F_{BM}	installation factor; the purchase cost of equipment multiplied by this factor yields bare module cost

F_C — contingency factor for unproven process steps (dimensionless; see Figure 5-2)

F_M^a — material factor; the ratio of purchased equipment cost for an item constructed of a special alloy or material relative to the price of the same item constructed of carbon steel

F_p — pressure factor; the cost of equipment designed for high pressure service relative to the cost of conventional equipment

i — annual interest rate, expressed either as a percentage or a fraction (yr^{-1}); also denotes rate of return based on annual cash income after tax: $i = (A_{NCI}/C_{TC}) \times 100 = (A_{NNP} + A_{BD})/C_{TC} \times 100$

i' — rate of return based on net profit after taxes (yr^{-1}; see Equation 6-9)

i'' — rate of return on average investment (yr^{-1}); $i'' = 2A_{NNP}/C_{TC} \times 100$

i_i — incremental return on incremental investment (yr^{-1}); $i_i = \Delta(A_{NNP} + A_{BD})/\Delta C_{TC} \times 100$

i_i' — incremental net profit after taxes divided by incremental investment (yr^{-1}); $i_i' = \Delta A_{NNP}/\Delta C_{TC} \times 100$

I_r, I_s — cost indices to correct equipment prices for inflation (dimensionless; see Equation 5-11)

(NPV) — net present value ($)

n — investment period (yr)

n_j — age of a process unit (yr)

(PBP) — payback period (yr)

r, s — year designated for escalation of cost data

S — scrap or salvage value ($)

s — lifetime remaining in a process at year j (yr)

t — fractional tax rate (dimensionless)

u, v — capacity designations pertaining to purchased equipment prices (dimensionless)

Chapter Five
CAPITAL COST ESTIMATION

After a flow sheet and a preliminary technical package have been prepared, the next logical and chronological step is to determine the price of a chemical plant. The most fundamental definition of price is what is known as fixed capital ("fixed" because it is invested in real equipment, which cannot be converted easily to any other form of capital). It is the price that would be paid if a processing plant could be bought, in the same way that a house, automobile, or washing machine is purchased.

Unlike consumer goods and appliances, which are bought directly and employed immediately, process equipment must be custom designed or at least identified to some extent by a professional and installed by specialists. This means, in most applications, that final plant cost is several times greater than the sum of bare equipment prices. When one considers further that a plant is contemplated for the future and an estimate must be based on the past, the need for inflation indices or escalation factors is evident. It is because of these uncertainties and complexities that a construction estimate approaching plus or minus 5 percent accuracy may cost as much as 10 percent of the plant itself. Fortunately, through the efforts of generations of cost engineers, an approximate translation of technical specifications into dollars is relatively quick and easy. Thus in predesign evaluation, where errors of 20 to 30 percent can be tolerated, the amounts of time and money involved in preparing an estimate are relatively small.

Unfortunately, in contrast to personal or domestic purchases, most projects that engineers explore do not already exist in the same form, having the same capacity, or using the same technology as that contemplated. Often, even if they do exist, the pertinent information is in the hands of competitors who are less than willing to share it. This situation usually calls for an original cost estimate.

As mentioned in Chapter Three, the process flow diagram is basic to an original capital cost estimate. One essentially defines the type of equipment, its size, and its construction material from information provided on the flow sheet. Then, the approximate purchase price of the particular equipment item can be obtained from a supplier, from a reference, or from past experience. Purchase prices of various types of process equipment are tabulated in the chemical engineering literature [1, 2, 8, 9]. A comprehensive collection of cost data also forms the bulk of this chapter.

Use of available data frequently requires scaling from one size or capacity to another and from the past to the present or future. Then, installation costs must be assessed and composite total capital determined. The purposes and uses of various factors and indices to accomplish this are described next.

VARIATION OF EQUIPMENT COST WITH SIZE

Escalation of cost with increasing capacity is obvious in everyday experience. If, for example, one were transporting people in taxicabs and one cab held four people (excluding driver), the capital cost would double for conveying eight people. For 20 to 30 people, a bus or van might be employed where the equipment cost is not necessarily directly proportional to the number of people. This can be expressed mathematically as follows:

$$C_{P,v,r} = C_{P,u,r} \left(\frac{v}{u} \right)^a \tag{5-1}$$

where $C_{P,v,r}$ is the purchase price of the equipment in question, which has a size or capacity of v in the year r, and $C_{P,u,r}$ is the purchase price of the same type of equipment in the same year but of capacity or size u. A size exponent a is applied to the capacity ratio to relate the cost of one size to another. In the twin taxicab example, u is 4, v is 8, and the size exponent is unity. Although a is 1 for multiple taxicabs, in comparing the cost per person on a bus to that in a taxicab, the exponent may be less than or greater than 1. For chemical process equipment, a is usually less than unity. The reason for this can be easily demonstrated by considering costs of storage tanks of different sizes. Assume two spherical tanks, made of identical materials, but of different capacity. The volume of tank u is

$$V_u = \frac{4}{3} \pi R_u^3 \tag{5-2}$$

The volume of tank v can be similarly expressed, giving for the volume or capacity ratio:

$$\frac{V_v}{V_u} = \left(\frac{R_v}{R_u} \right)^3 \tag{5-3}$$

However, costs of tanks are proportional not to the volume, but to surface areas or the quantities of metal plate used in their fabrication. The area of tank u is

$$A_u = 4\pi R_u^2 \tag{5-4}$$

and the area ratio is

$$\frac{A_v}{A_u} = \left(\frac{R_v}{R_u} \right)^2 \tag{5-5}$$

At a 1982 selling price of C_S dollars per cubic meter of tank surface, the purchase costs of the tanks would be

$$C_{P,v} = 4\pi R_v^2 C_S \tag{5-6}$$

$$C_{P,u} = 4\pi R_u^2 C_S \tag{5-7}$$

or

$$C_{P,v} = C_{P,u} \left(\frac{R_v}{R_u} \right)^2 \qquad (5\text{-}8)$$

Equation 5-3 can be employed to replace the radius ratio with a capacity ratio.

$$\frac{V_v}{V_u} \equiv \frac{v}{u} = \left(\frac{R_v}{R_u} \right)^3 \qquad (5\text{-}9)$$

Substituting $R_v/R_u = (v/u)^{1/3}$ in Equation 5-8 yields:

$$C_{P,v} = C_{P,u} \left(\frac{v}{u} \right)^{2/3} \qquad (5\text{-}10)$$

Thus, comparing Equations 5-1 and 5-10, the size exponent a, in this case, is $2/3$. Table 5-1 (taken primarily from Perry, p. 25-18), contains representative size exponents for numerous types of equipment. Note that the exponents for tanks are grouped near the value $2/3$ as anticipated.

The exponential size–capacity relationship is useful not only for individual

TABLE 5-1
TYPICAL EXPONENTS FOR EQUIPMENT COST AS A FUNCTION OF CAPACITY
[1, 2, 8, 9]

	Size Range	Capacity Unit	Exponent a
Agitator, turbine	4–40	kW	0.50
Blower, single-stage 14 kPa	0.05–0.4	m^3/s	0.64
Centrifugal pump	10–20	kW	0.50
Compressor, reciprocating	200–3000	kW	0.70
Conveyor, belt	5–20	m^2	0.50
Crusher			
Gyratory	12–50	kg/s	0.83
Jaw	7.5–25	kg/s	1.15
Dryer			
Drum	5–40	m^2	0.63
Vacuum shelf	10–100	m^2	0.53
Dust collector,			
Cloth	0.0001–0.5	m^3/s	0.70
Cyclone	0.0001–0.33	m^3/s	0.61
Electrostatic precipitator	0.5–2.0	m^3/s	0.68
Evaporator, agitated falling-film	3–6	m^2	0.55
Filter, plate and frame	1–60	m^2	0.58
Heat exchanger, shell and tube	5–50	m^2	0.41
Kettle, glass-lined, jacketed	3–10	m^3	0.65
Motor, 440 v, totally enclosed	0.75–15	kW	0.59
Refrigeration unit	25–14,000	kW	0.72
Screen, vibrating, single-deck	3–5	m^2	0.65
Stack	6–50	m	1.00
Tank			
API, storage	1000–40,000	m^3	0.80
Vertical	0.75–40	m^3	0.52
Horizontal	5–20	m^3	0.60
Tower, process	10–60	m^3	0.60

equipment items but for entire processing plants where the fixed capital cost of a plant having one capacity is scaled to that of another by use of Equation 5-1. This scaling practice is known prominently in the literature as the "sixth-tenths rule" because of the common usage of 0.6 as the exponent. Guthrie [2] correlated total capital investments for 59 process units, ranging from an alkylation module to a vinyl chloride plant, using this relationship. Exponents varied from 0.38 to 0.90 with a mean value of 0.64.

Although quick and easy to use, the practice of scaling entire plants using the six-tenths rule has limitations. It is necessary, for example, to know at least the capital cost of one plant of a given capacity. Also any changes in technology that affect the flow sheet will distort the accuracy of this interpolation. The possible range in exponents is quite large, requiring either prior experience or a few detailed cost estimates to define a correct size exponent for the process in question. Finally, maximum capacity of a plant is limited by sizes of equipment that can be fabricated. Beyond those extremes, capacity is increased merely by building multiple units. In this case, the exponent becomes unity.

The six-tenths rule is valuable for rapid order-of-magnitude feasibility estimates where substantial errors can be tolerated. It is also useful in extrapolating results from a detailed estimate at one capacity to another when one is testing the sensitivity of the product cost to capacity. It serves another function in correlating cost data. Guthrie, for example [1, 2], reports a number of equipment costs in terms of Equation 5-1, giving $C_{P,1,r}$, the capital cost equivalent to unit capacity. Included with this is the value of a, which allows one to determine the cost from Equation 5-1 for capacity equal to v. *There are pitfalls, however, if one extrapolates beyond the limits of this correlation.* For example, according to Equation 5-1, the cost decreases as the cost of materials or size of equipment declines. However, with some precision items, when a certain *minimum size* is reached, costs level off and even begin to rise again as savings in materials are offset by increasing labor required to meet the more stringent tolerances of miniature equipment. In this situation, the exponent can actually change to a negative number. Extrapolation in the opposite direction is limited by the maximum size of equipment that can be fabricated and conveniently shipped. When this maximum is exceeded, multiple units become necessary and the exponent becomes unity. This is illustrated graphically in Figure 5-36 for double-pipe heat exchangers. Unfortunately, some sources do not specify the range of applicability for size exponents reported.

Another advantage of the six-tenths rule is suggested by the form of Equation 5-1. If, for instance, capacity is plotted versus cost on log-log graph paper, the relationship yields a straight line of slope a. This is the format usually employed for graphical cost–capacity correlations found in standard sources [1, 2, 4, 8, 9], and later in this chapter. (A perceptive observer will note that the slope of the curve in Figure 5-61, for spherical storage tanks, is very near $2/3$, as predicted by Equation 5-10. It is slightly larger, however, because the wall thickness of the tank increases with size.) These plots have several advantages over an equation.

1　Limits of applicability can be defined easily by the length of the curve.

2　Changes in slope, which may occur over a wide capacity range, can be shown.

3　Costs can be read directly from the charts without computation.

ESCALATION OF EQUIPMENT COSTS DUE TO INFLATION

Predesign capital cost estimates are normally assembled from old price data. Thus, because of inflation (or, in rare instances, deflation), corrective indices are needed to adjust older data to current or future status. One hears frequently, through the news media, of changes in the Consumer Price Index, the Wholesale Price Index, and the Salary Survey Index. These are indicators of inflationary trends in the economy. By the same token, similar factors, specific to process plant construction, are available to chemical engineers.

An index is employed as a direct correction to Equation 5-1. For example, if the purchase price of equipment of capacity u in year r is $C_{P,u,r}$, the estimated purchase price of the same equipment of capacity v in year s is given by:

$$C_{P,v,s} = C_{P,v,r} \left(\frac{I_s}{I_r} \right) = C_{P,u,r} \left(\frac{v}{u} \right)^a \left(\frac{I_s}{I_r} \right) \tag{5-11}$$

Thus, one needs merely to find or estimate the cost index for the year s and divide it by the corresponding value for year r to evaluate the escalation ratio.

A number of different indices are used commonly by cost engineers in the chemical industry. The oldest and probably least specific is the *Engineering News Record* (ENR) index. Like all indices, it is based on a value of 100 at a specific year. The original datum year for the ENR index is 1913, but current values have grown so large that other bases (1949 = 100 and 1967 = 100) are also used. Historic trends and values of ENR indices are illustrated in Figure 5-1. The two ENR curves based on 1913 and on 1967, respectively, differ by a constant factor of 10.7.

The ENR index is published frequently in *Engineering News Record* so that current values are readily available. A major disadvantage of the ENR index is its lack of specificity to chemical process plants. Accepted as the standard in civil construction, it is influenced more by materials and techniques used in this industry and is less sensitive to process equipment fabrication and installation costs, which constitute the bulk of chemical plant capital expense.

Three other indices, developed specifically for chemical construction, are also included in Figure 5-1. The Nelson index applies particularly to petroleum refinery construction, whereas the M & S (Marshall & Swift, formerly Marshall & Stevens) and CE (*Chemical Engineering*) indices are overall chemical industry averages. Up-to-date Nelson and M & S values are reported frequently in *The Oil and Gas Journal*. M & S and CE values are published biweekly on the "Economic Indicators" page of *Chemical Engineering*, which also displays trends, historic values, and other economic data (including the most recent ENR index). The page is reproduced in each issue with minor changes as new data accumulate.

In general chemical process estimating, the CE Plant Cost Index has several advantages [4, 10]. First, it is based on equipment and labor typically employed in chemical plant construction. Second, improvements in productivity within the fabrication and construction industry are considered. Third, engineering costs as well as those for materials, manufacturing, and installation are included. In reality, there are two CE indices, one for escalating costs of complete plants and one for the purchased equipment only. In practice, differences between the two are almost indistinguishable. The former is that shown in Figure 5-1 and reported graphically

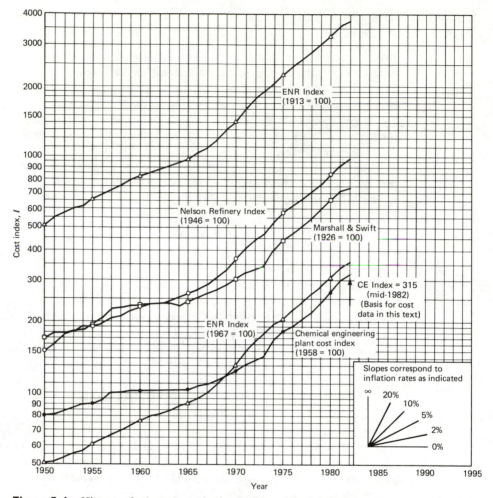

Figure 5-1 History of selected cost indices pertinent to chemical process construction. All cost data in this book are based on the *Chemical Engineering* Plant Cost Index Value of 315 (mid-1982).

in *Chemical Engineering*. M & S values are based on the installed cost of equipment but not a finished plant. Again, for predesign estimates, the distinction is academic. To illustrate, note that trends and slopes of the M & S and CE curves in Figure 5-1 are almost identical. In fact, from 1960 to 1980, their ratio has remained within plus or minus 3 percent of a constant value, namely 2.4.

Because of accessibility and accuracy, the *Chemical Engineering* Plant Cost Index is used in this text. Further information on its composition can be found in the articles by Kohn [4], Matley [5], and references cited therein. There is still a tendency in the profession to use the ENR index. This may be attributable to its more rapid ascent, providing higher predicted costs and a more conservative estimate. It seems wiser, however, to use a more accurate cost index and to allow for contingencies consciously and deliberately by other means. For escalation over short time periods, only slopes of the curves in Figure 5-1 are important. These are nearly the same for all indicators and correspond to inflation rates illustrated by the insert. To maintain currency, the reader is encouraged to add to the curves of Figure 5-1 as data become available. For study estimates, escalation by cost factors such as these is considered reliable for time jumps of 10 years or less.

INSTALLATION COSTS

From a catalog of basic equipment prices such as that presented later in this chapter, one can determine the purchase cost $C_{P,v,s}$ for each major item of equipment that appears on a process flow diagram. Not only must this equipment be transported to the site and placed on a foundation, but, as a visit to an operating plant dramatically reveals, it becomes shrouded with piping, structural steel, insulation, instruments, and other paraphernalia. Thus, the installed cost is usually several times greater than the purchase price.

One approach to assembling a total plant cost is to add purchase prices of all the equipment on the flow sheet and multiply the sum by a factor (usually between 3 and 5) to obtain an overall plant cost. This multiplier, known by the name of its originator as a Lang factor, depends markedly on the nature of the process. For example, in a paper mill, which contains expensive, precise, high speed machinery, a larger fraction of the cost is invested in the original equipment. Installation is relatively less expensive. Thus, the Lang factor is small. In an oil refinery, process vessels and equipment themselves are somewhat more simple, but installation of piping, insulation, and instruments is more expensive, creating a larger Lang factor. To allow for this variability, Peters and Timmerhaus [9] suggest factors of 3.9 for a plant that processes primarily solids (e.g., a cement plant), 4.1 for a process containing both solid and fluid streams (a fertilizer plant, perhaps), and 4.8 for a fluid processing plant such as an oil refinery.

Guthrie [1, 2] proposed the use of "module factors," which vary with equipment type. Since his technique is accurate, direct, and relatively easy to employ, it is advocated, illustrated, and employed in this text. In essence, the Guthrie approach is an efficient method for synthesizing Lang factors specific to processes under consideration.

As an introduction to this technique, steps in constructing a chemical plant should be reviewed. Consider the evolution of a grass-roots (beginning from scratch) project. Construction steps are examined according to category as follows.

Site Development

Land is located and surveyed, a price is negotiated, and the sale consummated. Civil engineering evaluation and design of the site are accomplished, and the land is drained, cleared, graded, and excavated. Sewers, water lines, roads, walkways, and parking lots are constructed. Grounds are landscaped and the site fenced.

Auxiliary Building Construction

Auxiliary buildings for administrative offices, laboratory, maintenance shop, garages, and warehouses are designed and constructed. Service buildings to house a cafeteria, personnel lockers, dressing rooms, and medical facilities are also necessary in a large modern plant.

Battery-Limits Process Equipment Installation

The battery limit is a geographic boundary that defines the manufacturing area of the plant. This includes process equipment and buildings or structures to house it but excludes auxiliary and service buildings and offsite facilities.

Purchased equipment prices are f.o.b. (free on board), that is, they pertain to equipment that is placed aboard the shipping truck, railcar, or barge at the fabrication plant. This equipment must be shipped, at the purchaser's expense, to the site where foundations and, in some cases, buildings, are constructed. It is hoisted or set into place. Piping, auxiliary steelwork such as ladders and walkways, instruments, and electrical equipment are installed. Units are then insulated, painted, and tested.

Offsite Facilities

In addition to battery-limits equipment displayed on a flow sheet, "offsite" process facilities are necessary to supply utilities and for convenience, safety, and pollution control. Examples include steam boilers, electrical power generators, cooling towers, fuel supply systems, stacks, flares, raw material and final product storage tanks and warehouses, pollution control facilities, fire protection, utilities, yard lighting, communications networks, railroad sidings, and unloading docks.

From f.o.b. purchase prices, which form the nucleus of a preliminary chemical process capital estimate, direct installation costs can be calculated, using appropriate factors. The sum of purchase and installation costs is what Guthrie terms a process module and represents the capital cost of an addition to a plant where site, auxiliary buildings, and offsite facilities already exist and are adequate.

For a module or battery-limits estimate, the capital expenses for offsite utilities such as steam plants, electric generators, and cooling towers are shared by the new unit through the price charged against the process per unit of steam, electricity, or water used. Because of savings in site development, auxiliary buildings, and offsite facilities, the fixed capital for an addition is generally less than it would be for a grass-roots plant. To derive the total fixed capital for an addition, a contingency allowance and contractor's commission are added to yield the total module cost. A contingency allowance provides latitude for unforeseen expenses, unexpected delays, disruptive weather, strikes, and other such natural or social phenomena. The contractor's fee is a commission granted to the construction firm and is based on bare module cost.

In estimating the fixed capital cost of a grass-roots plant, site preparation and auxiliary facilities are assessed as fractions of the module estimate. The sum of these yields the total plant cost.

For estimates of predesign accuracy, one merely needs the appropriate overall factors. However, it is both edifying and educational to review their derivation.

INSTALLATION FACTORS

Installation involves direct materials and labor, indirect expenses, contingency and fee allowances, and capital for auxiliary facilities. Derivation of installation factors is illustrated by the following analysis applied to a heat exchanger.

Direct Materials and Labor

According to Guthrie [1, 2], materials required for installation of a shell and tube heat exchanger represent the following percentages of the f.o.b. purchase price of a carbon steel unit.

Material	*Percent*
Piping	45
Concrete (foundations)	5
Steel (structural support)	3
Instruments	10
Electrical materials	2
Insulation	5
Paint	1
Subtotal	71

These figures, based on cost data from 42 different projects, indicate that materials for installing a carbon steel heat exchanger **amount to approximately 70** percent of the purchase price. Next, labor associated with installation must be considered. Again, experience reveals that direct wages paid to laborers and craftspeople who install heat exchangers are 37 percent of the total materials involved (including the exchanger itself). Thus, if the f.o.b. purchase price of a carbon steel exchanger C_P is $10,000, materials associated with installation, C_M, will cost $7100, and the wages paid to the workers for installation will be:

$$C_L = 0.37(C_P + C_M) = 0.37(C_P + 0.71C_P) = 0.63C_P \qquad (5\text{-}12)$$

Thus, the direct installation cost is:

$$C_P + C_M + C_L = C_P + 0.71C_P + 0.63C_P = 2.34C_P \qquad (5\text{-}13)$$

Indirect Costs

Now one must consider a number of indirect costs. Some of these are associated with equipment. *Freight, insurance*, and sales *taxes* on the purchased materials are examples. Others (*construction overhead*) pertain to labor: fringe benefits for workers (health insurance, vacation and holiday pay, sick leave, retirement benefits), so-called burden (mandatory fees such as social security, unemployment insurance, workmen's compensation), plus salaries, fringe benefits, and burden for supervisory and advisory personnel, including, in some cases, travel and subsistence expenses. There are additional costs for temporary buildings, roads, parking areas, and other facilities required only during the construction period. Special construction equipment such as cranes or other machinery must be purchased or rented and transported to the job site. Miscellaneous small tools and equipment also wear out and must be replaced. Finally, construction overhead also includes miscellaneous items such as jobsite cleanup and security costs, warehousing, and vendor services.

A third and final element of indirect cost is that associated with *engineering*. This includes our salaries (project and process engineering), those of other designers and draftspeople, procurement expenses, home office expenses, and the associated overhead.

These various elements of indirect cost are related to direct labor and materials as follows. *Freight, insurance,* and *taxes* are directly proportional to materials, typically 8 percent in the United States. *Construction overhead* is a function of the direct labor employed during installation, amounting to approximately 70 percent of wages. *Engineering* costs depend on the complexity of design and are thus proportional to materials; typically, they represent 15 percent of equipment and installation material costs. These expenses and their relation to purchased equipment cost are illustrated for a heat exchanger in Table 5-2. As indicated, the bare module cost of a heat exchanger is approximately three times the purchase price.

Contingency and Fee

To obtain the total module cost, that is, the total expense required to procure and install the heat exchanger in the battery limits and to make it ready for operation, *contingency* and *fee* must be added. These, according to Guthrie's data [1, 2], are 15

TABLE 5-2

TYPICAL COSTS ASSOCIATED WITH PURCHASE AND INSTALLATION OF A
HEAT EXCHANGER[a] IN A PROCESS MODULE

	Cost	Fraction of f.o.b. Equipment	
Direct Project Expenses			
Direct materials			
Equipment f.o.b. price, C_P	$10,000	1.0	C_P
Materials used for installation, $C_M = 0.71 C_P$	7,100	0.71	C_P
Direct labor, $C_L = 0.37(C_P + C_M) = 0.63 C_P$	6,300	0.63	C_P
Total direct materials and labor	$23,400	2.34	C_P
Indirect Project Expenses			
Freight, insurance, taxes, $C_{FIT} = 0.08(C_P + C_M)$	1,400	0.14	C_P
Construction overhead, $C_O = 0.70 C_L$	4,400	0.44	C_P
Contractor engineering expenses, $C_E = 0.15(C_P + C_M)$	2,600	0.26	C_P
Total indirect project costs	$ 8,400	0.84	C_P
Bare module capital, C_{BM}	$31,800	3.18	C_P
Contingency and Fee			
Contingency, $C_C = 0.15 C_{BM}$	4,800	0.48	C_P
Fee, $C_F = 0.03 C_{BM}$	1,000	0.10	C_P
Total contingency and fee	$ 5,800	0.58	
Total module capital, $C_{TM} = 1.18 C_{BM}$	$37,600	3.76	C_P
Auxiliary Facilities			
Site development, $C_{SD} = 0.05 C_{TM} = 0.19 C_P$	1,900		
Auxiliary buildings, $C_{AB} = 0.04 C_{TM} = 0.15 C_P$	1,500		
Offsite facilities, $C_{OS} = 0.21 C_{TM} = 0.79 C_P$	7,900		
Total grass-roots capital, $C_{GR} = 1.3 C_{TM}$	$48,900		

[a]Purchase price, $10,000.

and 3 percent of bare module capital, respectively. Thus, as illustrated in Table 5-2, the cost of a $10,000 heat exchanger, after installation, is $37,600, or 3.76 times the purchase price.

Auxiliary Facilities

To derive the contribution of a heat exchanger to a grass-roots plant, its share of site development, auxiliary building, and offsite capital must be assessed. Guthrie reports [2] the following ranges for use in "order to magnitude" estimates.

Item	Percentage of Total Module Cost
Site preparation	4–6
Auxiliary buildings	2–6
Offsite facilities	17–25
Total	23–37

For our purposes, an average value was employed to yield the contribution shown in Table 5-2. The capital cost associated with a $10,000 heat exchanger in a grass-root plant is estimated, accordingly, to be approximately $49,000. The appropriate Lang factor is thus 4.9, near the number recommended by Peters and Timmerhaus [9] for a fluid processing plant.

Using this technique, Guthrie [1, 2] estimated costs associated with battery-limits installations for three markedly different types of process: a typical chemical module with compressors, pumps, columns, and exchangers; a paper mill circuit (conveyors, mills, etc.); and a mechanical circuit (motors, punches, presses, and other mechanical equipment). For each item on the flow sheet, installation costs were tabulated and combined to give an overall Lang factor for the entire module. Guthrie's analysis yielded grass-roots Lang factors of 4.5 for the typical chemical plant, 2.5 for the paper mill, and 3.2 for the solids processing plant.

UNUSUAL CONSTRUCTION MATERIALS, EXTREME CONDITIONS, AND TECHNICAL UNCERTAINTY

Three extraneous influences—materials of construction, extreme conditions, and technological maturity—can dramatically alter capital costs. Guidance in their appraisal is provided in this section.

Corrosive, High Pressure, and High Temperature Service

With compatible process fluids and solids, carbon steel is normally the most economical material for the construction of chemical equipment. Because of corrosion, erosion, and other extreme conditions, however, more expensive materials of construction are often required. Selection of the appropriate material is often dictated by experience. For the novice, Table 4-28 should be of value. Aside from corrosion, extreme temperatures and pressures require extra-heavy construction or other costly modifications beyond conventional equipment design. What is needed at this point is a technique for adjusting capital cost estimates to compensate for alternate construction materials and extreme conditions.

The method recommended here for assessing unusual equipment costs follows that advocated by Guthrie, but the computational technique is slightly different. At first glance, it might seem appropriate to merely multiply the purchase price of carbon steel equipment by a ratio of purchase prices. However, installation costs are not a simple function of purchase cost. Although process piping and fittings made for the same unusual conditions are proportionally more expensive, other installation materials, labor, and indirect expenses are not. Furthermore, only about 70 percent of the piping is directly exposed to process fluid. The balance is auxiliary or utility piping made of conventional materials.

To account for expensive alloys or extreme conditions, special correction factors have been developed for use with the cost charts in this chapter. To prepare a capital estimate, the price of equipment fabricated from the most common or base material (usually carbon steel) is obtained from the appropriate cost chart. This is multiplied by a unique bare module factor F_{BM}^a, which applies to the material in question. Returning to the heat exchanger of Table 5-2, we note that the bare module factor for a carbon steel shell and tube exchanger F_{BM}^{cs} is 3.18. For a unit constructed entirely of stainless steel, purchased equipment cost is increased by a materials ratio F_M^{ss}. According to Figure 5-36, this is 3.0. Thus, the stainless steel exchanger would cost \$30,000 to purchase. For reasons already given, the bare module capital cannot be calculated simply by multiplying \$30,000 times the bare module factor for carbon steel. Rather, a special factor F_{BM}^{ss}, particular to stainless steel, must be employed. For convenience of use, the increased purchase price is also included in this factor so that bare module capital can be determined directly from the base (carbon steel) price. For example, the bare module factor for $F_M = 3.0$ is found from Figure 5-38 to be 5.8. Thus the bare module capital associated with this exchanger is $5.8 \times \$10,000 = \$58,000$.

Derivation of bare module factors for alloy construction is straightforward. As mentioned above, there is a direct increase in price of the purchased equipment.

$$C_P^a = C_P^{cs} F_M^a \qquad (5\text{-}14)$$

Next there are additional costs due to higher purchase prices for process piping, which would also be constructed of the more expensive alloy. For a heat exchanger, process piping or tubing represents about 70 percent of total piping; therefore, the associated materials cost is

$$C_t^a = 0.70 C_P^{cs} F_M^a f_t \qquad (5\text{-}15)$$

where f_t is the ratio of costs for piping or tubing materials relative to those for purchased equipment. Values of f_t reported by Guthrie range from near zero for conveyors and crushers to more than 0.6 for some process vessels (representative values are listed in Table 5-3).

Assuming that installation labor is not a function of materials cost, the increment of added bare module cost, above that for carbon steel, is:

$$\Delta C_{BM}^a = C_P^{cs}(1 + 0.7f_t)(F_M^a - 1) \qquad (5\text{-}16)$$

Equation 5-16 is simply the sum of Equation 5-14 and 5-15 with F_M^a replaced by $F_M^a - 1$. This yields the cost added to that for carbon steel or the base material whose F_M^a value is unity. The total bare module cost is thus that for the base material

TABLE 5-3

REPRESENTATIVE PIPING OR TUBING FACTORS FOR PROCESS EQUIPMENT

Generic Equipment Type	Piping or Tubing Factor,[a] f_t
Conveyors	0.03
Crushers, mills, grinders	0.03
Evaporators, vaporizers	0.40
Furnaces	0.15
Gas movers, compressors, exhausters	0.15
Gas–solid contacting equipment	0.18
Heat exchangers	
Shell and tube	0.45
Air-cooled	0.14
Mixers	0.20
Process vessels	
Horizontal	0.40
Vertical	0.60
Pumps	0.30
Separators	0.25
Size-enlargement equipment	0.25
Storage vessels	0.15

[a]This is the ratio of purchase price of piping materials to purchase price of equipment.

plus ΔC_{BM}^a:

$$C_{BM}^a = C_{BM}^{cs} + \Delta C_{BM}^a = C_P^{cs} F_{BM}^{cs} + C_P^{cs}(1 + 0.7f_t)(F_M^a - 1) \qquad (5\text{-}17)$$

or

$$C_{BM}^a = C_P^{cs} F_{BM}^a \qquad (5\text{-}18)$$

where F_{BM}^a is given by:

$$F_{BM}^a = F_{BM}^{cs} + (1 + 0.7f_t)(F_M^a - 1) \qquad (5\text{-}19)$$

This is an equation for a straight line with slope steeper than 45 degrees and having a value F_{BM}^a equal to F_{BM}^{cs} when F_M^a is 1. The slope increases, or course, with increasing values of f_t. Equation 5-19 is the basis for Figures 5-38, 5-46, and 5-51, which correlate F_{BM}^a with F_M^a. For the stainless steel exchanger, with $F_{BM}^{cs} = 3.18$, $F_M^{ss} = 3.0$, and $f_t = 0.45$, F_{BM}^{ss} is 5.8 from Equation 5-19 or from Figure 5-38. This yields a bare module cost of $58,000.

If the pressure were greater than 10 barg, another correction factor F_p would be required for the heat exchanger in question. Assume a shell-side pressure of 100 barg. The purchase price ratio of high pressure versus low pressure units, F_p, is found to be 1.22 (Figure 5-37). Assuming that F_p is the same for process piping as well as for the exchanger and for alloys as well as carbon steel, an equation analogous to Equation 5-19 can be derived

$$F_{BM}^a = F_{BM}^{cs} + (1 + 0.7f_t)(F_M^a F_p - 1) \qquad (5\text{-}20)$$

where F_M^a is multiplied by F_p. Equation 5-19 is, in fact, a more limited form of Equation 5-20 with $F_p = 1$. For the stainless steel exchanger, $F_p \times F_M^a = 1.22 \times 3.0 = 3.66$. Accordingly, F_{BM}^{ss}, from either Equation 5-20 or Figure 5-38, is 6.7. Consequently, the bare module capital associated with a stainless steel exchanger rated for 100 barg pressure is $6.7 \times \$10,000 = \$67,000$. If the exchanger were of carbon steel designed for 100 barg pressure, $F_M^{cs} = 1$, $F_p = 1.22$, and from Figure 5-38, F_{BM}^{cs} is approximately 3.5, increasing the price by about \$3000 over that of a low pressure unit.

Nonconventional materials create secondary complications in the estimation of grass-roots capital. Since contingency and fee are proportional to actual bare module capital, they are based on the new value of C_{BM}^a.

$$C_c = 0.15 C_{BM}^a \tag{5-21}$$

$$C_F = 0.03 C_{BM}^a \tag{5-22}$$

On the other hand, site development, auxiliary buildings, and offsite facilities are not affected by pressure or materials, and according to the earlier discussion:

$$C_{SD} + C_{AB} + C_{OS} = 0.30 C_{TM}^{cs} \tag{5-23}$$

Thus, the revised grass-roots contribution of a stainless steel exchanger designed for 100 barg service is

$$C_{GR}^{ss} = 1.18\, C_{BM}^{ss} + 0.30\, C_{TM}^{cs} \tag{5-24}$$

With C_{BM}^{ss} equal to \$67,000 and C_{TM}^{ss} equal to \$37,600, the grass-roots capital is approximately \$90,000. Alloy, pressure, and other appropriate cost factors are documented in the cost charts given later.

Technical Uncertainty

Although Lang factors of 3 to 5 are typically found using the foregoing technique, some engineers use values as large as 8 or 9 for certain projects. The demand for accuracy notwithstanding, these engineers point out that management accepts, cheerfully, economic surprises in only one direction—actual costs that are *less* then predicted. Thus, safety or contingency factors are introduced to provide a cushion. One danger of this practice, however, is that a viable project will be squelched because of excessively conservative predictions.

Within tolerance limits of a predesign estimate, the technique described earlier should be sufficient with no additional contingency margin *if the process steps are commercially established.* For operations that have not been demonstrated in full plant production, experience has revealed the need for an additional safety margin. In the past, the magnitude of a contingency allowance depended on the experience and judgment of the estimator. Recently, based on multiple historic cost estimates for 44 commercial chemical plants, a more suitable technique for quantifying the safety margin has been proposed [6]. A study by the Rand Corporation found that each innovative process step acted to decrease the total effective plant capacity. For fluids processing, the reduction was near 10 percent for each pioneering step. For solids-handling plants, the typical reduction was 20 percent per step. Waste-handling innovations created reductions falling between these limits. Using the six-tenths rule, the change in cost per unit of capacity can be predicted from these tentative conclusions. The results are shown in Figure 5-2, which illustrates the contingency factor as a function of innovative process steps. An innovative step is defined as a process sequence that has not been employed successfully by those associated with the prospective project. Practice by others who will not share their experience and data does not constitute successful commercial operation in this

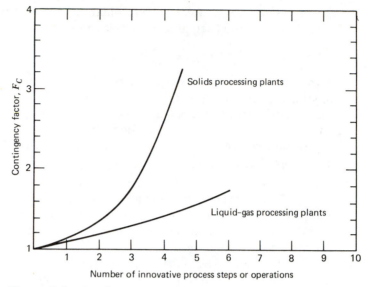

Figure 5-2   Contingency margins for prospective plants containing innovative process steps or operations. Conventionally estimated total module capital C_{TM} is multiplied by the appropriate contingency factor taken from this chart.

sense. If a project has one or more such unproven process steps, the capital cost estimated by preceding techniques is multiplied by the appropriate factor or factors from Figure 5-2. The soundness and accuracy of these safety factors can certainly be improved with experience. Meanwhile, these numbers provide theoretically defensible values as a beginning.

SUMMARY AND REVIEW

Capital Estimation Procedure

Predesign fixed capital estimates can be prepared by following eight steps.

Step 1   Determine the purchase price of each major flow sheet equipment item from charts in this chapter or from other appropriate sources.

Step 2   Escalate prices, using appropriate cost indices, to the date at which the cost estimate will apply.

Step 3   Multiply each updated purchase price $C_{P,i}$ by the installation factor $F_{BM,i}^{cs}$ (included with cost charts) to determine the base bare module contribution of each equipment item: $C_{BM,i}^{cs} = F_{BM,i}^{cs} \times C_{P,i}$.

Step 4   If components are constructed of special alloys or designed for unusually high pressure or temperature service, determine material and pressure factors from charts provided, and construct an actual bare module factor $F_{BM,i}^{a}$.

Step 5   Multiply updated purchase prices $C_{P,i}$ by actual bare module factors to obtain bare module costs $C_{BM,i}^{a}$ for alloy equipment or severe service.

$$C_{BM,i}^{a} = F_{BM,i}^{a} \times C_{P,i}$$

Step 6   Sum bare module prices to obtain total base bare module capital and, where different, total actual bare module capital.

$$C_{TBM}^{cs} = \sum_i C_{BM,i}^{cs} \qquad C_{TBM}^{a} = \sum_i C_{BM,i}^{a}$$

Step 7   Multiply C_{TBM}^{a} by 1.18 to account for contingency and fee, yielding the total module capital C_{TM}. This is the fixed capital required for installation of battery-limits equipment in an existing plant where auxiliary facilities are existing and adequate. If unconventional processing is involved, C_{TM} should be multiplied by an additional contingency factor. Recommended values are illustrated in Figure 5-2.

Step 8   If the estimate is for a totally new plant, multiply the total base bare module cost C_{TBM}^{cs} by 0.3 and add this to C_{TM}, yielding fixed capital for a grass-roots plant.

The foregoing procedure is applied in Illustration 5-1 at the end of this chapter. With experience, it may become possible to eliminate some of the steps, using overall factors rather than individual ones for each equipment item. Also, unusual circumstances, foreign locations, corporate experience, or other conditions may invalidate numerical values of the various multipliers employed here. In these cases, the same procedure can be followed using more appropriate factors.

A process design package traditionally includes an equipment list, a capital cost summary, and a manufacturing expense summary in addition to the flow sheet and material balance. Equipment lists and the capital cost summary sheet are discussed here. The manufacturing cost summary sheet is introduced in Chapter Six.

Equipment List

An equipment list is merely a summary of pertinent data and results employed and derived in the design procedure. Different organizations employ different formats. In fact, since both equipment and capital cost summaries relate to the same process items, they are often combined on the same sheet. They are separate here. An equipment list that I find effective is categorized according to generic types in keeping with the organization of this text. In the all-inclusive list of Table 5-4, each item of major process equipment heads a column of data that includes design parameters and equipment specifications. I refer to Table 5-4 as a composite equipment list because it is too general to be useful *per se*. In practice, there will often be a separate page or pages for each generic type. The data columns differ somewhat from case to case, since exchangers usually have two flow streams; distillation towers, three; absorbers, two; conveyors, crushers, and pumps, one; and electric motors, none. The data column in Table 5-4 includes all the items mentioned in Table 4-2. It should be amplified or reduced for each equipment type in a way that will be obvious to the designer. It is important that specifications employed for capital cost estimation be included in the list. This includes *area* for heat exchangers, *height* and *diameter* for towers, and internal *volume* for driers. These specifications are easily identified from cost charts in this chapter. Of course, the *construction material* is required as well.

Capital Cost Summary

A capital cost summary, like an equipment list, includes each major item of process equipment. The purpose of the cost summary is to document and illustrate the derivation of total estimated plant capital. It is organized according to steps outlined earlier in this chapter. Its use is best explained by reference to Table 5-5.

After the identification data in columns 1 and 2 of Table 5-5, capacity or size specifications are listed in column 3. These correspond to abscissa data in the respective cost charts. Succeeding columns include purchase costs of equipment fabricated from base material (usually carbon steel), module factors (from cost charts), material and pressure factors (from figures associated with cost charts), and actual bare module factors. The contribution of each equipment item to the overall bare module cost can then be entered both for base material and for special or alloy construction. Bare module and base bare module costs are easily totaled as indicated at the bottom of Table 5-5. Contingency and fee as a fraction of the total actual bare module sum is calculated and added as illustrated to yield total module cost. This, as noted earlier, is the capital required for an addition to an operating plant where auxiliary facilities already exist and are adequate. To compute the additional costs for a grass-roots process, auxiliary facilities are evaluated. As mentioned earlier, such represents about 30 percent of the base bare module cost even when alloy construction is employed. Individual steps should be clear from the table, prior discussion, and equations in this chapter.

While developing equipment lists and capital cost summaries, the usefulness of the practice of including equipment numbers and names in the margins of the flow sheet will be appreciated. That is why it was emphasized in Chapter Three.

COST DATA

Purchase costs (f.o.b. vendor's factory) for major process equipment are illustrated in the collection of charts that follows. Data are expressed graphically, with charts organized according to generic categories that were introduced in Chapter Four. To serve as an index, Table 5-6 contains figure numbers to identify the location of price data for each particular item.

As suggested by the sixth-tenths rule, logarithmic coordinates are used in the charts. In this format, data tend to fall in straight lines having slopes equal to size exponents. In most cases, a capacity or size parameter is plotted as the abscissa and the purchase price as the ordinate. With some types of equipment, unique correlation techniques are employed. In keeping with modern engineering practice and conventions of this text, all capacities are in SI units. The U.S. dollar is the price unit.

Wherever feasible and rigorous, motor or driver prices are included with the overall equipment. Otherwise, the absence of a driver is noted on the ordinate label. In these cases, the additional cost can be assessed from Figure 5-20 or 5-21.

Charts were assembled from information contained in references 1–3, 7–9, 11, and 12, and data provided by manufacturers. Some simplifications were necessary to confine the data and to resolve discrepancies. The charts, I feel, are accurate enough for preliminary design estimates and are certainly adequate for classroom use. Employment for more rigorous estimates is done at the user's risk.

All prices are based on a CE Plant Cost Index of 315 (mid-1982). Application of module factors, escalation, and contingency correctors has been explained. For convenience, bare module capital rather than purchased equipment price is illustrated for most auxiliary facilities (Figures 5-3 and 5-5 through 5-13).

TABLE 5-4
EQUIPMENT LIST[a]

Job title: _____ Location: _____ Capacity: _____

(Flow Sheet Page Number:) Equipment Identification Number Name	Auxiliary Facilities			Heat Exchangers			Process Vessel
	A-	**A-**		**E-**	**E-**		**D-**

Process Streams
  Stream Number
    Name
  Process orientation
  Phase
  Temperature (°C)
    In
    Out
  Pressure (barg)
    In
    Out
  Flow rate
    Mass (kg/s)
    Volumetric (m³/s)
  Concentration
  Viscosity (Pa · s)
  Enthalphy (kJ/kg)
    In
    Out							

Stream Number
    Name
  Process orientation
  Phase
  Temperature (°C)
    In
    Out
  Pressure (barg)
    In
    Out
  Flow rate
    Mass (kg/s)
    Volumetric (m³/s)
  Concentration
  Viscosity (Pa · s)
  Enthalpy (kJ/kg)
    In
    Out							

Heat Transfer Coefficient
    $(J/m^2 \cdot s \cdot K)$
  Efficiency
  Heating duty (kJ/s)
  LMTD (°C)
  MTD (°C)
  F_T
  $\Delta T_m = \text{MTD} \times F_T$ (°C)							

Utilities
  Electricity (kW)
  Cooling water (m³/s)
  Fuel
  Steam (kg/s)
    [Pressure (barg)]							

Equipment Size
  Length or height L (m)
  Width or diameter, D (m)
  Surface area, A (m²)
  Volume, V (m³)
  Design pressure (barg)
  Shaft power, $\dot{w}_s$ (kW)
  Orientation
  Material of construction
  Other specifications							

[a]See Table 5-7 for an example of a completed equipment list.

TABLE 5-4   (Continued)

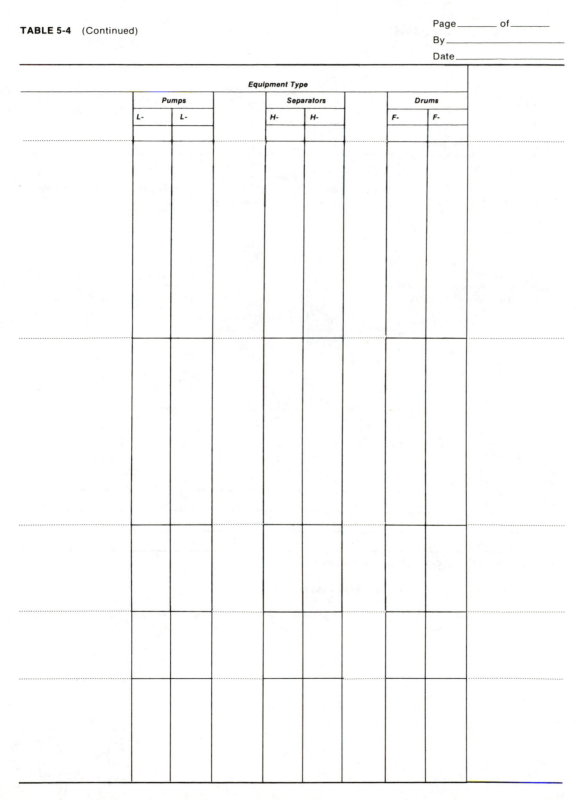

Equipment Type							
Pumps			Separators			Drums	
L-	L-		H-	H-		F-	F-

TABLE 5-5
CAPITAL COST SUMMARY[a]

Job title: _____

Location: _____

Effective date to which estimate applies: _____

Equipment Identification	Number	Capacity or Size Specifications	Purchased Equipment Cost (base material), $C_P^{cs,b} \times$ (CE Index/315)	Base Bare Module Factor, F_{BM}^{cs}	Base Bare Module Cost, C_{BM}^{cs}
Conveyors					
Storage lift	J-313				
....	....				
....	....				
Crushers, grinders, mills					
Product mill	C-358				
....	....				
....	....				
(Other items as taken from the equipment list)					
....	....				
....	....				
....	....				

Total bare module cost   Base materials, $C_{TBM}^{cs} = \sum_i C_{BM}^{cs} =$ ....

Contingency and fee

Total Module Cost[c]

Auxiliary facilities

Grass-roots capital

[a] For an example of a completed capital cost summary, see Table 5-8.

[b] Mid-1982 purchase price for carbon steel or base material construction.

[c] If unconventional processing is involved, an additional contingency allowance should be made using a factor such as is illustrated in Figure 5-2.

TABLE 5-5
CAPITAL COST SUMMARY[a]

Page _____ of _____

By _____

Date _____

Capacity _____

Cost Index Type _____

Cost Index Value _____

	Material Factor, F_M^a	Pressure or Other Factors, F_p	Actual Bare Module Factor, F_{BM}^a	Actual Bare Module Cost, C_{BM}^a

Actual materials, $C_{TBM}^a = \sum_i C_{BM}^a =$ ....

$C_C + C_F = C_{TBM}^a \times 0.18 =$ _____

$C_{TM} =$ ....

$C_{TBM}^{cs} \times 0.30 =$ _____

$C_{GR} =$

285

TABLE 5-6
INDEX TO COST CHARTS

Equipment	Figures
Auxiliary Facilities	5-3 to 5-13
Air plants	5-3
Boilers and steam generators	5-4
Chimneys or stacks	5-5
Water cooling towers	5-6
Water demineralizers	5-7
Electrical generating plants, substations	5-8, 5-9
Incinerators and incinerator–steam generators	5-10
Flares	5-10
Mechanical refrigeration units	5-11
Wastewater treatment plants	5-12
Water treatment plants	5-13
Thermal fluid heaters	5-26*b*
Conveyors	5-14, 5-15
Apron and auger	5-14*a*
Belt and bucket	5-14*b*
Chain and vibratory	5-15*a*
Pneumatic	5-15*b*
Crushers, Mills, and Grinders	5-16 to 5-19
Crushers	5-16
Grinders and cutters	5-17
Mills	5-18
Fluid energy (jet) mills	5-19
Drives and Power Recovery Machines	5-20 to 5-22
Electric motors and generators	5-20
Internal combustion engines, gas turbines, steam turbines	5-21
Power recovery machines	5-22
Evaporators and Vaporizers	5-23, 5-24
Vaporizers	5-23
Evaporators	5-24
Furnaces	5-25 to 5-28
Steam boilers	5-25, 5-26*a*
Thermal fluid heaters	5-26*b*
Process heaters	5-27
Reactive	5-27a
Nonreactive	5-27b
Industrial ovens	5-28
Gas Movers and Compressors	5-29 to 5-31
Fans	5-29
Blowers and compressors	5-30
Ejectors	5-31
Gas–Solid Contacting Equipment	5-32 to 5-35
Tunnel and vibratory conveyor dryers	5-32
Rotary and vertical tower contactors	5-33
Drum dryers	5-34*a*
Auger conveyor and spouted-bed contactors	5-34*b*
Fluid-activated contactors	5-35
Heat Exchangers	5-36 to 5-40
Shell and tube and double-pipe	5-36
Plate and spiral	5-39
Air-cooled (fin-fan)	5-40
Mixers	5-41 to 5-43
Motionless	5-41
Agitators	5-42
Heavy-duty	5-43
Process Vessels	5-44 to 5-48
Horizontally oriented	5-44*a*
Vertically oriented	5-44*b*
Tower packings	5-47
Sieve trays and mist eliminators	5-48
Pumps	5-49 to 5-51
Reactors	(see Table 4-21)
Separators	5-52 to 5-59
Clarifiers and thickeners	5-52
Rake and spiral classifiers	5-53
Centrifugal Separators	5-54, 5-55
Dust collectors	5-56
Liquid filters	5-57
Expression equipment	5-58
Vibratory screens	5-59
Size-Enlargement Equipment	5-60
Storage Vessels	5-61

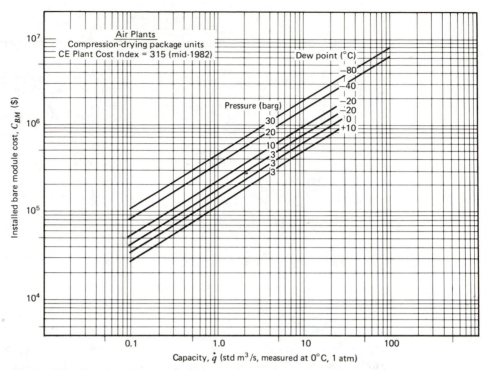

Figure 5-3 Bare module costs of packaged plants to produce compressed, dried air.

Figure 5-4 Purchase costs of boilers or steam generators. The factors in the bare module cost, $C_{BM} = C_P \times F_p \times F_T \times F_{BM}$ were taken from Guthrie [1, 2]. Note that they are applied somewhat differently in this instance compared with the general technique outlined in the text.

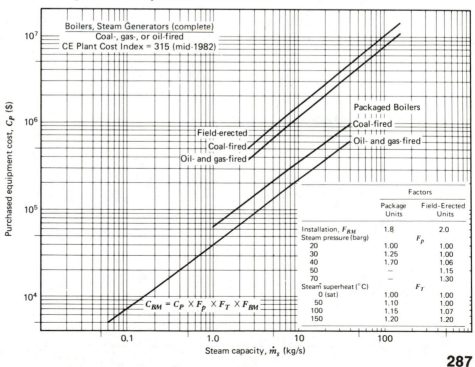

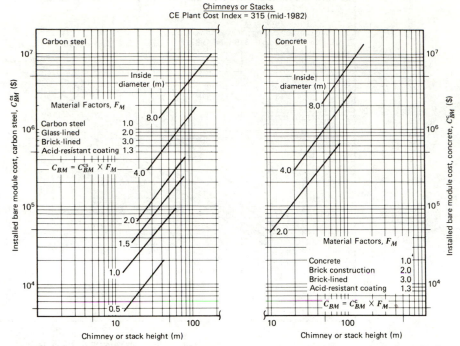

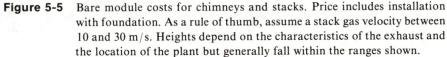

Figure 5-5 Bare module costs for chimneys and stacks. Price includes installation with foundation. As a rule of thumb, assume a stack gas velocity between 10 and 30 m/s. Heights depend on the characteristics of the exhaust and the location of the plant but generally fall within the ranges shown.

Figure 5-6 Bare module costs for cooling towers. Price includes delivery, erection, foundation, basin, pumps, and drives. To obtain total grass-roots capital, use standard factors as outlined in the text and in Table 5-2. Costs apply for 45°C water inlet temperature, 30°C outlet temperature, and 25°C wet bulb temperature. These are representative U.S. conditions. For definitive estimates, where higher accuracy is necessary, correlation factors for more specific atmospheric conditions can be found in reference 12.

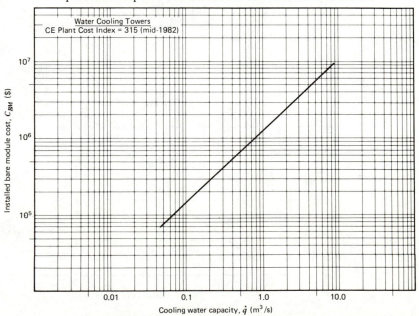

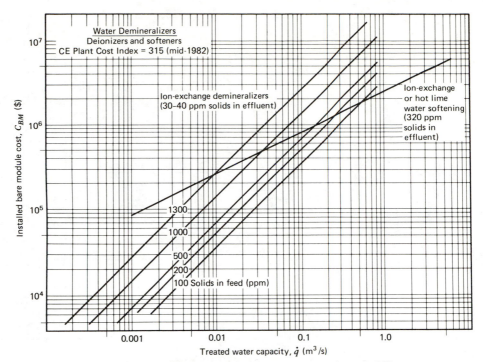

Water Demineralizers
Deionizers and softeners
CE Plant Cost Index = 315 (mid-1982)

Ion-exchange demineralizers
(30–40 ppm solids in effluent)

Ion-exchange
or hot lime
water softening
(320 ppm
solids in
effluent)

1300
1000
500
200
100 Solids in feed (ppm)

Installed bare module cost, C_{BM} ($)

Treated water capacity, $\dot{q}$ (m^3/s)

Figure 5-7   Bare module costs for water dimineralizers and softeners. Effluent concentrations of 30 to 40 ppm are suitable for boiler feed water. Up to 320 ppm is tolerable for cooling tower feedwater. Supply water, if not of normal potable quality must be processed by a conventional water treatment plant (see Figure 5-13).

Figure 5-8   Bare module costs for field-erected electrical power plants.

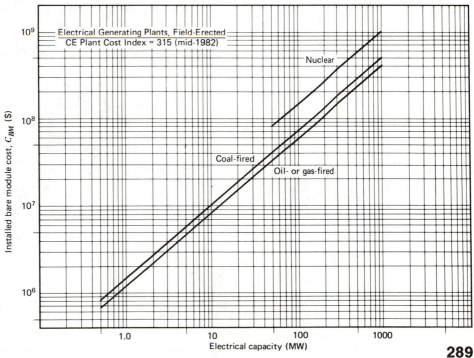

Electrical Generating Plants, Field-Erected
CE Plant Cost Index = 315 (mid-1982)

Nuclear

Coal-fired

Oil- or gas-fired

Installed bare module cost, C_{BM} ($)

Electrical capacity (MW)

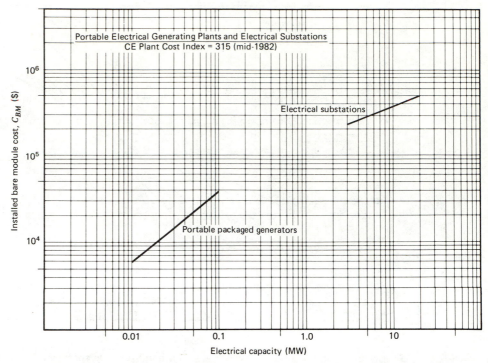

Figure 5-9 Bare module costs for electrical substations and portable packaged power plants.

Figure 5-10 Bare module costs for incinerators, incinerator–steam generators, and flares—liquid, solid, or gaseous fuel. For more details, see Lepeau, M., "Evaluating Heat Recovery from Burning Refuse," *Chem. Eng.*, pp. 81–84 (March 21, 1983).

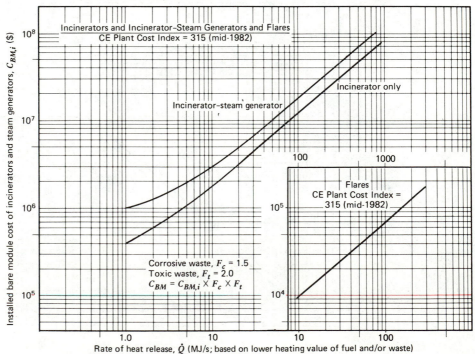

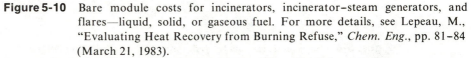

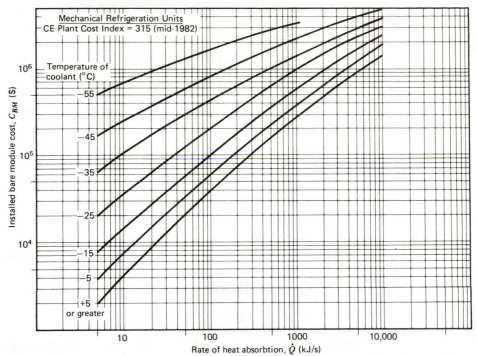

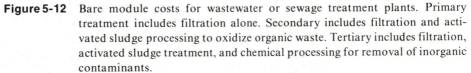

Figure 5-11 Bare module cost for air-cooled mechanical refrigeration units, complete except for the absorptive heat exchanger, which is part of main process module.

Figure 5-12 Bare module costs for wastewater or sewage treatment plants. Primary treatment includes filtration alone. Secondary includes filtration and activated sludge processing to oxidize organic waste. Tertiary includes filtration, activated sludge treatment, and chemical processing for removal of inorganic contaminants.

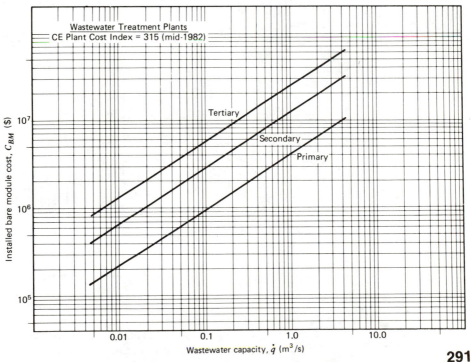

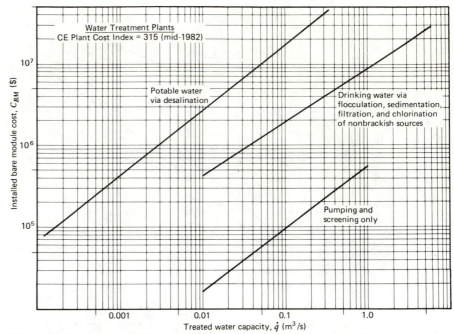

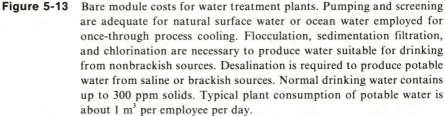

Figure 5-13 Bare module costs for water treatment plants. Pumping and screening are adequate for natural surface water or ocean water employed for once-through process cooling. Flocculation, sedimentation filtration, and chlorination are necessary to produce water suitable for drinking from nonbrackish sources. Desalination is required to produce potable water from saline or brackish sources. Normal drinking water contains up to 300 ppm solids. Typical plant consumption of potable water is about 1 m^3 per employee per day.

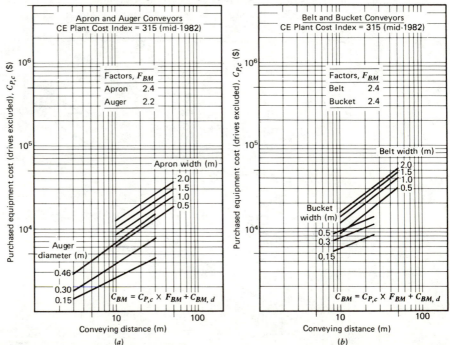

Figure 5-14 Purchased equipment costs for solids-conveying equipment: (a) apron and auger, (b) belt and bucket. Motor drives are not included. To determine bare module cost, multiply by the appropriate bare module factor F_{BM} and add the bare module cost of the drive $C_{BM,d}$ as illustrated.

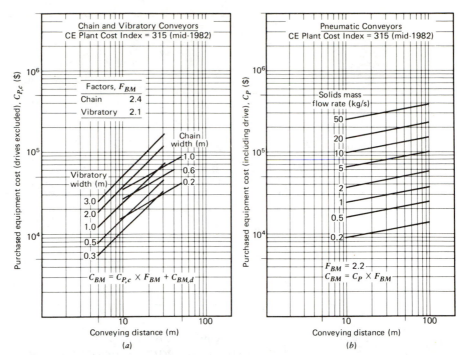

Figure 5-15 Purchased equipment costs for solids-conveying equipment. Note that the parameter for pneumatic conveyors is solids flow rate (kg/s). Drives are not included for chain and vibratory units (*a*) but are included for pneumatic conveyors (*b*). Total installed bare module costs are computed as illustrated.

Figure 5-16 Purchased equipment costs for crushers, including electrical motor drives. In the equations for electrical power consumption, $\dot{m}$ is solids flow rate (kg/s) and R is the size-reduction ratio. Prices are for closed-circuit packages including auxiliary classification equipment and controls. Custom provisions (e.g., for simultaneous drying and cooling, heating, dust collection, or controlled atmosphere) are not included.

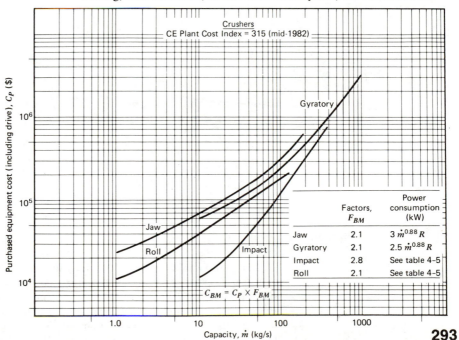

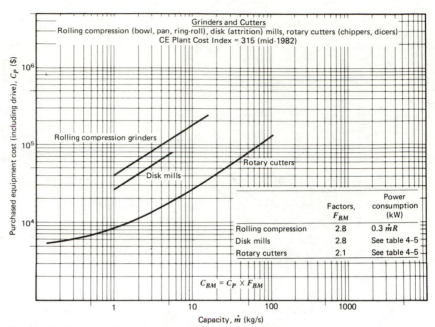

Figure 5-17 Purchased equipment costs for grinders and cutters, including electrical motor drives. In the equation for power consumption, $\dot{m}$ is solids flow rate (kg/s) and R is the size-reduction ratio. Prices are for closed-circuit packages including auxiliary classification equipment and controls. Custom provisions (e.g., for simultaneous drying and cooling, heating, dust collection, or controlled atmospheres) are not included.

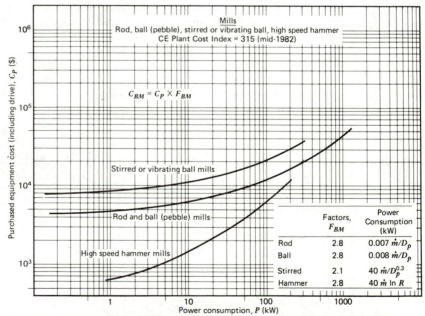

Figure 5-18 Purchased equipment costs for mills, including electrical motor drives. Note that the size parameter is power consumption because for fine grinding, this is a more accurate index of equipment size than is solids capacity. In the equations for power consumption, $\dot{m}$ is solids flow rate (kg/s) and D_p is mean product particle size (m). Prices are for closed-circuit packages including auxiliary classification equipment and controls. Custom provisions (e.g., for simultaneous drying and cooling, heating, dust collection, or controlled atmospheres) are not included.

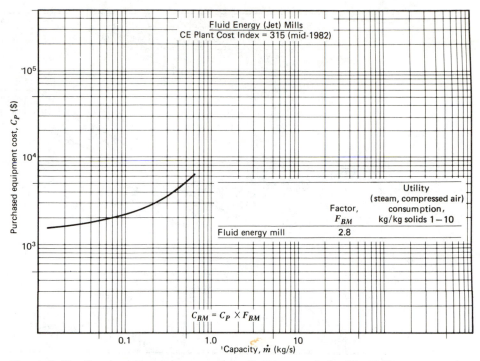

Figure 5-19 Purchased equipment costs for fluid energy (jet) mills. Utility consumption depends on solids properties, size-reduction ratio, and capacity (see Table 4-5). Motive fluid can be either steam or compressed air. Normal utility pressure is 8 bara or greater.

Figure 5-20 Purchased equipment costs for electric motors, 30 and 60 revolutions per second (rps). Electric generators are essentially equivalent in cost.

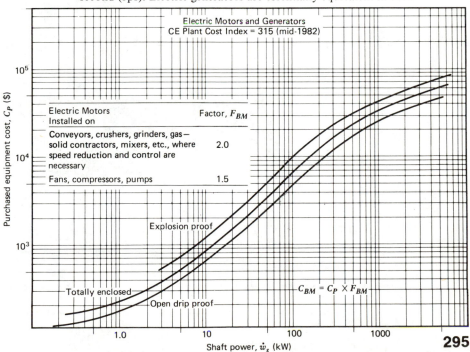

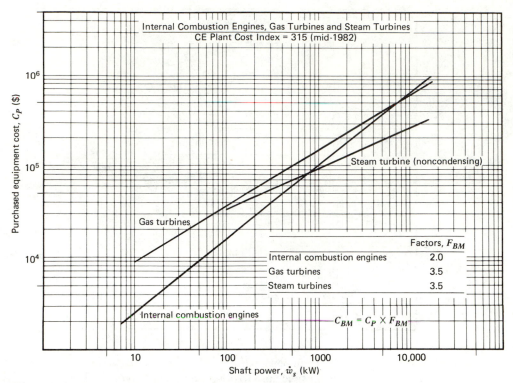

Figure 5-21 Purchased equipment costs for internal combustion engines and gas and steam turbines.

Figure 5-22 Purchased equipment costs for power recovery machines.

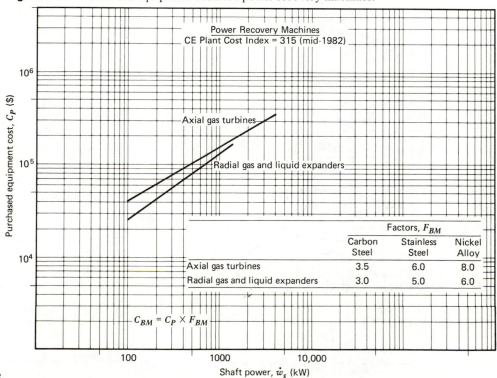

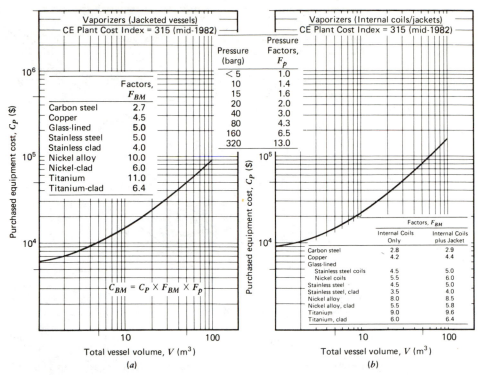

Figure 5-23 Purchased equipment costs for vaporizers having (*a*) jackets and (*b*) internal coils alone or internal coils plus jackets.

Figure 5-24 Purchased equipment costs and module factors for evaporators of various types. Costs are for single effects. Prices for multiple effects are directly proportional to the number of effects.

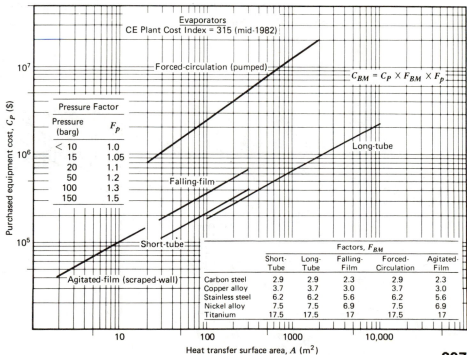

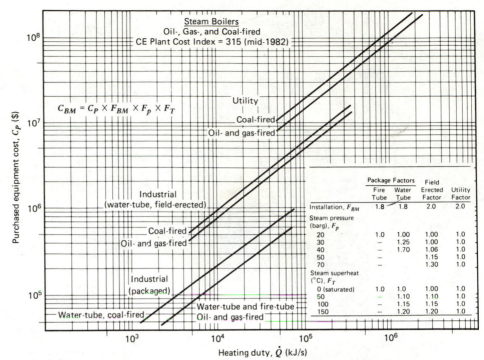

The chart (Figure 5-25) contains the equation:

$$C_{BM} = C_P \times F_{BM} \times F_p \times F_T$$

Labels on chart: Steam Boilers, Oil-, Gas-, and Coal-fired, CE Plant Cost Index = 315 (mid-1982); Utility, Coal-fired, Oil- and gas-fired; Industrial (water-tube, field-erected); Coal-fired; Oil- and gas-fired; Industrial (packaged); Water-tube, coal-fired; Water-tube and fire-tube Oil- and gas-fired.

Axes: Purchased equipment cost, C_P ($); Heating duty, $\dot{Q}$ (kJ/s)

	Package Factors		Field	
	Fire Tube	Water Tube	Erected Factor	Utility Factor
Installation, F_{BM}	1.8	1.8	2.0	2.0
Steam pressure (barg), F_p				
20	1.0	1.00	1.00	1.0
30	—	1.25	1.00	1.0
40	—	1.70	1.06	1.0
50	—		1.15	1.0
70	—		1.30	1.0
Steam superheat (°C), F_T				
0 (saturated)	1.0	1.0	1.00	1.0
50	—	1.10	1.10	1.0
100	—	1.15	1.15	1.0
150	—	1.20	1.20	1.0

Figure 5-25 Purchased equipment costs, module, pressure, and temperature factors for steam boilers. These data are comparable to those in Figure 5-4 except that duty is expressed there as steam capacity.

Figure 5-26 Purchase costs of (a) steam boilers and (b) thermal fluid heaters. In thermal fluid heaters, a temperature differential of 50°C is typical between fluid entering and leaving the heater. When the liquid is vaporized, vapor leaves at the saturation temperature.

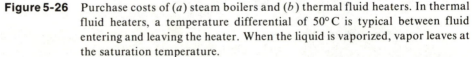
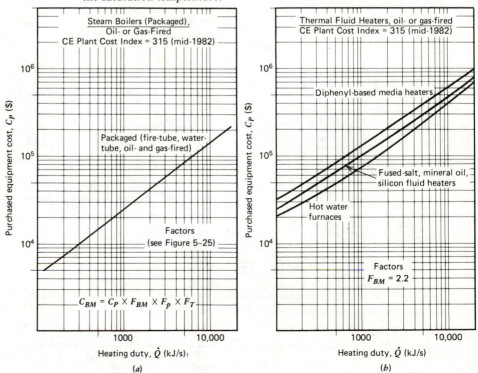

(a) chart labels: Steam Boilers (Packaged), Oil- or Gas-Fired, CE Plant Cost Index = 315 (mid-1982); Packaged (fire-tube, water-tube, oil- and gas-fired); Factors (see Figure 5-25);

$$C_{BM} = C_P \times F_{BM} \times F_p \times F_T$$

Axes: Purchased equipment cost, C_P ($); Heating duty, $\dot{Q}$ (kJ/s)

(b) chart labels: Thermal Fluid Heaters, oil- or gas-fired, CE Plant Cost Index = 315 (mid-1982); Diphenyl-based media heaters; Fused-salt, mineral oil, silicon fluid heaters; Hot water furnaces; Factors $F_{BM} = 2.2$

Axes: Purchased equipment cost, C_P ($); Heating duty, $\dot{Q}$ (kJ/s)

(a)   (b)

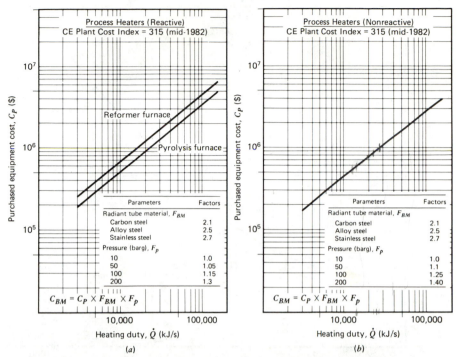

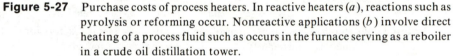

Figure 5-27 Purchase costs of process heaters. In reactive heaters (*a*), reactions such as pyrolysis or reforming occur. Nonreactive applications (*b*) involve direct heating of a process fluid such as occurs in the furnace serving as a reboiler in a crude oil distillation tower.

Figure 5-28 Purchase costs of industrial ovens. The heat loss expression is merely approximate for use in preliminary evaluation only. As indicated in Table 4-8 for oil-or gas-fired ovens, approximately 70 percent of the power heating value (LHV) of the fuel is equal to oven heat loss plus heat absorbed in the process. In electric ovens, 100 percent of the power is directed to these sinks.

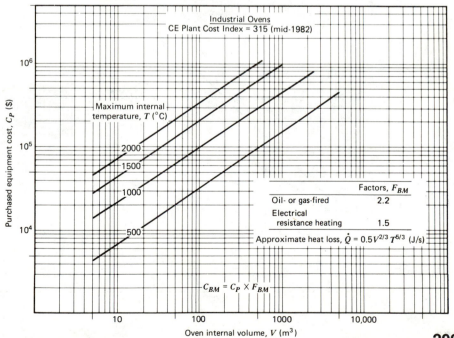

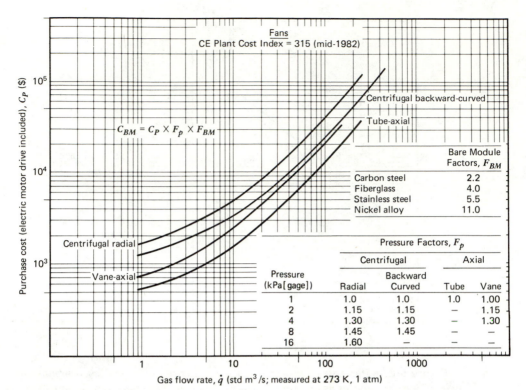

Figure 5-29 tables:

	Bare Module Factors, F_{BM}
Carbon steel	2.2
Fiberglass	4.0
Stainless steel	5.5
Nickel alloy	11.0

Pressure Factors, F_p

Pressure (kPa[gage])	Centrifugal Radial	Centrifugal Backward Curved	Axial Tube	Axial Vane
1	1.0	1.0	1.0	1.00
2	1.15	1.15	—	1.15
4	1.30	1.30	—	1.30
8	1.45	1.45	—	—
16	1.60	—	—	—

$$C_{BM} = C_P \times F_p \times F_{BM}$$

Figure 5-29 Purchased equipment costs and module and pressure factors for centrifugal and axial fans. Costs of electric motor drives are included.

Figure 5-30 Purchase costs of blowers and compressors. Costs of drives are excluded. To determine the bare module cost of drives $C_{BM,d}$, refer to Figures 5-20 and 5-21.

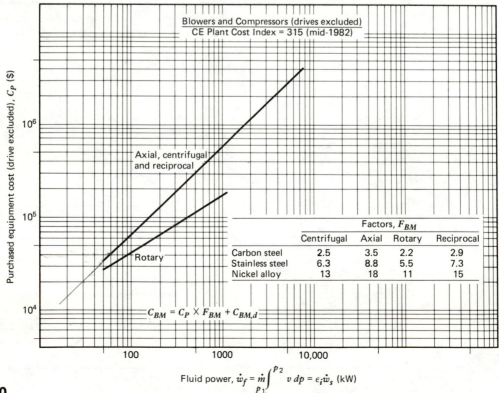

Factors, F_{BM}

	Centrifugal	Axial	Rotary	Reciprocal
Carbon steel	2.5	3.5	2.2	2.9
Stainless steel	6.3	8.8	5.5	7.3
Nickel alloy	13	18	11	15

$$C_{BM} = C_P \times F_{BM} + C_{BM,d}$$

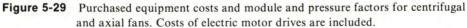

$$\text{Fluid power, } \dot{w}_f = \dot{m} \int_{p_1}^{p_2} v\, dp = \epsilon_i \dot{w}_s \text{ (kW)}$$

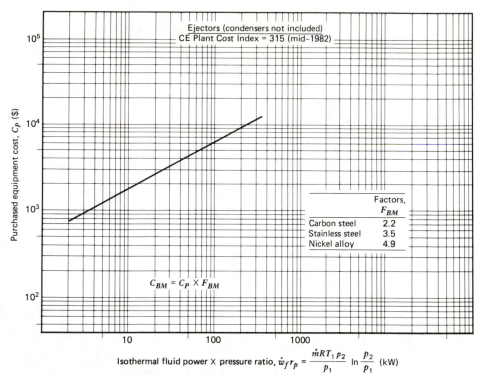

Figure 5-31 Purchase costs of ejectors. Condensers, where used, are not included and must be evaluated separately.

Figure 5-32 Purchased equipment costs for tunnel and vibratory conveyor dryers. Costs are for complete units, including drives.

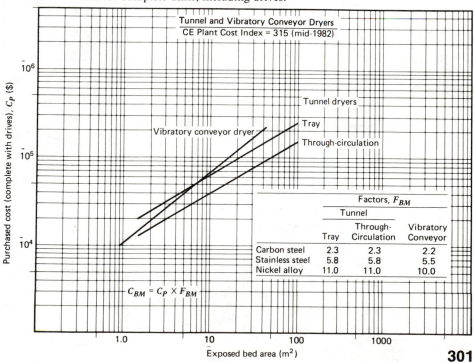

301

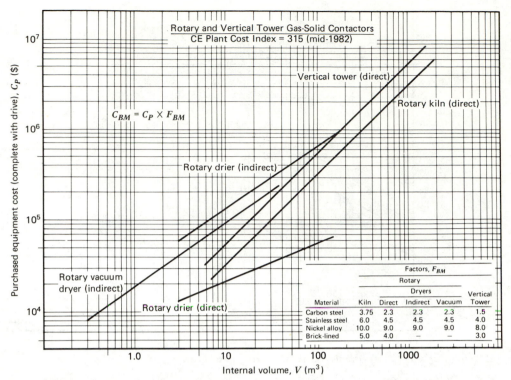

Figure 5-33 Purchased equipment costs for rotary and vertical tower contactors. Costs are for complete units, including drives.

Figure 5-34 Purchased equipment costs for (a) drum dryers and (b) auger conveyor and spouted-bed contactors. Costs are for complete units, including drives.

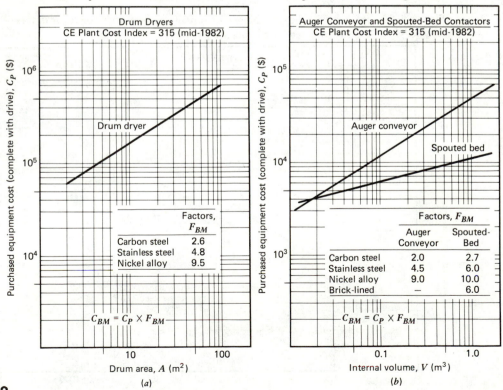

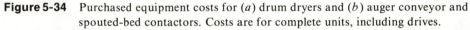

302

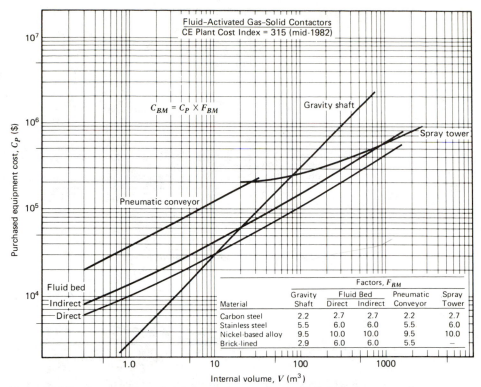

Figure 5-35 Purchased equipment costs for gravity shaft, fluid bed, pneumatic conveyor, and spray tower contactors.

Figure 5-36 Purchased equipment costs for shell and tube and double-pipe heat exchangers. Bare module factors F_{BM}^a are derived from Figure 5-38 using material factors given here and pressure factor F_p from Figure 5-37.

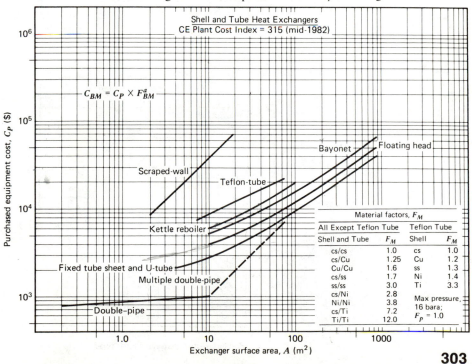

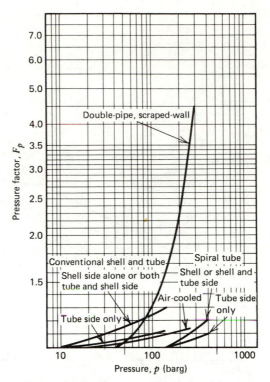

Figure 5-37 Pressure factors (ratio of purchase price of a high pressure heat exchanger to one designed for conventional pressures).

Figure 5-38 Bare module factors as a function of materials and pressure factors for various classes of heat exchanger.

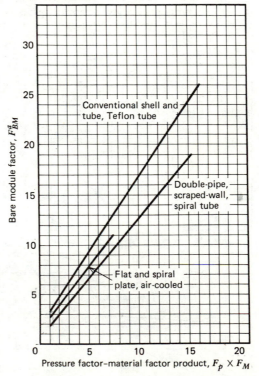

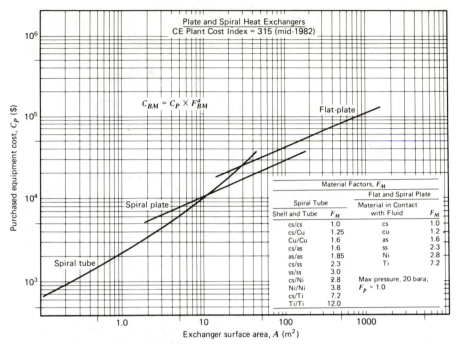

Figure 5-39 Purchased equipment costs of plate, spiral plate, and spiral tube heat exchangers. Area is that calculated from the basic heat transfer equation. Obtain F_{BM}^a from Figure 5-38, using F_M given here and F_p from Figure 5-37 for spiral tube exchangers. $F_P = 1.0$ for flat and spiral plate units.

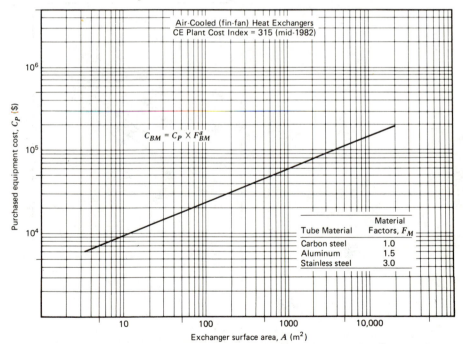

Figure 5-40 Purchased equipment costs for air-cooled exchangers. Area is outside area of equivalent bare tubes excluding fins. This is the parameter used to designate exchangers and in the basic heat transfer equation. The heat transfer coefficients shown in Table 4-15 are adjusted to this basis. Actual total outside areas including fins are approximately 15 to 20 times greater than the bare-tube values. Obtain F_{BM}^a from Figure 5-38, using F_M given here and F_p from Figure 5-37.

305

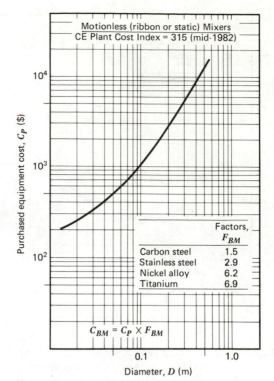

Figure 5-41 Purchased equipment costs for motionless mixers.

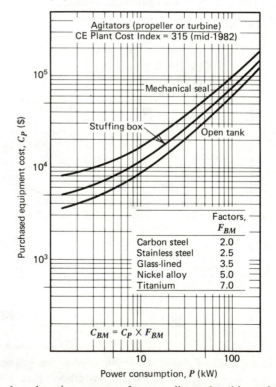

Figure 5-42 Purchased equipment costs for propeller and turbine agitators. Cost includes motor, speed reducer, and impeller ready for installation in a vessel. Stuffing-box seals can contain pressures up to 10 barg. Mechanical seals are employed for toxic or critical fluids at pressures up to 80 barg. (For information on other mixers not included in this chart see p. 316.)

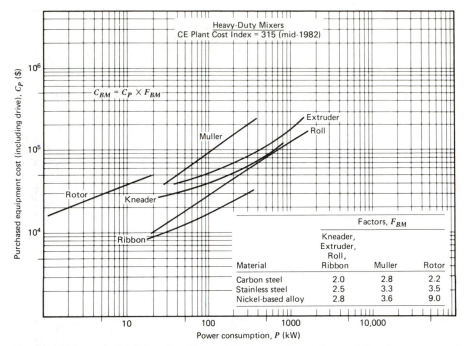

Figure 5-43 Purchased equipment costs for heavy-duty mixers of doughs, pastes, and powders. (For information on other mixers not included in this chart see p. 316.)

Figure 5-44 Purchased equipment costs for (*a*) horizontally oriented and (*b*) vertically oriented process vessels. Bases for costs are carbon steel construction and internal pressure less than 4 barg. Installation factors F_{BM}^a for higher pressures and different construction are found in Figure 5-46. For jacketed or internally heated vessels or autoclaves, see Figure 5-23. For packed or tray towers, add bare module costs of packing or trays from Figure 5-47 or 5-48.

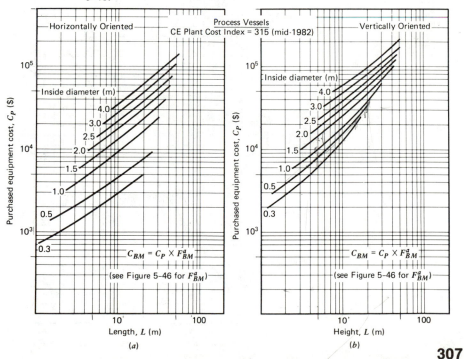

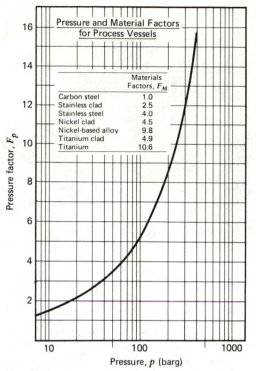

Figure 5-45 Vessel pressure and materials factors (ratio of purchase price of a high pressure or noncarbon steel vessel to one designed for carbon steel construction and pressures less than 4 barg).

Figure 5-46 Bare module factors as a function of materials and pressure factors for both horizontally and vertically oriented vessels.

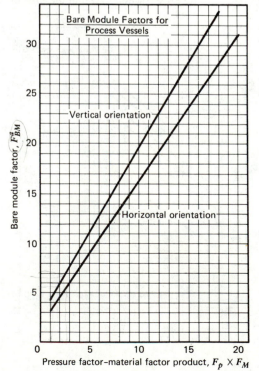

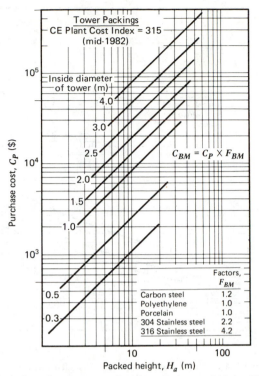

Figure 5-47 Purchased equipment costs for slotted-ring and high efficiency saddle tower packings (price includes tower internal supports and distributors).

Figure 5-48 Purchased equipment costs for sieve trays and mist eliminators. (Notice that costs shown are per tray.) In quantities fewer than 20, tray prices should be multipled by the factor f_Q as illustrated.

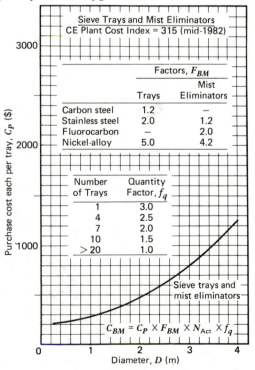

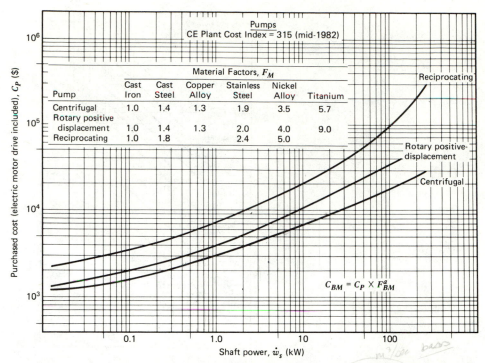

Figure 5-49 shown with:

Pumps
CE Plant Cost Index = 315 (mid-1982)

Pump	Material Factors, F_M					
	Cast Iron	Cast Steel	Copper Alloy	Stainless Steel	Nickel Alloy	Titanium
Centrifugal	1.0	1.4	1.3	1.9	3.5	5.7
Rotary positive displacement	1.0	1.4	1.3	2.0	4.0	9.0
Reciprocating	1.0	1.8		2.4	5.0	

$$C_{BM} = C_P \times F_{BM}^a$$

Figure 5-49 Purchased equipment costs for pumps. Shaft power $\dot{w}_s = \dot{q}\Delta p/\epsilon_i$. For unusual service and low capacities, use efficiencies near the lower extremes of ranges given in Table 4-20. Within the sensitivity allowed by a predesign cost estimate, this will compensate for higher priced pumps employed in severe service. Prices are complete with electric motor drives. To substitute other drives, use data in Figures 5-20 and 5-21. For values of F_{BM}^a, see Figure 5-51.

Figure 5-50 Pump pressure factor (ratio of purchase price of high pressure pump to that of one designed for 10 barg).

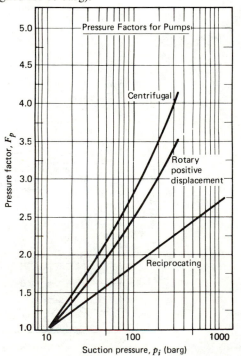

310

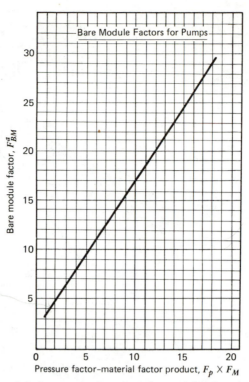

Figure 5-51 Bare module factors as a function of material and pressure factors for pumps.

Figure 5-52 Purchased equipment costs for thickeners and clarifiers.

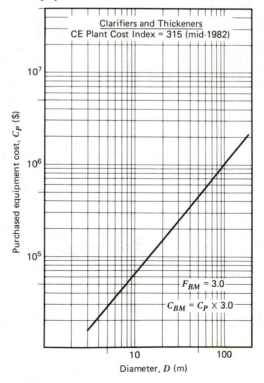

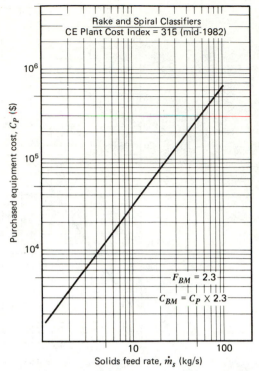

Figure 5-53   Purchased equipment costs for rake and spiral classifiers.

Figure 5-54   Purchased equipment costs for centrifugal filters and helical conveyor centrifugal separators.

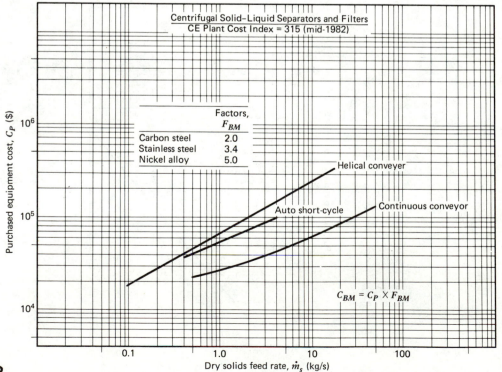

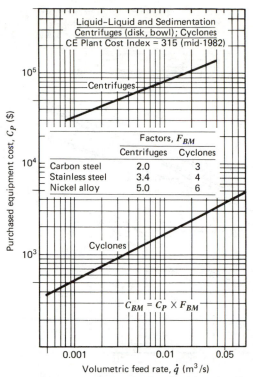

Figure 5-55 Purchased equipment cost for liquid–liquid and sedimentation centrifuges and cyclone separators.

Figure 5-56 Purchased equipment costs for high efficiency cyclones, bag filters, electrostatic precipitators, and venturi scrubbers. Bare module factors include valving, hoppers, and supports, as well as other installation expenses.

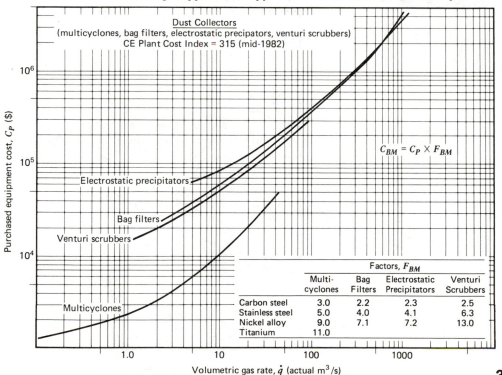

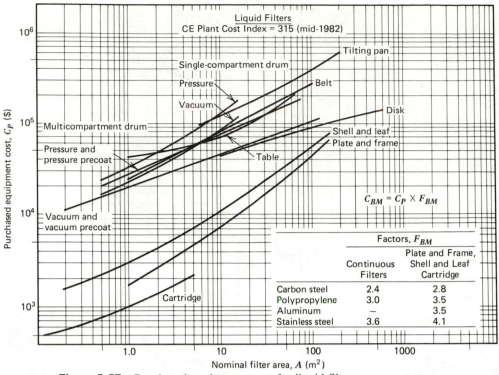

Figure 5-57 Purchased equipment costs for liquid filters.

Figure 5-58 Purchased equipment costs for screw and roll presses.

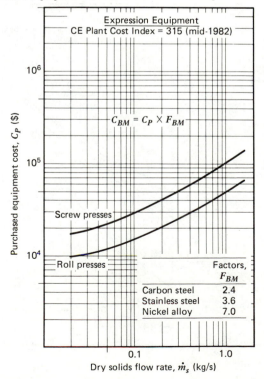

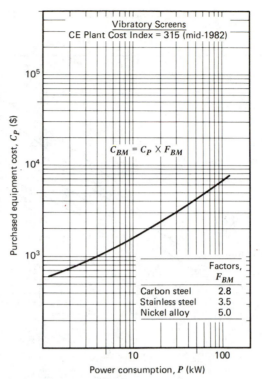

Figure 5-59 Purchased equipment costs for vibratory screens. The equation for power consumption ($P = 16{,}000\,\dot{m}_s/D_P$) is based on solids feed rate $\dot{m}_s$ (kg/s) and mesh size D_p (μm).

Figure 5-60 Purchased equipment costs for size-enlargement equipment.

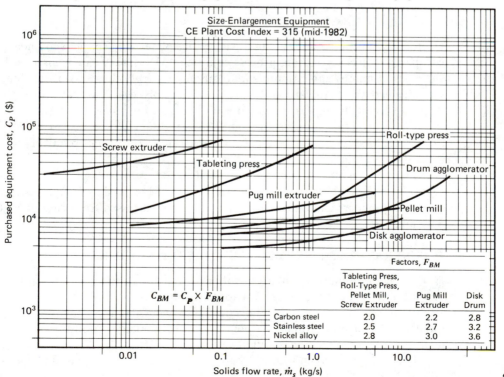

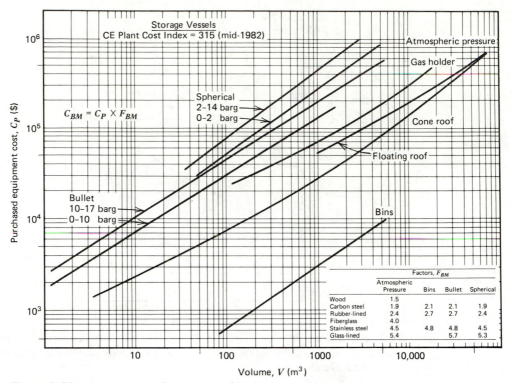

Figure 5-61 Purchased equipment costs for storage vessels.

Capital Cost Data for Mixers

Costs for some mixers such as fluid jets, orifice plates, and spargers are negligible compared with costs of other process equipment. Thus, they can be disregarded if the costs of pumps and compressors required to supply the pressurized fluid are included in the cost tabulation. When a vessel is involved such as that with a sparger or pump-mixing loop, it is evaluated using the data for process vessels.

To estimate the cost of a pump mixer, use the data for centrifugal pumps. For an agitated line mixer, assume a cost 1.5 times that of a centrifugal pump of the same capacity.

Data for motionless mixers, propeller and turbine agitators, kneaders, extruders, roll-muller-rotor; and ribbon mixers are included in Figures 5-41, 5-42, and 5-43. Grinders and mills used as mixers can be assessed from data in Figures 5-17 and 5-18.

ILLUSTRATION 5-1   ALKYLATE SPLITTER MODULE

A petroleum refinery distillation column, known as the alkylate splitter, is employed to separate a mixture of various C_4 through C_{14} hydrocarbons into two streams (see Figure 5-62). The split is made between C_8 and C_{10} fractions to yield a distillate suitable for automotive fuel (motor alkylate) and a bottoms product for aviation or furnace fuel (heavy alkylate).

Using techniques outlined in Chapter Four, the distillation column specifications have been established. It is 3 m in diameter and 30 m tall, having 44 sieve trays. Specific heats and latent heats are essentially independent of temperature and pressure in the range of interest. These data are also listed in Figure 5-62. For ease of discussion, detailed stream compositions are not shown, although they are known and were required for the distillation tower analysis.

Prepare an equipment list for the alkylate splitter module, and estimate its capital cost for mid-1984.

The process flow diagram and material balance have already been assembled in an acceptable form as presented in Figure 5-62. An equipment list is begun by identifying equipment items and assembling known information according to the format of Table 5-4. The list includes four heat exchangers, the distillation tower, three pumps, and two drums.

Exchanger duties can be determined from energy balances based on flow sheet data. Overall heat transfer coefficients can be extracted from Table 4-15 and areas calculated from Equation 4-72. The distillation tower has already been specified adequately for a predesign estimate. Pump shaft power is calculated from Equation 4-94 using intrinsic efficiencies based on Table 4-20. Driver efficiencies for electric motors are found in Figure 4-2. Drum volumes, lengths, and diameters are determined from residence times recommended in Tables 4-18 and 4-27. A length to diameter ratio of 3 as recommended in Table 4-24 is employed. Review of Table 4-28 indicates that carbon steel is adequate for this service. With this information as a basis, the equipment list in Table 5-7 was assembled.

To obtain purchased equipment costs, specifications listed in Table 5-7 were employed in concert with relevant equipment cost charts. Results are listed in Table 5-8. Prices were escalated to mid-1984 using a projected CE Plant Cost Index of 350, extrapolated at a 5 percent inflation rate in Figure 5-1. Since pressures and temperatures are modest and components are constructed of the base material, there were no corrections for special alloy construction or severe service. This process is conventional, and no additional contingency allowance is necessary. As shown in Table 5-8, total module cost is estimated to be $1.17M, and grass-roots capital is $1.52M. Further use is made of these figures in Chapter Six.

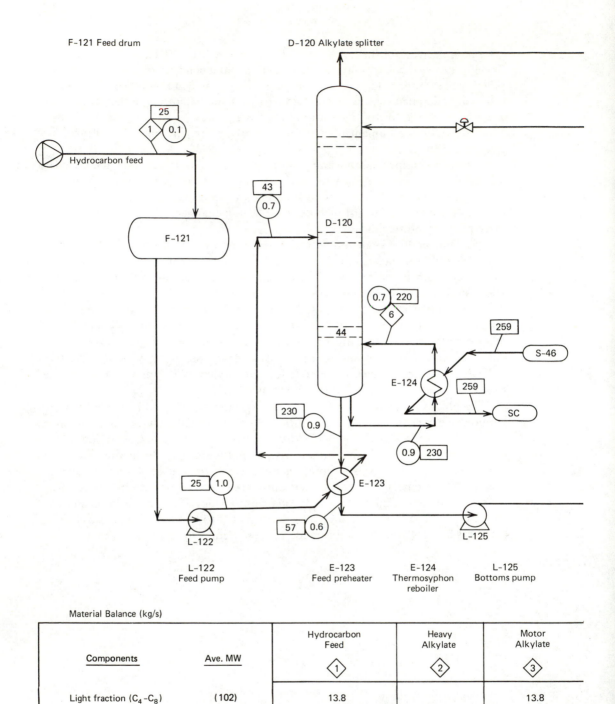

F-121 Feed drum   D-120 Alkylate splitter

Material Balance (kg/s)

Components	Ave. MW	Hydrocarbon Feed ◇1	Heavy Alkylate ◇2	Motor Alkylate ◇3
Light fraction (C_4–C_8)	(102)	13.8		13.8
Heavy fraction (C_{10}–C_{14})	(165)	1.65	1.65	
Total	(106)	15.45	1.65	13.8
Density at 25°C (kg/m^3)		706	760	700
Heat Capacity (kJ/kg · °C)		2.2	2.1	2.2
Latent Heat (kJ/kg)		294	246	300

Figure 5-62 Process flow diagram for alkylate splitter module.

E-127 Condenser   F-126 Condensate drum   E-129 Distillate cooler

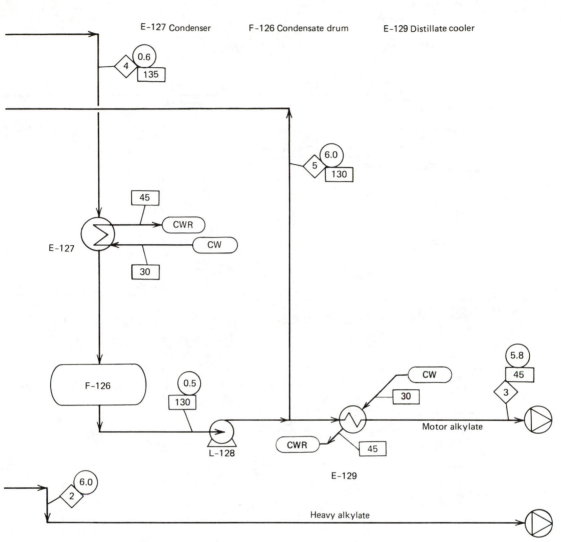

L-128
Distillate and
reflux pump

	Overhead Vapors ⬦4	Reflux ⬦5	Reboiler Flow ⬦6
	21.2	7.4	
			466
	21.2	7.4	**466**
	700	700	760
	2.2	2.2	2.1
	300	300	246

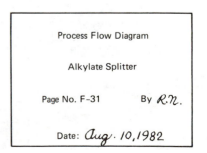

Process Flow Diagram

Alkylate Splitter

Page No. F-31   By *R.N.*

Date: *Aug. 10,1982*

TABLE 5-7
EQUIPMENT LIST

Job Title: Alkylate Splitter Module
Location: Eastern U.S.
Capacity: 3.92×10^8 kg/yr motor alkylate

(Flow Sheet Page Number: F-31)

Equipment Identification Number Name	Heat Exchangers				Process Vessel	
	E-123	E-124	E-127	E-129	D-120	
	Feed Preheater	Thermosyphon Reboiler	Condenser	Distillate Cooler	Alkylate Splitter	
Process Streams						
Stream number	1	6	4	3	1	
Name	Hydrocarbon feed	Reboiler flow stream	Overhead vapors	Motor alkylate	Hydrocarbon feed	
Process orientation	Tube side	Tube side	Shell side	Tube side	—	
Phase	Liquid	Saturated liquid → 7% vapor	Saturated vapor → saturated liquid	Liquid	Liquid	
Temperature (°C)						
In	25	230	135	130	43	
Out	43	230	135	45		
Pressure (barg)						
In	1.0	0.9	0.6	6.0	0.7	
Out	0.7	0.7	0.5	5.8		
Flow rate						
Mass (kg/s)	15.45	600	21.2	13.8	15.45	
Volumetric (m³/s)	0.022	0.79	0.030	0.020	0.022	
Concentration					$z_{lk} = 0.65$, $z_{hk} = 0.02$	
Enthalphy (kJ/kg)						
In	55	483	597	286	94.6	
Out	94.6	500	297	99		
Stream number	2	—			4	—
Name	Splitter bottoms	Steam-46 bara	Cooling water	Cooling water	Overhead vapors	Splitter bottoms
Process orientation	Shell side	Shell side	Tube side	Shell side		
Phase	Liquid	Vapor → Liquid	Liquid	Liquid	Saturated vapor	Saturated Liquid
Temperature (°C)						
In	230	259	30	30		
Out	57	259	45	45	135	230
Pressure (barg)						
In	0.9	45	3	3		
Out	0.6	~45	2.7	2.7	0.6	0.9
Flow Rate						
Mass (kg/s)	1.65	6.5	100	40	21.2	1.65
Volumetric (m³/s)	0.00217		0.10	0.04	0.030	0.00217
Concentration					$x_{hk} = 0.0001$	$x_{lk} = 0.01$
Enthalphy (kJ/kg)						
In	483	2797	126	126		
Out	120	1129	188	188	598	483
Heat transfer coefficient (J/m²·s·K)	400	480	450	425		
Efficiency	98%	95%	100%	100%	Tray efficiency 6.5 = 50%	
Heating Duty (kJ/s)	600	10,300	6360	2580	~1000 (losses)	
LMTD (°C)	88	29	97	40		
MTD (°C)	88	29	97	40		
F_T	0.90	1.0	1.0	1.0		
$\Delta T_m = \text{MTD} \times F_T$ (°C)	79	29	97	40		
Utilities						
Electricity (kW)						
Cooling water (m³/s)			0.10	0.04	(see E-127)	
Steam (kg/s) [pressure (barg)]		6.5 [45]			(see E-124)	
Equipment Size						
Length or height, L (m)					30	
Width or diameter, D (m)					3	
Surface area, A (m²)	20	740	150	150		
Volume, V (m³)						
Design pressure (barg)	<10	<50, tubes; <10, shell	<10	<10	<4	
Shaft power, w_s (kW)						
Orientation	Horizontal	Vertical	Horizontal	Horizontal	Vertical	
Material of construction	Carbon steel	Carbon steel	Carbon steel	Carbon steel	Carbon steel	
Other specifications	Two-pass Internally bolted floating head shell and tube exchanger	Single pass Internally bolted floating head shell and tube exchanger	Single pass Internally bolted floating head shell and tube exchanger	Single pass Fixed tube shell and tube exchanger	44 ss sieve trays Mist eliminator (D = 3 m)	

Equipment Type				
Pumps			Drums	
L-122	L-125	L-128	F-121	F-126
Feed Pump	Bottoms Pump	Distillate and Reflux Pump	Feed Drum	Condensate Drum
1	2	4	1	4
Hydrocarbon feed	Splitter bottoms	Overhead condensate	Hydrocarbon feed	Overhead condensate
—	—	—	—	—
Liquid	Liquid	Liquid	Liquid	Liquid
25	57	130	25	130
0.1 1.0	0.6 6.0	0.5 6.0	0.1	0.5
15.45 0.022	1.65 0.00217	21.2 0.030	15.45 0.022	21.2 0.030
55	120	297		
$\epsilon_i = 0.70$ $\epsilon_d = 0.77$	$\epsilon_i = 0.70$ $\epsilon_d = 0.73$	$\epsilon_i = 0.70$ $\epsilon_d = 0.88$		
3.6	2.3	27		
			8	4.7
			2.7	1.6
			60	9
<10	<10	<10	<4	<4
2.8	1.7	24		
Horizontal			Horizontal	
Cast iron			Carbon steel	
Standard low pressure centrifugals			$\theta = 2000s$ $L/D = 3$	$\theta = 300 s$ $L/D = 3$

TABLE 5-8
CAPITAL COST SUMMARY

by G. U.

Date: Sept. 3, 1983

Job Title: Alkylate Splitter Module

Location: Eastern United States

Effective Date to Which Estimate Applies: mid-1984

Capacity: 3.92×10^8 kg/yr Motor Alkylate

Cost Index Type: CE Plant Cost

Cost Index Value: 350 projected

Equipment Identification	Number	Capacity or Size Specifications	Purchase Equipment Cost (base material) $C_P^{cs}, 1982 \times 350/315 = C_P^{cs}, 1984$	Base Bare Module Factor, F_{BM}^{cs}	Base Bare Module Cost, C_{BM}^{cs}	Total
Heat Exchangers						
Feed preheater (floating head)	E-123	20 m^2	$5000 \times 1.11 = 5500$	3.2	18k	
Thermosyphon reboiler (floating head)	E-124	740 m^2	$42,000 \times 1.11 = 46,200$	3.3[a]	152k	
Condenser (floating head)	E-127	150 m^2	$16,000 \times 1.11 = 17,600$	3.2	56k	
Distillate cooler (fixed tube)	E-129	150 m^2	$12,000 \times 1.11 = 13,200$	3.2	42k	
Total exchangers						$ 268k
Process Vessel						
Alkylate splitter	D-120	$L = 30$ m, $D = 3$ m	$110,000 \times 1.11 = 121,000$	4.2	508k	
		44 ss sieve trays	$44 \times 800 \times 1.11 = 38,700$	1.2	46k	
		Mist eliminator	$800 \times 1.11 = 1,000$	1.2	1k	
Total process vessel						$ 555k
Pumps						
Feed pump (centrifugal)	L-122	$\dot{w}_s = 2.8$ kW	$4200 \times 1.11 = 4,600$	3.2	15k	
Bottoms pump (centrifugal)	L-125	$\dot{w}_s = 1.7$ kW	$3500 \times 1.11 = 3,900$	3.2	12k	
Distillate and reflux plump (centrifugal)	L-128	$\dot{w}_s = 24$ kW	$10,000 \times 1.11 = 11,000$	3.2	35k	
Total pumps						$ 62k
Drums						
Feed drum	F-121	$L = 8$ m, $D = 2.7$ m	$21,000 \times 1.11 = 23,100$	3.1	72k	
Condensate drum	F-126	$L = 4.7$ m, $D = 1.6$ m	$9,000 \times 1.11 = 9,900$	3.1	31k	
Total drums						$ 103k
Total bare module (carbon steel)						$ 988k
Contingency and fee					$988k \times 0.18 =$	178k
Total module cost						$1166k
Auxiliary facilities					$1166k \times 0.30 =$	350k
Grass-roots capital						$1516k

[a]Bare module factor is slightly larger for the reboiler because of higher internal tube pressure.

REFERENCES

1   Guthrie, K.M., "Data and Techniques for Preliminary Capital Cost Estimating," *Chem. Eng.*, pp. 114–142 (Mar. 24, 1969).

2   Guthrie, K.M., *Process Plant Estimating, Evaluation and Control*, Craftsman, Solano Beach, Calif. (1974).

3   Hall, R.S., J. Matley, and K.J. McNaughton, "Current Costs of Process Equipment," *Chem. Eng.*, pp. 80–116 (Apr. 5, 1982).

4   Kohn, P.M., "CE Cost Indexes Maintain 13-Year Ascent," *Chem. Eng.*, pp. 189–190 (May 8, 1978).

5   Matley, J., "CE Plant Cost Index—Revised," *Chem. Eng.* pp. 153–156 (Apr. 19, 1982).

6   Merrow, E.W., K.E. Phillips, and C.W. Myers, *Understanding Cost Growth and Performance Shortfalls in Pioneer Process Plants*, Rand Corporation, Santa Monica, Calif. (1981). See also *Chem. Eng.*, pp. 41–45 (Feb. 9, 1981).

7   Meyer, W.S., and D.L. Kime, "Cost Estimation for Turbine Agitators," *Chem. Eng.*, pp. 109–112 (Sept. 27, 1976).

8   Perry, J.H., and C.H. Chilton, *Chemical Engineers' Handbook*, 5th edition, McGraw-Hill, New York (1973).

9   Peters, M.S., and K.D. Timmerhaus, *Plant Design and Economics for Chemical Engineers*, 3rd edition, McGraw-Hill, New York (1980).

10   Pikalik, A., and H.E. Diaz, "Cost Estimation for Major Process Equipment," *Chem. Eng.*, pp. 107–122 (Oct. 10, 1977).

11   Vatavuk, W.M., and R.B. Neveril, *Chem. Eng.*, pp. 129–132 (July 12, 1982). This is Part 12 of a series. Preceding articles are indexed in this one.

12   Woods, D.R., S.J. Anderson, and S.L. Norman, "Evaluation of Capital Cost Data: Offsite Utilities," *Can. J. Chem. Eng.,* **57,** pp. 533–565 (October 1979).

PROBLEMS

Time frame for all estimates to be defined by instructor or reader.

5-1   Estimate the contribution that the steam boiler of Illustration 3-1 would make to total capital in a plant where it is located.

5-2   Estimate the grass-roots capital for the maple syrup process of Problem 3-2.

5-3   Estimate capital cost of the hydrogen chloride module of Problem 3-3.

5-4   Estimate the grass-roots capital for the multiple-effect maple syrup plant of Problem 3-4.

5-5   Estimate grass-roots capital for the coal-fired power plant of Problem 3-5. Do not use Figure 5-8 to arrive at an answer, but compare your result with that source. Note that electric generators are essentially equivalent to electric motors in cost. Assume that cost curves for turbines and generators can be extrapolated at constant slope.

5-6   Estimate capital cost of the nitric acid module of Problem 3-6.

5-7   Estimate grass-roots capital for the synthesis gas plant of Problem 3-7.

5-8   Estimate grass-roots capital for the Kraft pulping plant of Problem 3-8.

5-9   Estimate grass-roots capital for the coal gasification plant of Problem 3-9.

Chapter Six
MANUFACTURING COST ESTIMATION

In the taxicab illustration of Chapter Five (p. 266), it was taken for granted that a successful transportation business could not be conducted without rolling stock such as cabs or buses. By the same token, it should be obvious that operators, fuel, equipment, maintenance, garages, management, and other factors are also necessary elements of a healthy operation.

The translation of operating expenses such as gasoline, oil, tires, and labor into an annual cost is a straightforward application of arithmetic. The relation between capital cost and operating expense is not as obvious. Why, for example, does it cost 30 cents to travel one kilometer in a taxi when the operating expenses amount to only 15 or 20 cents? It doesn't take much experience with capitalism to realize where this difference lies. The original investment must be repaid by the owner so that vehicles can be replaced when they wear out. In addition, such expenses as maintenance, repairs, taxes, and insurance are related to the initial cost of the vehicle. In a typical chemical process, one fifth to one third of original investment must be recovered annually to satisfy these demands.

To illustrate various manufacturing expenses and how they are determined, refer to the cost summary sheet of Table 6-1, which is patterned after that of Holland et al. [3]. Although this traditional balance sheet used in elementary accounting has limitations as a sophisticated index of profitability, it is certainly adequate for defining most engineering alternatives, and it is a basic tool in refined higher order accounting procedures. Engineering aspects are emphasized here and in Chapter Seven. Less tangible economic influences and their implications are discussed in Chapter Eight.

FIXED CAPITAL, WORKING CAPITAL, AND TOTAL CAPITAL

Preliminary information, as given at the top of Table 6-1, is required to define the project and the time, usually future, to which an estimate applies. Next are listed the so-called capital expenditures. The dictionary alternative that most appropriately defines "capital" as used here is "accumulated possessions calculated to bring an income." *Fixed capital* is financially immobile. Usually, it is also physically immobile (although not necessarily, as illustrated by the taxicab). "Fixed" means

TABLE 6-1
MANUFACTURING COST SUMMARY

Page _____ of _____

By _____

Date _____

Job Title _____

Location _____ Annual Capacity (kg/yr) _____

Effective Date to Which Estimate Applies _____ Cost Index Type _____

Cost Index Value _____

	$/yr	$/kg
Fixed capital, C_{FC}		
Working capital (10–20% of fixed capital), C_{WC} _____		
Total capital investment, C_{TC}		
Manufacturing Expenses		
Direct		
Raw materials	....	....
By-product credits	(....................)	(....................)
Catalysts and solvents	....	....
Operating labor	....	....
Supervisory and clerical labor (10–20% of operating labor)	....	....
Utilities		
Steam ____ barg @ ____ $/kg		
____ barg @ ____ $/kg	....	....
Electricity @ ____ $/kWh	....	....
Process Water @ ____ $/m^3	....	....
Demineralized water @ ____ $/m^3	....	....
Cooling water @ ____ $/m^3	....	....
Waste disposal @ ____ $/kg^3	....	....
Maintenance and repairs (2–10% of fixed capital)	....	....
Operating supplies (10–20% of maintenance and repairs)	....	....
Laboratory charges (10–20% of operating labor)	....	....
Patents and royalties (0–6% of total expense)	_____	_____
Total, A_{DME}	....	....
Indirect		
Overhead (payroll and plant), packaging, storage (50–70% of the sum of operating labor, supervision, and maintenance)	....	....
Local taxes (1–2% of fixed capital)	....	....
Insurance (0.4–1% of fixed capital)	_____	_____
Total, A_{IME}	....	_____ _____
Total manufacturing expense (excluding depreciation), A_{ME}		....
Depreciation (approximately 10% of fixed capital), A_{BD}		....
General Expenses		
Administrative costs (25% of overhead)	....	....
Distribution and selling costs (10% of total expense)	....	....
Research and development (5% of total expense)	_____	....
Total, A_{GE}	....	_____ _____
Total expense, A_{TE}	....	....
Revenue from sales (_____ kg/yr @ _____ $/kg), A_S	_____	_____ _____
Net annual profit, A_{NP}	....	....
Income taxes (net annual profit times the tax rate), A_{IT}	(_____)	(_____)
Net annual profit after taxes ($A_{NP} - A_{IT}$) A_{NNP}	....	....
Aftertax rate of return, $i = ([A_{NNP} + A_{BD}]/C_{TC})$ $\times 100 =$ ____ %		

that the money, once spent, cannot be quickly converted back into cash or some other asset. To be more concrete, this is either the total module C_{TM} or grass-roots cost C_{GR}, depending on whether the project is for a new plant or an addition. For the alkylation splitter of Illustration 5-1, fixed capital is $1.52M, the total grass-roots price as derived in the capital cost summary sheet.

Working capital is what must be invested to get the plant into productive operation that is, money invested before there is a product to sell. Theoretically, in contrast to fixed capital, this money is not lost forever and can be recovered when the plant is closed down. For this reason, it is treated separately. Typically, working capital can be approximated as the value of 1 month's raw materials inventory and 2 or 3 months' product inventory. For predesign estimation, an assumed value between 10 and 20 percent of fixed capital is acceptable. This value should be nearer 10 percent when raw materials are cheap or the plant is unusually expensive (i.e., constructed of expensive alloy materials). The figure will be closer to 20 percent for opposite conditions and about 15 percent for a typical operation. For the alkylate splitter, working capital is assumed to be 15 percent of fixed capital or $230k.

Total capital, the simple sum of fixed and working capitals, represents the amount of money that must be provided by investors. Land, which is obviously a capital item, has been disregarded. Compared with fixed and working capital for a manufacturing plant, land costs are usually negligible.

Capital cost figures serve not only to identify and characterize the project, but they influence operating costs in several direct and indirect ways. This will be obvious as we continue our examination of the balance sheet.

MANUFACTURING EXPENSES
Direct Manufacturing Expenses

Costs in the direct manufacturing category are comprised of those due to materials or labor that either are physically in the product or have come in tangible contact with it during its evolution.

Raw materials comprise the most obvious direct manufacturing cost. They can be identified quickly by scanning the left-hand margin of a properly executed flow sheet. Quantities are taken directly from the material balance table. Since material balance flow rates are instantaneous values, a correction factor must be applied to convert from a one-second to a one-year basis, to acknowledge that no plant will operate continuously without interruption. This multiplier, known as the *operating factor* f_o, represents the fraction of time that a plant is in equivalent full-scale production. Competitive influences and high capital costs dictate the need for large operating factors in modern process plants. The devastating effect of this requirement is quite apparent in economic evaluations of small seasonal operations such as many in the food-processing industry. Process plants usually are designed for continuous, 24-h-day operation. With current engineering technology, operating factors falling between 85 and 95 percent are common. For academic purposes, 90 percent is a logical number. (In any economic evaluation, be certain that you have management concurrence on the factor that you use.)

Raw material prices usually are provided as part of the project assignment from management. Otherwise, prices for most chemical commodities and intermediates can be found in the *Chemical Marketing Reporter* (issued weekly by Schnell

Publishing Co., New York). Prices for refined hydrocarbons are reported weekly in the "Statistics" section of *The Oil and Gas Journal*. Since raw materials often represent the largest single expense on a balance sheet, magnitudes and sources of prices should be clearly explained when results are reported.

In the balance sheet, there are two columns, one for annual dollar values and one for unit values, usually in dollars per kilogram. The latter is derived from the former, dividing by the annual capacity noted at the top of the sheet.

By-product credits are assessed by the same technique as that employed for raw materials except that by-products exit with products, at the right-hand margin of the flow sheet. As with raw materials, multiplication by an operating factor f_o and the number of seconds in a year (31.5 million) is necessary for conversion to an annual credit. This and the unit credit are enclosed by parentheses in the balance sheet because they are opposite in sign from prevailing expenses. The difference between a product and a by-product is not always clear. Since both appear on the balance sheet, albeit in different places, profitability is unaffected by the distinction, and the question is an academic one, which you may decide.

ILLUSTRATION 6-1*a* ALKYLATION UNIT RAW MATERIALS AND CREDITS

Estimate raw materials costs and by-product credits for the alkylate splitter of Illustration 5-1.

The purpose of this process module is to produce automotive motor fuel as the primary product and heavy alkylate as a by-product. Hydrocarbon feed is the only raw material. Assuming that the feed is equivalent in price to aviation fuel, we find the mid-1982 price is \$253/m^3 (from the "Statistics" section of *The Oil and Gas Journal*). Assuming that prices will rise 10 percent per year, the projected mid-1984 price is \$306/m^3. This yields an annual raw materials cost of

$$A_{RM} = \frac{C_{S,1}\,\dot{m}_1\,(31.5 \times 10^6\ \text{s/yr})\,f_o}{\rho_1} = \frac{(\$306/\text{m}^3)(15.45\ \text{kg/s})(31.5 \times 10^6\ \text{s/yr})(0.90)}{(706\ \text{kg/m}^3)}$$

$$= \$190\text{M/yr}$$

Divided by the production rate, the unit raw materials cost per kilogram of product is:

$$\frac{\$190 \times 10^6/\text{yr}}{392 \times 10^6\ \text{kg/yr}} = \$0.485/\text{kg}$$

Credit for heavy alkylate is based on a mid-1982 price of \$243/m^3 reported for distillate fuel in *The Oil and Gas Journal*. Escalated at 10 percent per year, the estimated mid-1984 price is \$294/m^3. By-product credit is calculated similarly as follows.

$$A_{BPC} = \frac{C_{S,2}\,\dot{m}_2\,(31.5 \times 10^6\ \text{s/yr})\,f_o}{\rho_2} = \frac{(\$294/\text{m}^3)(1.65\ \text{kg/s})(31.5 \times 10^6\ \text{s/yr})(0.90)}{(760\ \text{kg/m}^3)}$$

$$= \$18.1\text{M/yr}$$

The unit credit, per kilogram of motor alkylate product, is:

$$\frac{\$18.1 \times 10^6/\text{yr}}{392 \times 10^6\ \text{kg/yr}} = \$0.046/\text{kg}$$

Catalysts and solvents, where used, must be replaced or regenerated. Adequate funds are provided in this category to pay for losses or depletion. The accounting technique is identical to that used for raw materials. There are no expenses in this category for the alkylation unit.

Operating labor refers to people who actually run the equipment. A census of the plant cafeteria or parking lot would reveal a significantly larger staff than is indicated by this category of the cost summary. The difference results from supervisory, maintenance, laboratory, and support personnel who are accounted for elsewhere. Enumeration of operating people is greatly aided by past experience with the same or similar processes. In the absence of more definitive data, a good estimate can be derived from the flow sheet, since each major equipment item requires a certain amount of operator attention. Recommended operator requirements are listed, according to generic equipment type, in Table 6-2. With modern, highly instrumented equipment, operating labor is insensitive to the size of a given device or vessel but is directly proportional to the number of such units. Unless a module conveniently adjoins other sections of the plant, it is impractical to employ a fraction of an operator, and the personnel estimate must be rounded up to the next highest integer value.

Values listed in Table 6-2 are based on continuous operation. Even though machines work 24 h/day, people, generally, do not. In a typical modern, intensive operation, each operating slot requires four people (three shifts per day plus the equivalent of one shift on the weekend). Since operators work even harder when equipment will not, salaries and wages are not multiplied by an operating factor.

Labor expenses entered under this category in the cost summary are direct wages based on an hourly rate. Fringe benefits and indirect labor reimbursement are entered later. The hourly wage rate varies from time to time and place to place. In the absence of more specific data, a typical U.S. rate of $12/h, $480/wk, or $25,000/yr, can be assumed, on a mid-1982 basis. To escalate for inflation, labor cost indices, like equipment cost indices, are tabulated and reported by a number of organizations. One of the most convenient and accurate is the Hourly Earnings Index for Chemicals and Allied Products prepared by the U.S. Bureau of Labor Statistics. It is reported on the "Economic Indicators" page of *Chemical Engineering* magazine. For obvious reasons, equipment and labor costs escalate at similar rates. Thus, the CE Plant Cost Index can also be used with reasonable confidence, to correlate inflationary changes in wage rates.

ILLUSTRATION 6-1b   ALKYLATION UNIT LABOR COSTS

Calculate annual and unit costs of operating labor for the alkylate splitter module of Illustration 5-1.

This module is essentially equivalent to a single distillation tower and its auxiliary equipment. Based on Table 6-2, the equipment is assumed to require one third of the attention of an operator. If the module were isolated from other modules, one full-time person would be required. However, this operation is integrated within a refinery so that "one third of an operator" is available. Considering operation is continuous (round the clock), four times this, or the equivalent of 1.3 persons, must be employed on an annual basis for this task. The annual cost, corrected to mid-1984, is thus estimated to be

$$\left(\frac{4}{3}\text{ person years}\right)(\$25,000/\text{yr})\left(\frac{350}{315}\right) = \$37,000/\text{yr}$$

TABLE 6-2

OPERATOR REQUIREMENTS FOR VARIOUS TYPES OF PROCESS EQUIPMENT

Generic Equipment Type	Operators per Unit per Shift
Auxiliary Facilities	
Air plants	1
Boilers	1
Chimneys or stacks	0
Cooling towers	1
Water demineralizers	0.5
Electric generating plants	3
Portable electric generating plants	0.5
Electric substations	0
Incinerators	2
Mechanical refrigeration units	0.5
Wastewater treatment plants	2
Water treatment plants	2
Conveyors	0.2
Crushers, mills, grinders	0.5–1
Drives and power recovery machines	—
Evaporators	0.3
Vaporizers	0.05
Furnaces	0.5
Gas Movers and Compressors	
Fans	0.05
Blowers and compressors	0.1–0.2
Gas–solids contacting equipment	0.1–0.3
Heat exchangers	0.1
Mixers	0.3
Process Vessels	
Towers (including auxiliary pumps and exchangers)	0.2–0.5
Drums	—
Pumps	—
Reactors	0.5
Separators	
Clarifiers and thickeners	0.2
Centrifugal separators and filters	0.05–0.2
Cyclones	—
Bag filters	0.2
Electrostatic precipitators	0.2
Rotary and belt filters	0.1
Plate and frame, shell and leaf filters	1
Expression equipment	0.2
Screens	0.05
Size-enlargement equipment	0.1–0.3
Storage vessels	—

The unit cost is:

$$\frac{\$37,000/\mathrm{yr}}{392 \times 10^6 \ \mathrm{kg/yr}} = \$0.0001/\mathrm{kg}$$

a rather low labor cost, which is typical of refineries and fluid-processing plants in general.

Estimates of *supervisory* and *clerical labor* are quite logically based on operating labor. The factor as indicated in Table 6-1 generally falls within the range of 0.10 to 0.20, depending on process complexity.

Utilities are identified by an oval-shaped flow sheet symbol (see Table 3-2). Electrical power consumption rates of motor-driven items, although sometimes not included on the flow sheet, are tabulated, along with other utilities, on the equipment list. Typical utilities include electricity, process steam, refrigerants, compressed air, cooling water, heated water, hot oil or molten salt, process water, demineralized water, municipal water, and river, lake, or ocean water. Waste disposal cost is also treated as a utility expense in Table 6-1. Unit costs depend on whether the plant is a customer or owner of the utility source. For example, most plants purchase electricity from an outside supplier. In this case, the price is easily defined by the specified rate. Cooling water, on the other hand, is usually processed on site, in an auxiliary tower. For a process module, where auxiliary facilities are not included in the fixed capital, the module is essentially a customer, purchasing utilities from the grass-roots plant.

Utility costs generally contain two separately escalating components. One is that comprised of materials and labor which inflate at a rate typified by the CE Plant Cost Index. The other is energy, which as we learned in the mid-1970s, can escalate at a much different rate. To account for these two factor, I have chosen to represent a utility unit cost as follows.

$$C_{S,u} = a \times \text{CE Plant Cost Index} + b \times C_{S,f} \tag{6-1}$$

The first coefficient a correlates capital- and labor-related expenses; b relates fuel price, $C_{S,f}$ to utility price $C_{S,u}$. Representative coefficients for common utilities are listed in Table 6-3. To illustrate, consider mid-1973: the CE Plant Cost Index was 144, energy from residual oil was $0.60/GJ ($4/barrel), and the cost of purchased power, estimated from Equation 6-1 with coefficients taken from Table 6-3, was:

$$C_{S,e,1973} = (\$1.3 \times 10^{-4}/\mathrm{kWh})(144) + (0.010 \ \mathrm{GJ/kWh})(\$0.60/\mathrm{GJ}) = \$0.025/\mathrm{kWh}$$

Using an index of 350 and an energy price of $5.90/GJ (number 6 fuel oil projected to mid-1984), the price of purchased electricity is estimated to be:

$$C_{S,e,1984} = (\$1.3 \times 10^{-4}/\mathrm{kWh})(350) + (0.010 \ \mathrm{GJ/kWh})(\$5.90/\mathrm{GJ}) = \$0.108/\mathrm{kWh}$$

Generated from coal at $1.90/GJ ($56/U.S. ton, eastern bituminous coal), the projected cost of power in 1984 would be $0.065/kWh.

In reality, costs of onsite power should be the same to the company for a module as for a grass-roots plant even though different values of a appear in Table 6-3. The reason for this is an artifact of bookkeeping techniques that assign additional capital to a grass-roots assessment and yield the same net result in the end.

TABLE 6-3

UTILITY COST COEFFICIENTS

The price of a given utility in dollars per unit, $C_{S,u}$, is calculated from the equation $C_{S,u} = a \times$ (CE Plant Cost Index) $+ b \times C_{S,f}$ where coefficients a and b are taken from this table, CE Plant Cost Index is the value (basis, 1958 = 100) at the effective date of the estimate and $C_{S,f}$ is the price of fuel used to generate the utility.[a] Representative price data are shown in Figure 6-1.

	Cost Coefficients	
	a	*b*
Electricity ($/kWh)		
Purchased from outside	1.3×10^{-4}	0.010
On-site power charged to process module	1.4×10^{-4}	0.011
On-site power charged to grass-roots plant	4.0×10^{-5}	0.011
Compressed and Dried Air ($/std m³)[b]	$0.1 < \dot{q} < 100$ std m³/s; $1 < p < 35$ bara	
Process module	$5.0 \times 10^{-5}\, \dot{q}^{-0.30} \ln p$	$9.0 \times 10^{-4} \ln p$
Grass-roots plant	$4.5 \times 10^{-5}\, \dot{q}^{-0.30} \ln p$	$9.0 \times 10^{-4} \ln p$
Instrument Air ($/std m³)[b]		
Process module	1.25×10^{-4}	1.25×10^{-3}
Grass-roots plant	1.10×10^{-4}	1.25×10^{-3}
Process Steam ($/kg)[c]	$1 < p < 46$ barg; $0.06 < \dot{m}_s < 40$ kg/s	
Process module	$\dfrac{2.7 \times 10^{-5}}{\dot{m}_s^{0.9}}$	$0.0020 p^{0.14}$
Grass-roots plant	$\dfrac{2.3 \times 10^{-5}}{\dot{m}_s^{0.9}}$	$0.0020 p^{0.14}$
Cooling Water ($/m³)[d]	$0.01 < \dot{q} < 10$ m³/s	
Process module	$0.0001 + \dfrac{3.0 \times 10^{-5}}{\dot{q}}$	0.0056
Grass-roots plant	$0.00007 + \dfrac{2.5 \times 10^{-5}}{\dot{q}}$	0.0056
Demineralized (boiler feed) Water ($/m³)[d]	$0.001 < \dot{q} < 1.0$ m³/s	
Process module	$0.007 + \dfrac{1.3 \times 10^{-5}}{\dot{q}}$	0.0022
Grass-roots plant	$0.00035 + \dfrac{1.2 \times 10^{-5}}{\dot{q}}$	0.0022
Drinking Water ($/m³)[d]	$0.01 < \dot{q} < 10$ m³/s	
Process module	$0.0007 + \dfrac{2.0 \times 10^{-6}}{\dot{q}}$	0.003
Grass-roots plant	$0.00035 + \dfrac{2.0 \times 10^{-6}}{\dot{q}}$	0.003
Natural Water, Pumped and Screened Only ($/m³)[d]	$0.01 < \dot{q} < 10$ m³/s	
Process module	$1 \times 10^{-4} + \dfrac{2 \times 10^{-7}}{\dot{q}}$	0.002
Grass-roots plant	$5 \times 10^{-5} + \dfrac{2 \times 10^{-7}}{\dot{q}}$	0.002
Waste Disposal ($/kg)		
Conventional solid or liquid wastes		
Process module	1×10^{-4}	—
Grass-roots	0.5×10^{-4}	—
Toxic or hazardous solids and liquids		
Process module	4×10^{-4}	—
Grass-roots	3×10^{-4}	—
Refrigerant ($/kJ cooling capacity)[e]	$200 < T < 300$ K; $1 < \dot{Q}_c < 1000$ kJ/s	
Process module	$\dfrac{6.0 \times 10^{6}}{\dot{Q}_c^{0.7}\, T^5}$	$\dfrac{1.1 \times 10^{6}}{T^5}$
Grass-roots plant	$\dfrac{180}{\dot{Q}_c^{0.9}\, T^3}$	$\dfrac{1.1 \times 10^{6}}{T^5}$

TABLE 6-3
UTILITY COST COEFFICIENTS (continued)

	Cost Coefficients	
	a	**b**
Wastewater Treatment ($/m³)[d]	$0.01 < \dot{q} < 10$ m³/s	
Primary (filtration)		
Process module	$1 \times 10^{-4} + \dfrac{2 \times 10^{-7}}{\dot{q}}$	0.002
Grass-roots plant	$5 \times 10^{-5} + \dfrac{2 \times 10^{-7}}{\dot{q}}$	0.002
Secondary (filtration and activated sludge processing)		
Process module	$7 \times 10^{-4} + \dfrac{2 \times 10^{-6}}{\dot{q}}$	0.003
Grass-roots plant	$3.5 \times 10^{-4} + \dfrac{2 \times 10^{-6}}{\dot{q}}$	0.003
Tertiary (filtration, activated sludge, and chemical processing)		
Process module	$1 \times 10^{-4} + \dfrac{3 \times 10^{-5}}{\dot{q}}$	0.005
Grass-roots plant	$5 \times 10^{-5} + \dfrac{3 \times 10^{-5}}{\dot{q}}$	0.005
Hot Water, Hot Oil, or Molten Salt Heat Transfer Media ($/kJ heating capacity)[f]	$350 < T < 850$ K; $100 < \dot{Q}_H < 20{,}000$ kJ/s	
Process module	$\dfrac{7.0 \times 10^{-7}\, T^{0.5}}{\dot{Q}_H^{0.9}}$	$6.0 \times 10^{-8} \sqrt{T}$
Grass-roots plant	$\dfrac{6.0 \times 10^{-7}\, T^{0.5}}{\dot{Q}_H^{0.9}}$	$6.0 \times 10^{-8} \sqrt{T}$

[a] $C_{S,f}$ is the price of fuel ($/GJ) based on the higher or gross heating value. For electrical power, compressed air, refrigerant, cooling water, and other auxiliary facilities where electricity is used to drive pumps and compressors, it is the price of fuel at the electric power station. For steam, it is the price of boiler fuel at the plant. Higher heating values of some representative fuels are as follows.

Fuel	Higher (Gross) Heating Value	Density
Bituminous and anthracite coals	26–32 MJ/kg	—
Lignite	17–19 MJ/kg	—
Wood	19–22 MJ/kg	—
Number 2 fuel oil	38 GJ/m³	870 kg/m³
Number 6 (residual) fuel oil	42 GJ/m³	970 kg/m³
Gasoline	37 GJ/m³	700 kg/m³
Natural gas	38 MJ/std m³	0.715 kg/std m³

For more extensive information, see Perry Section 9 and Ganapathy [1].

[b] Coefficients apply to ranges of $\dot{q}$ and p indicated where $\dot{q}$ is total auxiliary air plant capacity (std m³/s) and p is delivered pressure of air (bara).

[c] Use price of fuel burned in the boiler for $C_{S,f}$; $\dot{m}_s$ is total auxiliary boiler steam capacity (kg/s).

[d] $\dot{q}$ is total auxiliary water capacity (m³/s).

[e] $\dot{Q}_C$ is total auxiliary cooling capacity (kJ/s).

[f] $\dot{Q}_H$ is total auxiliary heating capacity (kJ/s).

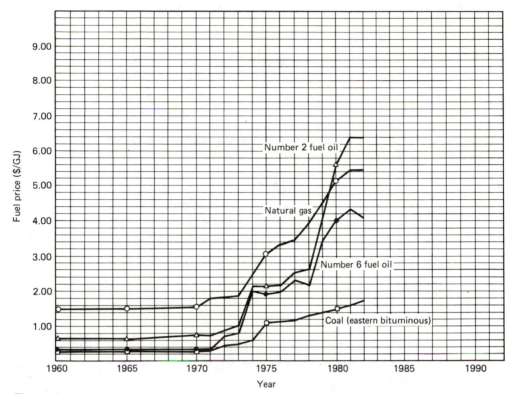

Figure 6-1 Representative prices of fuels on an energy-equivalent basis. Prices are for delivered fuels in the eastern United States. (*Source*: *The Oil and Gas Journal*, Boston Gas Company, and Public Service Company of New Hampshire.)

According to my analysis, purchased power is cheaper than onsite power if fuel prices are equivalent. The profit and general expenses required by an outside utility are more than balanced by other economies available to a large central power station. (This comparison was based on a 10 MW onsite plant and a 400 to 1000 MW central power station.) This supports the rule of thumb mentioned in Chapter Four; that is, self-generation of electricity is not advantageous unless cheap fuel is available or electricity can be cogenerated with process steam.

Costs for numerous common utilities can be estimated from appropriate coefficients taken from Table 6-3. In the absence of more specific energy prices, data for number 6 fuel oil, number 2 fuel oil, natural gas, and eastern bituminous coal are illustrated in Figure 6-1. These are typical of the northeastern United States. Lower fuel prices are available at other U.S. plant sites.

ILLUSTRATION 6-1*c*   UTILITY COSTS FOR ALKYLATE SPLITTER

Calculate the annual and unit costs of utilities for the alkylate splitter module of Illustration 5-1.

Quantities of utilities can be extracted from the equipment list of Table 5-7. Summing the numbers, the plant consumes 33 kW of electricity, 0.14 m^3/s of cooling water, and 6.5 kg/s of 46 bara steam.

Assuming that electricity can be generated on site with fuel energy equivalent to that contained in the by-product heavy alkylate, $C_{S,f}$ is calculated as follows.

$$C_{S,f} = \frac{\$243/m^3}{38 \text{ GJ}/m^3} = \$6.40/\text{GJ}$$

This is a higher price than is projected for residual fuel oil. Therefore, assume electricity is purchased offsite at $0.108/kWh as estimated above. Annual utility costs for electricity in the alkylate splitter module are given by:

$$A_e = (31.5 \times 10^6 \text{ s/yr}) f_o P C_{S,e} (\text{h}/3600 \text{ s})$$

$$= (31.5 \times 10^6 \text{ s/yr})(0.90)(33 \text{ kW})(\$0.108/\text{kWh})(\text{h}/3600 \text{ s})$$

$$= \$2800/\text{yr}$$

Since this is part of a large refinery, total cooling water consumption will probably be near the maximum for a single tower. In effect, the refinery may have several towers. To establish utility price, $\dot{q}$ is given a value of $10 \text{ m}^3/s$ to establish the first coefficient in Equation 6-1. Using figures from Table 6-3, $C_{S,cw}$ is calculated as follows.

$$C_{S,cw} = \left(0.00007 + \frac{2.5 \times 10^{-5}}{10} \right)(350) \text{ \$/m}^3 + (0.0056 \text{ GJ}/m^3)(\$5.90/\text{GJ})$$

$$= \$0.060/m^3$$

(The grass-roots coefficient was used, since grass-roots capital will be used for this illustration on the cost summary sheet.) Annual cooling water cost is calculated as for electricity.

$$A_{cw} = (3.15 \times 10^6 \text{ s/yr})(0.90)(0.14 \text{ m}^3/s)(\$0.060/m^3)$$

$$= \$2400/\text{yr}$$

For steam, auxiliary plant capacity is assumed to be the maximum and residual oil at $5.70/GJ is the postulated energy source. From data in Table 6-3, we can write:

$$C_{S,s-45} = (2.3 \times 10^{-5})(40)^{-0.9}(350) \text{ \$/kg} + (0.0020)(45)^{0.14}(\text{GJ}/\text{kg})(\$5.90/\text{GJ})$$

$$= \$0.020/\text{kg}$$

and

$$A_s = (31.5 \times 10^5 \text{ s/yr})(0.90)(6.5 \text{ kg/s})(\$0.020/\text{kg})$$

$$= \$3,690,000/\text{yr}$$

This result supports our general impression that reboiler energy is the overwhelming utility expense in a distillation operation.

Maintenance and repairs constitute an important and necessary budget item in any healthy manufacturing operation. Assuming that the plant is well designed and constructed, maintenance expenses will be proportional to the operation's size, scale, and complexity. Accordingly, annual maintenance expenses are assessed as a fraction, usually 2 to 10 percent, of fixed capital. This automatically yields higher maintenance expenses, as one would expect, for corrosive environments constructed of special materials. The range, 2 to 10 percent, is rather broad. Low maintenance costs pertain to well-established, relatively simple processes. Higher percentages should be assumed with unconventional or speculative processes. For most estimates, 6 percent is a reasonable assumption.

It is important to know that a large fraction of maintenance money, usually more than half, goes toward salaries. This compensates people one sees around a plant who are not accounted for elsewhere on the cost summary sheet. The remainder for maintenance is for new parts, tools, and equipment to replace worn-out components. This is the first of several annual manufacturing expenses that depend on fixed capital. It can be conveniently computed from the value of C_{FC} heading the cost summary sheet. Annual maintenance expense is entered in the first column of the sheet. Unit costs in the second column are derived from this figure when divided by annual capacity.

Operating supplies include replaceable materials such as instrument charts, lubricants, custodial supplies, and other items not considered as part of regular maintenance. These are found to vary with fixed capital and are, thus, proportional to maintenance and repair expenses. A value between 10 and 20 percent of maintenance is recommended in Table 6-1.

Laboratory expenses result from quality control testing and chemical or physical analyses necessary to certify product purity and viability or to identify faulty processing. The extent and cost of laboratory operations depend primarily on the complexity and sophistication of the process. Another cost that depends on the same variables is operating labor. Hence, laboratory expenses are conveniently factored from this item, which precedes it on the cost summary. Typical laboratory costs range from 10 to 20 percent of operating labor yielding a figure similar to that for supervisory and clerical labor.

Expenses for patents and royalties are accrued in any process that is licensed from another firm. Since many companies have a patent base of their own or use technology that is more than 17 years old (i.e., the patents have expired), royalty expenses are not incurred in many cases. Where necessary, the fee is usually based on sales income and may represent as much as 6 percent of total expense. Usually, the exact fee is known at the outset. When it is not but is anticipated, a value equal to 3 percent of total expense is reasonable for preliminary estimation. Since A_{TE} has not been assessed at this stage of the estimate, this item must be passed over until other costs have been itemized. Then a subtotal is computed that represents a known fraction of total expense, total expense is then computed and a royalty fee back-calculated.

Indirect Manufacturing Expenses

Summation of the costs enumerated above yields a total for *direct* manufacturing expenses. So-called *indirect* expenses are best identified by defining the items themselves as listed in Table 6-1. In many accounting techniques depreciation is included in this category, but because of it unusual impact on cash flow and taxes, it is listed separately here. This agrees with the approach recommended by Holland et al. [3].

Overhead and other miscellaneous expenses form an important component of indirect manufacturing expenses. Prominent in this category are fringe benefits, social security, unemployment insurance, and other compensation paid indirectly to plant personnel. This amounts to about 60 percent of direct salaries. Since a large fraction of maintenance expense is also salaries, overhead costs are based on the sum of operating labor, supervision, and maintenance. A range of 50 to 70 percent of this sum is found to represent overhead and miscellaneous expenses adequately.

Other indirect expenses include *local property taxes* and *insurance*. These are, in fact, assessed on the cost of the plant. They usually fall within the percentage ranges shown in Table 6-1. A total for these two items equivalent to 2 pecent of fixed capital is typical.

For facility in cash flow analysis, as discussed in Chapter Eight, the sum of total direct and indirect manufacturing expenses (excluding depreciation) is necessary. This is entered in the middle column of Table 6-1 for convenient access later.

Depreciation, according to Webster, denotes loss in value. In chemical processing, a well-designed and properly maintained plant does not necessarily wear out and decline in value like an automobile. Instead, the equipment is usually maintained in good condition, components are replaced periodically, and the plant retains its value as a production tool until advancements in technology justify replacement by a more modern and efficient design. Although rapid depreciation may not reflect physical reality, as a financial element, it is extremely important in process economics.

Holland et al. [3] present four common chemical engineering definitions of depreciation.

1　A tax allowance.

2　A cost of operation.

3　A means of building up a fund to finance plant replacement.

4　A measure of falling value.

The fourth, in effect, is that of Webster and least reflects process economics. Depreciation, as a means of building up a replacement fund, is likewise inaccurate in chemical plants because an obsolete plant is seldom replaced with one of the same design. In a conventional balance sheet such as Table 6-1, depreciation appears as an operating cost, in keeping with the second definition. This is the concept that we employ for preliminary analysis and optimization purposes because of its simplicity and utility. To financial managers, depreciation has more significance as a tax allowance. As you will note, depreciation, like profit, returns directly to the investor's purse. The difference is that depreciation is not slashed by taxation, which diverts approximately half of net profit to the government. Given no legal constraints, an investor would prefer to attribute all excess income to depreciation and generate no taxable profit. This, of course, is not allowed by the government, and strict rules regarding depreciation are in effect throughout the world.

In essence, only the fixed capital investment can be claimed as depreciation, and this must be spaced over a set period of time. With straight line depreciation, so-called, the annual amount A_{BD} (where BD denotes *book-value depreciation*) is merely the difference between fixed capital C_{FC} and salvage value S divided by the depreciating time period s.

$$A_{BD} = \frac{C_{FC} - S}{s} \qquad (6\text{-}2)$$

Traditionally, the U.S. chemical industry has used a value of $s = 10$ yr. Since no one wants an obsolete plant, the salvage value is generally negligible, and annual straight line depreciation over a 10 yr period is essentially 10 percent of fixed capital.

Aware of potential income from their capital and fearing loss of value due to inflation, entrepreneurs usually want to recoup their money as quickly as possible. To encourage investment, governments often allow accelerated depreciation schedules that are nonlinear. Two, known as the declining balance and double-declining balance methods, are in common use but are not discussed here. Another, the "sum-of-the-years'-digits" method, yields a similar result and is conceptually and computationally more tangible. In this method, the annual depreciation is calculated as the ratio of years left in a project divided by the sum of all the years. This is best described by an example.

Assume that project lifetime s_0 is 10 years. The remaining lifetime s_1, during year 1 is, thus, 10. For year two, s_2, it is 9, for year 3 s_3, it is 8, and so on, till year 10, when its remaining lifetime is 1 yr, or

$$s_j = s_0 + 1 - n_j \tag{6-3}$$

where n_j is the age. The sum of years digits is:

$$\sum_s = s_0 + (s_0 - 1) + (s_0 - 2) + \ldots + 1 = \frac{s_0(s_0 + 1)}{2} \tag{6-4}$$

$$\sum_{10} = 10 + 9 + 8 + \ldots + 1 = \frac{10(11)}{2} = 55 \tag{6-5}$$

During a given year, the fractional depreciation would be

$$\frac{s_j}{\Sigma_s} = \frac{s_0 + 1 - n_j}{\Sigma_s} \tag{6-6}$$

For the first year of 10, this is

$$\frac{S_1}{\Sigma_{10}} = \frac{10 + 1 - 1}{55} = \frac{10}{55} = 18\%$$

This drops to 16 percent the second year, 15 percent the third, 13 percent the fourth, and so on: 11, 9, 7, 5, 4, and finally 2 percent in the final year to yield 100 percent recovery.

With this brief discussion, we return to the cost summary sheet where a simple straight line value of 10 percent per year is the normal assumption. If capital and operating costs are assessed properly by us, financial people, with their computer programs, can readily manipulate the figures for more sophisticated investment analysis.

GENERAL EXPENSES

In addition to direct and indirect manufacturing expenses, a certain portion of corporate management cost, sales expense, and research effort must be financed from plant revenue. These activities, supported from plant general expenses, are usually conducted in central corporate headquarters, often remote from the plant. *Administrative costs* are proportional to the plant staff and can be scaled conveniently as a fraction, approximately 25 percent, of overhead. *Research and development (R & D), distribution, and selling* budgets are usually based on product value. Ten percent of total expense for distribution and selling costs plus 5 percent for R & D represent realistic estimates for most chemical products. As with patents and royalties, these are assessed last on the summary sheet. Assuming 3 percent of total expense for patents and royalties, 10 percent for distribution and selling, and 5 percent for R & D, all other manufacturing and general expenses would amount to 82 percent of the total expense. Division of this subtotal by 0.82 yields A_{TE}, from which the unknown expenses can be factored.

In caution, note that the percentages just cited and listed in Table 6-1 are typical but not universal values. Distribution and selling expenses for electricity may differ greatly, for example, from those of penicillin or gasoline. If the aid of experts or experience is lacking, these percentages should be used with care.

SALES REVENUE, PROFIT, AND TAXES

With total production expenses estimated, the profitability or sales price of a product can be determined. If the sales price is defined, either by existing markets or projections, the annual revenue is determined by multiplying unit price by annual capacity. The difference between this and total expense is net annual profit (or loss) A_{NP}. Assuming that there is a profit, income taxes, A_{IT} are estimated from the tax rate t.

$$A_{IT} = tA_{NP} \tag{6-7}$$

The real profit (net after taxes A_{NNP}) is:

$$A_{NNP} = A_{NP} - A_{IT} = A_{NP}(1 - t) \tag{6-8}$$

Historically, in recent U.S. experience except for wartime, corporate tax rates have ranged within a few percentage points of 50 percent. In effect, except for very small companies that receive a tax incentive, net profit after taxes and income taxes are the same, equal to half the net profit before taxes.

Profitability

Quite often, the engineer is not given a selling price but is asked what the selling price should be for a reasonable profit. This implies another question, What is a reasonable profit? There are numerous ways to assess profit, some having less credibility than others. The most meaningful answer is derived from cash flow analysis as discussed in Chapter Eight. Another measure of profitability is derived simply from the manufacturing cost summary as a rate of return. In a direct way, one should ask what is the rate of return or simple annual interest on the investment. The initial total investment is clearly identified as total capital C_{TC}. Net annual profit after taxes is the return. Simple annual interest or aftertax rate of return is given by:

$$i' = \frac{A_{NNP}}{C_{TC}} \times 100 \tag{6-9}$$

This yields a desceptively low result, since because of depreciation credits, the total capital investment declines with time.

Holland et al. [4] identify four different methods used by accountants to define the rate of return on original investment. All employ total capital as the denominator. For the numerator, one definition uses net profit after taxes, a second employs net profit before taxes A_{NP}, and a third considers the sum of net profit before taxes plus depreciation $A_{NP} + A_{BD}$ as the return. The fourth uses cash income after tax. (This is equal to net profit after taxes plus depreciation: $A_{NCI} = A_{NNP} + A_{BD}$.) The authors recommend the last definition. This is the one employed in Table 6-1.

$$i = \left(\frac{A_{NNP} + A_{BD}}{C_{TC}} \right) 100 \qquad (6\text{-}10)$$

Viewed as a traditional financial investment like a mortgage or loan, one would consider depreciation as repayment on the loan. The loan principal would, therefore, decline with time, and net profit after taxes would represent the interest payment. To simulate the traditional interest rate, rate of return on *average* investment is more appropriate, where average investment is approximated as half the total capital. Such a rate of return is given by:

$$i'' = \left(\frac{2A_{NNP}}{C_{TC}} \right) 100$$

For various reasons, any simple rate of return is limited as a measure of profitability. A project normally extends over a period of many years, during which economic parameters change. Time value of money is not strictly considered. The true significance of depreciation as a tax benefit is not appreciated. Nonetheless, rate of return is still employed by many in industry.

To establish selling price, suppose a 20 percent rate of return were required. From a rearrangement of Equation 6-10, the net profit after taxes is calculated from total capital.

$$A_{NNP} = \frac{i}{100} C_{TC} - A_{BD} = 0.20 C_{TC} - A_{BD} \qquad (6\text{-}10a)$$

But net profit before taxes is the sum of A_{NNP} and income taxes:

$$A_{NP} = A_{NNP} + A_{IT} \qquad (6\text{-}11)$$

where

$$A_{IT} = t A_{NP} \qquad (6\text{-}12)$$

and

$$A_{NNP} = (1 - t) A_{NP} \qquad (6\text{-}13)$$

With a 50 percent tax rate, $t = 0.50$, and net profit before taxes is:

$$A_{NP} = A_{NNP} + A_{IT} = A_{NNP} + \frac{t}{1-t} A_{NNP} = 2A_{NNP} = \frac{2i}{100} C_{TC} - 2A_{BD}$$

$$= 0.40 C_{TC} - 2A_{BD} \qquad (6\text{-}14)$$

Annual sales income must equal the sum of this and total expense.

$$A_S = A_{NP} + A_{TE} \qquad (6\text{-}15)$$

Division by annual capacity yields the unit sales price.

TABLE 6-4
MANUFACTURING COST SUMMARY

Page	15 of	25
By	G.U.	
Date	Sept. 4, 1982	

Job Title   Alkylate Splitter Module

Location   Eastern United States   Annual Capacity 3.92×10^8 kg/yr motor alkylate

Effective Date to Which Estimate Applies   mid-1984   Cost Index Type   CE Plant Cost

    Cost Index Value   350 (projected)

Fixed capital, C_{FC}	$1.52 M
Working capital (10–20% of fixed capital), C_{WC}	$0.23 M
Total capital investment, C_{TC}	$1.75 M

Cost	k$/yr		$/kg
Manufacturing Expenses			
Direct			
Raw materials (hydrocarbon feed at $306/m³)	$190,000		0.485
By-product credits (heavy alkylate at $294/m³)	(18,100)		(0.046)
Catalysts and solvents (none)			
Operating labor (4/3 equivalent at $27,350 per year)	37		0.001
Supervisory and clerical labor (15% of operating labor)	6		0.00002
Utilities			
Steam, 45 barg @ 0.02 $/kg	3,690		0.0094
Electricity @ 0.108 $/kWh	3		0.00001
Cooling water @ 0.060 $/m³	2		0.00001
Maintenance and repairs (6% of fixed capital)	91		0.00023
Operating supplies (15% of maintenance and repairs)	14		0.00004
Laboratory charges (15% of operating labor)	6		0.00002
Patents and royalties (3% of total expense)	6,440		0.0165
Total, A_{DME}	182,190	182,190	0.465
Indirect			
Overhead (payroll and plant), packaging, and storage (60% of the sum of operating labor, supervision, and maintenance)	80		0.00020
Local taxes (1.5% of fixed capital)	23		0.00006
Insurance (0.5% of fixed capital)	8		0.00002
Total, A_{IME}	111	111	0.00028
Total manufacturing expense (excluding depreciation), A_{ME}		182,301	0.4653
Depreciation (10% of fixed capital), A_{BD}		152	0.0004
General Expenses			
Administrative costs (25% of overhead)	20		0.0001
Distribution and selling expenses (10% of total expense)	21,470		0.0548
Research and development (5% of total expense)	10,734		0.0274
Total, A_{GE}	32,224	32,224	0.0823
Total expense, A_{TE}		214,677	0.5480
Revenue from sales (3.92×10^8 kg/yr @ 0.551 $/kg), A_S		215,073	0.5490
Net annual profit, A_{NP}		396	0.0010
Income taxes (50% of net annual profit), A_{IT}		(198)	(0.0005)
Net annual profit after taxes, A_{NNP}		198	0.0005

After rate of return, $i = ([A_{NNP} + A_{BD}]/C_{TC}) \times 100 = 20\%$

ILLUSTRATION 6-1*d*.   SELLING PRICE OF MOTOR ALKYLATE

Assuming a 20 percent rate of return on original investment, recommend a selling price for the motor alkylate of Illustration 5-1.

This answer is a direct consequence of completing a manufacturing cost analysis. Using results from Illustrations 5-1 and 6-1*a* through 6-1*c*, and following the step-by-step discussion above, total expense is easily computed. In this case, it is $214,700k/yr or $0.548 per kilogram of motor alkylate. Details are given in Table 6-4. Since 20 percent return on total capital is required, annual net profit after taxes is computed from Equation 6-10a.

$$A_{NNP} = \frac{i}{100} C_{TC} - A_{BD} = 0.20(\$1.75\text{M}) - \$152\text{k} = \$198\text{k}$$

At a rate of 50 percent, income taxes are equivalent to A_{NNP}. The net annual profit A_{NP} must then be double this as illustrated in Table 6-4 and Equation 6-14. Annual sales revenues, equal to net profit plus total expense, must be $215,100k/yr or $0.549 per kilogram of gasoline to provide the desired return on investment.

This concludes the discussion of balance sheet economics. For a more extensive treatment the series by Holland et al. and references 2, 5, and 6 should prove helpful. Material contained in this chapter is as much as many chemical engineers will need to know about process costs. It is also a sufficient basis for elementary economic optimization and cash flow analysis, which are discussed in Chapters Seven and Eight, respectively.

REFERENCES

1   Ganapathy, V., *Oil Gas J.* pp. 84–86 (June 9, 1980).

2   Grant, E.L., W.G. Ireson, and R.S. Leavenworth, *Principles of Engineering Economy*, 7th edition, Wiley, New York (1982).

3   Holland, F.A., F.A. Watson, and J.K. Wilkinson, "Capital Costs and Depreciation," *Chem. Eng.*, pp. 118–121 (July 23, 1973); also Part 2 of *Chem. Eng. Reprint* No. 215 (1975).

4   Holland, F.A., F.A. Watson, and J.K. Wilkinson, "Profitability of Invested Capital," *Chem. Eng.*, pp. 139–144 (Aug. 20, 1973); also Part 3 of *Chem. Eng. Reprint* No. 215 (1975).

5   Jelen, F.C., Editor, *Cost and Optimization Engineering*, McGraw-Hill, New York (1970).

6   Schweyer, H.E., *Process Engineering Economics*, McGraw-Hill, New York (1960).

PROBLEMS

Time frame for estimates and rate of return on capital to be defined by instructor or reader.

6-1   Estimate the price of steam generated in the equipment of Problems 3-1 and 5-1.

6-2 Based on problems 3-2 and 5-2, estimate what the selling price should be for maple syrup. (Assume sap value to be half the retail price of milk.)

6-3 Estimate the price of hydrogen chloride generated in the plant defined in Problems 3-3 and 5-3.

6-4 Based on Problems 3-4 and 5-4, estimate the selling price for maple syrup produced by multiple-effect evaporation. (Assume sap value to be half the retail price of milk.)

6-5 What should be the price charged for electricity produced in the plant described in Problems 3-5 and 5-5?

6-6 Estimate the selling price of nitric acid produced by the plant described in Problems 3-6 and 5-6.

6-7 Estimate the price for synthesis gas produced by the plant described in Problems 3-7 and 5-7.

6-8 Estimate the price for paper pulp as produced in the plant described in Problems 3-8 and 5-8.

6-9 Estimate the price for gas produced from coal by the plant described in Problems 3-9 and 5-9.

Chapter Seven
ECONOMIC OPTIMIZATION

"Optimum" denotes the quantity or condition that is most favorable. Economic optimization, as applied to manufacturing, is the process of finding the condition that maximizes financial return or, conversely, minimizes expenses. For this reason, it is convenient to think of optimization as a process that maximizes the bottom line on the balance sheet in Table 6-1.

CONVENTIONAL OPTIMIZATION
Functions of a Single Variable

Consider the selection of an optimum thickness of insulation to place around a steam pipe. Two economic factors are obvious to lay observer and engineer alike. One is the cost of steam, which decreases with insulation thickness. The second is cost of the insulation itself, which increases with thickness. Numerous other items, raw materials and laboratory charges for instance, are essentially independent of thickness.

While the layman realizes there must be a favorable balance between insulation purchase price and utilities costs, formulating the balance offers problems. First of all, these quantities are not immediately comparable. They are in different economic units. Utilities costs are in dollars per unit time, for instance, whereas capital costs are merely in dollars with no relationship to time. In Table 6-1, the impact of purchase price is manifest in fixed capital C_{FC}, which appears above the annualized expenses. Utilities costs, on the other hand, are integrated within direct manufacturing expenses.

Reconciliation is obvious as one notes the impacts of capital cost on various items of annual total expense A_{TE}. For example, the relation between insulation and fixed capital is expressed as follows.

$$C_{FC} = C_{CFC} + 1.18 \, F_{BM} C_{P,I} \, \pi \left[\left(\frac{D}{2} + t \right)^2 - \left(\frac{D}{2} \right)^2 \right] L \tag{7-1}$$

or

$$C_{FC} = C_{CFC} + C_{VFC} \tag{7-2}$$

The independent variable in Equation 7-1 is insulation thickness t. Pipe

343

diameter D is a parameter; that is, there will be an optimum value of t for each diameter. In addition, $C_{P,I}$ is the purchase price per unit volume of insulation, L is pipe length, F_{BM} is a bare module or installation factor, and the multiplier 1.18 is that employed in Table 5-2 to convert bare module costs to total module capital. (It is assumed here that insulation thickness has no effect on auxiliary facilities. Otherwise, another multiplier such as the 1.3 employed in Table 5-2 would be employed to denote the additional effect.) Fixed capital associated with everything in the plant except insulation, C_{CFC}, will certainly be many orders of magnitude larger than C_{VFC}, the variable capital—that dependent on insulation thickness. As shown later, C_{CFC} exerts no influence on the result because it is independent of t.

Conversion of fixed capital to an annual operating cost, is easily accomplished by reviewing individual categories in Table 6-1. The capital-related items are as follows.

Capital-Related Cost Item	*Typical Fractions of Fixed Capital*
Maintenance and repairs	0.06
Operating supplies	0.01
Overhead, etc.	0.04
Taxes and insurance	0.02
Depreciation	0.10
General	0.01
Total	0.24

Thus we find that about one fourth of the fixed capital investment must be spent each year to own and maintain a process plant. In mathematical terms,

$$A_{FC} = f_a C_{FC} \tag{7-3}$$

where A_{FC} represents annual expenses due to fixed capital and f_a is the annuity factor. For the typical case above, f_a is 0.24. Using extreme values from Table 6-1, factors as low as 0.1 or as high as 0.45 can be derived.

Referring again to Table 6-1, the annual total expense can be expressed by

$$A_{TE} = A_{FC} + A_{OE} \tag{7-4}$$

Annual operating expenses A_{OE} can be divided into two components:

$$A_{OE} = A_{COE} + A_{VOE} \tag{7-5}$$

where A_{COE} includes those independent of insulation thickness and A_{VOE} those that vary with t. A survey of items in Table 6-1 reveals only one noncapital-related operating expense affected by insulation thickness—the steam cost:

$$A_{VOE} = A_S = C_{S,s} \frac{\dot{Q}}{\lambda} (31.5 \times 10^6 \text{ s/yr}) f_o \tag{7-6}$$

where $\dot{Q}$ is heat loss rate, λ is latent heat, $C_{S,s}$ is the unit cost of steam, and f_o the operating factor. Insulating thickness enters Equation 7-6 by way of a heat transfer analysis:

$$\dot{Q} = U\pi DL \, \Delta T_m \tag{7-7}$$

where

$$\frac{1}{U} = \frac{1}{h_i} + \frac{D \ln[(D + 2t)/D]}{2k} + \frac{D}{h_o(D + 2t)} \qquad (7\text{-}8)$$

One versed in thermal science will recognize Equations 7-7 and 7-8 as those defining the rate of heat transfer through a cylindrical wall. The overall coefficient U is a composite value based on internal and external film values h_i and h_o, external pipe diameter D, and insulation thermal conductivity k. An assumption of negligible pipe wall resistance has been integrated into Equation 7-8. Invariabley, h_i will be large, allowing us to discard $1/h_i$ relative to the last two terms.

In abbreviated form, annual total expense can be expressed as a function of insulation thickness by substitution into Equation 7-4.

$$A_{TE} = A_{FC} + A_{OE} \qquad (7\text{-}9)$$

$$A_{TE} = A_{CFC} + A_{VFC}(D, t, L) + A_{COE} + A_{VOE}(D, t, L, k, h_o, \Delta T_m) \qquad (7\text{-}10)$$

$$A_{TE} = A_C + A_{VFC}(D, t, L) + A_{VOE}(D, t, L, k, h_o, \Delta T_m) \qquad (7\text{-}11)$$

Costs that are independent of insulation are lumped into a constant term A_C. Variable capital-related and operating costs are expressed by A_{VFC} and A_{VOE}, which are functions of indicated parameters and the single independent variable t. Since A_C has no effect on the optimum thickness, it can be eliminated by considering variable annual expenses only.

$$A_{VTE} = A_{TE} - A_C = A_{VFC}(D, t, L) + A_{VOE}(D, t, L, k, h_o, \Delta T_m) \qquad (7\text{-}12)$$

In addition, the length L is a constant multiplier, allowing us to factor it from the right-hand side of Equation 7-12. This amounts to optimizing the annual variable expenses per unit length of pipe.

$$\frac{A_{VTE}}{L} = \frac{A_{VFC}}{L} + \frac{A_{VOE}}{L} \qquad (7\text{-}13)$$

The final optimization relationship is assembled by manipulating Equations 7-1, 7-2, 7-3, 7-6, 7-7, and 7-8 and substituting the results into Equation 7-13.

$$\frac{A_{VTE}}{L} = f_a \frac{C_{VFC}}{L} + \frac{A_{VOE}}{L} \qquad (7\text{-}14)$$

$$\frac{A_{VTE}}{L} = 1.18 F_{BM} C_{P,I} \, \pi \left[\left(\frac{D}{2} + t \right)^2 - \left(\frac{D}{2} \right)^2 \right] f_a$$

$$+ \frac{C_{S,s} \, \pi (\Delta T_m/\lambda) \, (31.5 \times 10^6 \text{ s/yr}) f_o}{\ln[(D + 2t)/D]/2k + 1/[h_o(D + 2t)]} \qquad (7\text{-}15)$$

Equation 7-15, containing two functions, is typical of expressions that yield an optimum. One function increases with the variable of interest. The other decreases. This behavior is illustrated in Figure 7-1, which gives curves for increasing capital and decreasing operating costs. The sum of these curves A_{VTE}/L is also plotted. Optimum insulation thickness is that corresponding to the minimum value of A_{VTE}/L. To evaluate t_{opt}, one could obtain the various captial cost factors and insulation properties from a supplier. For a given steam pressure and ambient temperature, ΔT_m and λ are fixed. Unit steam price $C_{S,s}$ can be computed from Table 6-3 and h_o obtained from a reference source. This would yield numerical

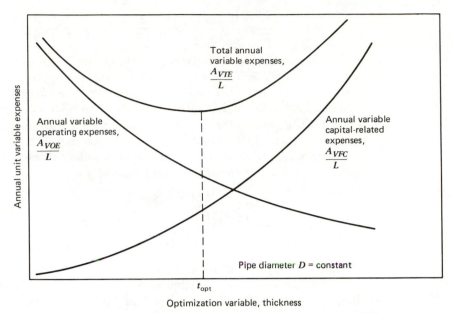

Figure 7-1   Optimization relationship for determining the most efficient insulation thickness for a steam pipe. These curves and the position of the optimum vary with pipe diameter.

coordinates for plotting in Figure 7-1. (For a more sophisticated treatment, see reference 3).

In practice, one wishes to maximize profit. In this illustration, I have assumed implicitly that sales revenues will not change with t. Given this, one can easily verify from Table 6-1 that A_{NNP} reaches a maximum at the same value of t where A_{VTE} is minimum. In some situations, a continuous minimum is not achieved. For example, at modest pressures, the most economic flash drum diameter is that of a single unit capable of processing the total stream. As stream quantity increases, drum diameter increases until a practical limit is achieved. This is near 4 m. Above this, conventional shop equipment and transportation devices are inadequate, and the optimum shifts discontinuously from a single drum to double or multiple units of 4 m diameter or less.

The optimum is shifted dramatically in seasonal or short-term processes, where operating time is limited. Mathematically, this effect is reflected in the operating factor f_o. In Figure 7-1, for example, a decrease in f_o lowers variable operating expenses by a constant multiplier. This, in effect, rotates that curve downward and counterclockwise, shifting the minimum in total annual expenses toward thinner pipe insulation.

From calculus, an alternate method of defining minimum expense is obvious. The derivative of annual expense is taken relative to the independent variable and the result set equal to zero. (One should also take the second derivative to verify minimum rather than maximum.) This analytical approach identifies the optimum with precision and, if the function is simple, quickly. It is also more convenient for computer calculation than the graphical approach. However, precision is usually less important than accuracy. A graph reveals not only location of the minimum but the shape of the curve. Real materials such as insulation are provided in incremental sizes. Usually the exact numerical optimum does not agree with a readily available size, and a value either side of the optimum must be selected. If the curve is markedly steeper on one side than on the other, the appropriate choice is obvious from a graphical correlation. Only the optimum point itself is computed from the mathematical derivative.

In an unfamiliar situation, I recommend using the graphical approach for the first time, at least. Once economic behavior has been defined, mathematical techniques can be employed safely for repetitive iterations.

The balance sheet approach employed in this section is a distorted image of true process optimization. Its weakness stems from a simplified view of depreciation as an operating cost and failure to consider the time value of money. (For rigorous optimization, trial-and-error cash flow analysis is necessary. Using a computer, this economic refinement is easily employed once cash flow calculations are understood. Such refinement is usually not necessary for preliminary economic evaluations.) To improve accuracy using the simplified approach, I recommend a lower value, between 0.15 and 0.20, for the annuity factor f_a. This compensates somewhat for the weaknesses.

In contrast to the insulation case, analyzed minutely for illustration, painstaking consideration of the total cost summary sheet is unnecessary. You merely pass quickly through annual cost items to identify those that are functions of the independent variable. Then, you can develop an equation such as 7-14 that considers only those cost elements. Variable fixed capital usually can be assessed rather easily, and variable operating expenses can be similarly derived. These, with an appropriate annuity factor, are the only terms necessary for optimization.

A number of chemical engineering operations contain opposing cost functions that lead to classical behavior such as that demonstrated for pipe insulation. These include selection of approach temperatures in heat exchangers, definition of distillation reflux ratio, selection of optimum pipe size, and definition of liquid and gas flow rates in various process vessels. Another type of optimum is encountered in cyclic operations such as cleaning of heat exchangers, discharging filter presses, or replacing a reactor catalyst. Selection of optimum reactor size is another classical problem. In all cases, optimization can be accomplished by developing relationships between anticipated variables and annual cash income. In some operations, particularly cyclic ones, annual sales as well as labor, utilities, or raw materials are affected and must be considered. When faced with a specific problem, formulation of an analytical cost relationship should follow naturally from procedures illustrated.

ILLUSTRATION 7-1 MULTIPLE-EFFECT EVAPORATION

A multiple-effect evaporator is to be designed for concentrating Kraft liquor. Boiling point elevation is negligible, and 20 kg/s of water is to be removed. Using reasonable economic assumptions and data, determine the optimum number of effects for this process module. Base this evaluation on costs in mid-1982.

Since sales income is unaffected by this module, our goal is to minimize expenses. Based on discussion in Chapter Four and other background information, we recognize that sizes of individual effects will be independent of number but steam and cooling water rates will be strongly influenced by number. Increasing the number of effects increases capital cost and labor while decreasing utility expenses. Examination of the balance sheet (Table 6-1) reveals various elements that depend on the number of effects, including all capital-related items plus labor and utilities.

A flow sketch is presented in Figure 7-2. Reference to Chapter Four and Table

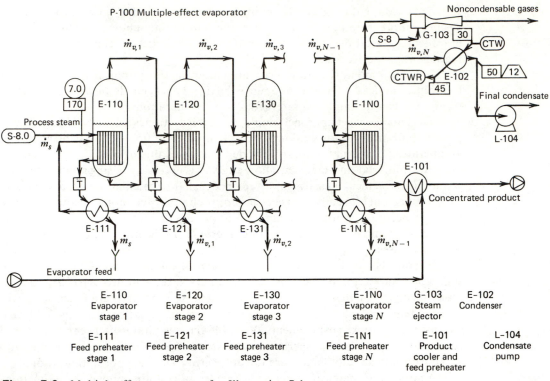

P-100 Multiple-effect evaporator

E-110	E-120	E-130	E-1N0	G-103	E-102
Evaporator	Evaporator	Evaporator	Evaporator	Steam	Condenser
stage 1	stage 2	stage 3	stage N	ejector	

E-111	E-121	E-131	E-1N1	E-101	L-104
Feed preheater	Feed preheater	Feed preheater	Feed preheater	Product	Condensate
stage 1	stage 2	stage 3	stage N	cooler and	pump
				feed preheater	

Figure 7-2 Multiple-effect evaporator for Illustration 7-1.

4-7 suggests the use of long-tube units. Based on a cooling water exit temperature of 45° C and a 5° C approach temperature, the last effect will be at approximately 50° C. Employing conventional process steam at 8 bara, temperature of steam in the first effect will be 170° C.

Assuming a ΔT of 20°C, U is 2000 J/s·m^2·K from Figure 4-4 and is assumed constant. Since there is no boiling point elevation, vapor yield is estimated at 0.85. Total vapor yield, based on Equations 4-27 and 4-28, is

$$\sum_{i=1}^{n} \dot{m}_{v,i} = \dot{m}_s \, n \, \exp \, (y^{2/y}) = \dot{m}_s n^{0.68}$$

where $\sum_{i=n}^{n} \dot{m}_{v,i}$ is also given by the problem statement as 20 kg/s.

Annual costs, which depend on n, are:

LABOR

$$A_L = n(0.3 \times 4 \text{ operators}) (\$25,000/\text{ operator yr}) = 30,000n \ (\$/\text{yr})$$
(0.3 operator per evaporator per shift taken from Table 6-2)

SUPERVISION

$$A_{Su} = 0.15A_L = 4500n \ (\$/\text{yr})$$

UTILITIES

Steam

$$A_S = \dot{m}_s \, (31.5 \times 10^6 \text{s/yr}) \, f_o \, C_{S,s}$$

$$= \frac{\Sigma \dot{m}_{v,i}}{n^{0.68}} \, (31.5 \times 10^6 \text{s/yr}) \, f_o \, C_{S,s}$$

Cooling Water

$$\dot{m}_{v,n} = \dot{m}_s y^n = \dot{m}_s \, (0.85)^n = \frac{\Sigma \dot{m}_{v,i}}{n^{0.68}} \, (0.85)^n$$

$$\dot{m}_{cw} = \frac{\lambda \, \dot{m}_{v,n}}{C_{P,cw}(45 - 30° \text{C})}$$

$$A_{cw} = \dot{m}_{cw}(31.5 \times 10^6) f_o C_{S,cw}$$

OVERHEAD, LABORATORY, ADMINISTRATIVE

$$A_{OH} = 30{,}000n \quad (\$/\text{yr})$$

Including capital-related expenses, annual variable total expenses are:

$$A_{VTE} = A_{VFC} + A_{VOE} = f_a C_{VFC} + A_L + A_{Su} + A_s + A_{cw} + A_{OH} \qquad (7\text{-}16)$$

From Table 6-3, steam costs are assessed at $0.017/kg. Assuming an operating factor of 90 percent, we have:

$$A_s = (20 \text{ kg/s}) \, \frac{(31.5 \times 10^6 \text{ s/yr})}{n^{0.68}} \, (0.90)(\$0.017/\text{kg})$$

$$= \frac{9.6 \times 10^6}{n^{0.68}}$$

Based on a cooling water unit cost of 6×10^{-5} \$/kg (Table 6-3), this utility amounts to less than 10 percent of the stream cost and is disregarded.

Assembly of these and previous data within Equation 7-16 yields

$$A_{VTE} = f_a C_{VFC} + 65{,}000 \, n + \frac{9.6 \times 10^6}{n^{0.68}}$$

Basically, variable fixed capital is total capital for the evaporator module of Figure 7-2. If there were only one evaporator, ΔT would be 120° C and evaporator area could be calculated as follows.

$$\dot{Q} = \dot{m}_{s,1} \, (h_v - h_l)_{8 \text{ bara, sat}} = \frac{\dot{m}_v}{0.85} \, (2769 - 721 \text{ kJ/kg})$$

$$Q = \frac{20 \text{ kg/s}}{0.85} \, (2048 \text{ kJ/kg}) = 48 \times 10^6 \text{ J/s}$$

$$A = \frac{\dot{Q}}{U \Delta T} = \frac{48 \times 10^6 \text{J/S}}{(200 \text{ J/s} \cdot m^2 \cdot \text{K}) \, (120° \text{C})} = 200 \text{ m}^2$$

From Figure 5-24,

$$C_P = \$250,000$$

$$F_{BM}^{cs} = 2.9$$

$$C_{TM} = (\$250,000)\,(2.9)\,(1.18) = \$850,000$$

Since each effect is the same size regardless of number, for n units

$$C_{TM} = \$850,000n$$

If the annuity factor were assumed to be 0.18 yr^{-1}, we would have

$$A_{VTE} = 0.18\ \text{yr}^{-1}\ (\$850,000n) + 65,000n + \frac{9.6 \times 10^6}{n^{0.68}}$$

or

$$A_{VTE} = 220,000\ n\ (\$/\text{yr}) + \frac{9.6 \times 10^6}{n^{0.68}}\ \$/\text{yr}$$

Cost components tabulated for various values of n are as follows.

n	220,000n ($/yr)	$9.6 \times 10^6/n^{0.68}$($/yr)	A_{VTE} ($/yr)
2	440,000	6,000,000	6,440,000
4	880,000	3,740,000	4,620,000
6	1,320,000	2,840,000	4,160,000
8	1,760,000	2,330,000	4,090,000
10	2,220,000	2,006,000	4,226,000
12	2,640,000	1,772,000	4,412,000

Results are plotted in Figure 7-3. Seven or eight effects seems to be the optimum number. Analytically,

$$\frac{dA_{VTE}}{dn} = 220,000 - \frac{0.68(9.6 \times 10^6)}{n^{1.68}}$$

$$\frac{d^2 A_{VTE}}{dn^2} = 1.68\,\frac{(0.68)(9.6 \times 10^6)}{n^{2.68}}$$

The second derivation is postive, denoting a minimum, and the first derivative is zero at

$$n = \left[\frac{0.68(9.6 \times 10^6)}{220,000}\right]^{0.60} = 7.6$$

True average ΔT's are near those assumed. Seven or eight evaporators will be proposed.

ILLUSTRATION 7-2   THE GOLDENROD

The Goldenrod, a candy factory and restaurant in York Beach, Maine, has a dishwashing system that operates 100 days/yr, 7 h/day. It consumes hot rinse water at a rate of 0.42 liter/s. Water enters the factory at 15°C, is heated in an oil-fired hot water heater to 82°C, and goes to the dishwasher. Water leaves the dishwater at about 75°C and is discharged to the drain. One logical scheme for energy recovery

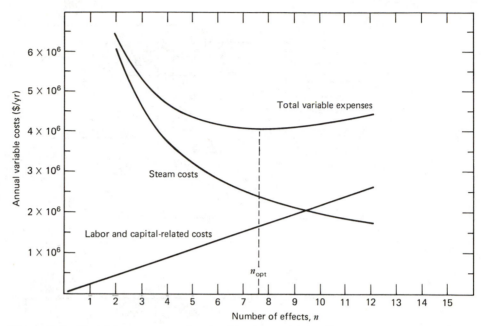

Figure 7-3 Variable costs associated with the multiple-effect evaporator module of Illustration 7-1.

involves use of effluent water to preheat the feed water, thus reducing the cost of oil. Design the optimum system to accomplish this. Assume operation is based on mid-1982 costs. A preliminary flow sheet with material balance is shown in Figure 7-4.

Based on an energy balance around the water heater, oil consumption is calculated as follows.

$$\frac{\dot{m}_2 (\text{HHV}) \epsilon_{\text{P-110}}}{\rho_2} = \dot{m}_1 C_{p,1} (82°\text{C} - T_1)$$

$$\dot{m}_2 = \frac{\dot{m}_1 C_{p,1} (82°\text{C} - T_1) \rho_2}{(\text{HHV}) \epsilon_{\text{P-110}}}$$

Based on Table 6-1, annual variable costs are as follows.

UTILITIES

$$\text{Fuel oil} = A_{fo} = m_2 \, C_{s,fo} (100 \times 7 \times 3600) \, \text{s/yr}$$

$$\text{From Figure 6-1, } C_{s,fo} = \frac{\$6/\text{GJ}}{\rho_2} (\text{HHV}) \, (\$/\text{kg})$$

$$A_{fo} = \frac{\dot{m}_1 \, C_{p,1} \, (82 - T_1)(\$6/\text{GJ})(2.5 \times 10^6 \, \text{s/yr})}{\epsilon_{\text{P-110}}}$$

$$= \frac{(0.42 \, \text{kg/s})(4190 \, \text{J/kg°C})(82 - T_1)(\$6 \times 10^{-9}/\text{J})(2.5 \times 10^6 \text{s/yr})}{\epsilon_{\text{P-110}}}$$

$$= \frac{26.6 \, (82 - T_1)}{\epsilon_{\text{P-110}}} \, \$/\text{yr}$$

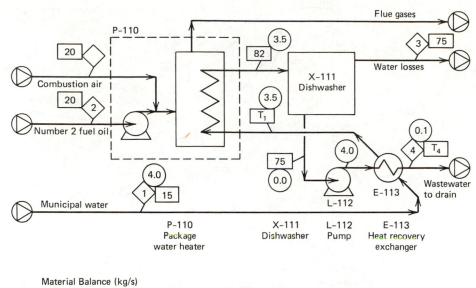

Figure 7-4 Tentative flow sheet and material balance for Goldenrod energy recovery scheme.

Material Balance (kg/s)

	◇1 Municipal Water	◇2 Number 2 Fuel Oil	◇3 Water Losses	◇4 Waste-water
Fuel oil Water	0.42		0.04	0.38
Total	0.42		0.04	0.38

ELECTRICITY

$$P_{\text{L-112}} = \frac{\dot{w}_p}{\epsilon_{o,\text{L-112}}} = \frac{\dot{m}_4 \, \Delta p}{\rho_4 \, \epsilon_i \epsilon_d} \cong \frac{(0.38 \text{ kg/s})(4 \times 10^5 \text{J/m}^3)}{(1000 \text{ kg/m}^3)(0.60)(0.75)} \cong 400 \text{ W}$$

$$A_e = \frac{P(100 \times 7 \text{ h/yr})(C_{s,e})}{(1000 \text{ W/kW})}$$

$$= (0.4 \text{ kW})(700 \text{ h/yr}) \, (0.08 \text{ \$/kWh}) = \$22/\text{yr}$$

Capital-related expenses are estimated by techniques described in Chapter Five.

New Equipment

Pump L-112

$$P = 400 \text{ W}$$

$$\dot{w}_s = 0.75P \cong 300 \text{ W}$$

$$C_P = \$1300 \text{ (centrifugal, from Figure 5-49)}$$

For this service, a plastic pump seems appropriate. Assume F_M for such is the same as for cast iron. Suction pressure is below 10 barg, so $F_p = 1$ and

$$F_{BM} = 3.2$$

$$C_{BM} = 1300 \times 3.2 = \$4200.$$

Exchanger E-113

$$\dot{Q} = UA \, \Delta T_m = \dot{m}_1 C_{p,1}(T_1 - 15) = \dot{m}_4 C_{p,4}(75 - T_4)$$

$$(75 - T_4) = \frac{\dot{m}_1}{\dot{m}_4}(T_1 - 15) = \frac{0.42}{0.38}(T_1 - 15) = 1.11(T_1 - 15)$$

$$T_4 = 75°C - 1.11\,(T_1 - 15°C)$$

$$\Delta T_m = \Delta T_{lm} = \frac{(T_4 - 15) - (75 - T_1)}{\ln\left[(T_4 - 15)/(75 - T_1)\right]}$$

From Table 4-15, $U = 1200 \text{ J}/\text{m}^2 \cdot \text{s} \cdot \text{K}$.

$$A = \frac{(0.42 \text{ kg}/\text{s})(4190 \text{ J}/\text{kg})\,(T_1 - 15)}{(1200 \text{ J}/\text{m}^2 \cdot \text{s} \cdot \text{K})\,\Delta T_{lm}}$$

$$A = \frac{1.5 \text{ m}^2\,(T_1 - 15)}{\Delta T_{lm}}$$

Examination of Figure 5-36 indicates that double-pipe exchangers will be the most inexpensive units for this service. The corrosion guide of Table 4-28 indicates copper for both shell and tube to be the most favorable construction material. With information assembled thus, capital costs can be expressed for various values of T_1, as follows.

$T_1(°C)$	$T_4(°C)$	ΔT_{lm}	$A\,(m^2)$	$C_{P,E\text{-}113}$	$F_{BM}^{Cu/Cu}$	$C_{BM,E\text{-}113}$	ΣC_{BM}	$\Sigma C_{TM}^{a} = C_{VFC}$
15	75	—	0	0	2.5	0	$ 4,200[b]	$ 4,950[b]
40	47	33	1.1	$ 850	2.5	$ 2,125	6,325	7,460
60	25	12	5.5	1000	2.5	2,500	6,700	7,910
64	20.6	8.0	9.0	1100	2.5	2,750	6,950	8,200
66	18.4	5.8	12.9	1300	2.5	3,250	7,450	8,800
68	16.2	3.3	23.6	2300	2.5	5,750	9,950	11,700
69	15.1	1.3	61.1	6000	2.5	15,000	19,200	22,700

[a] $\Sigma C_{TM} = \Sigma C_{BM} \times 1.18$ (from Table 5-2).

[b] Pump only.

The optimization expression is:

$$A_{VTE} = f_a C_{VFC} + A_{VOE}$$

where $C_{VFC} = \Sigma C_{TM}$ denoted above

$$A_{VOE} = \text{fuel oil expense} = A_{fo} = \frac{26.6(82 - T_1)}{\epsilon_{P\text{-}110}} \ (\$/\text{yr})$$

(A_e = electricity cost = \$22/yr is a constant)

Assuming $\epsilon_{P\text{-}110} = 0.70$, $A_{VOE} = 38\,(82 - T_1)$.

Use $f_a = 0.18$.

$$A_{VTE} = 0.18\,C_{VFC} + 38(82 - T_1)$$

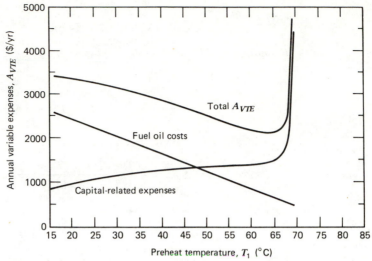

Figure 7-5   Optimization curve for Goldenrod energy recovery scheme.

Now, A_{VTE} can be evaluated for various values of T_1. Results are tabulated below and plotted in Figure 7-5. The optimum preheat temperature is apparently 64° C. Capital costs escalate sharply as the approach temperature in E-113 becomes smaller.

$T_1(°C)$	A_{fo} ($/yr)	C_{VFC}	$0.18\,C_{VFC}$ ($/yr)	A_{VTE} ($/yr)
15	2546	$ 4,950	890	3440
40	1596	7,460	1340	2940
60	836	7,910	1420	2260
64	684	8,200	1480	2160
66	608	8,800	1580	2190
68	532	11,700	2110	2640
69	494	22,700	4090	4580

This result is examined further in Chapters Eight and Nine.

Functions of Multiple Variables

When more than one undetermined property is present in the optimization equation, the expression for A_{VTE} contains two or more independent variables rather than one. The optimum under such conditions is a composite minimum for the whole system. To find the optimum mathematically, A_{VTE} is differentiated with respect to each variable and each partial derivative is set equal to zero. This yields a number of equations equal to the number of variables. The optimum set of variable values is found by solving this system of equations simultaneously.

Graphical solution of multivariable functions is somewhat more complicated. If there are only two variables, cost curves are plotted versus one for fixed values of the other. This yields a family of curves, each having a minimum. Then, a locus is defined by a line through these minima. The lowest value of the locus or composite curve represents the optimum combination of twin variables. For three variables, multiple charts are employed. Peters and Timmerhaus [4] have illustrated graphical

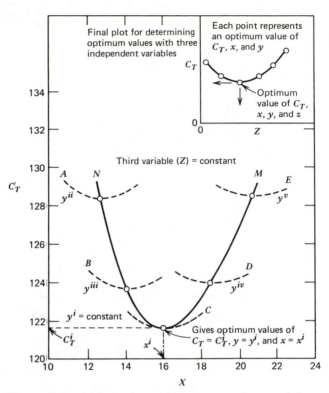

Figure 7-6   Optimization within a system of two and three independent variables [4]. (From *Plant Design and Economics for Chemical Engineers*, 3rd edition, by M. S. Peters, and K. D. Timmerhaus, Copyright © 1980. Used with the permission of McGraw-Hill Book Company.)

techniques applied to multiple variable problems (see Figure 7-6). Complexity and tedium increase markedly with an increase in the number of variables.

There is an optimum, it seems, for every human action. Because of its mathematical foundation and commercial significance, optimization has attracted the attention of many mathematicians and technologists. It has, in fact, developed as a subdiscipline in science and engineering. Its growth has certainly not been inhibited by the development of modern computer search techniques. With such sophisticated tools available, there is a danger that the process of solution can become more intriguing than the problem. Because of this hazard, I recommend that you also engage common sense and approximate hand calculations whenever more sophisticated techniques are employed. For an introduction to optimization science applied specifically to chemical engineering situations, references 1, 2, 4, and 5 are recommended.

INCREMENTAL RETURN ON INCREMENTAL INVESTMENT

To a scientist, the maximum in a net profit curve is mathematically and intuitively most satisfying. To an economist, it is less so. The economist, more aware of interest rates, is concerned with alternate investments that may provide more lucrative

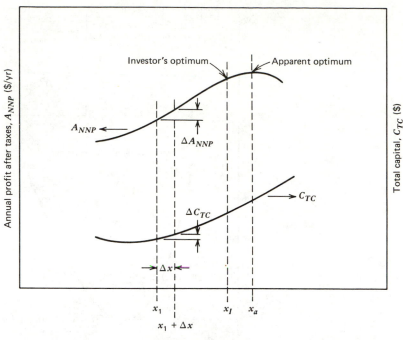

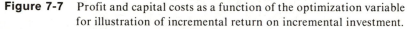

Figure 7-7 Profit and capital costs as a function of the optimization variable for illustration of incremental return on incremental investment.

returns. To appreciate this, one should plot both annual profit and total capital in the same chart, as in Figure 7-7. (Note that two ordinate scales must be employed because the units, $ and $/yr, are different.) The apparent optimum, based on prior discussion in this chapter, is located at the maximum of the net profit curve. The investor's optimum occurs lower on the curve. To understand the difference, we must consider incremental return on incremental investment.

Examine the effect of a small change in the optimization variable Δx in Figure 7-7. The total capital increases by an amount ΔC_{TC}, and, at the point illustrated, profit also increases by an amount ΔA_{NNP}. The ratio

$$i_i' = \frac{\Delta A_{NNP}}{\Delta C_{TC}} \qquad (7\text{-}17)$$

is termed *incremental* return on *incremental* investment[1] and has important economic significance. It reveals the return on each additional dollar of invested capital. Thus, if one were considering the wisdom of increasing the optimization variable from x_1 to $x_1 + \Delta x$, i_i' measures the profitability on the additional capital required.

Note that i_i' becomes smaller as x increases from x_1, eventually becoming zero at the apparent optimum. It then goes negative at larger values of x. Faced with alternate investment opportunities, one would stop adding money to this project

---

[1]To be strictly consistent with Table 6-1, the quantity

$$i_i = \frac{\Delta(A_{NNP} + A_{BD})}{\Delta C_{TC}}$$

should be used. The simpler form (Equation 7-17) is employed here for the sake of illustration.

when i'_i dropped below the return available from the alternatives. This point is designated as the investor's optimum x_I in Figure 7-7. It is as if we were climbing a mountain with a point other than the peak as our goal. That point would be the location, near the peak, where the slope declined to a set value. In most mountains, this point is not far removed from the peak, but it depends on i'_i. In periods of low interest rates, i'_i is small, and the difference between investor's optimum and apparent optimum is negligible. When interest rates are high, the distinction is important.

Negative values of i'_i occur when x exceeds x_a. This is obviously an undesirable economic range. You should be careful, however, not to dismiss all negative results automatically. This practice is especially dangerous when using a mathematical relationship without the benefit of a graph such as Figure 7-7. The hazard, which occurs in many real situations, lies where capital declines, as shown for lower values of x in this figure. Note that i'_i is negative in this range; not because of ΔA_{NNP} but because ΔC_{TC} is negative. Continuing to increase x under these circumstances is highly profitable and should be done.

ILLUSTRATION 7-3 INVESTOR'S OPTIMUM—THE GOLDENROD

Determine the optimum equipment for Illustration 7-2 if an incremental return on investment of 10 percent is required.

Examination of Table 6-1 reveals that annual net profit after taxes is given by:

$$A_{NNP} = (A_C - A_{VTE})(1 - t) \qquad (7\text{-}18)$$

where A_C is a constant and A_{VTE} represents variable total expenses from Illustration 7-2. Similarly, the total capital can be expressed by

$$C_{TC} = C_{\text{const}} + 1.15 C_{VFC} \qquad (7\text{-}19)$$

where the additional 0.15 comes from working capital and C_{const} represents capital that is independent of preheat temperature. Since constant terms drop out of an incremental analysis, i'_i, given by either Equation 7-17 or 7-20 is the same.

$$i'_i = \frac{\Delta(A_{NNP} - A_C)}{\Delta(C_{TC} - C_{\text{const}})} \qquad (7\text{-}20)$$

With the substitution of Equations 7-18 and 7-19, Equation 7-20 becomes

$$i'_i = \frac{\Delta[-A_{VTE}(1 - t)]}{\Delta(1.15 C_{VFC})} \qquad (7\text{-}21)$$

Assuming a tax rate of 50 percent and 10 percent straight line depreciation, the terms in Equation 7-21 can be computed from the following data, taken from the solution to Illustration 7-2.

Preheat Temperature, T_1 $(°C)$	$-0.5 A_{VTE}$ $(\%/yr)$	$1.15\ C_{VFC}$	i'_i (yr^{-1})
15	−1720	$5,690	0.01
40	−1470	8,580	1.00
60	−1130	9,100	0.44
64	−1080	9,430	0.01
66	−1095	10,120	−0.04
68	−1320	13,450	—
69	−2290	26,100	—

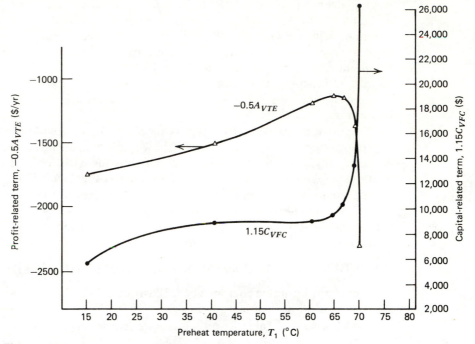

Figure 7-8 Profit and capital-related terms for evaluation of incremental return on investment from Equation 7-21.

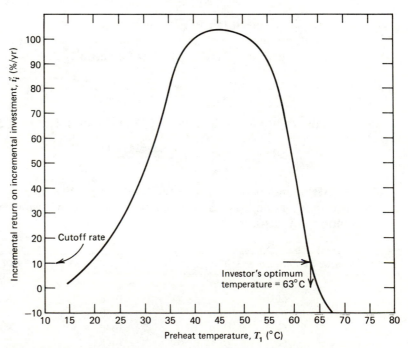

Figure 7-9 Incremental return on incremental investment (based on net profit after taxes) for the Goldenrod energy recovery scheme.

Plots of $-0.5\,A_{VTE}$ and $1.15\,C_{VFC}$ appear in Figure 7-8. From these curves, i'_i is determined by taking ratios of the two slopes. This was done graphically to yield numerical values as listed in the tabulation above. These are plotted in Figure 7-9. (Incremental rate values here are larger than true rates of return because of pump capital, which, being constant, does not enter this calculation.)

We can see that the 10 percent cutoff yields an investor's optimum temperature of 63° C. This is only slightly less than the apparent optimum found in Illustration 7-2. As revealed in Figure 7-9 and mentioned above, the distinction becomes greater as the cutoff rate of return becomes larger. Investor's optimum, like apparent optimum, is influenced by the time value of money and the tax implications of depreciation. These complications are illustrated after an explanation of cash flow techniques in Chapter Eight.

REFERENCES

1 Beveridge, G.S., and R.S. Schechter, *Optimization Theory and Practice*, McGraw-Hill, New York (1970).

2 Jelen, F.C., Editor, *Cost and Optimization Engineering*, McGraw-Hill, New York (1970).

3 McChesney, M. and P. McChesney, "Insulation Without Economics," *Chem. Eng.*, pp. 70–79 (May 3, 1982). See also D.E. McConnell and B.F. Blackwell, *Chem. Eng.*, p. 5 (Aug. 9, 1982).

4 Peters, M.S., and K.D. Timmerhaus, *Plant Design and Economics for Chemical Engineers*, 3rd edition, p. 361, McGraw-Hill, New York (1980).

5 Wilde, D.J., and C.S. Beightler, *Foundations of Optimization*, Prentice-Hall, Englewood Cliffs, New Jersey (1967).

PROBLEMS

7-1 Economic Pipe Diameter

Derive an equation that expresses the optimum diameter of pipe for a given flow rate in process piping. Consider both carbon steel and stainless steel. In the size range of interest (0.02–0.20 m inside diameter), the January 1983 price of carbon steel pipe (dollars per meter of length) is given by $C_P = 1000\,d^{1.6}$, where d is inside diameter in meters.

Assume the bare module factor for pipe is the same as that for horizontal process vessels. Assume friction losses due to fittings are 50 percent of those in straight pipe. Employ after assumptions that seem appropriate and necessary. See Peters and Timmerhaus, p. 377 and Perry, p. 5-31 for additional help if necessary.

7-2 Maple Sap Preheating

Design optimum pipe diameter and length for the sap preheater described in Problem 3-2. Use the price quoted for pipe in Problem 7-1 modified by appropriate materials cost factors.

7-3 Evaporator Fouling

In a given service employing short-tube evaporators, heating surfaces become coated with residue, causing the heat transfer coefficient to decrease with time according to the relation

$$U = U_o(1 - 3 \times 10^{-7}\theta)$$

where U_o is the heat transfer coefficient in a freshly cleaned unit, and θ is time in seconds since the last cleaning.

A single evaporator is designed to evaporate 50 million kg of water per year in a plant that has an operating factor of 0.90. Vapor yield is 0.9 kg of water vapor per kilogram of steam provided. Cleaning requires an evaporator to be pulled off-line for 24 h. Cleaning labor is equivalent to the continuous time of two extra persons during the off-line period. Temperature difference in the evaporator is 20° C. Steam temperature is 170° C. Liquor in the evaporator contains 20 percent sucrose.

Determine the optimum length of time between cleanings and the requisite surface area.

7-4 Multiple-Effect Evaporation of Maple Sap

In a central processing plant, multiple-effect evaporation has been proposed for concentrating maple sap. Processing rates are 100 times greater than in Problems 3-2 and 3-4 but concentrations are similar. Feed-forward processing is proposed. Boiling point elevation is negligible except in the final effect, where it is 4° C. Operation is continuous except for minor interruptions over a 50-day season. Estimate capital and operating costs for an optimum evaporating plant.

7-5 Filter Press Cleaning

In a water treatment plant, a filter press is employed to polish 100 kg/s of effluent (average flow). Because of residue accumulation, actual flow rates decrease with time according to the equation

$$\dot{m} = \frac{\dot{m}_0}{1 + 0.05\,\theta^{1/3}}$$

where m_0 is flow rate through a clean filter and θ is time since cleaning, in seconds. Cake discharge is automatic so that cleaning can be done without additional operators, but flow is interrupted for 10 min during each cleaning cycle. The treatment plant has an operating factor of 0.85.

How long should the filter press be operated between cleanings?

7-6 Alkylation Unit Heat Pump Fractionation[2]

An oil refiner requires additional i-butane feed for a new alkylation unit that produces gasoline from butylene and i-butane. To meet this need, a system to produce an i-butane-rich steam containing 800 m³ (measured at 15° C) a day of i-butane is needed. The propane content of this stream is acceptable, since it will be removed in the alkylation unit facilities, but the n-butane content of this stream may not exceed 64 m³ a day. The refinery has available a maximum of 4800 m³ a day of mixed butanes to supply this need of the following composition.

Percentage of Liquid
Volume at 15° C

Propane	2.5
i-Butane	21.5
n-Butane	76.0
Total	100.0

[2]Based on the 1980 AIChE Student Contest Problem. (The American Institute of Chemical Engineers, by permission.)

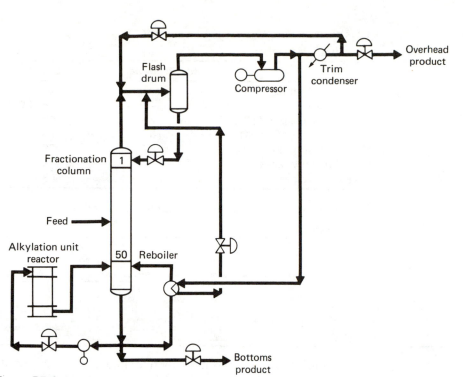

Figure P7-6-1 Heat pump fraction column.

There is an existing fractionation column with 50 actual trays and adequate diameter and design specifications (temperature and pressure) which the refiner wishes to use for this system. The alkylation unit, which must be maintained at 40°C, provides 8.8 MJ/s of heat from the exothermic alkylation reaction. This is to be used in the fractionation column reboiler. It is expected that a vapor compression condensation cycle (heat pump) can be economically employed to provide the additional energy requirements for reboiling this system. All compressors, pumps, and heat exchange equipment will be purchased new.

You, as a design engineer for the engineering company licensing the alkylation process unit, are to decide the most profitable feed rate, operating conditions, and equipment configuration for this system. A preliminary flow diagram for a heat pump fractionation column is shown in Figure P7-6-1. Capital costs for the alkylation unit exchanger, piping, instrumentation, and tray modifications (if required) may be considered to be constant for all schemes, therefore, only the compressor, pumps, and fractionator heat exchange equipment need be considered for capital cost analysis.

For the fractionation calculations, all the propane in the feed may be assumed to go overhead, and the split between *i*-butane and *n*-butane may then be treated as a *binary system*.

For this system, constant molal overflow is a reasonable assumption; therefore, the use of a McCabe–Thiele diagram will give valid results.

Exchange with the alkylation unit (8.8 MJ/s) is attractive, since the temperature rise upon vaporization is minimal. Hence, 32°C is used as the inlet and outlet temperature of this stream to and from the reactor exchanger. This temperature controls the operating pressure of the fractionation column.

Technical Data
1.   Physical Properties

	Propane	i-Butane	n-Butane
Molecular formula	C_3H_8	C_4H_{10}	C_4H_{10}
Molecular weight	44.094	58.120	58.120
Normal boiling point (°C)	−42	−12	0
Critical temperature (K)	370	408	425
Critical pressure (bara)	42.6	36.5	38
Liquid density (kg/m³ at			
15°C)	507	563	584

2.   Equilibrium Vaporization Constants See Perry Figure 13-6.
3.   Enthalpy

Temperature (°C)	Liquid Enthalpy (kJ/kg)			Vapor Enthalpy (kJ/kg)								
				1.7 bara			3.5 bara			7.0 bara		
	Propane	i-Butane	n-Butane	Propane	i-Butane	n-Butane	Propane	i-Butane	n-Butane	Propane	i-Butane	n-Butane
15	0	0	0	374	346	372	367	337	363	356	318	342
27	30	26	26	393	365	390	388	357	383	376	342	365
38	63	53	53	411	383	409	407	379	404	397	365	388
49	95	84	84	432	404	430	428	397	423	418	386	409
60	132	114	113	453	425	451	448	421	446	439	409	432
71	172	144	142	474	446	472	469	442	467	462	430	455

Figure P7-6-2 may be used to find liquid volumes for all streams when temperatures differ from 15°C.

From a Mollier diagram of overhead vapor, Figure P7-6-3 was prepared to show the enthalpy change for isentropic compression for various suction pressures from 1.4 to 4.15 barg.

Basic Design Engineering Data
4.   Battery-Limit Conditions

	Temperature (°C)	Pressure (barg)
Feed	27	Adequate (no feed pump required)
Net overhead product	43 maximum	7.0 minimum
Net bottoms product	43 maximum	7.0 minimum

Figure P7-6-2   Thermal expansion factor.

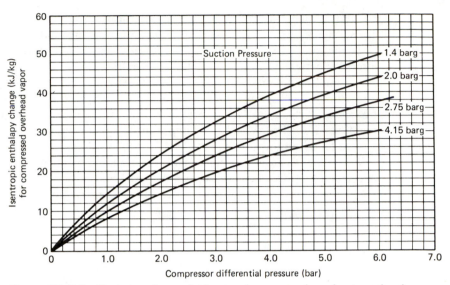

Figure P7-6-3   Enthalpy change for isentropic compression of net overhead vapor.

5. *Economic Guidelines*

The refiner uses the following guidelines. A \$1/day reduction in operating utilities will be equivalent to a \$1300 capital outlay. The value of the fractionation column bottoms product is equal to that of the fractionation column feed.

7-7 **Dicyanobutene Reactor System[3]**

Nylon 66 is produced by the condensation polymerization of adipic acid, $HOOC(CH_2)_4COOH$, and hexamethylenediamine, $H_2N(CH_2)_6NH_2$. Both monomers are produced by multistep syntheses with a high overall yield. We are concerned with one of the steps in making hexamethylenediamine (HMD), namely, the cyanation of dichlorobutene (DCB) to dicyanobutene (DNB).

The reactions in the synthesis of HMD are as follows.

$$C_4H_6 + Cl_2 \rightarrow C_4H_6Cl_2 \tag{P7-7-1}$$
$$\text{dichlorobutene (DCB)}$$

$$C_4H_6Cl_2 + 2NaCN \rightarrow C_4H_6(CN)_2 + 2\,NaCl \tag{P7-7-2}$$
$$\text{dicyanobutene (DNB)}$$

$$C_4H_6(CN)_2 + H_2 \rightarrow C_4H_8(CN)_2 \tag{P7-7-3}$$
$$\text{adiponitrile (ADN)}$$

$$C_4H_8(CN)_2 + 4H_2 \rightarrow NH_2(CH_2)_6NH_2 \tag{P7-7-4}$$
$$\text{hexamethylenediamine (HMD)}$$

The cyanation reaction (Equation P7-7-2) is carried out in an aqueous medium using a copper cyanide complex catalyst. Pilot plant studies have shown that control of pH and temperature is crucial.

[3]Based on the 1981 AIChE Student Contest Problem. (American Institute of Chemical Engineers, by permission.)

Also, materials of construction are important; only glass-lined steel or Hastelloy C appears to be adequate. Since both the raw materials and materials of construction are expensive, it will be necessary to carry out a thorough evaluation of alternate designs to find the economic optimum.

You are an engineer in a chemical engineering consulting, design, and construction company. Your recent work has been in nylon intermediates processes. A client has delivered basic data from a pilot plant study of the cyanation reaction. The client would like your company to provide the design of a reactor system based on these data.

The client's basic data contain information on the rate of reaction, materials of construction, investment and operating costs, and physical properties.

You are to design a continuous reactor system to convert DCB to DNB.

Letter of Commission

J. Q. Engineer
DOALL Chemical Engineering Co.
Richtown, Texas 98765

Dear J. Q.:

To increase our nylon intermediates capacity, we are planning to install a facility for producing dicyanobutene (DNB) from dichlorobutene (DCB). The process (see Figure P7-7-1) involves cyanation of DCB with aqueous sodium cyanide in the presence of a soluble copper cyanide complex catalyst.

Our Technical Division will design the feed preparation, product recovery, refining, and catalyst recovery areas of the process, and the supporting services. We request your services in designing a continuous reactor system. The annual production of crude DNB (100% DNB basis) shall be 96,000 metric tons, based on 8000 operating hours per year.

The high cost of the raw materials and materials of construction requires that you provide an optimum economic design. Our Economics Division will provide a guide for optimizing in accord with our internal evaluation practices. We plan to begin construction next year. The expected midpoint of construction should be in two years, with startup three years from today. The proof year in which the project economics will be evaluated is three years after startup.

Very truly yours,

Jane D. Manager

Jane D. Manager
Magic Monomers, Inc.

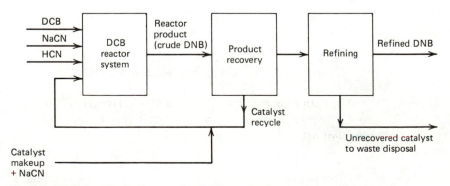

Figure P7-7-1   Block diagram for DNB synthesis process.

Basic Data Report

Process for the Manufacture of
Dicyanobutene from Dichlorobutene
by Aqueous Cyanation

Prepared by Elizabeth B. Basic
MAGIC MONOMERS, INC.
Corporateville, U.S.A.

During the past two years, the Research and Technical divisions of Magic Monomers, Inc., have made extensive studies on alternate routes for making dicyanobutene, an intermediate in manufacturing polyamide resins. Special circumstances related to supply and production of other company products show that cyanation of dichlorobutene with aqueous sodium cyanide offers the most favorable economics.

The process chemistry has been worked out in considerable detail, and processing problems including corrosion have been studied in the pilot plant. The results are recorded in several reports filed under the general title, "Dicyanobutene from Dichlorobutene." This report provides only the basic data for developing reactor designs and economic evaluations.

PHYSICAL PROPERTIES

The physical properties of the feed and product streams are presented in Table P7-7-1.

The DCB feed and the DNB produced are mixtures of isomers; however, these mixtures can be treated as single entities with the average properties given in Table P7-7-1. A number of by-products are formed, but these can be treated as inerts. For sizing the reactor, use the average reactor mixture properties.

OVERALL REACTION

The cyanation of DCB is carried out by reacting DCB with aqueous sodium cyanide in the presence of a soluble copper cyanide complex.

$$C_4H_6Cl_2 + 2NaCN \rightarrow C_4H_6(CN)_2 + 2NaCl$$
$$\text{(DCB)} \qquad\qquad Cu(CN)_x \quad \text{(DNB)}$$

The kinetics of this reaction show an interesting behavior. The catalyst is soluble only in the aqueous phase; therefore the reaction proceeds only in that phase. Since DCB is only slightly soluble in the aqueous phase, the reaction starts out slowly. However, the DNB produced enhances the solubility of the DCB and as soon as the DNB begins to form, the reaction rate increases until the solubility of both DCB and DNB are at a maximum. When the reaction is run by the batch or semibatch method, the reaction starts slowly, increases to a maximum, and finally, when the DCB feed is depleted, the rate falls off and the reaction stops.

PILOT PLANT OPERATION

A schematic flowsheet of the pilot plant operation is shown in Figure P7-7-2.

The reactor, a 25 liter, jacketed, glass-lined vessel with an agitator was operated in a semibatch manner. The reactor was charged initially with DCB and catalyst solution, and brought to temperature. Sodium cyanide solution was then fed at a rate that held the pH in the range of 5.0–5.5. An aqueous solution of HCN was added continuously to neutralize excess alkali (NaOH, NH$_3$, and Na$_2$CO$_3$) in the NaCN feed. Exothermic reaction heat was removed by circulating cooling water through the jacket.

The reactor effluent was decanted. The crude DNB (organic phase) was sent to product recovery, and the heavier aqueous phase was sent to catalyst recovery.

TABLE P7-7-1
PHYSICAL PROPERTIES

	Weight Percent	Molecular Weight
Refined DCB		
DCB	99.25	125
Miscellaneous organics	0.50	125
H_2O	0.25	18
Heat capacity (liq)	1.5×10^3 J/kg $\cdot$ K	
Density (liq)	1.16×10^3 kg/m³	
Viscosity (liq)	0.65×10^{-3} Pa $\cdot$ s	
Catalyst Solution (combination of recycle and makeup)		
$NaCu(CN)_2$	6.5	138.6
NaCN	17.3	49.0
H_2O	76.2	18.0
Heat capacity (liq)	3.8×10^3 J/kg $\cdot$ K	
Density (liq)	1.15×10^3 kg/m³	
Viscosity (liq)	1.5×10^{-3} Pa $\cdot$ s	
Sodium Cyanide Solution		
NaCN	26.0	49
Na_2CO_3	1.0	106
NH_3	0.3	17
NaOH	0.2	40
H_2O	72.5	18
Heat capacity (liq)	4.0×10^3 J/kg $\cdot$ K	
Density (liq)	1.13×10^3 kg/m³	
Viscosity (liq)	1.5×10^{-3} Pa $\cdot$ s	
Hydrogen Cyanide Solution		
HCN	9.0	27
H_2O	91.0	18
Heat capacity (liq)	4.0×10^3 J/kg $\cdot$ K	
Density (liq)	1.00×10^3 kg/m³	
Viscosity (liq)	1.0×10^{-3} Pa $\cdot$ s	
Average Reactor Mixture		
Heat capacity (liq)	3.4×10^3 J/kg $\cdot$ K	
Density (liq)	1.13×10^3 kg/m³	
Viscosity (liq)	1.3×10^{-3} Pa $\cdot$ s	

pH Control   Control of pH was found to be most important. It governs the concentration of the active catalyst species and must be held within the range of 5.0–5.5. If the pH drops to 4.5, or increases to 6.0, the reaction rate drops by an order of magnitude. Since the NaCN feed is controlled by pH measurement, it is necessary to neutralize small amounts of NaOH, NH₃, and Na₂CO₃ contaminants present in the NaCN. This excess alkali is not consumed in the reaction, and if not neutralized by addition of aqueous HCN, it would interfere with control of the NaCN addition. The HCN addition can be controlled accurately with a millivolt readout from a special electrode system. The cyanide salts formed in the neutralization react with the DCB.

Catalyst   The optimum catalyst ratio was found to be 0.038 kg of copper per kilogram of DCB in feed.

Temperature Control   Temperature control is also important. The preferred temperature is $80 \pm 2°$C. At higher temperatures, yield loss increases from hydrolysis and irreversible

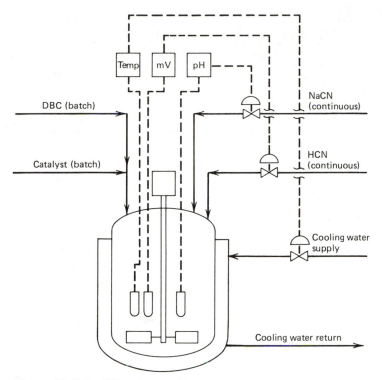

Figure P7-7-2   Pilot plant semibatch reactor.

polymerization of the DCB. At lower temperatures the reaction rate slows appreciably; and at 76°C the DNB starts to crystallize on the heat transfer surfaces.

Agitation   Tests to determine the degree of agitation needed showed that the rate of reaction increased with increasing agitation, but became constant at a power input of 0.6 kW/m³. The data suggested that only enough agitation was required to disperse the organic and aqueous phases to the extent that the aqueous phase was always saturated with DCB.

Summary of Experimental Results   Numerous runs were made in the pilot plant reactor. For convenience, the results of several runs under the most favorable conditions are summarized, along with the conditions, in Table P7-7-2. The results are expressed as percentage conversion of DCB versus time from the start of sodium cyanide addition.

In a reaction of this type, the overall rate is a complex relationship between the concentration of reactants (DCB, cyanide, and catalyst) in the aqueous phase. Since the DCB is only slightly soluble in the aqueous phase, the overall rate involves the mass transfer of the DCB into the aqueous phase. The rate for design purposes can be determined directly from the conversion data in Table P7-7-2.

REACTOR DESIGN SPECIFICATION

The reactor conditions and specifications are summarized in Tables P7-7-3 and P7-7-4. Note that the reaction fluid is severely corrosive. Only glass-lined steel or Hastelloy C was satisfactory. Glass-lined piping is unacceptable because it is prone to leak and is more expensive to maintain.

The instrumentation needs are also shown in Table P7-7-4. Again, because of the corrosive material, installed spare instrument loops are needed for the pH and millivolt control.

TABLE P7-7-2

DCB REACTION KINETICS: PILOT PLANT DATA

Time (min)	DCB Conversion (%)
0	0.00
10	6.21
20	13.33
30	21.48
40	30.82
50	41.52
60	53.78
70	67.82
71	69.33
72	70.86
73	72.42
74	73.99
75	75.59
76	77.21
77	78.85
78	80.52
79	82.21
80	83.77
82.5	87.38
85.0	90.26
87.5	92.54
90.0	94.31
92.5	95.68
95.0	96.73
97.5	97.53
100.0	98.14
102.5	98.60
105.0	98.95
107.5	99.21
110.0	99.41
112.5	99.56
115.0	99.67
Catalyst ratio	0.038 kg Cu/kg DCB in initial charge
Reaction temperature	$80°C \pm 2$
Cyanide consumption	8% of the CN^- in the catalyst stream is consumed in the reaction; 100% of CN^- added in sodium cyanide and hydrogen cyanide streams reacts.

The heat transfer specifications are in Table P7-7-5. Included are the heat transfer coefficients measured in the pilot plant. We have also included the formulas for estimating coil and jacket areas in a vessel.

Possible designs included jacketed vessels, internal coils, and/or external heat exchangers with pumped recirculation. The pilot plant had serious pump maintenance and heat exchanger fouling problems with an external loop. Any external pumping and heat exchange system must have an installed duplicate system.

TABLE P7-7-3
DESIGN CONDITIONS

Operating temperature 80°C $\pm$ 2
Design pressure 3.45 barg
Feed temperatures (°C)
  Refined DCB, 60
  Catalyst solution, 60
  Sodium cyanide solution, 40
  Hydrogen cyanide solution, 40
Reaction
The heat of reaction for the DCB to DNB reaction was determined experimentally at the reaction temperature of 80°C, and found to be -1.023×10^8 J/mol DCB consumed.
Yield
The yield loss of DCB to by-products in the reactor is 6 percent regardless of conversion when the reaction is carried out at 80°C $\pm$ 2. Any unconverted DCB constitutes an additional yield loss.
Catalyst
The catalyst is composed mainly of recycled material. The catalyst to DCB ratio should be the same as for the pilot plant tests. Catalyst loss is 0.5 percent through the reactor system.

ECONOMIC CRITERIA

Magic Monomers management has agreed to use incremental return on incremental investment as the basis for optimizing process equipment. This is based on net profit after taxes. A value of i'_i equal to 15 percent has been specified as the cutoff limit. The proof year should be the third year of operation.

Cost of Raw Materials	
DCB solution	$0.62/kg
Catalyst solution	$0.30/kg
NaCN solution	$0.082/kg
HCN solution	$0.0192/kg

TABLE P7-7-4
REACTOR DESIGN SPECIFICATIONS

Instrumentation
pH control for NaCN addition to reactor (provide duplicate installation)
Millivolt control for HCN addition to reactor (provide duplicate installation)
Temperature control for cooling water to any coils, jackets, and heat exchangers
Flow rate control for maintaining the proper ratio of catalyst to DCB in the feed
Materials of Construction
Reactor vessels may be glass-lined steel or Hastelloy C; piping must be Hastelloy C
See Table P7-7-5 for materials for heat transfer equipment.

TABLE P7-7-5
HEAT TRANSFER SPECIFICATIONS

Heat of reaction must be removed by one or more of the following
   Jacket on glass-lined steel vessel
   Internal helical coils of Hastelloy C using 50.8mm Sch 10 tubing
   External heat exchanger system, including pump, piping, and heat exchanger
Heat Transfer Area—Standard Reactor Vessel
Jackets
   Area of jacket is related to the working volume of reactor according to:

$$A = 3.7 \, V^{2/3}$$

where A = jacket area (m²)

   V = reactor working volume (m³)

Coils
   Maximum area of the coil is related to working volume of reactor according to:

$$A = 4.6 \, V^{2/3}$$

where A = coil area (m²)

   V = reactor working volume (m³)

External Heat Exchanger
There are no restrictions on the area that can be supplied in an external loop.
Heat Transfer Coefficients
The overall heat transfer coefficients U_0 for the various modes outlined above are as follows.

Jacketed glass lined steel	80 J/m² · s · K
International Hastelloy C coils	120 J/m² · s · K
External heat exchanger Hastelloy	150 J/m² · s · K

Cooling Water
Cooling water is available at 30°C. Assume a maximum rise of 10°C.

Direct cost for each control loop will be $7000, including equipment, material, and labor.

Assume that operating labor, operating supervision, and overheads will not vary with the size of the process equipment.

Assume straight line depreciation over a 12-yr period.

Chapter Eight
PROFITABILITY (CASH FLOW) ANALYSIS

Two factors limit the traditional balance sheet accounting method outlined in Chapter Six: the time value of money and the tax impact of depreciation.

TIME VALUE OF MONEY

We tend to think of a monetary unit—the dollar, for example—as a fixed standard like the international meter. In reality, even when inflation and currency fluctuations are nonexistent, dollars should be thought of as perishable. The lesson is as old as the Bible, where Jesus told of the unwise servant (Matthew 25:14–30). This man buried money left in his charge (measured in "talents" at the time) rather than risk loss and the wrath of a stern master. A wise servant, on the other hand, invested his portion in the money market and generated a profit for the master. Profound in a spiritual sense, the literal meaning is also genuine. Money should be thought of as a commodity capable of growth and having time-dependent value. Money to the entrepreneur is much like cattle to a rancher. An ambitious rancher would rather have 10 cows now than 10 cattle 3 yrs from now, because with wise management and hard work, 10 today could conceivably multiply to as many as 50 in 3 yrs. Similarly, one dollar today is more valuable to any competent investor than one dollar in the future.

The actual difference in time value depends on interest rate. An old object lesson involved asking a child if he or she would prefer a nickel today or a dime tomorrow. When the child requested the nickel immediately, it created an opportunity to explain the value of future rewards versus immediate gratification. The child, in truth, may have been the wise one, preparing for a future when, with skyrocketing interest rates, a nickel today might indeed be the better economic choice.

Since we, as planners, are vitally concerned with future projects, we must know the numerical difference between current and future dollars. To derive a quantitative relationship, consider the value in 5 yr of one dollar invested at an interest rate i (compounded annually). At the end of year 1, capital plus interest is:

$$C_1 = C_0 + iC_0 = C_0(1 + i) \tag{8-1}$$

Compounding means that interest is added to principal so that both receive interest in subsequent years. At the end of year 2, the investment is worth:

$$C_2 = C_1 + iC_1 = (1 + i)C_1 = (1 + i)(1 + i)C_0 = (1 + i)^2 C_0 \tag{8-2}$$

At the end of year 5, capital will be:

$$C_5 = C_0(1 + i)^5 \tag{8-3}$$

Given $C_0 = \$1$ and an interest rate of 10 percent,

$$C_5 = \$1(1 + 0.1)^5 = \$1.62 \tag{8-4}$$

A more interesting question is the converse, "How much money must one invest today to have one dollar in 5 yr?" Returning to Equation 8-3, C_5 is now known, and we solve for C_0:

$$C_0 = \frac{C_5}{(1 + i)^n} \tag{8-5}$$

Where C_5 is $1, n is 5, and i is 0.1,

$$C_0 = \frac{\$1}{(1 + 0.1)^5} = \frac{\$1}{1.62} = \$0.62$$

Thus, 62 cents, invested today at 10 percent annual interest, will be worth $1 in 5 yr. Hence, to convert future dollars to current value, one multiplies by a discount factor f_d

$$C_0 = C_n f_d \tag{8-6}$$

where, as illustrated,

$$f_d = \frac{1}{(1 + i)^n} \tag{8-7}$$

In cash flow analysis discussed later, future incomes or expenditures are multiplied by appropriate discount factors to reveal a transaction's real value in current dollars.

Anyone with money in the bank is exposed to another means of compounding interest—so-called continuous compounding, whereby interest is added continuously to principal rather than at year's end. In this case, the discount factor is given by [1] the following.

$$f_d' = [\exp(ni)]^{-1} \tag{8-8}$$

For 5 yr at 10 percent interest, f_d' is 0.61. Compared with 0.62 for annual compounding, this difference is too slight to consider. Even though the world of commerce uses continuous compounding, because of convenience and simplicity, annual compounding remains the standard for process accounting. For a detailed discussion of alternate compounding techniques, see Holland et al. [1].

ILLUSTRATION 8-1 AUTOMOBILE PURCHASE

Consider the economics of purchasing a new car. Assume that the price is $6000 and that you have the cash in hand. You have two alternatives: to pay cash or to invest

TABLE 8-1

BEHAVIOR OF PRINCIPAL AND INTEREST COMPOUNDED ANNUALLY

Completion of Year	Principal During Year	Interest at Year End
1	6000	600
2	6600	660
3	7260	726
4	7986	799
5	8785	879
6	9664	966
	Final principal = $9664 + $966 = $10,630	

$6000 at 10 percent interest while taking out an automobile loan at 15% interest. In both investment and loan, assume that interest is compounded annually and that all transactions occur at the end of the year.

1 Compute the value of $6000 at the end of year 6, if not spent at all.

2 Compute the cost of the automobile in current dollars if you elect to retain access to emergency capital by investing the principal at 10% interest. The loan will be paid from that $6000 investment in installments of $1000 plus interest until the bank account is depleted.

The problem is illuminated by examining cash flow as projected over the 6-yr period. For part 1, it is illustrated in Table 8-1.

This could also have been calculated from Equation 8-3 or 8-5.

$$C_n = C_0(1 + i)^n = \frac{C_0}{f_d}$$

$$C_6 = \$6000(1 + 0.1)^6 = \frac{\$6000}{0.564} = \$10,630$$

For part 2, the projected cash flow is illustrated in Table 8-2. At the end of year 6, $1192 plus $279 or $1471 must be provided from other resources to complete the transaction. This means an additional cost of $1471 to maintain access to emergency

TABLE 8-2

ECONOMICS OF SIMULTANEOUS LOAN AND INVESTMENT METHOD FOR AUTOMOBILE PURCHASE

	(1)	(2)	(3)	(4)	(5)	
Completion of Year	Invested Principal Throughout Year	Interest Received on Investment (i = 0.10)	Outstanding Loan Balance	Interest Owed on Loan[a]	Loan Payment[a]	Net Cash Flow (2) − (4) − (5)[a]
1	$6000	$600	$6000	($900)	($1000)	($1300)
2	4700	470	5000	(750)	(1000)	(1280)
3	3420	342	4000	(600)	(1000)	(1258)
4	2162	216	3000	(450)	(1000)	(1234)
5	928	93	2000	(300)	(1000)	(1207)
6	(279)	(42)	1000	(150)	(1000)	(1192)

capital. In current dollars, the difference is considerably less, as found by Equation 8-5.

$$C_0 = 1471f_d = \frac{1471}{(1 + 0.1)^6} = 1471(0.564) = \$830$$

This means that $6830 invested today would cover the complete purchase under the financial conditions stated.

CASH FLOW ANALYSIS

To define the economic performance of a manufacturing venture, an analyst must predict various sources and sinks of money throughout the lifetime of a project. The result is converted to a numerical index of the project's worth to an investor. Not only is time a factor, but the economic behavior of depreciation and other allowances must be considered.

To visualize the economic behavior of a project, let us return to a modified form of Figure S2-1 as shown in Figure 8-1. As described in Chapter Six, depreciation, in modern times, has unique significance to investors because of federal tax policies. In the traditional balance sheet approach, depreciation is treated as a manufacturing expense. In reality, it returns to investors the same as net profit after taxes. More significantly, as illustrated in Figure 8-1, depreciation passes through a separate conduit thus bypassing the profit pipeline which suffers from a big leak leading to the federal treasury. Given a choice, investors would prefer to have all their money return as depreciation. Governments, wary of this, limit depreciation to a total no greater than the original fixed capital investment. Because the value of money is perishable, investors would prefer to recover invested capital through depreciation as quickly as possible. To encourage new investments, governments often compromise by allowing accelerated depreciation schedules such as that computed by the sum-of-years'-digits method mentioned in Chapter Six. For the same reason, various other annual adjustments A_A are sometimes allowed. These also pass through the tax-free conduit along with depreciation, as illustrated in Figure 8-1.

Figure 8-1   Cash flow considering the behavior of depreciation and other allowances.

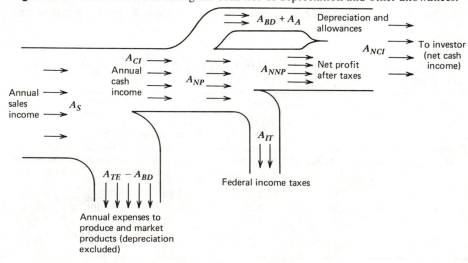

To an engineer, versed in complex material balance calculations, cash flow analysis should be elementary. One merely prepares an economic balance for each year of projected operation and assesses the cumulative cash flow to and from the investor's purse. To compensate for money's time value, cash flows are corrected by appropriate discount factors. Annual cash flows can be accessed simply by treating Figure 8-1 as a steady-state system with no accumulation, similar to a common pipeline system. The following example should clarify the calculation procedure.

ILLUSTRATION 8-2   CASH FLOW ANALYSIS

The balance sheet for a proposed project reveals a fixed capital investment C_{FC} of $350M. Annual total project expenses A_{TE} are estimated to be $150M and sales income A_S is projected to be $240M. Calculate annual and cumulative cash flows for an estimated project lifetime of 10 yr. Compute payback period (PBP), the discounted break-even period (DBEP), and net present value (NPV) for a 10 percent rate of return. What is aftertax rate of return? What is the discounted cash flow rate of return (DCFRR)? Assume that the plant is constructed in 3 yr with annual investments of $100M in the first and second years and $150M the third. Assume that working capital is $50M invested at the end of year 3. Because of startup problems, sales income for year 4 is only two thirds of normal. Disregard any extraneous allowances or credits. Assume straight line depreciation with zero scrap value.

Annual cash flow data can be assembled by executing a cash balance for each year of anticipated operation using Figure 8-1 as the system. Results are listed in Table 8-3. Annual cash investments A_I are required for construction costs and working capital during the first 3-yr. These show as negative cash flows in Table 8-3, although working capital appears again as a positive influx when the plant is shut down. Sales income, except for startup in year 4, is constant. Sources of other numbers leading to cash flow values should be obvious. Cumulative cash flow is merely the algebraic sum of annual cash flows. The first column of cumulative values shown in Table 8-3 is undiscounted; that is, no compensation is made for the time at which money is spent or received. For a more realistic analysis, annual cash flows are multiplied by discount factors appropriate to the year and interest rate in question, yielding a cumulative discounted cash flow in terms of current dollars. The result depends, of course, on the interest rate that one assumes. Thus, discounted cash flows are shown for two interest or "discount" rates, 10 and 15 percent, in Table 8-3.

All the economic questions posed in this illustration can be answered from the set of cash flow curves in Figure 8-2. These represent cumulative cash flow data taken from Table 8-3. Note that there are three curves, one undiscounted, one for a discount rate of 10 percent, and a third discounted at 15 percent.

Both DBEP and NPV depend on the discount factor chosen. *Pay back period* is the time that must elapse after startup until cumulative undiscounted cash flow repays fixed capital investment. In this illustration PBP is the point where undiscounted cash flow rises to the level of negative working capital (4.8 yr, as indicated in Figure 8-2). *Discounted break-even period* is the time from the decision to proceed until discounted cumulative cash flow becomes positive. For this case with $i = 0.10$, DBEP is 12 yr. *Net present value*, is the final cumulative discounted cash flow value at project conclusion. As illustrated in Figure 8-2, for $i = 0.10$, NPV = $32.7M, for $i = 0.15$, NPV is negative, $-$41.9M. *Discounted cash flow rate*

TABLE 8-3

CASH FLOW DATA FOR ILLUSTRATION 8-2 (MILLIONS OF DOLLARS)[a]

Completion of Year	Annual Capital Investment, A_I	Sales Income, A_S	Total Expenses Less Depreciation, $A_{TE} - A_{BD}$	Cash Income, A_{CI}	Depreciation, A_{BD}	Allowances, A_A	Net Profit, A_{NP}	Federal Income Taxes, A_{IT}	Net Profit After Taxes, A_{NNP}
1	(100)								
2	(100)								
3	(200)								
(Startup)									
4		160	(115)	45	35	—	10	5	5
5		240	(115)	125	35	—	90	45	45
6		240	(115)	125	35	—	90	45	45
7		240	(115)	125	35	—	90	45	45
8		240	(115)	125	35	—	90	45	45
9		240	(115)	125	35	—	90	45	45
10		240	(115)	125	35	—	90	45	45
11		240	(115)	125	35	—	90	45	45
12		240	(115)	125	35	—	90	45	45
13	50	240	(115)	125	35	—	90	45	45
(Shutdown)									

[a]Negative cash quantities in parentheses.

Figure 8-2   Cash flow profiles for Illustration 8-2 (based on Table 8-3).

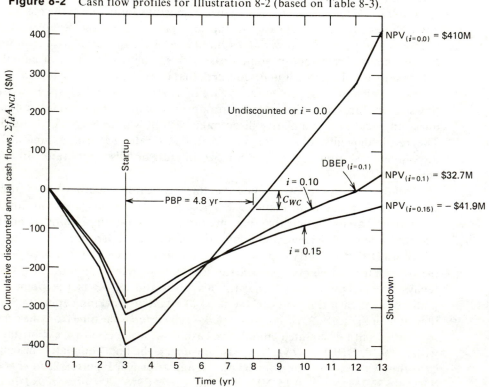

TABLE 8-3
CASH FLOW DATA FOR ILLUSTRATION 8-2 (MILLIONS OF DOLLARS)[a] (continued)

		$i = 0.10$			$i = 0.15$		
Cash Flow or Net Cash Income, $A_{NCI} = A_I + A_{BD} + A_A + A_{NNP}$	Cumulative Cash Flow, ΣA_{NCI}	Discount Factor, f_d	Discounted Cash Flow, $f_d A_{NCI}$	Cumulative Discounted Cash Flow, $\Sigma f_d A_{NCI}$	Discount Factor, f_d	Discounted Cash Flow, $f_d A_{NCI}$	Cumulative Discounted Cash Flow $\Sigma f_d A_{NCI}$
(100)	(100)	0.909	(90.9)	(90.9)	0.870	(87.0)	(87.0)
(100)	(200)	0.826	(82.6)	(173.5)	0.756	(75.6)	(162.6)
(200)	(400)	0.751	(150.2)	(323.7)	0.658	(131.6)	(294.2)
40	(360)	0.683	27.3	(296.4)	0.572	22.9	(271.3)
80	(280)	0.621	49.7	(246.7)	0.497	39.8	(228.5)
80	(200)	0.564	45.2	(201.5)	0.432	34.6	(194.0)
80	(120)	0.513	41.1	(160.5)	0.376	30.1	(163.9)
80	(40)	0.467	37.3	(123.2)	0.327	26.2	(137.9)
80	40	0.424	33.9	(89.2)	0.284	22.7	(115.0)
80	120	0.386	30.8	(58.4)	0.247	19.8	(95.2)
80	200	0.350	28.0	(30.4)	0.215	17.2	(78.0)
80	280	0.319	25.2	(4.9)	0.187	15.0	(63.1)
130	410	0.290	37.6	32.7	0.163	21.1	(41.9)

of return is the discount rate that yields an NPV of zero. In this cash, DCFRR falls between the two cases at approximately 12 percent. *Aftertax rate of return* is calculated from key balance sheet values for a typical operating year.

$$i = \frac{A_{NNP} + A_{BD}}{C_{TC}} \times 100 = \frac{\$45M + \$35M}{\$400M} \times 100 = 20\%$$

This illustrates how a few years' separation between investment and profits reduces the true or discounted cash flow of return from the simple balance sheet number.

In discussing economic attractiveness, decision makers usually speak in terms of three parameters: rate of return i, DCFRR, and NPV. As illustrated above, simple rate of return can be quite deceptive as an index of time profitability because it depends so much on when investments occur and when income is realized. For these reasons, DCFRR and NPV are the favorite indicators used by modern entrepreneurs [4–7]. Both result from the same calculation process and can be identified directly from cash flow curves of the form shown in Figure 8-2.

Discounted cash flow rate of return is designed to reflect the actual interest on investment. Because of this, it is useful for comparing alternate possibilities. If, for example, the DCFRR for a manufacturing project were less than interest available from safer, strictly financial ventures, one would be foolish to take the risk. Thus, corporate managers usually set minimum acceptable DCFRR values based on prevalent commercial interest rates.

Beyond their dependence on future and thus uncertain numbers, DCFRR methods have several other inherent weaknesses. One is the need for trial-and-error calculations to find a value of i that yields an NPV of zero. With advanced computational methods, this is a minor problem. Another deficiency exists when unusually high rates are found. Even though such values indicate high profitability, they are not, because of implicit assumptions in the procedure, true interest rates [6].

This discrepancy becomes greater as differences between DCFRR values and prevalent interest rates increase.

Both NPV and DCFRR data are essential when examining alternate investments. Superficially, one might be interested in the rate of return only. This, however, does not reflect investment scale. Assume, for example, that commercial interest rates are 15 percent and we have a choice of two ventures of similar risk. One offers 20 percent return and the second, 30 percent. With no other information, we would be inclined to favor the latter. If, however, the first project involved a capital investment of $1000 and the second, $1 billion, the rate of return is an incomplete measure of venture worth. To place these two projects in perspective, NPVs should be calculated at a 15 percent discount rate. This would reveal the overwhelming economic importance of the larger project. Again, however, with a complete cash flow profile like Figure 8-2, these questions, plus those concerning depth of financial exposure, payback, and others, can be answered.

ILLUSTRATION 8-3   THE GOLDENROD—ECONOMIC PROFILE

In Illustration 7-2, the proposed heat exchange improvement yields a savings in fuel cost of $1860/yr. It requires a fixed capital investment of $8200 and increased electricity costs of $22/yr. Assuming depreciation at sum-of-the-years'-digits rates and zero salvage value, evaluate the economics of this project. Estimate aftertax return on investment, PBP, and DCFRR. Assuming a discount rate of 10 percent, compute NPV and DBEP. Assume that construction can be completed within a year and that operating lifetime will be 10 yr.

A decision by Goldenrod managers can be based on a cash flow analysis of the project in isolation. The project, for instance, causes no change in sales income. Rather, a positive cash flow is created by a reduction in manufacturing expenses. During the first year there will, of course, be a negative cash flow of $8200 for fixed

TABLE 8-4

CASH FLOW DATA FOR INSTALLATION OF A HEAT EXCHANGE MODULE IN THE GOLDENROD[a]

Completion of Year	Annual Capital Investment, A_I	Total Expenses Less Depreciation, $A_{TE} - A_{BD}$[b]	Cash Income, A_{CI}	Depreciation, A_{BD}	Allowances, A_A	Net Profit, A_{NP}	Federal Income Taxes, A_{IT}
1	(9000)						
2		(690)	690	1270	—	(580)	(290)
3		(690)	690	1145	—	(455)	(228)
4		(690)	690	1020	—	(330)	(165)
5		(690)	690	890	—	(200)	(100)
6		(690)	690	765	—	(75)	(38)
7		(690)	690	635	—	55	28
8		(690)	690	510	—	180	90
9		(690)	690	380	-	310	155
10		(690)	690	255	—	435	218
11	800	(690)	690	130	—	560	280

[a]Unless otherwise noted, quantities in parentheses are negative cash flows.

[b]This is a negative expense or a positive contribution to cash income.

capital and about \$800 due to working capital. By reference to Figure 8-1, we note a change in annual cash income A_{CI} of \$1860 less \$22 for increased electricity and minus about 14 percent of fixed capital for additional maintenance, overhead, property taxes, and supplies. We assume no significant differences in operating labor with this change. Annual net savings are thus:

$$-(A_{TE} - A_{BD}) = \$1860 - \$22 - 0.14(\$8200) = \$690$$

This is actually less than depreciation during early years of the project. Thus, to supply the cash balance, net profit after taxes from other parts of the corporation can be channeled to the depreciation conduit, picking up taxes as a credit in the process. The cash flow scheme is enumerated in Table 8-4 and illustrated in Figure 8-3. Depreciation rates are based on Equation 6-6.

Annual fuel oil savings are essentially swamped by capital-related expenses. Positive cash flows are insufficient to repay capital within the project lifetime, even with zero discount rate. Aftertax rate of return varies from year to year because of variable depreciation. In year 3 it is

$$i = \frac{A_{NNP} + A_{BD}}{C_{TC}} \times 100 = \frac{-227 + 1145}{9000} \times 100 = 10\%$$

In year 7, $i = (27 + 635)/9000 \times 100 = 7.4$ percent, and in year 10, $i = (218 + 255)/9000 \times 100 = 5.3$ percent. The PBP is larger than estimated project lifetime, since undiscounted cumulative cash flow is still negative in year 11. DCFRR is also negative and was not determined. NPV, based on a 10 percent discount rate, is negative (−\$3743), representing a loss on the project.

ALTERNATE INVESTMENTS

Economic comparison of process alternatives can be accomplished in an accurate and effective way using *incremental return on incremental investment*. This is done

TABLE 8-4
Continued

			$i = 0.10$		
Net Profit After Taxes, A_{NNP}	Net Cash Flow or Net Cash Income, $A_{NCI} = A_I + A_{BD} + A_A + A_{NNP}$	Cumulative Cash Flow, ΣA_{NCI}	Discount Factor, f_d	Discounted Cash Flow, $f_d A_{NCI}$	Cumulative Discounted Cash Flow, $\Sigma f_d A_{NCI}$
	(9000)	(9000)	0.909	(8181)	(8180)
(290)	980	(8020)	0.826	809	(7372)
(227)	918	(7102)	0.751	689	(6683)
(165)	855	(6247)	0.683	584	(6099)
(100)	790	(5457)	0.621	491	(5608)
(37)	728	(4729)	0.564	411	(5197)
27	662	(4067)	0.513	340	(4857)
90	600	(3467)	0.467	280	(4577)
155	535	(2932)	0.424	227	(4350)
218	473	(2459)	0.386	183	(4167)
280	1210	(1249)	0.350	424	(3743)

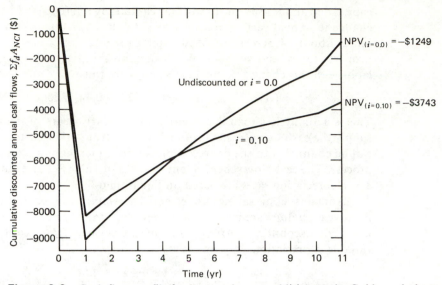

Figure 8-3   Cash flow profile for heat exchanger addition to the Goldenrod plant.

using the same mathematical concepts employed in Chapter Seven except that the optimization variable may be a discrete quantity such as number of reactors or some other function that changes in steps rather than continuously. To assess the viability of an alternative, one option is selected as a basis. Then other alternatives are compared with the base case. Incremental return on incremental investment is given by:

$$i_i' = \frac{A_{NNP} - A_{NNP,b}}{C_{TC} - C_{TC,b}} \tag{8-9}$$

or

$$i_i = \frac{(A_{NNP} + A_{BD}) - (A_{NNP} + A_{BD})_b}{C_{TC} - C_{TC,b}} \tag{8-10}$$

where subscript b denotes the base case. Either equation may be pertinent, depending on whether return is defined to include depreciation. Selection of the most attractive alternative involves a directed trial-and-error search. The base case and change increments are usually obvious from context. For example, if the question is one of well-stirred reactors in series, a single reactor might be chosen as the base case and a second reactor postulated. Values of i_i or i_i' would be calculated for this change. If the incremental return were greater than the cutoff value established by financial advisors, a third reactor would be postulated. Returns i_i or i_i' would be computed for this second change using two reactors as the base case. The procedure would be repeated until the rate of return dropped below the cutoff value.

CASH FLOW AND RATE OF RETURN CRITERIA

Cash flow and incremental return techniques can be combined to define the amount of annual income or savings that should be generated per dollar of capital expense. This is demonstrated with an illustration.

ILLUSTRATION 8-4   CASH FLOW–RATE OF RETURN CRITERIA

Analyze the feasibility of investing additional capital in a process having an anticipated lifetime of 5 yr. Construction will require 1 yr. Depreciation is assumed to be straight line over the operating years. Salvage value is zero. Financial advisors specify that any proposed investment must yield a net present value equal to 1 percent of total capital multiplied by the investment lifetime in years. They also estimate the prevailing interest rate over the period to be 12 percent. Establish the economic yield required of each dollar invested.

Assume the capital is ΔC_{TC}. As illustrated in Figure 8-1, the change in annual cash flow due to this incremental investment will be:

$$\Delta A_{NCI} = \Delta A_{NNP} + \Delta A_{BD} \tag{8-11}$$

Based on Table 6-1, annual expenses A_{TE} or annual sales A_S must be changed by the investment in a way that will yield the desired return. Assume, in this case, that sales price or volume is not affected. This means that the change in net annual profit will be:

$$\Delta A_{NP} = \Delta (A_S - A_{TE}) \tag{8-12}$$

or

$$\Delta A_{NP} = -\Delta A_{TE} \tag{8-13}$$

since ΔA_S is constant. At a 50 percent tax rate, incremental net profit after taxes is:

$$\Delta A_{NNP} = 0.5\Delta A_{NP} = -0.5\Delta A_{TE} \tag{8-14}$$

According to Equations 8-11 and 8-14, the change in annual cash flow is:

$$\Delta A_{NCI} = -0.5\Delta A_{TE} + A_{BD} \tag{8-15}$$

Since the proposed investment does not affect sales, it must yield a saving in raw materials, catalysts, utilities, labor, or some other component of manufacturing expense listed in Table 6-1. I term this the annual recovered expense ΔA_R. Annual total expense will be reduced by the recovered component but increased because of certain factors reflecting the change in capital.

$$\Delta A_{TE} = 0.04\Delta C_{FC} + 0.005\Delta C_{FC} + 0.02\Delta C_{FC} + 0.02\Delta C_{FC} + \Delta A_{BD}$$
$$+ 0.005\Delta C_{FC} - \Delta A_R \tag{8-16}$$

Those factors involving fixed capital are for maintenance, supplies, overhead, taxes, insurance, and administrative expense. These add up, in this case, to $0.09\Delta C_{FC}$. With these terms combined, Equation 8-16 substituted into Equation 8-15 yields:

$$\Delta A_{NCI} = -0.5(0.09) \, \Delta C_{FC} + 0.5\Delta A_{BD} + 0.5\Delta A_R$$
$$= -0.045\Delta C_{FC} + 0.5\Delta A_{BD} + 0.5\Delta A_R \tag{8-17}$$

(With high depreciation rates such as in this illustration, the second term on the right of Equation 8-17 more than compensates for the first. The result is a positive change in cash flow even with no contribution from ΔA_R. This consequence of tax credits for depreciation explains why government policy is so important in investment decisions.)

TABLE 8-5

CASH FLOW DATA FOR ILLUSTRATION 8-4[a]

$$\Delta C_{TC} = \$1000 \text{ (assume } \Delta C_{WC} = 0.15 \, \Delta C_{FC} = \$130)$$

$$\Delta C_{FC} = \$870$$

$$A_{BD} = \frac{\$870}{5} = \$174$$

Completion of Year	Annual Capital Investment, A_I	Capital-Related Contribution, $-0.045 \, \Delta C_{FC}$	Depreciation Contribution, $0.5 A_{BD}$	Net Cash Income, ΔA_{NCI}	Discount Factor $(i = 0.12)$, f_d
1	(1000)			(1000)	0.89
2		(39)	87	$48 + 0.5 \Delta A_R$	0.80
3		(39)	87	$48 + 0.5 \Delta A_R$	0.71
4		(39)	87	$48 + 0.5 \Delta A_R$	0.64
5		(39)	87	$48 + 0.5 \Delta A_R$	0.57
6	130	(39)	87	$178 + 0.5 \Delta A_R$	0.51

[a] Negative cash quantities in parentheses.

Based in Equation 8-17, a cash flow profile for the proposed investment can be constructed. Such, based on a total capital investment of $1000, is documented in Table 8-5. Net present value, the algebraic sum of annual discounted cash flows, is

$$\text{NPV} = \Sigma f_d \, \Delta A_{NCI} \tag{8-18}$$

or, in this illustration,

$$\text{NPV} = 0.89(-1000) + (48 + 0.5 \, \Delta A_R)(0.80 + 0.71 + 0.64 + 0.57 + 0.51)$$

$$+ 0.51(130)$$

$$= -824 + 3.23(48 + 0.5 \Delta A_R)$$

$$= -669 + 1.62 \Delta A_R$$

According to corporate policy, the NPV must be 1 percent times 6 yr or 6 percent of total capital. Thus,

$$NPV = 0.06 \Delta C_{TC} = -669 + 1.62 \Delta A_R$$

Based on $1000 total capital,

$$\Delta A_R = \frac{669 + 60}{1.62} = \$450$$

Thus, to satisfy investment criteria, each dollar of capital investment must yield an annual process saving or recovered expense of 45 cents.

INFLATION, RISK, AND OTHER VARIABLES

In reading the preceding cash flow analyses, you probably wondered why inflationary impacts were not included. Basically, inflation can be considered to be a component of interest. That is, to compensate for inflation or deflation, investors vary the minimum percentage return that they will tolerate. In process analysis, we are usually given a minimum value of i by management. We can use this to pursue a cash flow profile. The acceptable rate of return does depend on inflation, but responsibility for its definition rests appropriately with financial specialists rather than with engineers.

Changes in such factors as utility costs, salaries, and tax rates could be anticipated in a cash flow analysis, but such variables affect the marketplace and one's competitors as well. Thus, if a process offers superior economic prospects under today's conditions, it will probably retain these advantages in the future and it is unnecessary to predict changes or trends. To compensate for variations in the balance sheet with time, future managers will adjust selling prices to retain earnings at the projected rate of return. Some processes, such as those with a large utilities component, are vulnerable to more capricious factors. Unexpected changes may shift their economic position relative to competitors. Anticipation of this is part of risk analysis.

Additional factors and adornments can be added to refine and complicate a cash flow analysis. Sophisticated sales projections, land costs, annual changes in working capital, and various tax allowances and other credits are examples. If these quantities are known, there is no computational reason for not including them. However, we are dealing with the future and its uncertainties. Based on past history filled with cyclic variations in the world economy, changing interest rates, inflation, and war, an infatuation with too much speculative detail seems futile [7].

There are numerous indices of profitability other than those emphasized here, but most can be derived simply from the balance sheet or cash flow data illustrated thus far [2]. One could also examine impacts of numerous contingencies and probabilities. This is known as sensitivity or risk analysis. It simply involves asking the question, What if? With computer techniques, answers to such questions can be provided quickly and a proliferation of contingencies examined [3]. These and other statistical and probability exercises extend further into the realm of financial analysis than most engineers care to or need to proceed. With the tools at hand, you are equipped to discuss investment possibilities intelligently with anyone, corporate directors included.

REFERENCES

1   Holland, F.A., F.A. Watson, and J.K. Wilkinson, "Time Value of Money," *Chem. Eng.*, pp. 123–126 (Sept. 17, 1973); also part 4 of *Chem. Eng.* Reprint No. 215 (1975).

2   Holland, F.A., F.A. Watson, and J.K. Wilkinson, "Methods of Estimating Project Profitability," *Chem. Eng.* pp. 80–86 (Oct. 1, 1973), also part 5 of *Chem. Eng.* Reprint No. 215 (1975).

3   Holland, F.A., F.A. Watson, and J.K. Wilkinson, "Sensitivity Analysis of Project Profitabilities," *Chem. Eng.*, pp. 115–119 (Oct. 29, 1973); also part 6 of *Chem. Eng.* Reprint No. 215 (1975).

4   Holland, F.A., F.A. Watson, and J.K. Wilkinson, "Time, Capital and Interest Affect Choice of Project," *Chem. Eng.*, pp. 83–89 (Nov. 26, 1973); also part 7 of *Chem. Eng.* Reprint No. 215 (1975).

5   Holland, F.A., F.A. Watson, and J.K. Wilkinson, "Engineering Economics, parts 8, 9, 10, 11, *Chem. Eng.* Reprint, No. 215 (1975).

6   Horwitz, B.A., "The Mathematics of Discounted Cash Flow Analysis," *Chem. Eng.,* pp. 169–174 (May 19, 1980).

7   Weaver, J.B., "Project Selection in the 1980s," *Chem. Eng. News*, pp. 37–46 (Nov. 2, 1981).

PROBLEMS

8-1   Repeat Part 2 of Illustration 8-1, but consider the effect of federal income taxes, assuming that a buyer is in the 30 percent tax bracket (i.e., interest received is taxable at 30 percent, and 30 percent of interest paid returns to the buyer as a tax credit).

8-2   Repeat Illustration 8-2 but assume a project lifetime of 15 operating years. Federal tax policies allow accelerated depreciation in this case. Hence assume that plant is depreciated to zero value in the first 10 operating years at sum-of-the-years'-digits rates.

8-3   Alkylation Plant Evaluation[1]

A large refinery is operating an alkylation plant constructed during World War II. Expanded over the years, the unit has been made more efficient by minor technological improvements. However, the increased demands on this unit due to the need for lead-free gasoline will require still more investment. Already throughput to the unit has been cut back in an attempt to improve the quality of the product as needed for lead-free gasoline blends. Instead of investing still more money in what is basically old hardware, the company may prefer to construct a new unit of a design optimized for the forthcoming clear-octane requirements. The decision will require the economics of an optimized new plant that would produce 1590 m^3 per stream day (332 stream days a year) of debutanized motor alkylate with a clear (lead-free) research octane number of at least 93. Higher octane numbers would, of course, be desirable if equally profitable.

You are to establish the economics of the new plant.

Assume that in addition to this motor alkylate, 177 m^3 per stream day of heavy alkylate, or 10 percent by volume of the total reaction product, will be produced as the major by-product, which will be separated for solvent production.

A simplified flow scheme is shown in Figure P8-3-1. Furthermore, design calculations may suggest changes in heat exchange or pump alignment. The following information provides additional assumptions and process particulars that may be used in the development of an optimum plant design.

Basis for Economic Calculations

Motor alkylate is used as a blending stock with other gasoline components, and the Economics Department has developed the following product values (RVP = Reid vapor pressure).

---

[1]Based on the 1977 AIChE Student Contest Problem. (The American Institute of Chemical Engineers, by permission.)

Product and Byproduct Values
(Basis: 1982)

Motor Alkylate	Value ($/$m^3$)
Research octane number	
89	253
92	254
95	261
98	265

By-products	Value ($/$m^3$)
Propane	138
Butanes	155
Heavy alkylate	216

Feed streams available to the alkylation plant are listed, on the basis of 1982 stream costs or values, in Table P8-3-1.

Because it is available from other company-operated facilities, sulfuric acid (H_2SO_4) is to be used as the alkylation catalyst. Favorable experience has been obtained by other refineries of the same company using horizontal, internally stirred loop reactors containing heat exchange surfaces (see Perry, Figure 21-8). The characteristics of the recommended contactor are as follows.

Heat Exchange Contactor

Net Operating Volume (m^3)	Heat Exchange Surface (m^2)	Power Required (kW)	Costs, Including Drive and Settler (mid-1982)	
			Equipment Only	Installed Cost
50	790	300	$415,000	$1,540,000
	Overall heat transfer coefficient: 350 $J/m^2 \cdot s \cdot K$			

The cost of sulfuric acid exchange, 100 percent H_2SO_4 exchanged for as low as 85 percent H_2SO_4, may be assumed to be 7.0¢/kg on the basis of 100 percent H_2SO_4. Fifty percent sodium hydroxide (NaOH) may be purchased for 28.3 ¢/kg based on 100 percent NaOH. Fresh acid makeup costs 8.8 ¢/kg on the basis of 100 percent H_2SO_4.

If the project looks attractive, the final design will be completed in one year. The actual commitment of project capital will occur at the end of that year with the signing of a construction contract for a firm price. Progress payments will be made to the contractor as follows

First year of construction	15 percent
Second year of construction	55 percent
Third year of construction	30 percent

Because of the high concentration of H_2SO_4 and the relatively low temperatures used, carbon steel may be used for all vessels, pumps, and piping.

A 20 yr process life, beginning at startup is to be used. Assume 13 yr depreciation at sum-of-years'-digits rates with no salvage value at the end of the process life.

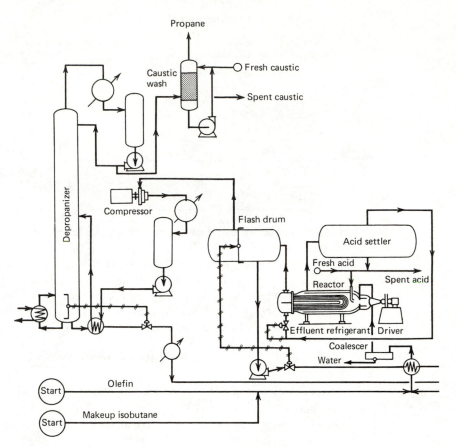

Figure P8-3-1 Flow sketch of refrigeration alkylation unit.

TABLE P8-3-1
HYDROCARBON FEED STREAMS AVAILABLE

Component	Composition (vol %)			
	Butylene from Catalytic Cracker	Isobutane from Gas Plant	Purchased Butylenes	Purchased Mixed Butanes
Propane	0.3	1.0	2.2	6.1
Isobutane	23.9	94.2	12.3	48.0
n-Butane	10.3	4.3	2.6	41.6
Butene-1	20.1	—	24.0	
Isobutene	12.7	—	33.1	
trans-Butene-2	25.2	—	19.3	
cis-Butene-2	7.5	—	6.5	
Isopentane		0.5	—	4.3
Availability (m³/day)	795	811	636	955
Cost/value ($/m³)[a]	156	149	162	157

[a] Basis: 1982 dollars.

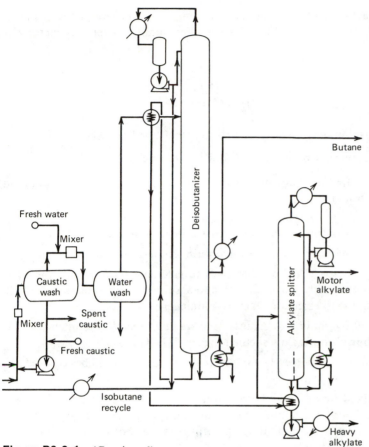

Figure P8-3-1 (Continued)

Calculate the discounted cash flow rate of return for the optimum process configuration and net present value at a 15 percent discount rate.

Process Notes

1 On the basis of the literature references and the operation of other alkylation plants in company facilities, the following range of variables will probably lead to an optimum plant design.

Reaction Temperature	4.5°C
iC_4 in reactor hydrocarbon effluent	75 vol %
Olefin space velocity, (volume of olefin)/(volume of acid) (hour)	0.25–0.45
Acid consumption	72 kg acid catalyst/m^3 *total* alkylate
F-1-0 (research octane number clear) debutanized motor alkylate	93–99

Note: Maximum olefin space velocity should be limited to 0.45 to ensure sufficient reaction time.

2 A high isobutane–butylene ratio is necessary to maintain isobutane concentration in the acid that will favor primary alkylation reactions and suppress

secondary reactions. The alkylate octane number can be computed from this ratio and other process variables. The so-called quality correlation factor F is given by

$$\frac{1}{F} = \frac{100 \, (SV)_o}{(I)_E \times (I/O)_F}$$

where $(I)_E$ = isobutane in reactor hydrocarbon effluent (vol %)

$\quad (I/O)_F$ = external isobutane–olefin ratio (volume)

$\quad (SV)_O$ = olefin liquid hourly space velocity, (volume olefin)/(volume acid) (hour)

Octane number as a function of this factor is displayed in Figure P8-3-2.

3 Use reactor effluent refrigeration to maintain temperature control.

4 Olefin space velocity SV_o should be used to calculate reaction volume. Use volumetric liquid hourly space velocity and assume that the reactors are half filled with catalyst to calculate reactor volume.

5 Feeds are dried by chilling to remove free water at reaction temperature. A coalescer is shown on the flow sketch (Figure P8-3-1).

6 If feed pumps are sized correctly, no interreactor pumps should be required for a multiple reactor system.

7 Purchased mixed butane may be required. Figure P8-3-1 does not indicate where it should be charged into the process; the optimum addition point may be chosen on the basis of capital or operating costs or both.

8 For this evaluation it should be assumed that acid life and alkylate quality are independent of acid strength.

Figure P8-3-2 Research octane number as a function of the correlation factor.

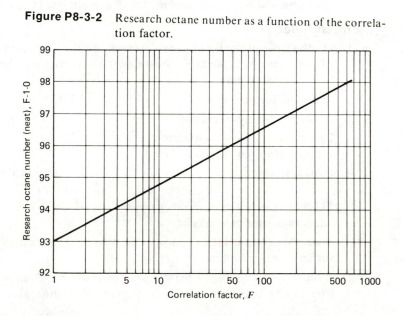

TABLE P8-3-2

TYPICAL REACTOR-PRODUCT COMPOSITIONS

Alkylate	Composition (Vol %)
Debutonized Motor alkylate	
iC_5	4.9
C_6	4.5
C_7	4.2
C_8	83.2
C_9	2.6
C_{10}	0.6
	100.0
Heavy alkylate	
C_{10}	0.8
C_{11}	19.9
C_{12}	76.2
C_{13}	2.5
C_{14+}	0.6
	100.0

Hint: The deisobutanizer may be calculated as two columns in series.

9   Acid carryover out of the acid settlers is assumed to be 100 ppmw. The base is hydrocarbon.

10   The true boiling point end temperature of the motor alkylate should be 166° C, and the RVP at 38° C of debutanized motor alkylate will be 0.24 bara. The yield of heavy alkylate is assumed to be 10 percent by volume of the total alkylate, or 177m³/stream day. The isobutane content of the liquid butane side stream is specified as 3 percent of the volume of the butanes. Assume that isopentane in the feed streams leaves in this stream. Table P8-3-2 contains the typical product compositions.

11   The recycle isobutane purity should be 95 percent by volume.

12   Propane recovery from the depropanizer should be 95 percent by volume, and the purity of the propane product stream should be 97 percent by volume.

13   Sufficient normal butane should be withdrawn in the deisobutanizer bottoms to yield a motor alkylate having an RVP at 38°C of 0.69 bara. Assume that RVP at 38°C is equivalent to the true vapor pressure and blend on a molar basis. The normal butane octane number is assumed to be 96 research, and the RVP may be assumed to be 3.6 bara. Assume that the octane numbers of 0.69 bara RVP blends may be calculated according to the volume fraction of butane in the motor alkylate.

14   Short-cut distillation routines may be used for fractionator calculations. Column loads are to be 75 percent of the maximum allowable vapor velocities at the point of highest internal column flows.

15   In order to simplify the calculations, the depropanizer bottoms stream is specified to be returned to the deisobutanizer feed. However, in some plants an additional refrigeration loop is set up, with this stream being returned directly to the reactor input and the size of the deisobutanizer being thereby reduced.

16   For flash drum calculations, assume an initial flash pressure of 0.07 barg.

Sulfuric Acid Alkylate Yield Data	
Isobutane consumed (volume/vol olefin)	1.10
Heat of reaction, (kJ/kg olefin)	1430
Total alkylate yield (vol/vol olefin)	1.72
Acid consumption, (kg acid/m³ total alkylate)	72

8-4   Methanation Unit Design[2]

Petroleum shortages in the United States have generated a great deal of interest in producing substitute or synthetic natural gas (SNG) from coal. This gas can be distributed through the natural gas pipeline network and burned in existing equipment without modification.

The major processing sequence of converting coal to SNG is shown in Figure P8-4-1. It starts with coal gasification (partial oxidation), where coal reacts with oxygen and steam to form a raw gas rich in carbon monoxide (CO) and hydrogen (H_2) and having a higher heating value of 8 to 16 MJ/std m³. Before it can be upgraded into pipeline-quality gas in the methanation unit, the raw gas is cooled, sent to the shift converter to produce sufficient H_2 for the methanation reaction, and then sent to the gas purification unit where sulfur compounds plus some carbon dioxide (CO_2) are removed. In the methanation unit, H_2 and CO are converted to methane. Any excess CO_2 may have to be removed so that the final gas will have a higher heating value comparable to that of natural gas (about 39 MJ/std m³). Finally, the gas is compressed, dehydrated, and delivered to the pipeline.

You are to develop a process design for the methanation unit in this SNG plant.

Theory

Methanation of coal-derived synthesis gases involves catalytic conversion of carbon oxides and hydrogen according to the following exothermic reactions.

[2]Based on the 1983 AIChE Student Contest Problem. (The American Institute of Chemical Engineers, by permission.)

Figure P8-4-1   Typical process sequence for coal conversion to SNG: scope of process design for the problem boxed by broken line.

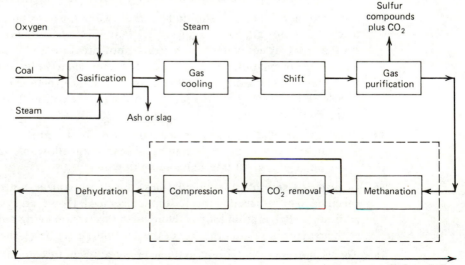
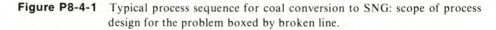

$$CO + 3H_2 = CH_4 + H_2O \qquad \text{(CO methanation)}$$

$$CO_2 + 4H_2 = CH_4 + 2H_2O \qquad \text{(CO}_2 \text{ methanation)}$$

The CO_2 methanation most probably proceeds through an endothermic intermediate stage as follows.

$$CO_2 + H_2 = CO + H_2O \qquad \text{(reverse shift)}$$

The methanation process is usually carried out in two steps: a bulk methanation step and a cleanup methanation step. Most of the carbon oxides are methanated in the bulk methanation step. The product gas from the bulk reactors generally does not meet the heating value and CO content specifications of substitute natural gas. This is because the extent of methanation is limited by the large amount of water by-product. To achieve additional methanation, the gas from the bulk methanator is cooled to condense out the water, then reheated for reaction in the cleanup methanator. The cleanup methanation step uses a lower temperature than the bulk methanation step because this increases the equilibrium constant.

For the purpose of this study, the bulk and cleanup methanators are adiabatic, fixed-bed reactors containing alumina- or silica-supported nickel oxide catalysts. Other systems, such as direct-cooled tubular or fluid bed reactors, have been proposed, but have not been commercially proved. There is an upper temperature limit as well as a lower temperature limit at which the methanation catalyst can operate. Below the lower temperature limit, the reactions will not initiate. Above the upper temperature limit, the catalyst sinters and loses its activity. Owing to the high heat of reaction, there is a potential for a large temperature rise in the methanator. The temperature rise can be controlled either by recycling cooled product gas or by injecting steam into the methanator. Both the recycled gas and the steam act as heat carriers for the heat released during methanation. Multiple reactors in series are often suggested for the bulk methanation step. This allows interreactor cooling, which reduces the quantity of recycled gas or stream injection.

Two empirical factors are specified for commercial reactor designs: the "space velocity" and "approach temperature difference." Space velocity is defined as the gas rate in standard cubic meters per second divided by catalyst volume. Approach temperature difference, or approach ΔT, indicates the degree of departure from chemical equilibrium. The approach ΔT is the difference between the actual reactor outlet temperature and the equilibrium temperature corresponding to the composition of the reactor effluent. The approach ΔT varies with catalyst activity for a given space velocity. At start of run, the fresh catalyst is very active and the reaction proceeds to a close approach to equilibrium conversion, or small approach ΔT. The catalyst activity declines with continued use. This results in less conversion or a larger approach ΔT at end of run. The approach ΔT usually recommended for carbon monoxide methanation is $5.6°C$ at start-of-run and $27.8°C$ at end-of-run.

The reverse shift reaction is considerably faster than CO methanation. Therefore, it has a closer approach to equilibrium in the methanators. The approach ΔT for this reaction is usually assumed to be zero at start and end of run.

Process Description

Figure P8-4-2 is a flow sketch for a methanation plant. Note that this sketch does not include all equipment and/or heat exchange services, and the flow sequence shown is not necessarily optimum. It is provided for guidance only.

Purified synthesis gas enters a guard reactor for removal of the last traces of

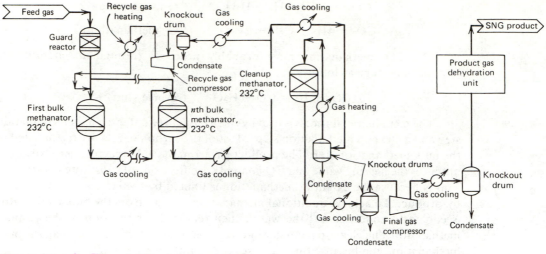

Figure P8-4-2   Flow sketch of the methanation of coal-derived synthesis gas for SNG production.

sulfur compounds by reaction with zinc oxide (ZnO) to protect the methanation catalyst from poisoning. Zinc oxide in the guard reactor is periodically replaced. The minimum temperature required to operate the guard reactor is 232°C.

The sulfur-free feed gas from the zinc oxide reactor then enters the bulk methanator system. Effluent gas from each reactor in the bulk methanation system is cooled before entering the next reactor. Part of the effluent gas from the last bulk methanator is compressed and recycled to join the fresh feed, while the rest of the gas is further cooled to condense out the water. After water separation in the knockout drum, the gas is reheated and sent to the cleanup methanator. The effluent gas from the cleanup methanator is then compressed and dried in a dehydration unit before delivery to the pipeline.

The dehydration unit consists of an absorber and a regenerator. In the absorber, the gas is dehydrated by a glycol solution. Wet glycol solution is regenerated by steam stripping and recycled to the absorber. The dehydration unit is not part of this assignment.

Problem Scope

Your company is considering installation of a coal-based SNG plant and you are a member of the study team assigned to the project. Other team members have been asked to design gasification, syngas shift, and purification sections. You are to develop an optimum methanation plant design based on the feed gas composition and product specifications listed in Table P8-4-1.

Many processing options for a methanation plant will produce a workable design. Any number of bulk methanators with interreactor cooling for heat removal may be used, along with steam injection and/or product gas recycle for temperature moderation.

In this case, previous studies indicate that steam injection need not be considered. The principal focus is on the tradeoff between the number of bulk methanators with interreactor cooling versus the recycle gas rate. A system consisting of one bulk methanator will require a very high recycle gas rate. On the other hand, a series of bulk methanators with interreactor cooling will require a

TABLE P8-4-1
FEED AND PRODUCT SPECIFICATIONS

	Feed	Product
Gas composition[a]		
CH_4	15.60 mol %	balance
CO	16.50 mol %	0.1 mol % max
CO_2	4.15 mol %	3.0 mol % max
H_2	63.40 mol %	5.0 mol % max
N_2	0.35 mol %	
H_2O	Nil	0.0085 mol % max
H_2S	0.1 ppm (molar)	
Rate	5.41 mol/s	
Pressure	24.8 bara	69 bara
Temperature	18° C	49° C max
Higher heating value		37.4 MJ/std m³ (minimum)

[a] After dehydration.

lower recycle gas rate, but will entail higher reactor capital costs and a higher system pressure drop. Thus, there is a best case from the standpoint of capital and operating cost.

In addition, optimization of the various heat exchange services is important. Various process streams may be heated with steam or hot process gas and they may be cooled by steam generation, cool process gas, boiler feed water, or cooling water.

Equipment should be designed to handle start-of-run operation with a fresh catalyst and end-of-run operation with reduced catalyst activity. Equipment necessary for startup of the plant should also be included. In preparing your report, use the information provided in the Design Guidelines and the Economic Guidelines.

Design Guidelines

Flow rate, composition, and properties of the feed and product gases are specified in Table P8-4-1. Other design related information is in Tables P8-4-2 through P8-4-4.

For the methanator mass and heat balance calculations, actual reactor outlet gas composition versus actual reactor outlet temperature is provided in Table P8-4-3. These compositions were computed using the equilibrium constant data from thermodynamic tables and the indicated feed pressures and approach ΔT as in Table P8-4-3. At a given reactor outlet temperature and pressure, the reactor outlet composition is a function of the atomic ratios, percentage inerts, and approach ΔT. Since atomic ratios are unaffected by chemical reactions, recycling some of the product gas to the feed does not affect the atomic ratios as long as no material is removed from the recycle stream. Tables P8-4-3a and P8-4-3b can be used for bulk methanators with and without recycle. Removal of water from bulk methanator products changes the atomic ratio. Tables P8-4-3c through P8-4-3f provide data for the cleanup methanation calculations. Neglect pressure effects for our range of operation. Linear interpolations of these tables will give valid results.

Compressors can be driven by either electric motors or steam turbines. The flow sketch is for guidance. It should not limit your optimization effort. For example, heat exchangers may be added or eliminated.

TABLE P8-4-2

MOLECULAR WEIGHT (MW) AND CRITICAL PROPERTIES OF PERTINENT GASES

	MW	$T_c(K)$	P_c (bara)
H_2O	18.02	647	221
H_2	2.01	33	13
N_2	28.01	126	34
CO	28.01	134	35.5
CO_2	44.01	304	74
CH_4	16.04	191	44

TABLES P8-4-3

METHANATION REACTOR EFFLUENT COMPOSITIONS

TABLE P8-4-3a Reactor Effluent Compositions for Bulk Methanation

Pressure 22 bara	Approach ΔT 5.6°C		Atom Ratio (C/H) 0.192			Atom Ratio (C/O) 1.462	
Composition of Feed Used as Basis							
Component Mol percent	H_2O 0.00	H_2 63.40	N_2 0.35	CO 16.50	CO_2 4.15	CH_4 15.60	Total 100.00

Reactor Effluent Temperature and Composition

Temperature (°C)	Mol Percent of Component						Total (mol per 100 mol feed)
	H_2O	H_2	N_2	CO	CO_2	CH_4	
404.5	35.42	5.23	0.57	0.03	2.38	56.37	61.67
410	35.24	5.50	0.57	0.04	2.44	56.22	61.76
415.6	35.01	5.82	0.57	0.04	2.52	56.04	61.86
421.1	34.79	6.15	0.56	0.05	2.59	55.86	61.97
426.7	34.57	6.47	0.56	0.06	2.66	55.68	62.08
432.2	34.34	6.80	0.56	0.06	2.74	55.49	62.18
437.8	34.12	7.13	0.56	0.07	2.81	55.31	62.29
443.3	33.89	7.45	0.56	0.08	2.88	55.13	62.40
448.9	33.67	7.78	0.56	0.09	2.96	54.94	62.51
454.4	33.40	8.17	0.56	0.11	3.04	54.72	62.64
460	33.14	8.56	0.56	0.12	3.12	54.50	62.77
465.5	32.87	8.95	0.56	0.14	3.21	54.28	62.91
471.1	32.60	9.34	0.56	0.15	3.29	54.05	63.04
476.7	32.33	9.74	0.55	0.17	3.37	53.83	63.18
482.2	32.06	10.15	0.55	0.20	3.46	53.59	63.32
487.8	31.74	10.61	0.55	0.22	3.55	53.33	63.49
493.3	31.50	10.96	0.55	0.24	3.62	53.12	63.61
498.9	31.19	11.43	0.55	0.27	3.71	52.85	63.78
504.4	30.86	11.92	0.55	0.31	3.80	52.56	63.96
510	30.57	12.34	0.55	0.34	3.88	52.31	64.12
515.6	30.28	12.78	0.54	0.38	3.96	52.06	64.28
521.1	29.94	13.28	0.54	0.43	4.05	51.75	64.47
526.7	29.60	13.80	0.54	0.48	4.14	51.45	64.66

TABLES P8-4-3

METHANATION REACTOR EFFLUENT COMPOSITIONS

TABLE P8-4-3*b*   Reactor Effluent Compositions for Bulk Methanation

Pressure 22 bara	Approach ΔT 27.8°C		Atom Ratio (C/H) 0.192		Atom Ratio (C/O) 1.462	
	Composition of Feed Used as Basis					
Component Mol percent	H_2O 0.00	H_2 63.40	N_2 0.35	CO 16.50	CO_2 4.15	CH_4 15.60	Total 100.00

	Reactor Effluent Temperature and Composition						
	Mol Percent of Component						**Total**
Temperature (°C)	H_2O	H_2	N_2	CO	CO_2	CH_4	(mol per 100 mol feed)
404.5	34.41	6.70	0.56	0.05	2.73	55.56	62.15
410	34.15	7.08	0.56	0.06	2.81	55.35	62.27
415.6	33.93	7.40	0.56	0.06	2.88	55.17	62.38
421.1	33.70	7.72	0.56	0.07	2.96	54.98	62.49
426.7	33.41	8.16	0.56	0.08	3.05	54.74	62.63
432.2	33.18	8.48	0.56	0.09	3.13	54.56	62.74
437.8	32.92	8.87	0.56	0.11	3.21	54.34	62.87
443.3	32.66	9.25	0.56	0.12	3.29	54.12	63.00
448.9	32.35	9.69	0.55	0.14	3.39	53.87	63.15
454.4	32.09	10.08	0.55	0.16	3.47	53.65	63.29
460	31.82	10.48	0.55	0.17	3.55	53.42	63.43
465.5	31.51	10.93	0.55	0.20	3.65	53.17	63.59
471.1	31.24	11.33	0.55	0.22	3.73	52.93	63.73
476.7	30.97	11.74	0.55	0.25	3.81	52.70	63.88
482.2	30.65	12.20	0.55	0.28	3.90	52.43	64.05
487.8	30.29	12.74	0.54	0.31	4.00	52.11	64.24
493.3	30.01	13.16	0.54	0.35	4.08	51.87	64.40
498.9	29.64	13.70	0.54	0.39	4.18	51.54	64.60
504.4	29.35	14.15	0.54	0.43	4.25	51.28	64.77
510	29.01	14.65	0.54	0.48	4.34	50.98	64.96
515.6	28.66	15.17	0.54	0.54	4.43	50.66	65.17
521.1	28.31	15.71	0.54	0.59	4.51	50.34	66.38
526.7	28.00	16.18	0.53	0.65	4.59	50.05	65.57

TABLES P8-4-3

METHANATION REACTOR EFFLUENT COMPOSITIONS

TABLE P8-4-3c Reactor Effluent Compositions for Cleanup Methanation

Pressure 21 bara	Approach ΔT 5.6° C		Atom Ratio (C/H) 0.243			Atom Ratio (C/O) 7.800	
	Composition of Feed Used as Basis						
Component	H_2O	H_2	N_2	CO	CO_2	CH_4	Total
Mol percent	0.56	14.42	0.81	0.22	5.01	78.99	100.00

Reactor Effluent Temperature and Composition

Temperature (°C)	Mol Percent of Component						Total (mol per 100 mol feed)
	H_2O	H_2	N_2	CO	CO_2	CH_4	
293.3	7.91	0.65	0.87	0.00	1.85	88.71	92.99
298.9	7.88	0.72	0.87	0.00	1.86	88.67	93.02
304.4	7.85	0.78	0.87	0.01	1.88	88.62	93.05
310	7.81	0.85	0.87	0.01	1.89	88.58	93.08
315.6	7.78	0.91	0.87	0.01	1.91	88.53	93.11
321.1	7.73	0.99	0.87	0.01	1.92	88.47	93.15
326.7	7.69	1.07	0.87	0.01	1.94	88.42	93.19
332.2	7.65	1.16	0.87	0.01	1.96	88.36	93.23
337.8	7.60	1.24	0.87	0.01	1.98	88.30	93.27
343.3	7.55	1.34	0.87	0.02	2.00	88.23	93.31
348.9	7.50	1.44	0.86	0.02	2.02	88.16	93.36
354.4	7.44	1.54	0.86	0.02	2.05	88.08	93.41
360	7.39	1.64	0.86	0.03	2.07	88.01	93.46
365.5	7.33	1.76	0.86	0.03	2.09	87.92	93.52
371.1	7.27	1.87	0.86	0.03	2.12	87.85	93.57
376.7	7.21	1.99	0.86	0.04	2.14	87.76	93.63
382.2	7.13	2.13	0.86	0.05	2.17	87.65	93.70
387.8	7.08	2.24	0.86	0.05	2.19	87.57	93.76
393.3	7.00	2.39	0.86	0.06	2.22	87.47	93.83
398.9	6.92	2.54	0.86	0.07	2.25	87.36	93.91
404.4	6.85	2.69	0.86	0.08	2.28	87.24	93.98
410	6.77	2.85	0.86	0.09	2.31	87.13	94.06
415.5	6.70	2.99	0.86	0.10	2.33	87.02	94.14
421.1	6.61	3.17	0.86	0.12	2.37	86.88	94.23

TABLES P8-4-3
METHANATION REACTOR EFFLUENT COMPOSITIONS

TABLE P8-4-3d Reactor Effluent Compositions for Cleanup Methanation

Pressure 21 bara	Approach ΔT 5.6°C		Atom Ratio (C/H) 0.236			Atom Ratio (C/O) 5.176	
			Composition of Feed Used as Basis				
Component Mol percent	H_2O 5.72	H_2 13.67	N_2 0.77	CO 0.21	CO_2 4.75	CH_4 74.89	Total 100.00

Reactor Effluent Temperature and Composition

Temperature (°C)	Mol Percent of Component						Total (mol per 100 mol feed)
	H_2O	H_2	N_2	CO	CO_2	CH_4	
293.3	12.91	0.83	0.82	0.00	1.80	83.64	93.45
298.9	12.87	0.91	0.82	0.00	1.81	83.59	93.49
304.4	12.83	0.98	0.82	0.00	1.83	83.54	93.52
310	12.77	1.07	0.82	0.00	1.85	83.48	93.56
315.6	12.72	1.16	0.82	0.01	1.87	83.42	93.61
321.1	12.67	1.25	0.82	0.01	1.89	83.36	93.65
326.7	12.61	1.36	0.82	0.01	1.92	83.28	93.70
332.2	12.56	1.45	0.82	0.01	1.94	83.22	93.74
337.8	12.50	1.57	0.82	0.01	1.97	83.14	93.80
343.3	12.43	1.69	0.82	0.01	2.00	83.06	93.86
348.9	12.36	1.81	0.82	0.01	2.02	82.97	93.92
354.4	12.29	1.93	0.81	0.02	2.05	82.89	93.98
360	12.21	2.09	0.81	0.02	2.09	82.79	94.05
365.5	12.41	2.21	0.81	0.02	2.11	82.70	94.11
371.1	12.05	2.37	0.81	0.03	2.15	82.59	94.19
376.7	11.97	2.51	0.81	0.03	2.18	82.50	94.26
382.2	11.88	2.68	0.81	0.04	2.22	82.38	94.34
387.8	11.79	2.84	0.81	0.04	2.25	82.27	94.42
393.3	11.68	3.03	0.81	0.05	2.29	82.13	94.52
398.9	11.59	3.19	0.81	0.05	2.33	82.02	94.60
404.4	11.51	3.36	0.81	0.06	2.36	81.91	94.68
410	11.39	3.57	0.81	0.07	2.41	81.76	94.79
415.5	11.29	3.75	0.81	0.08	2.44	81.63	94.88
421.1	11.18	3.96	0.81	0.09	2.48	81.48	94.98

TABLES P8-4-3
METHANATION REACTOR EFFLUENT COMPOSITIONS
TABLE P8-4-3e Reactor Effluent Compositions for Cleanup Methanation

Pressure 21 bara	Approach ΔT 27.8° C		Atom Ratio (C/H) 0.243		Atom Ratio (C/O) 7.800	

	Composition of Feed Used as Basis						
Component	H_2O	H_2	N_2	CO	CO_2	CH_4	Total
Mol percent	0.56	14.42	0.81	0.22	5.01	78.99	100.00

	Reactor Effluent Temperature and Composition						
	Mol Percent of Component						Total
Temperature (°C)	H_2O	H_2	N_2	CO	CO_2	CH_4	(mol per 100 mol feed)
293.3	7.74	0.98	0.87	0.01	1.92	88.48	93.14
298.9	7.69	1.06	0.87	0.01	1.94	88.43	93.18
304.4	7.65	1.14	0.87	0.01	1.96	88.37	93.22
310	7.60	1.23	0.87	0.01	1.98	88.31	93.26
315.6	7.55	1.34	0.87	0.01	2.01	88.23	93.31
321.1	7.50	1.42	0.86	0.01	2.02	88.17	93.35
326.7	7.45	1.52	0.86	0.02	2.05	88.10	93.40
332.2	7.40	1.62	0.86	0.02	2.07	88.03	93.45
337.8	7.33	1.75	0.86	0.02	2.10	87.94	93.51
343.3	7.27	1.85	0.86	0.02	2.12	87.86	93.56
348.9	7.20	1.99	0.86	0.03	2.15	87.77	93.63
354.4	7.14	2.11	0.86	0.03	2.18	87.68	93.69
360	7.08	2.23	0.86	0.04	2.20	87.59	93.75
365.5	7.00	2.37	0.86	0.05	2.23	87.49	93.82
371.1	6.93	2.51	0.86	0.05	2.26	87.39	93.89
376.7	6.85	2.66	0.86	0.06	2.29	87.28	93.96
382.2	6.78	2.81	0.86	0.07	2.32	87.17	94.03
387.8	6.70	2.96	0.86	0.08	2.35	87.06	94.11
393.3	6.62	3.11	0.86	0.09	2.37	86.95	94.19
398.9	6.53	3.30	0.86	0.10	2.41	86.80	94.29
404.4	6.46	3.44	0.86	0.12	2.43	86.69	94.36
410	6.36	3.63	0.85	0.13	2.47	86.55	94.46
415.5	6.28	3.80	0.85	0.15	2.49	86.42	94.55
421.1	6.19	3.98	0.85	0.17	2.52	86.28	94.65

TABLES P8-4-3
METHANATION REACTOR EFFLUENT COMPOSITIONS
TABLE P8-4-3f   Reactor Effluent Compositions for Cleanup Methanation

Pressure 21 bara	Approach ΔT 27.8° C		Atom Ratio (C/H) 0.236			Atom Ratio (C/O) 5.176	
			Composition of Feed Used as Basis				
Component Mol percent	H_2O 5.72	H_2 13.67	N_2 0.77	CO 0.21	CO_2 4.75	CH_4 74.89	Total 100.00

Reactor Effluent Temperature and Composition

Temperature (°C)	Mol Percent of Component						Total (mol per 100 mol feed)
	H_2O	H_2	N_2	CO	CO_2	CH_4	
293.3	12.68	1.25	0.82	0.00	1.90	83.36	93.65
298.9	12.62	1.35	0.82	0.01	1.92	83.29	93.70
304.4	12.56	1.45	0.82	0.01	1.94	83.22	93.75
310	12.50	1.56	0.82	0.01	1.97	83.15	93.80
315.6	12.43	1.68	0.82	0.01	2.00	83.07	93.85
321.1	12.37	1.80	0.82	0.01	2.02	82.99	93.91
326.7	12.30	1.92	0.81	0.01	2.05	82.90	93.97
332.2	12.21	2.07	0.81	0.01	2.09	82.80	94.04
337.8	12.14	2.21	0.81	0.02	2.12	82.71	94.11
343.3	12.06	2.34	0.81	0.02	2.15	82.61	94.17
348.9	11.97	2.50	0.81	0.02	2.18	82.51	94.25
354.4	11.89	2.65	0.81	0.03	2.22	82.40	94.32
360	11.79	2.84	0.81	0.03	2.26	82.28	94.41
365.5	11.70	2.99	0.81	0.03	2.29	82.17	94.49
371.1	11.61	3.15	0.81	0.04	2.33	82.06	94.57
376.7	11.49	3.37	0.81	0.05	2.38	81.91	94.68
382.2	11.39	3.56	0.81	0.05	2.42	81.77	94.77
387.8	11.28	3.76	0.81	0.06	2.46	81.64	94.87
393.3	11.17	3.95	0.81	0.07	2.50	81.50	94.97

TABLE P8-4-4
CATALYST PROPERTIES

	Bulk Methanation	Cleanup Methanation	Zinc Oxide
Form (cylindrical pellet)	Tablet	Tablet	Extrusion
Height (= diameter)	6.35 mm	6.35 mm	4.76 mm
Operating temperature range	232–482°C	232–454°C	232–427°C
Void fraction	0.4	0.4	0.4
Life	2 yr	2 yr	—
Bulk density	960 kg/m³	800 kg/m³	1120 kg/m³
Cost per cubic meter	$14,000	$18,300	$3875

Vessel Design

As an exercise in vessel design, use the following guidelines to specify process vessels and identify their cost.

- Maximum diameter, as stated in Chapter Four, is 4 m, because of shop fabrication and transportation limitations.
- For vessel wall thickness and weight calculations, see the section on materials of construction in Chapter Four.
- Design pressure (bara) = 1.1 × operating pressure in bara

  or

  2.5 bar + operating pressure in bara

  (use the greater value)
- Design temperature = 28°C + operating temperature (315°C, minimum). Add 3 mm for corrosion allowance.
- Total vessel weight = 1.1 × (weight of cylindrical section + weight of heads); this factor accounts for nozzles, flanges, skirts, and other parts of the vessel.
- For methanator and ZnO guard reactors, use gas density at reactor outlet temperature, as an average, for pressure drop calculations.
- Assume the pressure drop for nozzles, distributor, and supports is equivalent to 1 m of bed height.
- All reactors should be axial downflow vertical vessels.

- Parallel vessels may be used to minimize pressure drop.
- If hemispherical heads are used, the catalyst bed and supports may extend into the vessel heads up to 15 percent of the vessel diameter. If ellipsoidal heads are used, the catalyst bed may not extend into the heads.

Methanators

- Reactor space velocity is 4.0 std m^3/s of total reactor feed gas per cubic meter of catalyst.
- Minimum catalyst bed depth is 0.5 × bed diameter.

Zinc Oxide Reactor

- Assume that 1 kg of ZnO absorbs 0.15 kg of H_2S.
- Set amount of ZnO for 2 yr life, minimum.
- Use minimum ZnO bed depth of 0.5 × bed diameter.

Waste Heat Recovery

- Use kettle-type reboilers. An amount of water equal to 2 percent of feedwater should be blown down or discharged to keep mineral concentrations low.

Materials of Construction

- Use the following criteria for gas streams.

Temperature (°C)	H_2 (mol %)	Material
>315	All	Low alloy steel
<315	<5	Carbon steel
<315	>5	Low alloy steel

- Use carbon steel for shells of waste heat boilers and water-cooled gas coolers.

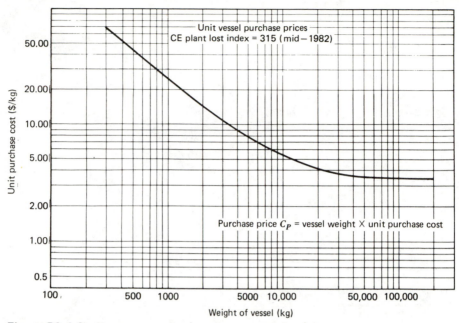

Figure P8-4-3 Pressure vessel unit cost versus vessel weight.

Economic Guidelines

Alternative designs are to be compared on the basis of a cash flow-incremental return on incremental investment guideline.

No alternative design that produces less than a 17 percent rate of return on incremental capital should be chosen. The return, in this case, is net present value (NPV) over 20 yr. The rate of return ROR is, therefore,

$$\text{ROR} = \frac{\Delta \text{NPV}}{\Delta C_{TC}} \times 100$$

For cash flow calculations, assume the following.

- Time value of money is 11 percent per year.
- Interest compounded annually.
- Expenses paid annually at end of year.
- Project lifetime is 20 yr from start of construction.
- Construction period of 3 yr.
- Sum-of-years'-digits depreciation over 10 yr with zero salvage value.
- 330 operating days/yr.
- Maintenance is 4 percent of fixed capital.
- Operating labor is constant regardless of design.
- Use Figure P8-4-3 to estimate purchase prices of process vessels.

Section 3
TECHNICAL
REPORTING

Chapter Nine   REPORT PREPARATION

Chapter Nine
REPORT PREPARATION

In the beginning, the story goes, God, after creating humankind, was defining the professions. "Anticipating that squabbles would ultimately develop between chemists and chemical engineers, He decided to settle the issue once and for all; dictating to His typist, 'All a chemical engineer does is *right*.' Unfortunately the typist mis-spelled the last word" [4]. At times, many of us might agree that all a chemical engineer does is *write*. Some feel we don't even do that very well and that we're getting worse. In a recent survey of educators and industrialists, for instance, some respondents complained that language skills among chemical engineering graduates had deteriorated severely in recent years. Others, according to the reporter, felt simply that the skills were not better than before—abominable [6].

I have heard some managers in industry claim that communcation skills are more imortant than technical competence. I do not agree. Communication would be unnecessary if there were no technical result to report. (It does, indeed, require exceptional writing or speaking skill to camouflage an inept or incomplete engineering job.) But I do agree with a variant of the managers' statement: "Many exceptional engineering jobs go unappreciated because of poor writing or speaking." With this in mind, let us consider the elements of effective writing.

PHILOSOPHY OF TECHNICAL REPORTING

Unlike politicians, engineers should write with the hope that readers will find their errors. It is much less embarrassing and painful for an engineering mistake to be found in print before it appears in fact. Thus, a technical report should be designed with clarity as the major goal.

Basic honesty is a key ingredient of clear writing. If there is no concrete result or recommendation, say so. Perhaps your most important contribution will be to expose a question or a mistake. Such honesty may not always pay off immediately, but a reputation for integrity is worth the wait. Reports intended to reveal rather than obscure will be better understood by others and, when deserving, will be defended by them.

MECHANICS OF REPORT WRITING

An outline does wonders to initiate the writing process. Professional or experienced writers often outline their work mentally, not formally. However, judging from the indictment in the first paragraph of this chapter, you should prepare a written outline if you are a student or an engineer. As an example, a skeleton outline of this chapter is shown in Figure 9-1. (Of course the real outline is scribbled on three sheets of paper with numerous insertions and marginal notations.) As you prepare your outline, think about the audience. Van Ness and Abbott [8] caution that readers of most technical reports:

1   Are busy or at least believe so.
2   Have a background similar to yours but know much less about the project in question.

Other reader characteristics may prevail under various circumstances. In fact, the abstract and summary of a report often are designed for administrators and business people with nonengineering backgrounds.

Some suggestions by Bolmer [1] for preparing a speech are also appropriate for prose. At each juncture, ask the magic questions: Who? What? When? Where? Why? How? These answers will usually lead you to the next step. Bolmer also suggests writing or identifying the conclusions first (asking the same questions) to provide focus in the outline. Next, review your notes and write prominent thoughts, quotations, and ideas on slips of paper. Do not then cast them into the air and pick them up randomly from the floor. Instead, organize them as your mind directs. In the shuffle, some ideas might appropriately land in the wastebasket.

Report Structure

Composing a report is much like baking cinnamon rolls. A cook does not put dough in one pile, raisins in another, cinnamon and sugar in a third, then bake the ingredients separately. Neither does one place all the materials in a blender and atomize them into a uniform mass. Instead, individual elements are assembled wisely and in proper proportion to yield an interesting, attractive, and tasty result. So is a report organize to provide mental nourishment, impetus, and satisfaction.

I see three primary purposes of a design report.

1   To present technical information.
2   To serve as a repository of data.
3   To promote or define action.

The first two might be viewed as the dough, the third as cinnamon and raisins. Unfortunately, I cannot give you an exact recipe for composing a report. A rigid outline for all reports and situations is stifling. However, for a beginner, a skeleton format may be helpful. The format illustrated in Table 9-1 is discussed in detail below. As you read about each section, think which of the foregoing purposes is satisfied. (I have even provided space in Table 9-1 to keep score. I will divulge my ratings later.)

OUTLINE
Chapter Nine
REPORT PREPARATION

I. Introduction
   (Attention) "Skills no better than before—abominable."
   A. Importance of communication skill.
      1. More important than technical skills? Hogwash.
      2. "All an engineer does is write (right?)."
   B. Philosophy of writing.
      1. Honesty.
II. Mechanics of report writing.
   A. Outline.
      1. Reader identification.
      2. Who, What, When, Where, Why, How?
      3. Write conclusions first.
      4. Review literature or calculations.
      5. Write thoughts on sheets of paper.
   B. Structure.
      1. Cinnamon roll.
      2. Dangers of rigid format.
      3. Sample outline.
         a. Purpose of section.
            i. Present technical information.
            ii. Define, recommend, encourage, promote action.
            iii. Data repository.
         b. Sample format (see Table 9-1).
   C. Length.
      1. Long enough to reach the ground.
      2. 50 mile hike.
III. Style and technique.
   (Interest) Hydrochloric acid to clean pipes.
   A. First person, humor, informal versus formal.
   B. Fog Index.
   C. How to improve.
      1. Practice, practice, practice.
      2. Invite criticism.
      3. Read good writing appreciatively.
      4. Read bad writing critically.
Illustration 9-1   Goldenrod Report

Figure 9-1   The outline employed to write this chapter.

I like to think of a report as containing four divisions: a beginning procedural segment, the summary, a body, and an end procedural segment.

Front Matter

The beginning procedural segment usually contains a *letter of transmittal*, *title page*, *table of contents*, and *abstract*. It is much like the pages at the beginning of this book numbered in lowercase roman numerals. (This section is known as "front matter" in the publishing business.) In many reports, especially brief ones, some of these components are unnecessary. In a short or letter report, the title, abstract, and beginning of body may appear on the first page.

TABLE 9-1
SAMPLE FORMAT FOR A TECHNICAL REPORT

Division	Section	Purpose		
		Present Information	Data Repository	Promote or Define Action
I. Beginning procedural section (front matter)	Letter of transmittal Title page Table of contents Abstract			
II. Summary	Summary			
III. Body	Introduction (background, literature survey, theory, etc.) Method of approach (procedure) Results Discussion of results Conclusions Recommendations			
IV. Concluding procedural section (back matter)	References Appendix			

Summary

The *summary* is an isolated section because it is often circulated separately to a wider audience that includes managers and nontechnical readers who are concerned with action and recommendations rather than computational detail. Because of its political impact and importance in decision making, the summary should be written most carefully, emphasizing vital conclusions and recommendations. Supporting data must be summarized and presented clearly and interestingly to a less sophisticated reader. Illustrations should be used effectively but sparingly. Since the summary is based on the broader report, it is, of course, written last. It appears, however, near the front of the finished document as shown in Table 9-1.

Body

Asking who, what, when, where, why, and how leads smoothly to an efficient outline for the report body. An *introduction* of some sort is necessary to bring the reader "up to speed." Historical or chronological structure is ofttimes effective in this section. If appropriate, *literature survey, theory*, and other topics may be folded into an introduction or inserted as separate sections thereafter.

To evaluate your report, a technical reader must understand how you derived the results. A section on *approach* or *procedure* serves this need. It should be written in a way that permits a reader to duplicate experiments or calculations independently if necessary. In a design project, pivotal assumptions and bases should be included and, where appropriate, explained. More common assumptions are listed in the appendix or not at all. Detailed calculations should not be placed here or in the appendix. They belong in your files. Representative sample calculations should be in the appendix.

A key structural role is played in the *results* section. Information vital to the final conclusions and recommendations is found here. Peripheral data should be in the appendix or in your files. Inclusion of unnecessary detail obscures vital results.

The results secton is followed by the *discussion* or *discussion of results*. This is where logical conclusions are exposed. Many authors fail to develop and manipulate their data enough. Table 4-28, for example, was assembled from about 10 different sources. I spent hours arriving at a format, days defining the details. This single table required more than a week's hard labor. The original 10 sources could easily have been reprinted directly, but I wanted focused data, not diffuse data. Most of the tables in this book required the same kind of selection and concentration. Many engineers do not invest enough energy in massaging results. They are satisfied with detailed tabulations of numbers when refined charts or curves would tell the story better.

The *conclusions* and *recommendations* sections represent the apex of your report. As you outline these sections, think, analyze, and ask the magic questions. Skilled technical writers, not unlike popular authors, often employ suspense to create a climax. Since preceding sections have created a focusing effect, this segment can be concentrated and brief. Often *conclusions* and *recommendations* are combined into a single section. Sometimes recommendations are presented as a list of numbered statements.

How you say it *does* make a difference. We could imagine someone walking down a corridor, stopping at each door, knocking, and politely stating, "My senses perceive a conflagration at the extremes of this structure. I advise you to depart with haste." A real messenger would, of course, race up and down the hall screaming "Fire! Fire!" Provocations and emotions created by screaming "Fire!" in a technical report sometimes cause regret. On the other hand, we want readers to sit up, take notice, and in many cases, act. Of the two approaches illustrated, a tone nearer "Fire!" is suggested.

Back Matter

The final procedural section will not be opened by many readers, yet it serves a fundamental role in supporting the report. Not only should we be considerate of our more technical readers who will read this section, but want to help them find any mistakes that might be present.

References can be presented in any logical consistent format so long as they are clear and unambiguous. The format used in this book should be acceptable in most reports. As a reader, I find article titles informative and recommend their inclusion. Sometimes, authors try to impress readers by citing exhaustive lists of nonpertinent references. This creates the same result as unnecessary detail in the text—foggy and misleading communication.

Efficiency and clarity are traits of an effective appendix. Sometimes, students dump their raw calculations here to prove the work was done and to impress the teacher. As a reader, I am confused, discouraged, and angered by this strategy. Writers often fail to separate the wheat from the chaff. In almost every case I have seen, computer printout is chaff and should not be included at all in the report. Raw calculations are also chaff and should remain in your files. They do serve nicely, nontheless, as a useful outline for preparing the appendix. Illustrative and sample calculations selected critically from your work provide effective support to more focused information found in the report body. An effective appendix demands the same kind of creativity as any other part of the report. Sometimes even good authors are careless with this section.

By the way, in my opinion, purpose 1 (to present information) applies to the letter of transmittal, title page, and table of contents in Table 9-1. The abstract, summary, discussion of results, and conclusions accomplish the same end and promote or define action (purpose 3) as well. Goals 1 and 2 generally suit the introduction and approach sections. Action is promoted and defined primarily in recommendations. References and the appendix serve as data repositories.

I reemphasize that the outline above is only a suggestion; the nature of the project—and your own personality—will shape the structure of the report. It reminds me of my first 50-mile backpacking trip. I had listened to a man who frequently hiked in California's Sierra Mountains. He stressed the importance of lightweight packing and illustrated it by telling how he took only three pair of socks. He wore two pair and carried the other. When camping for the evening, he changed socks and washed out the sweaty ones; laying them on a warm stone. The next morning, they were dry and ready for the day's hike.

I tried the same technique on a trip in the Appalachian mountains in New England. What succeeded in dry California failed miserably and odorously in the Northeast. (Where does one find warm rocks in the rain?) Consider the situation in designing a report. Is yours a three-sock or a nine-sock project?

Report Length

The question of report length might well be answered the same way Abraham Lincoln answered a similar query about a man's legs. He said they should be long enough to reach the ground. A report should be long enough to tell the story.

Length is also somewhat dependent on audience and other circumstances. Many of us, infatuated with our own writing, tend to inflate its length. The old saying "Length of a graduate thesis is inversely proportional to the data it contains" is boringly valid at times. It's as though there were some minimum weight limit. Even though I am considered sparse with words, a ruthless but respected critic eliminated about 20 percent of what was originally drafted for this chapter. The improvement was worth the pain.

STYLE AND TECHNIQUE

Some years ago, a New York plumber discovered that hydrochloric acid was dandy for cleaning clogged drains. He sent his suggestion to the National Bureau of Standards.

"The efficacy of hydrochloric acid is indisputable," the Bureau wrote back, "but the ionic residues are incompatible with metallic permanence."

"Thank you," replied the plumber. "I thought it was a good idea too."

Finally, someone at the Bureau wrote, "Don't use hydrochloric acid! It eats hell out of the pipes!" [2]

No doubt, crisp language communicates ideas efficiently. No one knows how many years scientific and technological progress has been retarded by foggy writing. Communications professionals have been criticizing the characteristically formal impersonal language of science for years. Yet, we still encounter unpleasant examples in our professional literature. Fortunately, promising trends are evident, and we find more humor and use of first person in modern technical prose. Van Ness and Abbott wrote [8]:

> *For many years the dominant attitude with respect to scientific and technical writing was that it should be impersonal, because science and technology were said to be impersonal. This forced adoption of the passive voice, and promoted the lifeless syntax, the witless style, to say nothing of the grammatical mistakes of technical prose. We repudiate the whole of it. Not only does habitual use of the passive voice make for dull writing; it forces a convoluted style almost impossible for an engineer to make concise, precise, and grammatical. I and we are not four-letter words; they are entirely acceptable in technical reports and publications. We do not suggest that every sentence start with I or we; one seeks variety. If you are too humble or shy to bring yourself to write I, use we, in the sence of you, the reader, and I, the writer. One also has its place. Do not think you can avoid responsibility for what you write by adopting an impersonal style. No way; your name is on the title page. Take some pride in it; you are the expert.*

The entire article is a useful guide for engineers.

I remember speaking, not long ago, with a student who went to work at duPont. As a new recruit, he spent his first month on the job in a writing course. Instructors emphasized informal personal style because it makes written communication so much more effective. If the largest U.S. chemical corporation believes in it, we should feel free to promote it.

In the following set of examples, the same information is written at different levels of formality.

> This experiment was designed to define the relationship between temperature, time, and location in the curing of a polyurethane automobile bumper. It was initiated because of failures in certain applications.

Alternately:

> About 5 percent of the bumpers we manufacture for the new Z-cars are dropping from the vehicles at subfreezing temperatures. In a crash program to salvage our contract with Studebaker, Jean Doe assigned Dan Jordan and me to analyze the curing process and isolate any flaws.

Unfortunately not all organizations tolerate informal technical documents. You may find the need to aim your language somewhere between that befitting an automobile purchase agreement and that in a letter to an intimate friend. However, grammar, punctuation, spelling, clarity, and precision of writing at any level of formality can be improved. As guides to the technical rules for good English expression, references 3 and 7 are recommended.

A recent article in *Science 82* [2] discusses the "Fog Index" employed by Douglas Mueller, a writing consultant. It is a measure of writing clarity. As reported in that article, big words and long sentences are the two major culprits. The Fog Index puts these factors into a simple formula that tells how many years of schooling are needed to read a sample easily. The first letter to the plumber quoted above has a Fog Index of 26. To understand it would require a Ph.D. and 7 yr of postdoctoral study. The second letter, with a Fog Index of 6, should be clear to a sixth grader.

The article continues to describe how to calculate a Fog Index. My 12-yr-old son Thatcher, intrigued with the challenge, computed indices for two important recent documents created in our family. A selection from the preface of this book scored 15, low enough for students with 12 yr of grammar school and 3 yr of college. My wife's recent book on colonial history rated 11. According to *Science 82*, she wins.

> *At what Fog Index should a writer write? "A low one," says Mueller. The nation's largest daily newspaper,* The Wall Street Journal, *got that way by lowering its Fog Index to 11.* Time *and* Newsweek *also averge 11. The New Yorker usually comes in under 12. Technical journals range a lot higher, but most are notoriously hard reading, even for specialists. Good technical memos, according to a recent study at Bell Laboratories, average only 14. "The truth is," says Mueller, "no matter what Fog Index your readers can tolerate, they prefer to get their information without strain." Mueller says he's never met anyone, in any field, who couldn't lower his Fog Index to 15. "Einstein could. It's easy. Just keep your average sentence length under 20, cross out every useless word, and never use a Big Word unless you absolutely need to. Remember: The less energy your reader wastes on decoding your language, the more he'll have left for your brilliant ideas." [2]*

Some examples, prominent and otherwise, were also given.

From a business letter

We might further mention that we would be glad to furnish any one of these whistles on a trial basis, to the extent that if the smaller size was not adequate enough, it could be returned in lieu of the purchase of a larger size, depending upon actual operation and suitability of your requirements for signal distance and audibility.   (Fog Index: 28)

Translation

"If your whistle isn't loud enough, send it back and we'll give you a bigger one."   (Fog Index: 6)

From the scientific journal *Nature*

The current fashion for environmental impact assessment (EIA) is partly explained by the continuing force of the environmental protection movement in Western countries. That movement is now under severe pressure from economic recession, and there are signs that impact assessments themselves will play a decreasing role in planning and development. Certainly, this is the message that emerges from the U.S.A., where the emphasis is switching back to the costs of environmental protection.   (Fog Index: 17)

Opening of the Gettysburg Address

Fourscore and seven years ago our fathers brought forth on this continent a new nation, conceived in liberty and dedicated to the proposition that all men are created equal. Now we are engaged in a great civil war, testing whether that nation or any nation so conceived and so dedicated can long endure. We are met on a great battlefield of that war. We have come to dedicate a portion of that field as a final resting place for those who here gave their lives that that nation might live. It is altogether fitting and proper that we should do this.   (Fog Index: 10)

Matthew 6:9–13 (King James Version)

Our Father which art in heaven, hallowed be thy name. Thy kingdom come. Thy will be done in earth as it is in heaven. Give us this day our daily bread. And forgive us our debts, as we forgive our debtors. And lead us not into temptation, but deliver us from evil: For Thine is the kingdom, and the power, and the glory, for ever. Amen.   (Fog Index: 4)

Knowing the facts of good style does not necessarily create good writing. A reporter is said to have asked a famous football coach the secret of his success. He said there were three reasons: (1) practice, (2) practice, (3) practice. (A bystander added, "But it helps if the players are big and fast.") By analogy, to improve writing skills, you should write, write, write. (But it helps if you have grown up in an articulate family, studied debate for 8 years, and taken a minor in English.)

Not only must you write, but you should swallow your ego and invite expert criticism. In a less threatening vein, read quality writing by others and try to understand why it is good. When it is necessary to read bad writing, read it critically, noting errors and problems in margins as you observe them. Rewrite passages to see if you can improve them. (If the writer is your professor or a corporate vice president, it might be wise to destroy the marked copy.) The following example is presented for your evaluation.

ILLUSTRATION 9-1   REPORT ON THE GOLDENROD PROJECT

Prepare a report on the Goldenrod energy conservation project.

This report is based on calculations included in Illustrations 7-2 and 8-3. It is brief and rather informal as suits the character of this project. My letter report follows.

Durham, New Hampshire 03824
2 April 1982

Mr. Richard Boston, Engineering Manager
The Goldenrod
York Beach, Maine

Dear Rick:

Recently, you and I discussed energy lost through the discharge of hot rinse water from your plant. Since our conversation, I have examined the technology and economics of this problem. My analysis is contained in the attached report.

This work revealed several pivotal bases we did not define adequately in our conversation. Now that I have some preliminary results, I think we need to meet again and discuss the project some more. As soon as you have read this report, will you please call and let me know how you feel.

Sincerely,

Gael D. Ulrich

Gael D. Ulrich
Consultant

Economics of Waste Heat Recovery from
Hot Rinse Water at
The Goldenrod

Prepared by
Gael D. Ulrich, Consultant
Durham, New Hampshire

ABSTRACT

A scheme for recovering energy from hot rinse water at the Goldenrod plant was examined. Calculations indicate that a heat exchanger to cool effluent and preheat fresh feed can reduce water heater fuel consumption by about 75 percent. This, however, is not enough to justify the required fixed capital investment of $8200. Reasons for the negative economic result are discussed. Alternate energy-saving possibilities are considered.

SUMMARY

Based on current production rates and fuel prices, the Goldenrod management spends about $2500 per year to heat rinse water for dishes. Discharge of this heated water directly to the municipal sewer represents an unnecessary waste of energy. Hot discharge water could be used, instead, to preheat fresh feed, saving fuel in the process. Such recovery could be accomplished with a heat exchanger and pump, simple modifications employing proven technology. This report is the result of an effort to define the economics of such an energy recovery scheme.

To estimate capital costs, the energy recovery loop was designed in a preliminary way. Since economic parameters depend on exchanger size (i.e., preheat temperature), the optimum exchanger area was also determined. Maximum return was found for a 10 m^2 exchanger and a corresponding preheat temperature of 64°C. At higher temperatures, exchanger areas and costs increase dramatically. A double-pipe exchanger constructed of copper is the recommended type. The proposed pump is of the centrifugal type, 400 W, with a plastic casing. The estimated installed cost, mid-1982, for this system is $8200. Annual fuel savings of $1860 are predicted.

Because of expenses associated with maintaining and owning new equipment, only about $700 is available as a clear cash savings. Cash flow analysis indicates that this is inadequate to repay the investment in 10 years even with a zero discount factor. With a 10 percent discount factor, the project has a negative net present value of −$3675.

Based on this analysis, the proposed modifications are economically unsound. On the other hand, potential fuel savings justify deeper examination. Perhaps other conservation measures can be employed. Economic assumptions designed for large industrial plants may be too conservative for this small project. Energy credits and other ways of saving capital should also be explored.

After this report has been examined, I suggest we meet to examine the problem in more depth and define future action.

INTRODUCTION

The Goldenrod, established in 1896, has grown from a small local enterprise to a well-known manufacturer of custom novelty confections for the New England tourist trade. In contrast with antique taffy-making machines one sees on display and in use, the dishwashing operation is sophisticated and modern. Because of dramatic price increases in the mid-1970s, fuel costs, once considered minor, have become a significant factor in plant economics. Hot water in particular represents energy that is literally poured down the drain.

Recently, plant engineering manager Rick Boston and I discussed ways of recovering energy from hot waste water. We decided that using waste water to preheat fresh wash water was a promising alternative. I have examined this possibility in a preliminary way. Conclusions from my analysis and its details are discussed in the paragraphs that follow.

APPROACH

The process to be examined was specified in preliminary discussions. Small size and a need for process simplicity dictate a flow sheet that is rather well defined (see Figure 9-2). The process as illustrated includes the current hot water heater and dishwasher. A heat exchanger and pump (E-113 and L-112) are the only new items shown in the flow sheet.

In concept, a pump might be unnecessary because municipal water is available at adequate pressure. However, since the dishwasher must be near atmospheric pressure, a pump was included to provide positive flow through the heat exchanger under all conditions. Steady state was assumed. I think this is one of the most vulnerable assumptions and one that must be examined further. Nonsteady behavior seems likely during certain periods. If the process is unsteady, most of the results will be invalid. In this event, the process must be modified with hold tanks or other equipment to compensate for unsteady behavior.

Flow rates are based on data provided by Rick Boston. According to his figures, rinse water is consumed at an average rate of 0.42 liter/s, 7 h/day, throughout a tourist season of 100 days per year. As illustrated, water is heated from 15 to 85°C. It exits to the drain at a temperature (estimated) of 75°C. Material balance figures include water losses, through evaporation and other means, of about 10 percent.

On examination, this process emerges as a classic optimization problem. Counteractive economic elements are fuel savings opposed to exchanger capital costs. The optimization variable is clearly preheat temperature T_1. This can vary anywhere from the present value 15°C to about 70°C, which requires an infinite heat exchanger.

For economic evaluation and optimization, a double-pipe heat exchanger was assumed. Copper seems to be the most appropriate material of construction for both inner and outer

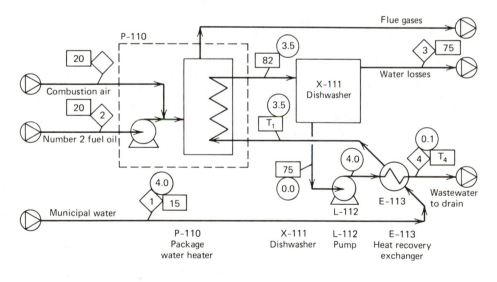

Material Balance (kg/s)

	◇ 1 Municipal Water	◇ 2 Number 2 Fuel Oil	◇ 3 Water Losses	◇ 4 Waste-water
Fuel oil Water	0.42		0.04	0.38
Total	0.42		0.04	0.38

Figure 9-2 Flow sheet and material balance for Goldenrod energy recovery scheme.

tubes. Conventional shell and multitube exchangers are considerably more expensive then double-pipe units in this duty range. Additional process details can be found in Appendix 1. (Illustration 7-2 of this book contains the details said for the purposes of this hypothetical report to be in Appendix 1. This is the type of material that is appropriately relegated to an appendix.)

Once the optimum temperature has been identified, a cash flow analysis can be prepared. This was done as outlined in Appendix 2. (Illustration 8-3, for the purpose of this example.) Results are reported and discussed in the following section.

RESULTS

As described above, the cost of fuel decreases, and expenses associated with capital increase, as the preheat temperature rises to its limiting value of 70° C. Based on a CE Plant Cost Index of 315 (mid-1982) and fuel prices of $6/ GJ ($0.86 per gallon of number 2 fuel oil), annual fuel prices drop from $2550 to $500 as preheat temperature rises from 15 to 69° C. Annual electricity cost for the pump remains constant at about $22. Capital cost increases gradually as preheat temperature rises. Near 69° C, capital costs rise sharply as ΔT_m approaches zero and calculated area goes to infinity. These trends are illustrated in Figure 9-3. Detailed calculations are outlined in Appendix 1.

For the optimum preheat temperature of 64° C, annual fuel oil savings are estimated to be $1860. This is balanced against an anticipated fixed capital investment of $8200 and increased electricity costs of $22/yr. Using depreciation at a sum-of-the-years'-digits rate over a 10 yr lifetime, a cash flow profile as generated. Details are found in Appendix 2. The results are illustrated in Figure 9-4.

DISCUSSION

The optimization curves of Figure 9-3 suggest an important conclusion about equipment design. Note that annual expense drops gradually, then increases sharply, as preheat temperatures approach 69° F. The most dramatic variable here is heat exchanger capital, which changes from a rather weak function of temperature to an extremely sensitive function. This has an interesting connotation, namely, that optimum temperature is relatively insensitive to fuel price *per se*. To elaborate, regardless of price, fuel expense is a linear function of T_1. Although fuel price dictates slope and intercept of this annual cost line, it has a relatively minor impact on the position of T_1, where total annual cost is a minimum. Thus, a

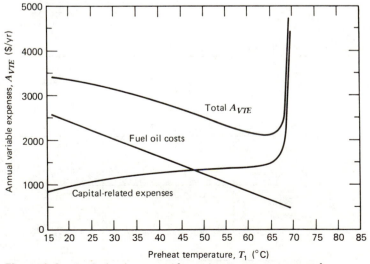

Figure 9-3 Optimization curve for energy recovery proposal.

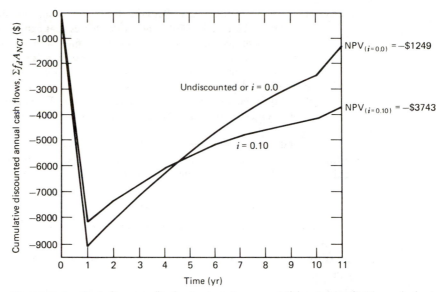

Figure 9-4   Cash flow profile for heat exchanger addition to the Goldenrod plant.

heat exchanger of about 10 m² area, where capital costs exhibit a plateau, is optimum regardless of fuel price. Trial calculations indicate that the optimum preheat temperature falls between 63 and 66° C whether the fuel price is half that assumed or double that assumed in this analysis. Thus, we can be reasonably comfortable with the equipment as specified.

Given a potential fuel savings of 75 percent, one is inclined to be optimistic. Unfortunately, a cash flow analysis is not encouraging. As described in Appendix 2, an investment of $8200 is required to realize an annual fuel savings of $1860. Considering typical rates for added maintenance, overhead, insurance, and taxes on the new equipment, annual net savings are reduced to $690. This relatively small amount is overwhelmed by capital cost, making the project unattractive. For example, positive cash flows are insufficient over the 10 yr lifetime to repay the invested capital, even with zero discount rate. With a minimum reasonable discount rate of 10 percent, net present value is minus $3676. This means that installation of the proposed change represents a loss in 1982 dollars of $3676, assuming that you can invest capital elsewhere at 10 percent interest.

These economic results are discouraging for one major reason: the low plant operation factor. When capital must be amortized on the basis of 700 h of annual operation (less than 10 percent of the hours available in a year), operating savings must be inordinately large to justify investment. If the plant functioned continuously throughout the year rather than over a short summer season, the economics would be more attractive. The same would be true if the plant operated around the clock rather than through only one shift.

Several major assumptions limit this analysis, and they should be remembered. One is the assumption of continuous operation. If water flows are cyclical or batchwise rather than continuous, a more sophisticated process is required. Such can only make the economics less attractive. Fuel savings were based on a constant price of $0.86 per gallon of number 2 fuel oil. This is certainly a questionable assumption, but one consistent with current market conditions.

CONCLUSIONS AND RECOMMENDATIONS

Economic results in this report do not justify investing $8200 for a heat exchanger loop to save $1860 per year in fuel. This is disappointing because the extra fuel is technically unnecessary and certainly seems wasteful to one having a frugal New England unbringing. As responsible citizens and business people, you should continue to pursue profitable alternatives for alleviating this waste. Perhaps direct water reuse or other conservation measures are possible. There may be other hot water losses that can be combined with the rinse water to make the project more attractive. Perhaps some of the economic assumptions were overly conservative. Maintenance costs, taxes, insurance, and overhead, for example, were based on rates characteristics of large industries. These should be reassessed in light of your own experience and situation. The potential for energy-saving tax credits should be explored. There may be federal subsidies or credits that change the economic outlook. In any event, this document is submitted as a first step toward a more enlightened discussion of the problem.

I recommend the foregoing course of action. After you have read this report and discussed it with others in the firm, we should meet and examine the problem further. In particular, we should explore the following questions.

1 Is the assumption of continuous operation a valid one?

2 Can fuel savings be increased by extending operating hours or including the other waste streams?

3 Are there other viable energy-saving alternatives that do not involve a heat exchanger with its relatively large capital cost?

4 Are the economic assumptions employed in this study valid? What about various cost factors? Are energy credits available? Were projected fuel prices realistic?

I hope you find this information worthwhile, and I look forward to discussing it with you.

Gael D. Ulrich

Gael D. Ulrich

REFERENCES

1 Bolmer, J., "Tips on Talking in Public," *Chem. Eng.*, pp. 143–146 (Sept. 21, 1981).

2 Dunkle, T., "Obfuscatory Scrivenery (Foggy Writing)," *Science 82*, pp. 82–84 (April 1982).

3 Hodges, John C., and Mary E. Whitten, *Harbrace College Handbook*, 7 edition, Harcourt Brace Jovanovich, New York (1972).

4 Leesley, M.E., and Williams, M.L., Jr., "All a Chemical Engineer Does Is Write," *Chem. Eng. Educ.*, pp. 188–192 (Fall 1978).

5 Peters, M.S., and K.D. Timmerhaus, *Plant Design and Economics for Chemical Engineers*, 3 ed., Chapter 12, McGraw-Hill, New York (1980).

6 Ricci, L.J. "Chemical Engineer's Education Goes Downhill," *Chem. Eng.*, pp. 94–98 (April 27, 1979).

7 Strunk, W., Jr., and E.B. White, *The Elements of Style*, 3rd ed., Macmillan, New York (1978).

8 Van Ness, H.C., and Abbott, M.M., "Technical Prose: English or Techlish?" *Chem. Eng. Educ.*, pp. 154–159 (Fall 1977).

APPENDIXES

Appendix A
UNITS AND CONVERSION FACTORS

The centimeter-gram-second (cgs) system of units, which has been used for many years, often has been called "the metric system." However a somewhat different system, the Système Internationale (SI) was adopted internationally by the ninth General Conference on Weights and Measures in 1948. Acceptance in this country has been slow, but in 1977 the American Institute of Chemical Engineers (AIChE) Council established schedules for the AIChE to enter into metric conversion, using SI.

Among SI principles are: the use of the kilogram for mass only, and the use of the newton for force or weight. Pressure is expressed in terms of newtons per square meter and is given the name pascal. The pascal (Pa) is a very small unit and the AIChE suggests the kilopascal (kPa) as a more logical unit for pressure. In this book, the bar, which is equal to 100 kPa and very nearly equal to one standard atmosphere, is advocated.

The main advantage of SI is coherence, which means that no conversion factors are needed when basic or derived SI units are used. Any exception to SI units destroys this coherence and is a serious detriment to efficient, error-free computation.

Tables A-1 through A-7 are from Oldshue and Buck [1], on which the foregoing introductory remarks also were based.

REFERENCE

1   Oldshue, J.Y., and E. Buck, "AIChE Goes Metric," *Chem. Eng. Prog.*, pp. 135–138 (August 1977).

TABLE A-1
ACCEPTABLE METRIC UNITS

| Quantity | Système Internationale | | | Metric Alternative[a] | |
	Unit	Symbol		Unit	Symbol
Energy	joule	J			
Force	newton	N			
Length	meter	m			
Mass	kilogram	kg		metric ton	t
Molar mass	kilogram mole	mol		gram mole	g mol
Pressure	pascal	Pa		bar	bar
				atmosphere	atm
Temperature	degree kelvin	K		degree Celsius	°C
Time	second	s		year	yr
				day	d
				hour	h
				minute	min
Volume	cubic meter	m^3		liter	liter
Viscosity	pascal-second	$Pa \cdot s$			

[a] National Bureau of Standards Special Publication 330, "The International System of Units (SI)," 1974 edition. Most of these are listed as acceptable alternatives in reference 1. Some units such as the bar, atmosphere, and minute are considered temporary alternatives. The bar, because of its convenient size for process design, is the pressure unit emphasized in this text. The gram mole, though not recommended, is mentioned here and denoted by g mol to avoid ambiguity and confusion with the preferred kilogram mole denoted mol.

TABLE A-2
EXAMPLES OF SI DERIVED UNITS

| Quantity | SI Unit | |
	Name	Symbol
Area	Square meter	m^2
Speed, velocity	Meter per second	m/s
Acceleration	Meter per second squared	m/s^2
Kinematic viscosity	Square meter per second	m^2/s
Density, mass density	Kilogram per cubic meter	kg/m^3
Concentration (of amount of substance)	Mole per cubic meter	mol^a/m^3

[a] Kilogram mole.

TABLE A-3
SI DERIVED UNITS WITH SPECIAL NAMES

Quantity	SI Unit		
	Name	Symbol	Expression in terms of Other Units
Frequency	hertz	Hz	s^{-1}
Force	newton	N	$kg \cdot m/s^2$
Pressure, stress	pascal	Pa	N/m^2
Energy, work, quantity of heat	joule	J	$N \cdot m$
Power, radiant flux	watt	W	J/s
Quantity of electricity, electric charge	coulomb	C	$A \cdot s$
Electric current	ampere	A	C/s
Electric potential, voltage, potential difference, electromotive force	volt	V	W/A
Capacitance	farad	F	C/V
Electric resistance	ohm	Ω	V/A

TABLE A-4
EXAMPLES OF SI DERIVED UNITS EXPRESSED BY MEANS OF SPECIAL NAMES

Quantity	SI Unit	
	Name	Symbol
Dynamic viscosity	pascal-second	$Pa \cdot s$
Moment of force	meter-newton	$N \cdot m$
Surface tension	newton per meter	N/m
Heat flux density, irradiance	watt per square meter	W/m^2
Specific heat capacity, specific entropy	joule per kilogram-kelvin	$J/(kg \cdot K)$
Specific energy	joule per kilogram	J/kg
Thermal conductivity	joule per meter-second-kelvin	$J/(m \cdot s \cdot K)$
Energy density	joule per cubic meter	J/m^3
Molar energy	joule per mole	J/mol[a]
Molar entropy, molar heat capacity	joule per mole-kelvin	$J/(mol^a \cdot K)$

[a] Kilogram mole

TABLE A-5
UNITS IN USE WITH THE INTERNATIONAL SYSTEM[a]

Name	Symbol	Value in SI Units
Minute	min	1 min = 60 s
Hour	h	1 h = 60 min = 3600 s
Day	d	1 d = 24 h = 86400 s
Year	yr	1 yr ≈ 365 d
Degree	°	$1° = (\pi/180)$ rad
Liter	liter	1 liter = 1 dm^3 = 10^{-3} m^3
Ton (metric)	t	1 t = 10^3 kg
Ångström	Å	1 Å = 0.1 nm = 10^{-10} m
Bar (absolute)	bara	1 bar = 0.1 MPa = 10^5 Pa
Bar (gage)	barg	(pressure in bar above atmospheric pressure)
Standard atmosphere	atm	1 atm = 1.01325 bara

[a]In SI units (where mol = kilogram mole) the universal gas constant is

$$R = 0.0832 \text{ m}^3 \cdot \text{bara/mol} \cdot \text{K}$$

$$= 8320 \text{ J/mol} \cdot \text{K}$$

$$= 0.08206 \text{ m}^3 \cdot \text{atm/mol} \cdot \text{K}$$

In addition to the thermodynamic temperature (symbol T), expressed in kelvins, use is also made of Celsius temperature (symbol t) defined by the equation

$$t = T - T_0$$

where $T_0 = 273.15$ K by definition. The Celsius temperature is expressed in degrees Celsius symbol (°C). The unit "degree Celsius" is thus equal to the unit "kelvin," and an interval or a difference of Celsius temperature may also be expressed in degrees Celsius.

TABLE A-6
SI PREFIXES[a]

Factor	Prefix	Symbol	Factor	Prefix	Symbol
10^{18}	exa	E	10^{-1}	deci	d
10^{15}	peta	P	10^{-2}	centi	c
10^{12}	tera	T	10^{-3}	milli	m
10^9	giga	G	10^{-6}	micro	μ
10^6	mega	M	10^{-9}	nano	n
10^3	kilo	k	10^{-12}	pico	p
10^2	hecto	h	10^{-15}	femto	f
10^1	deka	da	10^{-18}	atto	a

[a]SI symbols are not capitalized unless the unit is derived from a proper name (e.g., Hz for H. R. Hertz). Unabbreviated units are not capitalized (e.g., hertz, newton, kelvin). Only E, P, T, G, and M prefixes are capitalized. Except at the end of a sentence, SI units are not to be followed by periods. With derived unit abbreviations, use center dot to denote multiplication and a slash for division (e.g., newton-second/meter^2 = $N \cdot s/m^2$).

TABLE A-7
CONVERSION FACTORS TO SI FOR SELECTED QUANTITIES

To Convert from	To	Multiply by[a]
barrel (for petroleum, 42 gal)	meter^3 (m^3)	1.5898729×10^{-1}
British thermal unit (Btu, International Table)	joule (J)	1.0550559×10^3
Btu/lbm-°F (heat capacity)	joule/kilogram-kelvin (J/kg · K)	$4.1868000^* \times 10^3$
Btu/hour	watt (W)	2.9307107×10^{-1}
Btu/second	watt (W)	1.0550559×10^3
Btu/ft^2-hr-°F (heat transfer coefficient)	joule/meter^2-second-kelvin (J/m^2 · s · K)	5.6782633
Btu/ft^2-hour (heat flux)	joule/meter^2-second (J/m^2 · s)	3.1545907
Btu/ft-hr-°F (thermal conductivity)	joule/meter-second-kelvin (J/m · s · K)	1.7307347
calorie (International Table)	joule (J)	4.1868000^*
cal/g · °C	joule/kilogram-kelvin (J/kg · K)	$4.1868000^* \times 10^3$
centimeter	meter (m)	$1.0000000^* \times 10^{-2}$
centimeter of mercury (0°C)	pascal (Pa)	1.3332237×10^3
centimeter of water (4°C)	pascal (Pa)	9.80638×10^1
centipoise	pascal-second (Pa · s)	$1.0000000^* \times 10^{-3}$
centistoke	meter^2/second (m^2/s)	$1.0000000^* \times 10^{-6}$
degree Fahrenheit (°F)	kelvin (K)	$t_K = (t_F + 459.67)/1.8$
degree Rankine (°R)	kelvin (K)	$t_K = t_R/1.8$
dyne	newton (N)	$1.0000000^* \times 10^{-5}$
erg	joule (J)	$1.0000000^* \times 10^{-7}$
farad (International of 1948)	farad (F)	9.99505×10^{-1}
fluid ounce (U.S.)	meter^3 (m^3)	2.9573530×10^{-5}
foot	meter (m)	$3.0480000^* \times 10^{-1}$
foot (U.S. Survey)	meter (m)	3.0480061×10^{-1}
foot of water (39.2°F)	pascal (Pa)	2.98898×10^3
foot^2	meter^2 (m^2)	$9.2903040^* \times 10^{-2}$
foot/second^2	meter/second^2 (m/s^2)	$3.0480000^* \times 10^{-1}$
foot^2/hour	meter^2/second (m^2/s)	$2.5806400^* \times 10^{-5}$
foot-pound-force	joule (J)	1.3558179
foot^2/second	meter^2/second (m^2/s)	$9.2903040^* \times 10^{-2}$
foot^3	meter^3 (m^3)	2.8316847×10^{-2}
gallon (U.S. liquid)	meter^3 (m^3)	3.7854118×10^{-3}
gram	kilogram (kg)	$1.0000000^* \times 10^{-3}$
horsepower (550 ft · lbf/s)	watt (W)	7.4569987×10^2
inch	meter (m)	$2.5400000^* \times 10^{-2}$
inch of mercury (60°F)	pascal (Pa)	3.37685×10^3
inch of water (60°F)	pascal (Pa)	2.48843×10^2
inch^2	meter^2 (m^2)	$6.4516000^* \times 10^{-4}$
inch^3	meter^3 (m^3)	$1.6387064^* \times 10^{-5}$
kilocalorie	joule (J)	$4.1868000^* \times 10^3$
kilogram-force (kgf)	newton (N)	9.8066500^*
micrometer	meter (m)	$1.0000000^* \times 10^{-6}$
mil	meter (m)	$2.5400000^* \times 10^{-5}$
mile (U.S. Statute)	meter (m)	$1.6093440^* \times 10^3$
mile/hour	meter/second (m/s)	$4.4704000^* \times 10^{-1}$
millimeter of mercury (0°C)	pascal (Pa)	1.3332237×10^2
ohm (International of 1948)	ohm (Ω)	1.000495
ounce-mass (avoirdupois)	kilogram (kg)	2.8349523×10^{-2}
ounce (U.S. fluid)	meter^3 (m^3)	2.9573530×10^{-5}
pint (U.S. liquid)	meter^3 (m^3)	4.7317647×10^{-4}
poise (absolute viscosity)	pascal-second (Pa · s)	$1.0000000^* \times 10^{-1}$
poundal	newton (N)	1.3825495×10^{-1}
pound-force (lbf avoirdupois)	newton (N)	4.4482216
pound-force-second/ft^2	pascal-second (Pa · s)	4.7880258×10^1
pound-mass (lbm avoirdupois)	kilogram (kg)	$4.5359237^* \times 10^{-1}$
pound-mass/foot^3	kilogram/meter^3 (kg/m^3)	1.6018463×10^1
pound-mass/foot-second	pascal-second (Pa · s)	1.4881639
psi	pascal (Pa)	6.8947573×10^3
quart (U.S. liquid)	meter^3 (m^3)	9.4635295×10^{-4}
slug	kilogram (kg)	1.4593903×10^1
stoke (kinematic viscosity)	meter^2/second (m^2/s)	$1.0000000^* \times 10^{-4}$
ton (long, 2240lbm)	kilogram (kg)	$1.0160469^* \times 10^3$
ton (short, 2000 lbm)	kilogram (kg)	$9.0718474^* \times 10^2$
torr (mm Hg, 0°C)	pascal (Pa)	1.3332237×10^2
volt (International of 1948)	volt (absolute) (V)	1.000330
watt (International of 1948)	watt (W)	1.000165
watt-hour	joule (J)	$3.6000000^* \times 10^3$
yard	meter (m)	$9.1440000^* \times 10^{-1}$

[a]An asterisk after the seventh decimal place indicates that the conversion factor is exact and all subsequent digits are zero.

Appendix B
RULES OF THUMB[1]

GENERAL

Safety and overdesign philosophies vary among professionals. Some say that if a plant can be debottlenecked, it was overdesigned and too expensive in the first place. Others emphasize the risk to one's reputation if a plant cannot operate at the nameplate capacity and within product specifications. I recommend not including an overdesign factor because conservative allowances are built into most of the design data and equations. Pilot plant, laboratory, or research equipment, on the other hand, should be overdesigned to provide flexibility. Approximately 50 percent excess capacity is a reasonable allowance in such instances.

Pressure drops through pipelines connecting equipment are generally small compared with those in the equipment itself. If needed, typical values are in the vicinity of 0.03 bar. Average flow velocities normally range from 1 to 4 m/s for pipes carrying liquids and process vapors. Typical velocities are 3 to 10 m/s for lines conveying gases.

AUXILIARY FACILITIES

Compressed air is commonly provided at 3, 10, 20, or 30 barg pressure. Respective dew points are indicated in Figure 5-3. Instrument air is normally at 3 barg with a dew point of 0°C.

Process Steam

Traditional standard specifications are as follows.

Designation	Pressure (bara)	Temperature (°C)
High pressure	45	400 (superheated)
Medium–high pressure	33	239 (saturated)
Moderate pressure	17	204 (saturated)
Medium–low pressure	9	175 (saturated)
Low pressure	4.5	148 (saturated)

[1]Assembled with the aid of the University of New Hampshire Chemical Engineering Class of 1982, S. M. Hsieh assisting.

Cooling Water

Temperature of cooling water to process is normally 30°C. It is usually heated to about 45°C before returning to the cooling tower. Lower temperatures are possible in dry climates.

Dimineralized Water for Boilers

Contains 30 to 40 ppm solids. Up to 320 ppm is suitable for cooling tower feed water.

Electricity

Usually purchased offsite
See discussion of drives and power recovery machines for exceptions. Electricity costs can be estimated using Table 6-3.

Refrigerant

Usually a "Freon" fluid is traditionally available to temperatures from +5 to −60°C. Costs depend on temperature required; see Figure 5-11 and Table 6-3 for specifics.

Waste Treatment

Water and air pollution guidelines control levels of plant emissions. Legislation controlling U.S. plants is discussed in references 1 and 2. Actual allowable emissions are a complex function of politics and location. In the absence of better data, the approximate guidelines in Table B-1 may be consulted.

TABLE B-I
REPRESENTATIVE EMISSIONS LIMITS FOR U.S. PLANTS

Pollutant	Maximum Design Concentration in Effluent Gases (g/std m³)
Sulfur oxides	0.2
Particulate matter	0.2
Carbon monoxide	0.01
Hydrocarbons	0.4
Nitrogen oxides	0.2
Heavy metals and toxic elements	0.005

Pollutant	Maximum Design Concentration in Wastewater (g/m³)
Dissolved organics	10
General inorganic compounds	20
Heavy metals and toxic elements	5
Halogenated hydrocarbons	5
Total suspended solids	20

Solid and Liquid Wastes

Solid and nonaqueous liquid waste disposal costs can be estimated from data contained in Table 6-3.

CONVEYORS

Auger or Screw Conveyor. Flexible and capable of conveying sticky, gummy solids. Simultaneous mixing, heating, cooling, and drying can be accomplished.

Belt Conveyor. High capacity, long distance transportation, especially for minerals and ores. Maximum incline is 30 degrees. Number of turns is limited.

Apron Conveyor. Similar to belt conveyor. Used when conveyed materials incompatible with belts.

Bucket Elevator. Best for vertical lifting of nonsticky materials. Usually used for abrasive material transport such as minerals and ores.

Continuous Flow Conveyor. Convenient self-feeding, self-discharging. Frequently chosen for relatively low capacity operations where convenience and versatility are required. Unsuitable for highly abrasive, lumpy or sticky materials.

Pneumatic Conveyor. Employed for high volume, short distance transport. Avoid large lumps, abrasive solids, and explosive gas–solid mixtures.

Vibratory Conveyor. For nonsticky materials. Operation under vacuum or controlled atmospheres possible. Drying, sieving, heating, and cooling can be conducted simultaneously.

CRUSHERS, MILLS, GRINDERS

For *coarse crushing*, use primary crushers such as *jaw* and *gyratory units* which are capable of handling large quantities of friable materials *up to 2 m* in diameter.

To crush lumps of *0.5 m or smaller diameter*, *impact* and *roll devices* are usually employed.

For *fine crushing* or *coarse grinding*, use *pan, bowl, ring-roll,* or *attrition mill* (100 μm–1 cm).

For *fine grinding*, choose *ball, rod,* or high speed *hammer mill* (1–100 μm).

To produce *submicrometer* particles, use a fine or ultrafine grinder such as *fluid energy mill* (0.1–1 μm).

DRIVES AND POWER RECOVERY MACHINES

Drives

Electric motors are almost always used when less than 100 kW of shaft power is needed. They are competitive with steam turbines at higher power levels as well.

Steam turbines are frequently employed for driving centrifugal pumps, compressors, or generators where speeds and demands are relatively constant.

Internal combustion engines are usually reserved for remote locations or mobile applications. *Combustion gas turbines* are chosen for remote, mobile, or other nonconventional applications where high speed, high capacity, and fixed-load service are needed. Because of materials limitations, combustion gas temperatures normally fall below 750° C.

Power Recovery Machines

Avoid large irreversible or unnecessary drops in process stream pressures when possible. Otherwise, an expander should be considered if available power is greater than 100 kW. Throttling valves are usually used to drop the pressures of process streams if power available is less than 100 kW.

In new plants where single drives or continuous electric power needs of 1000 kW or greater are found, cogeneration should be considered. It will be economically feasible if there is a matching need for low pressure process steam. Otherwise, purchase of electric power from an offsite source will usually be less expensive.

Axial turbines are employed for power recovery where flow rates, inlet temperatures, or pressure drops are high.

Radial flow or *turboexpanders* are used to recover power in many applications where inlet temperatures do not exceed 550° C.

EVAPORATORS AND VAPORIZERS

1 Steps in designing vaporizers.
  (a) Determine superficial vapor velocity, u.

$$u = 0.06 \left(\frac{\rho_l - \rho_g}{\rho_l} \right)^{0.5}$$

  (b) Calculate cross-sectional area and diameter of a vertical, cylindrical vaporizer. (If diameter is larger than 4 m, use a horizontal pressure vessel, multiple vertical vessels, or special entrainment separation.)
  (c) Calculate heat duty of vaporizer from $\dot{Q} = \dot{m}\lambda$.
  (d) Calculate heat transfer area. Process steam saturated at 4.5, 9, 17, or 33 bara or superheated to 400° C at 45 bara is the likely heat source.
  (e) Calculate height of jacket required.
  (f) Allow one additional diameter for vapor disengagement, and determine total height of vaporizer. If height is more than four times diameter, internal heating coils should be employed.

2 Steps in designing single-effect evaporators.
  (a) Use material balance to determine amount of vapor and liquid leaving.
  (b) Use energy balance to determine heat load.
  (c) Calculate heat transfer area, $A = \dot{Q}/U\Delta T$.
  (d) Determine power consumption in forced-circulation and agitated-film units from Figure 4-5.

3 Steps in designing multiple-effect evaporators.
  (a) Determine liquid flow scheme. Use forward feed if the heat transfer coefficient of the first effect is greater than that in the last by less than 50 percent. Otherwise mixed feed is preferred.
  (b) Select number of effects: four to six effects if boiling point elevation (BPE) is significant, otherwise six to ten stages are normally used.

   (c)   Determine stage size.
      (1)   Select type of evaporator using criteria in Table 4-7.
      (2)   Estimate the vapor generated in each effect.
      (3)   Calculate the concentrations of intermediate liquid streams.
      (4)   Estimate temperatures in the system.
      (5)   Obtain overall heat transfer coefficient U. If any U does not exceed $500 \, J/m^2 \cdot s \cdot K$, use another type of evaporator or rearrange the feed sequence.
      (6)   Determine temperature of the system with better values of U.
      (7)   Calculate area of each evaporator (Equation 4-29).
      (8)   Determine power consumption.

FURNACES

Furnaces are usually purchased as packaged units.

In process heating, steam is normally used as the working fluid if the required heat is in the range of 100 to 250° C. A special thermal fluid such as Dowtherm A is usually employed in the temperature range of 250 to 400° C.

Steam generators or boilers are usually chosen in preference to other furnaces if they can perform the needed service.

Direct-fired heaters can be simply considered as process vessels that provide 10 to 60 seconds residence time for the process.

Indirect-fired furnaces are often employed to vaporize or heat process streams indirectly.

Incinerators are furnaces designed to dispose of unwanted wastes.

Ovens are designed to maintain solid objects at high temperatures for extended periods.

Fire-tube units are employed primarily for generating modest amounts of low pressure saturated steam.

Water-tube boilers are used to generate high pressure or superheated steam. Steam at 45 bara pressure superheated to 400° C is a typical maximum. Saturated steam at pressures ranging from 17 to 33 bara is also commonly generated.

GAS MOVERS, COMPRESSORS, AND EXHAUSTERS

Fans operate near 1 atm. Pressure drops are usually less than 15 kPa. Efficiencies vary from 60 to 80 percent.

Blowers operate with pressure drops of 3 kPa to 5 bar. They are more expensive than fans. Efficiencies range from 70 to 85 percent. For vacuum operation, efficiencies are lower, usually 50 to 70 percent.

Compressors are for high pressure operation, from 2 to 3000 bara. Typical efficiencies are 60 to 80 percent. Staged compressors are usually employed if the compression ratio is greater than 4. Reciprocal piston types are common when the pressure is above 1000 bara.

Ejectors are employed for corrosive gases and vapors at pressure ranges from 0.01 to 5 bara. Efficiencies are comparatively low (25 to 30 percent). With no moving parts, ejectors are almost maintenance-free. Capital cost is low. Ejectors are not recommended if the process stream should not be diluted with a motive fluid.

Selection Criteria

1 Rotary blowers or compressor are often chosen for pressures ranging from low vacuum to 10 bara and for flow rates of less than 20 std m^3/s.

2 Centrifugal blowers or compressors are usually preferred for high capacity and constant delivery pressure. They are favored when discharge pressures are below 15 bara.

3 Reciprocating positive displacement compressors are chosen at ultrahigh pressures and modest capacities.

GAS–SOLID CONTACTORS

Tunnel contactors are usually employed in small-scale operations. Temperatures can be varied along the tunnel length.

Rotary contactors are used in most large-scale bulk solid–gas processes because of their flexibility and large capacities.

Tower contactors are attractive for controlled atmosphere and vacuum operation. Capacities are limited. Maximum temperature is 300°C.

Vibrating conveyor contactors give excellent contacting efficiency for solids that are not easily suspended in a fluid bed.

Drum dryers are often used for slurries and pastes that will not flow through other contactors.

Screw conveyors are employed for handling gummy, sticky, or other problem solids.

Gravity shaft units provide high efficiency contact and large capacity operation. The solid must be relatively rigid and strong. Blast furnaces and limestone calciners are examples of gravity shaft contactors.

Fluid bed contactors are popular for particulate solids except for gummy, friable, or fragile materials.

Spouted-bed contactors are employed when solids are too large to fluidize but the advantages of a fluid bed are desired.

Pneumatic conveyor contactors are used prominently in drying, cooling, and reacting. They are limited to fine particle materials that dry or react in short times.

Spray dryer–coolers are relatively expensive to operate. They are usually employed when special characteristics such as dispersibility and pourability are desired in the final product or if heat-sensitive materials are involved.

HEAT EXCHANGERS

STEP 1. Identify Flow Path and Exchanger Type.

1 Flow path.

(a) Corrosive fluids are usually passed on the tube side.

(b) High pressure fluids usually pass on the tube side. Plate exchangers are not recommended for pressures above 10 bara.

(c) Fouling or scaling fluids are placed on the tube side of fixed-tube exchangers. If deposits can be removed by high velocity steam or water jets, fouling fluids may also pass on the shell side of exchangers that can be exposed for cleaning.

(d)   Highly viscous fluids are usually placed in the shell side of conventional shell and tube exchangers. Plate exchangers are attractive for such service. For viscosities greater than 1 Pa · s, scraped-wall exchangers are attractive.

(e)   Condensing vapors are usually placed on the shell side.

2   Exchanger type (see Table 4-12).

STEP 2.   Specify Heat Duty.

Determine exchanger duty from an energy balance on one side. Allow up to 10 percent losses depending on shell-side temperature.

STEP 3.   Determine ΔT_m (see text).

1   Approach ΔT's are approximately 10° C for liquids or systems with high heat transfer coefficients.

2   Approach ΔT's are approximately 50° C for gases or systems with low heat transfer coefficients.

STEP 4.   Determine Overall Heat Transfer Coefficient U (see Table 4-15).

STEP 5.   Calculate Heat Transfer Area $A = \dot{Q}/U\Delta T_m$.

STEP 6.   Choose Construction Materials.

Pressure drops are approximately 0.2 to 0.6 bar for liquid heating, cooling, or boiling. For condensation or heat transfer to or from gases, pressure drops are approximately 0.1 bar.

MIXERS

Mixing of Fluids having Moderate Viscosities

Combustion units, reactors or other equipment where gas uniformity is necessary use *fluid jet* mixing. Power consumption of a fluid jet increases linearly with the square of viscosity. For intense jet mixing of liquids having a viscosity of 0.01 Pa · s, a pressure differential of 300 bar is required.

Motionless mixers are ideal for viscous fluid mixing when turbulence is not necessary or desired.

Spargers are commonly employed in extremely corrosive situations requiring gas–liquid contact for brief periods and with mild agitation. The recommended gas rate is 0.004 m^3/s per square meter of tank cross section for mild agitation, 0.008 for relatively complete agitation, and 0.02 for violent motion.

Propeller and turbine agitators are usually employed in chemical process plants. Propellers seldom exceed 0.5 m in diameter. Turbine impeller diameters are limited to about 3 m. In general, turbines are best for suspension of solids, dispersion of immiscible liquids, heat transfer enhancement, and promotion of chemical reaction.

For mixing of miscible liquids and solutions, moderate agitation with either a propeller or a turbine impeller is recommended, requiring a specific power consumption of 0.3 to 0.7 kW/m^3.

For suspension of solid particles, if settling velocity is less than 0.01 m/s, mild agitation can be achieved with either a propeller or a turbine impeller. For up to 0.05

m/s settling velocity, either will serve under vigorous agitation. From 0.05 to 0.1 m/s, intense agitation with a propeller is recommended.

In agitated tanks, the liquid height is usually between 1 and 1.5 times the vessel diameter. Vessel heights are usually 1.5 to 2.0 times diameter.

To create an emulsion, propeller agitators are employed with high specific power.

Mixing of Solids and Pastes

Agitation along with simultaneous squeezing, dividing, and folding is necessary as viscosity of the paste increases above 200 Pa · s.

To mix heavy, stiff, and gummy materials such as clays, pastes, adhesives, light polymers, and doughs, use a *kneader*.

Extruders are ideal for pastes or nonabrasive semisolids that are reasonably well mixed but require high shear under pressure to give improved consistency.

Mixing rolls are employed commonly for dispersing additives and pigments into heavy polymer suspensions.

Rotor mixers are effective with modestly viscous nonsticky pastes and soft lumpy solids or cakes.

PROCESS VESSELS

Distillation

Distillation is usually more economical than liquid–liquid extraction or other competitive techniques. Unless there is a good reason for distillation not to work, it is usually the best technique for separating miscible liquid mixtures.

In distilling three-component mixtures, take the lightest overhead from the first tower and feed the bottoms to the second tower for separation.

For more than three components, separation one by one rather than splitting into multicomponent mixtures is generally more economical. The number of columns is the same, but energy requirements are substantially different. This is affected by other considerations.

- Sequences yielding equal molal division in distillate and bottoms are preferred.
- Separations of components having similar volatilities should be made in the absence of other components.
- High purity separations should be reserved until last.

Operate distillation towers at the lowest feasible pressure. This is usually dictated by cooling water, which can hold the condenser temperature as low as 40° C.

Consider using the latent heat of an overhead product to lower reboiler duty. Reboilers and partial condensers act as additional stages.

Gas Absorption

Gas temperatures leaving absorbers are usually within 2° C of the entering liquid temperature.

Equipment Specification

Drums are used to flash, mix, settle, and retain liquids or gases. Volumes are based on average residence times of process fluids. Drums can be oriented either horizontally or vertically. They commonly range up to 4 m in diameter and 20 m in length.

Flash drums are used to separate a liquid–vapor mixture. A liquid residence time of approximately 600 s is common.

Holdup drums have typical residence times of 300 to 600 s.

Mixer drums have length to diameter ratios of 1:1 to 2:1.

Settler drums have length to diameter ratios of 2 to 4. Residence times are approximately 300 to 600 s.

Towers are used to separate components that are "energically" combined, commonly, gas–liquid, liquid–liquid, and liquid–solid mixtures.

For *tray towers*, tower height = (number of theoretical stages) × (stage height)/(tray efficiency)

$$H_a = \frac{(N)(H_t)}{\epsilon_s}$$

$$H_t = 0.5D^{0.3} \qquad \text{for} \quad D > 1.0 \text{ m}$$

$$= 0.5 \, m \qquad \text{otherwise}$$

Pressure drop is usually near 1 kPa per stage.
Diameters range to 4 m, heights to 60 m.

For *packed towers*, tower height = (number of stages) × (height of a theoretical plate)

$$H_a = (N)(\text{HETP})$$

For distillation, $\text{HETP} = 0.5D^{0.3} \qquad \text{for} \quad D > 0.5 \text{ m}$

For gas absorption, $\text{HETP} = D^{0.3}$

For both distillation and absorption, $\text{HETP} = D \qquad \text{for } D < 0.5 \text{ m}$

Bubble and spray towers have low contacting efficiencies. They are used where the number of stages is relatively small. Heights range to 40 m. Gas residence time is approximately 60 s.

PUMPS

Shaft power ($\dot{w}_s$) and power consumption (P) are required in pump design and cost estimation.

Centrifugal radial pumps are employed widely in chemical processes where fluids of moderate viscosity are raised to modest pressures.

Axial flow pumps are used for moving large volumes at low differential pressures.

Regenerative or turbine pumps are often chosen for low flow, high pressure services.

Positive displacement pumps are used when liquid viscosities are large, flow rates are small, or carefully metered liquid rates are desired.

Reciprocating piston pumps are often required if extremely high pressure are required at moderate flow rates.

Volumetric displacement pumps can be used when direct contact pumping is not possible or fluid displacement pumping is merely more convenient (e.g., in laboratory or temporary situations).

REACTORS

Reactors usually fall under the categories of mixers, process vessels, furnaces, gas–solid contactors, heat exchangers, or other types of process equipment that have been adapted or modified for a specific reaction.

Inlet temperature, pressure, and concentration are necessary for specification of a reactor.

Equilibrium analysis should be made, to rule out impossible results and to define the limits of possible conversion.

In *isothermal reactors* or those where temperatures drop with conversion, the volume required for a given conversion is larger in a *well-stirred* than in a *plug flow* operation.

In adiabatic exothermic reactions, well-stirred reactors require the least volume except near equilibrium conversions.

Isothermal performance and general temperature and concentration control are accomplished more easily in well-stirred reactors.

Some well-stirred exothermic reactors have multiple stable operating modes.

Plug flow behavior can be approached in well-stirred reactors by placing multiple reactors in series.

Both energy balances and material balances are usually necessary to determine reactor size.

If pressure drops are large, as through a long plug flow or packed-bed reactor, it may be necessary to design the reactor in increments.

Fluid Beds and Packed Beds

Fluid beds give intimate fluid–solid contact and uniform temperature distribution (temperature differences less than 5°C from point to point). They are generally well-stirred when length to diameter ratios are less than 2. For ratios greater than 4, they approach plug flow behavior.

Packed beds are generally plug flow devices. To provide uniform flow distribution, bed height should be no less than half the diameter.

SEPARATORS
Centrifugal Separators

Employed where particles have very low settling velocities.

1 *Sedimentation centrifuges* are ideal for clarification of dilute liquid–solid systems and for separation of immisicible liquid mixtures.

2 *Helicial conveyor centrifuges* are suitable for producing relatively dry cakes from large flows of concentrated slurries containing moderately fine clays or ores.

3 *Continuous conveyor centrifugal filters* are used in applications involving sandlike particles or crystals having diameters larger than 150 μm.

4 *Cyclone separators* are widely used for removing solid or liquid particles from gases if entrainment of smaller particles can be tolerated. Gas cyclones are limited at subatmospheric pressure because of leakage through the solids discharge.

Electrostatic Precipitators

Used with corrosive or high temperature gases for collection of ultrafine particles.

Gravity and Impingement Separators

1 If settling velocity is less than 0.1 m/s and a pressure differential of 10 kPa is available, a *cyclone separator* will generally be more economical than a *settling drum*.

2 Liquid *settling drums* are not practical unless sedimentation velocity is greater than 0.001 m/s.

3 *Clarifiers* and *thickeners* are usually restricted to aqueous-based or other nontoxic, nonhazardous systems. Otherwise, use filters or centrifuge-type separators.

4 *Rake* and *spiral classifiers* are effective in separating coarse solids or sands from the overall flow of a mill.

Filters

1 *Bag filters* are highly efficient for collecting submicrometer particles from gas streams. They are, however, limited by maximum fabric service temperatures.

2 *Cartridge filters* are used primarily for final cleaning or polishing of a low concentration effluent stream.

3 *Sand filters* are used primarily in wastewater and culinary water treatment plants.

4 *Horizontal filters* are suitable for separating conventional slurries. Cake washability and flexibility are excellent.

5 *Rotary disk filters* are designed for relatively easy noncritical separations where flexibility and efficient cake washing are not essential.

6 *Rotary drum filters* are used in most liquid–solid chemical process filtrations.
   (a) Rotary drum precoat units are the only continuous filters capable of ultimate crystal-clear cleaning of effluent streams.
   (b) Multicompartment media-covered drums are popular because of flexibility and cost.

7 Batch filters
   (a) The *plate and frame filter press* is noted for flexibility of design, filter media, and operation. Leakage, dripping, and high labor costs are drawbacks.

(b) *Shell and leaf filters* are similar to plate and frame filters in performance. Slightly more expensive, they are usually required for filtration of toxic, odorous, or hazardous materials.

Sludges from liquid–solid process filters contain about 50 weight percent solids.

Presses

1 *Screw presses* are used to express vegetable oils and to dewater paper pulps, plastics, and rubber. They are also suitable for a variety of sludges and residues except those that contain coarse and abrasive solids.

2 *Roller presses* are widely used for dewatering paper and fabrics. They can accept abrasive materials.

Screens

1 *Grizzlies* are used to separate particles larger than 5 cm.

2 *Vibrating* or *rotary screens* are used if particles are smaller.

SIZE-ENLARGEMENT EQUIPMENT

For products demanding extreme uniformity in size and composition, choose a *tableting press*. Otherwise, use a *roll press* or *pellet mill*, which operates at higher capacity and is lower in cost.

For gummy pastes, use *pellet mills*.

For materials where heat is required, and shape of final product is a major consideration, a *screw extruder* is recommended.

Plug mill extruders are ideal when mixing is desired (e.g., of catalysts or fertilizers).

Agglomerators give a loosely packed product, and operating costs are low.

Prilling is a good choice for enlarging sticky and gummy materials if they can be melted to a free-flowing liquid. Energy costs are relatively high, however.

STORAGE VESSELS

Capacity, temperature, pressure, and exposure conditions must be known for one to specify storage vessels. Large *atmospheric pressure storage tanks* are usually chosen to contain raw materials for processing or products awaiting shipment. In them

1 Either an internal vent system or a floating roof is employed to compensate for ambient temperature and pressure changes.

2 Pressures must be within a few kilopascals of ambient.

3 Thirty days' capacity is usually specified for raw materials and products to assure uninterrupted plant operation. Such storage tanks should be at least 1.5 times the size of the transport vessel. (Capacities of various shipping vessels are listed in Table 4-27, note *a*.)

Bins are frequently employed in solids handling operations, between grinders, conveyors, and so on, to provide surge capacity.

Spherical or bullet-shaped tanks with thicker walls are used to accommodate higher pressure gas or liquid storage.

Large-scale storage of gases is often accomplished by liquefying the gas and storing it at *cryogenic* temperature and relatively modest pressure.

Day tanks are designed with about 8 h capacity. Residence times for feed tanks are about 1800 s.

For batch transfer within a storage system, provide for one transfer per shift and allow 15 min for it to occur.

REFERENCES

1 Buonicore, A.J., "Air Pollution Control," *Chem. Eng.*, pp. 81–101 (June 30, 1980).

2 Robertson, J.H., W.F. Cohen, and J.Y. Longfield, "Water Pollution Control," *Chem. Eng.*, pp. 102–119 (June 30, 1980).

Appendix C
JANAF THERMOCHEMICAL DATA[1]

A complete set of thermodynamic properties for a given element or compound in the ideal state can be developed from its entropy at a specified condition, heats of transformation, plus the specific heat and specific volume known as a function of temperature.[2] The mathematical procedures, however, are tedious and time consuming. With the task of finding and selecting accurate, consistent, fundamental data from the literature added, such computations become formidable. Fortunately, these tedious procedures have been performed for us by several organizations. Data from one of these, the JANAF thermochemical tables, are found in this appendix.

The JANAF (the acronym formed from Joint Army–Navy–Air Force) collection originated in the early days or rocket propulsion. More accurate data were desired by those estimating the performance of numerous prospective fuels. Using literature sources and statistical mechanics, absolute entropies, standard heats of formation, and heat capacities over a range of temperatures were assembled. From these, free energy, enthalpy, entropy, and equilibrium constants of formation were computed from absolute zero to 6000 K. The second edition of the *JANAF Thermochemical Tables* cited contains such data for more than 1000 chemical species. Tables C-1 to C-12 give data for 12 species prominent in chemical engineering calculations.

[1]Stull, D. R., and H. Prophet, *JANAF Thermochemical Tables*, 2nd edition, U.S. Government Printing Office, Washington, D.C. (1971). Tables C-1 through C-12, though similar to the charts contained in this reference, are updated versions expressed in SI units. They were provided by Malcolm W. Chase, Thermal Research, Dow Chemical Company, Midland, Michigan (January 1983).

[2]Generalized correlations can be employed to compensate for nonideal behavior when appropriate. These pressure–volume–temperature correlations are available in any modern chemical engineering thermodynamics text. A more exhaustive reference is that of R. C. Reid, J. M. Prausnitz, and T. K. Sherwood, *Properties of Gases and Liquids*, 3rd edition, McGraw-Hill, New York (1977).

TABLE C-1
METHANE (CH_4); (IDEAL GAS); M = 16.043

T(K)	$C_p°$	$S°$	$-(G°-H°_{298})/T$	$H°-H°_{298}$	$\Delta H_f°$	$\Delta G_f°$	$Log_{10}K$
	J/(gmol·K)			kJ/gmol			
0	0.000	0.000	INFINITE	-10.025	-66.906	-66.906	INFINITE
100	33.259	149.394	216.380	-6.699	-69.990	-64.434	33.656
200	33.476	172.473	189.313	-3.368	-72.032	-58.195	15.198
298.15	35.639	186.146	186.146	0.000	-74.873	-50.815	8.902
300	35.710	186.368	186.146	0.067	-74.931	-50.668	8.822
400	40.501	197.250	187.598	3.862	-77.973	-42.116	5.500
500	46.342	206.911	190.510	8.201	-80.818	-32.823	3.429
600	52.229	215.882	194.000	13.129	-83.329	-22.983	2.001
700	57.794	224.354	197.736	18.636	-85.475	-12.744	0.951
800	62.932	232.413	201.568	24.673	-87.266	-2.230	0.146
900	67.601	240.099	205.426	31.204	-88.730	8.489	-0.493
1000	71.797	247.446	209.267	38.179	-89.881	19.351	-1.011
1100	75.530	254.467	213.058	45.551	-90.776	30.321	-1.440
1200	78.835	261.182	216.790	53.271	-91.437	41.367	-1.801
1300	81.747	267.609	220.455	61.304	-91.927	52.446	-2.107
1400	84.308	273.763	224.045	69.609	-92.257	63.576	-2.372
1500	86.559	279.659	227.555	78.153	-92.483	74.722	-2.602
1600	88.538	285.311	230.990	86.910	-92.621	85.856	-2.803
1700	90.287	290.729	234.346	95.855	-92.667	97.023	-2.981
1800	91.826	295.934	237.626	104.960	-92.650	108.173	-3.139
1900	93.190	300.938	240.827	114.215	-92.579	119.336	-3.281
2000	94.399	305.750	243.952	123.595	-92.462	130.486	-3.408
2100	95.479	310.382	247.007	133.089	-92.320	141.633	-3.523
2200	96.441	314.846	249.990	142.687	-92.157	152.762	-3.627
2300	97.303	319.151	252.902	152.373	-91.969	163.900	-3.722
2400	98.077	323.310	255.751	162.143	-91.776	175.029	-3.809
2500	98.776	327.327	258.534	171.988	-91.579	186.117	-3.889
2600	99.403	331.214	261.253	181.895	-91.374	197.238	-3.962
2700	99.972	334.979	263.914	191.866	-91.169	208.326	-4.030
2800	100.491	338.624	266.521	201.891	-90.964	219.409	-4.093
2900	100.964	342.159	269.069	211.961	-90.768	230.509	-4.152
3000	101.391	345.586	271.563	222.083	-90.579	241.567	-4.206
3100	101.784	348.920	274.002	232.241	-90.383	252.634	-4.257
3200	102.144	352.155	276.395	242.438	-90.211	263.701	-4.304
3300	102.479	355.305	278.738	252.668	-90.056	274.759	-4.349
3400	102.780	358.368	281.035	262.931	-89.906	285.805	-4.391
3500	103.064	361.351	283.286	273.224	-89.784	296.859	-4.430
3600	103.324	364.259	285.495	283.541	-89.676	307.896	-4.467
3700	103.562	367.092	287.663	293.888	-89.596	318.951	-4.503
3800	103.788	369.857	289.792	304.256	-89.525	330.000	-4.536
3900	103.993	372.556	291.880	314.645	-89.492	341.042	-4.568
4000	104.186	375.192	293.930	325.055	-89.479	352.084	-4.598
4100	104.366	377.765	295.943	335.481	-89.483	363.108	-4.626
4200	104.533	380.284	297.922	345.925	-89.525	374.171	-4.653
4300	104.692	382.744	299.863	356.389	-89.588	385.192	-4.679
4400	104.838	385.154	301.775	366.866	-89.680	396.225	-4.704
4500	104.977	387.510	303.654	377.355	-89.801	407.250	-4.727
4600	105.106	389.819	305.503	387.861	-89.948	418.329	-4.750
4700	105.228	392.083	307.323	398.375	-90.123	429.383	-4.772
4800	105.341	394.300	309.110	408.902	-90.328	440.441	-4.793
4900	105.449	396.472	310.871	419.442	-90.558	451.504	-4.813
5000	105.550	398.601	312.603	429.994	-90.818	462.550	-4.832
5100	105.646	400.693	314.310	440.554	-91.115	473.620	-4.851
5200	105.738	402.748	315.992	451.123	-91.433	484.691	-4.869
5300	105.822	404.760	317.649	461.700	-91.784	495.808	-4.886
5400	105.901	406.739	319.281	472.286	-92.169	506.871	-4.903
5500	105.981	408.685	320.888	482.880	-92.584	517.975	-4.919
5600	106.052	410.593	322.473	493.482	-93.027	529.063	-4.935
5700	106.123	412.471	324.034	504.093	-93.504	540.180	-4.950
5800	106.186	414.316	325.574	514.707	-94.010	551.292	-4.965
5900	106.248	416.132	327.097	525.330	-94.550	562.447	-4.979
6000	106.311	417.919	328.595	535.958	-95.111	573.547	-4.993

TABLE C-2
CHLORINE, DIATOMIC (CL₂); (REFERENCE STATE—IDEAL GAS); M = 70.914

	J/(gmol·K)			kJ/gmol			
T(K)	$C_p°$	$S°$	$-(G°-H°_{298})/T$	$H°-H°_{298}$	$\Delta H_f°$	$\Delta G_f°$	$Log_{10}K$
0	0.000	0.000	INFINITE	-9.180	0.000	0.000	0.000
100	29.292	188.908	251.584	-6.268	0.000	0.000	0.000
200	31.698	209.853	226.003	-3.230	0.000	0.000	0.000
298.15	33.936	222.961	222.961	0.000	0.000	0.000	0.000
300	33.970	223.170	222.961	0.063	0.000	0.000	0.000
400	35.300	233.149	224.313	3.535	0.000	0.000	0.000
500	36.083	241.116	226.903	7.104	0.000	0.000	0.000
600	36.572	247.743	229.840	10.740	0.000	0.000	0.000
700	36.907	253.404	232.810	14.414	0.000	0.000	0.000
800	37.146	258.349	235.701	18.121	0.000	0.000	0.000
900	37.330	262.738	238.467	21.845	0.000	0.000	0.000
1000	37.472	266.676	241.095	25.585	0.000	0.000	0.000
1100	37.593	270.253	243.584	29.338	0.000	0.000	0.000
1200	37.698	273.529	245.944	33.104	0.000	0.000	0.000
1300	37.790	276.550	248.187	36.878	0.000	0.000	0.000
1400	37.869	279.353	250.312	40.660	0.000	0.000	0.000
1500	37.945	281.968	252.337	44.451	0.000	0.000	0.000
1600	38.016	284.420	254.266	48.250	0.000	0.000	0.000
1700	38.083	286.730	256.149	52.053	0.000	0.000	0.000
1800	38.146	288.905	257.872	55.865	0.000	0.000	0.000
1900	38.212	290.972	259.559	59.681	0.000	0.000	0.000
2000	38.279	292.934	261.178	63.509	0.000	0.000	0.000
2100	38.351	294.800	262.734	67.337	0.000	0.000	0.000
2200	38.426	296.587	264.236	71.178	0.000	0.000	0.000
2300	38.505	298.298	265.471	75.023	0.000	0.000	0.000
2400	38.589	299.938	267.073	78.877	0.000	0.000	0.000
2500	38.681	301.516	268.420	82.743	0.000	0.000	0.000
2600	38.777	303.035	269.722	86.613	0.000	0.000	0.000
2700	38.882	304.499	270.981	90.496	0.000	0.000	0.000
2800	38.991	305.917	272.203	94.391	0.000	0.000	0.000
2900	39.104	307.286	273.391	98.295	0.000	0.000	0.000
3000	39.221	308.612	274.542	102.211	0.000	0.000	0.000
3100	39.342	309.901	275.663	106.140	0.000	0.000	0.000
3200	39.463	311.152	276.751	110.081	0.000	0.000	0.000
3300	39.585	312.369	277.813	114.035	0.000	0.000	0.000
3400	39.706	313.553	278.847	117.997	0.000	0.000	0.000
3500	39.823	314.704	279.855	121.976	0.000	0.000	0.000
3600	39.940	315.829	280.838	125.964	0.000	0.000	0.000
3700	40.053	316.925	281.801	129.959	0.000	0.000	0.000
3800	40.158	317.992	282.738	133.972	0.000	0.000	0.000
3900	40.258	319.038	283.654	137.993	0.000	0.000	0.000
4000	40.355	320.059	284.554	142.026	0.000	0.000	0.000
4100	40.443	321.055	285.432	146.063	0.000	0.000	0.000
4200	40.522	322.030	286.290	150.114	0.000	0.000	0.000
4300	40.593	322.984	287.131	154.168	0.000	0.000	0.000
4400	40.660	323.921	287.960	158.231	0.000	0.000	0.000
4500	40.719	324.833	288.767	162.302	0.000	0.000	0.000
4600	40.765	328.486	289.562	166.373	0.000	0.000	0.000
4700	40.811	326.607	290.340	170.452	0.000	0.000	0.000
4800	40.844	327.465	291.106	174.536	0.000	0.000	0.000
4900	40.869	328.310	291.855	178.623	0.000	0.000	0.000
5000	40.894	329.134	292.591	182.711	0.000	0.000	0.000
5100	40.911	329.946	293.319	186.799	0.000	0.000	0.000
5200	40.932	330.741	294.031	190.891	0.000	0.000	0.000
5300	40.949	331.519	294.729	194.987	0.000	0.000	0.000
5400	40.961	332.285	295.420	199.083	0.000	0.000	0.000
5500	40.970	333.038	296.093	203.179	0.000	0.000	0.000
5600	40.978	333.774	296.763	207.275	0.000	0.000	0.000
5700	40.982	334.498	297.415	211.376	0.000	0.000	0.000
5800	40.991	335.214	298.064	215.472	0.000	0.000	0.000
5900	40.991	335.912	298.700	219.572	0.000	0.000	0.000
6000	40.995	336.603	298.947	223.672	0.000	0.000	0.000

TABLE C-3
CARBON MONOXIDE (CO); (IDEAL GAS); M = 28.011

T(K)	J/(gmol·K)			kJ/gmol			
	$C_p°$	$S°$	$-(G°-H_{298}°)/T$	$H°-H_{298}°$	$\Delta H_f°$	$\Delta G_f°$	$Log_{10}K$
0	0.000	0.000	INFINITE	-8.669	-113.805	-113.805	INFINITE
100	29.104	165.741	223.430	-5.770	-112.449	-120.252	62.809
200	29.108	185.916	200.209	-2.858	-111.290	-128.524	33.566
298.15	29.142	197.543	197.543	0.000	-110.529	-137.164	24.029
300	29.142	197.723	197.543	0.054	-110.516	-137.331	23.910
400	29.342	206.125	198.690	2.975	-110.115	-146.335	19.109
500	29.794	212.719	200.857	5.929	-110.022	-155.410	16.235
600	30.443	218.204	203.305	8.941	-110.173	-164.477	14.318
700	31.171	222.953	205.777	12.021	-110.495	-173.502	12.946
800	31.899	227.162	208.192	15.175	-110.935	-182.473	11.914
900	32.577	230.957	210.514	18.397	-111.449	-191.393	11.108
1000	33.183	234.421	212.735	21.686	-112.010	-200.242	10.459
1100	33.710	237.609	214.853	25.033	-112.608	-209.041	9.926
1200	34.175	240.563	216.873	28.426	-113.227	-217.773	9.479
1300	34.572	243.316	218.802	31.865	-113.880	-226.463	9.099
1400	34.920	245.889	220.647	35.338	-114.541	-235.095	8.771
1500	35.217	248.312	222.413	38.848	-115.215	-243.680	8.485
1600	35.480	250.592	224.103	42.384	-115.897	-252.228	8.234
1700	35.710	252.751	225.727	45.940	-116.587	-260.726	8.011
1800	35.911	254.797	227.283	49.522	-117.286	-269.186	7.811
1900	36.091	256.743	228.785	53.124	-117.993	-277.604	7.631
2000	36.250	258.600	230.229	56.739	-118.708	-285.989	7.469
2100	36.392	260.370	231.622	60.375	-119.424	-294.328	7.321
2200	36.518	262.065	232.965	64.019	-120.160	-302.650	7.185
2300	36.635	263.692	234.266	67.676	-120.892	-310.917	7.061
2400	36.740	265.253	235.526	71.346	-121.646	-319.164	6.946
2500	36.836	266.755	236.747	75.023	-122.399	-327.385	6.840
2600	36.924	268.203	237.927	78.714	-123.169	-335.565	6.741
2700	37.003	269.596	239.074	82.408	-123.943	-343.728	6.649
2800	37.083	270.943	240.191	86.115	-124.725	-351.845	6.563
2900	37.150	272.249	241.275	89.826	-125.524	-359.941	6.483
3000	37.217	273.508	242.325	93.542	-126.332	-368.012	6.407
3100	37.279	274.730	243.354	97.270	-127.143	-376.050	6.336
3200	37.338	275.914	244.350	100.998	-127.972	-384.070	6.269
3300	37.392	277.064	245.325	104.734	-128.809	-392.070	6.206
3400	37.443	278.182	246.274	108.479	-129.654	-400.028	6.145
3500	37.493	279.265	247.203	112.223	-130.507	-407.978	6.088
3600	37.543	280.324	248.107	115.976	-131.373	-415.890	6.034
3700	37.589	281.353	248.994	119.734	-132.248	-423.781	5.982
3800	37.631	282.357	249.856	123.495	-133.127	-431.638	5.933
3900	37.673	283.332	250.701	127.261	-134.018	-439.483	5.886
4000	37.715	284.286	251.530	131.026	-134.921	-447.303	5.841
4100	37.756	285.219	252.341	134.800	-135.829	-455.110	5.798
4200	37.794	286.131	253.136	138.578	-136.750	-462.876	5.756
4300	37.832	287.018	253.914	142.361	-137.679	-470.629	5.717
4400	37.869	287.888	254.676	146.147	-138.616	-478.369	5.679
4500	37.903	288.742	255.421	149.934	-139.562	-486.085	5.642
4600	37.941	289.575	256.157	153.724	-140.515	-493.762	5.607
4700	37.974	290.391	256.877	157.523	-141.478	-501.431	5.573
4800	38.007	291.190	257.584	161.322	-142.448	-509.076	5.540
4900	38.041	291.976	258.274	165.122	-143.428	-516.711	5.508
5000	38.074	292.742	258.956	168.929	-144.415	-524.318	5.477
5100	38.104	293.499	259.626	172.736	-145.415	-531.920	5.448
5200	38.137	294.240	260.287	176.548	-146.419	-539.489	5.419
5300	38.171	294.964	260.935	180.364	-147.432	-547.020	5.391
5400	38.200	295.679	261.571	184.184	-148.448	-554.556	5.364
5500	38.233	296.378	262.199	188.004	-149.482	-562.062	5.338
5600	38.263	297.068	262.814	191.832	-150.515	-569.564	5.312
5700	38.296	297.746	263.420	195.656	-151.565	-577.053	5.288
5800	38.325	298.411	264.019	199.489	-152.616	-584.496	5.264
5900	38.355	299.068	264.605	203.321	-153.678	-591.923	5.240
6000	38.388	299.712	265.186	207.162	-154.745	-599.354	5.218

TABLE C-4
CARBON DIOXIDE (CO$_2$); (IDEAL GAS); M = 44.011

	J/(gmol·K)			kJ/gmol			
T(K)	$C_p°$	$S°$	$-(G°-H°_{298})/T$	$H°-H°_{298}$	$\Delta H_f°$	$\Delta G_f°$	$Log_{10}K$
0	0.000	0.000	INFINITE	-9.364	-393.150	-393.150	INFINITE
100	29.209	178.899	243.459	-6.456	-393.283	-393.714	205.645
200	32.359	199.865	216.936	-3.414	-393.413	-394.095	102.922
298.15	37.129	213.685	213.685	0.000	-393.522	-394.405	69.095
300	37.221	213.915	213.685	0.067	-393.526	-394.413	68.670
400	41.325	225.225	215.200	4.008	-393.589	-394.698	51.540
500	44.627	234.814	218.187	8.314	-393.677	-394.965	41.260
600	47.321	243.199	221.673	12.916	-393.815	-395.212	34.405
700	49.564	250.663	225.287	17.761	-394.003	-395.430	29.506
800	51.434	257.408	228.890	22.815	-394.208	-395.622	25.830
900	52.999	263.559	232.404	28.041	-394.426	-395.790	22.970
1000	54.308	269.215	235.806	33.405	-394.639	-395.924	20.680
1100	55.409	274.445	239.086	38.894	-394.848	-396.049	18.806
1200	56.342	279.307	242.237	44.484	-395.049	-396.145	17.243
1300	57.137	283.847	245.266	50.158	-395.258	-396.229	15.920
1400	57.802	288.106	248.174	55.907	-395.451	-396.292	14.785
1500	58.379	292.114	250.973	61.714	-395.647	-396.342	13.801
1600	58.886	295.901	253.663	67.580	-395.836	-396.388	12.940
1700	59.317	299.482	256.253	73.492	-396.016	-396.417	12.180
1800	59.701	302.884	258.751	79.442	-396.208	-396.434	11.504
1900	60.049	306.122	261.157	85.429	-396.401	-396.438	10.898
2000	60.350	309.210	263.483	91.450	-396.593	-396.442	10.353
2100	60.622	312.160	265.734	97.500	-396.785	-396.417	9.860
2200	60.865	314.988	267.906	103.575	-396.999	-396.409	9.411
2300	61.086	317.695	270.014	109.671	-397.212	-396.371	9.001
2400	61.287	320.302	272.056	115.788	-397.442	-396.325	8.625
2500	61.471	322.808	274.035	121.926	-397.681	-396.283	8.280
2600	61.647	325.222	275.960	128.085	-397.928	-396.216	7.960
2700	61.802	327.549	277.826	134.256	-398.191	-396.154	7.664
2800	61.952	329.800	279.642	140.444	-398.463	-396.066	7.388
2900	62.095	331.975	281.412	146.645	-398.756	-395.970	7.132
3000	62.229	334.084	283.131	152.862	-399.057	-395.869	6.892
3100	62.350	336.126	284.809	159.092	-399.367	-395.752	6.668
3200	62.467	338.109	286.441	165.331	-399.698	-395.639	6.458
3300	62.576	340.034	288.039	171.586	-400.036	-395.518	6.260
3400	62.685	341.904	289.596	177.849	-400.392	-395.367	6.074
3500	62.785	343.720	291.114	184.121	-400.760	-395.229	5.898
3600	62.886	345.490	292.600	190.405	-401.137	-395.057	5.732
3700	62.982	347.213	294.052	196.698	-401.530	-394.882	5.574
3800	63.074	348.895	295.474	202.999	-401.932	-394.681	5.425
3900	63.166	350.536	296.867	209.313	-402.342	-394.493	5.283
4000	63.258	352.134	298.227	215.635	-402.764	-394.288	5.149
4100	63.342	353.699	299.562	221.965	-403.200	-394.074	5.020
4200	63.425	355.226	300.867	228.304	-403.643	-393.840	4.898
4300	63.509	356.719	302.152	234.647	-404.103	-393.597	4.781
4400	63.584	358.180	303.407	241.003	-404.568	-393.359	4.670
4500	63.664	359.611	304.641	247.366	-405.040	-393.104	4.563
4600	63.739	361.012	305.850	253.734	-405.526	-392.815	4.460
4700	63.823	362.380	307.039	260.115	-406.015	-392.535	4.362
4800	63.898	363.728	308.206	266.500	-406.517	-392.233	4.268
4900	63.973	365.046	309.352	272.893	-407.024	-391.949	4.178
5000	64.040	366.338	310.478	279.295	-407.538	-391.635	4.091
5100	64.128	367.606	311.587	285.704	-408.066	-391.321	4.008
5200	64.220	368.853	312.675	292.123	-408.593	-390.995	3.927
5300	64.312	370.079	313.750	298.549	-409.124	-390.622	3.850
5400	64.404	371.280	314.800	304.984	-409.664	-390.284	3.775
5500	64.496	372.464	315.842	311.428	-410.208	-389.907	3.703
5600	64.588	373.627	316.863	317.884	-410.756	-389.547	3.633
5700	64.680	374.769	317.867	324.348	-411.308	-389.183	3.566
5800	64.773	375.895	318.858	330.821	-411.865	-388.769	3.501
5900	64.865	377.004	319.833	337.302	-412.425	-388.359	3.438
6000	64.957	378.096	320.796	343.791	-412.990	-387.957	3.377

TABLE C-5
HYDROGEN CHLORIDE (HCl); (IDEAL GAS); M = 36.465

T(K)	$C_p°$	$S°$	$-(G°-H°_{298})/T$	$H°-H°_{298}$	$\Delta H_f°$	$\Delta G_f°$	$Log_{10}K$
		J/(gmol·K)			kJ/gmol		
0	0.000	0.000	INFINITE	-8.640	-92.127	-92.127	INFINITE
100	29.116	154.980	212.689	-5.770	-92.215	-93.236	48.701
200	29.125	175.163	189.456	-2.858	-92.169	-94.286	24.625
298.15	29.137	186.795	186.795	0.000	-92.312	-95.303	16.696
300	29.137	186.975	186.795	0.054	-92.316	-95.320	16.597
400	29.175	195.359	187.937	2.971	-92.588	-96.282	12.573
500	29.305	201.882	190.096	5.891	-92.914	-97.169	10.151
600	29.573	207.246	192.523	8.837	-93.253	-97.985	8.530
700	29.987	211.832	194.962	11.811	-93.579	-98.751	7.369
800	30.493	215.869	197.326	14.836	-93.889	-99.470	6.495
900	31.054	219.493	199.589	17.912	-94.161	-100.148	5.812
1000	31.627	222.794	201.748	21.046	-94.399	-100.805	5.265
1100	32.188	225.836	203.803	24.238	-94.604	-101.433	4.817
1200	32.715	228.660	205.757	27.481	-94.780	-102.048	4.442
1300	33.204	231.296	207.623	30.778	-94.931	-102.646	4.124
1400	33.652	233.777	209.401	34.121	-95.060	-103.232	3.852
1500	34.062	236.111	211.104	37.505	-95.178	-103.813	3.615
1600	34.430	238.321	212.740	40.932	-95.274	-104.387	3.408
1700	34.765	240.417	214.304	44.392	-95.362	-104.947	3.225
1800	35.070	242.413	215.811	47.886	-95.445	-105.512	3.062
1900	35.346	244.316	217.263	51.405	-95.517	-106.069	2.916
2000	35.602	246.136	218.660	54.953	-95.588	-106.621	2.785
2100	35.828	247.881	220.011	58.526	-95.655	-107.173	2.666
2200	36.041	249.550	221.317	62.120	-95.722	-107.721	2.558
2300	36.233	251.157	222.580	65.731	-95.784	-108.261	2.459
2400	36.409	252.705	223.802	69.367	-95.847	-108.801	2.368
2500	36.577	254.195	224.990	73.015	-95.918	-109.340	2.285
2600	36.727	255.630	226.141	76.680	-95.985	-109.876	2.207
2700	36.869	257.019	227.258	80.358	-96.060	-110.412	2.136
2800	37.003	258.362	228.346	84.052	-96.140	-110.939	2.070
2900	37.125	259.663	229.400	87.759	-96.224	-111.470	2.008
3000	37.242	260.927	230.434	91.479	-96.307	-111.993	1.950
3100	37.355	262.149	231.434	95.207	-96.404	-112.512	1.896
3200	37.459	263.337	232.413	98.947	-96.504	-113.035	1.845
3300	37.556	264.492	233.367	102.700	-96.613	-113.545	1.797
3400	37.652	265.613	234.300	106.458	-96.726	-114.056	1.752
3500	37.740	266.705	235.212	110.227	-96.847	-114.570	1.710
3600	37.828	267.768	236.099	114.006	-96.977	-115.068	1.670
3700	37.907	268.805	236.969	117.792	-97.111	-115.570	1.632
3800	37.987	269.818	237.823	121.587	-97.253	-116.073	1.595
3900	38.062	270.805	238.655	125.390	-97.404	-116.562	1.561
4000	38.137	271.772	239.471	129.202	-97.563	-117.052	1.529
4100	38.208	272.713	240.270	133.018	-97.722	-117.541	1.497
4200	38.275	273.634	241.053	136.842	-97.893	-118.022	1.468
4300	38.342	274.537	241.823	140.674	-98.069	-118.499	1.439
4400	38.409	275.420	242.576	144.511	-98.249	-118.964	1.412
4500	38.472	276.282	243.316	148.356	-98.437	-119.441	1.386
4600	38.530	277.127	244.040	152.206	-98.629	-119.905	1.362
4700	38.593	277.960	244.756	156.063	-98.826	-120.361	1.338
4800	38.652	278.772	245.454	159.925	-99.027	-120.821	1.315
4900	38.706	279.571	246.141	163.791	-99.232	-121.269	1.293
5000	38.765	280.353	246.818	167.665	-99.441	-121.721	1.272
5100	38.819	281.119	247.484	171.544	-99.650	-122.156	1.251
5200	38.874	281.876	248.136	175.427	-99.864	-122.595	1.231
5300	38.928	282.617	248.781	179.318	-100.081	-123.035	1.213
5400	38.978	283.345	249.412	183.213	-100.299	-123.466	1.194
5500	39.028	284.060	250.036	187.113	-100.521	-123.888	1.177
5600	39.083	284.763	250.651	191.021	-100.738	-124.315	1.160
5700	39.129	285.453	251.258	194.928	-100.964	-124.742	1.143
5800	39.179	286.135	251.852	198.845	-101.182	-125.148	1.127
5900	39.229	286.805	252.437	202.765	-101.403	-125.566	1.112
6000	39.275	287.466	253.015	206.690	-101.629	-125.972	1.097

TABLE C-6
HYDROGEN, DIATOMIC (H_2); (REFERENCE STATE—IDEAL GAS); M = 2.0158

	J/(gmol·K)			kJ/gmol			
T(K)	$C_p°$	$S°$	$-(G°-H_{298}°)/T$	$H°-H_{298}°$	$\Delta H_f°$	$\Delta G_f°$	$Log_{10}K$
0	0.000	0.000	INFINITE	-8.468	0.000	0.000	0.000
100	28.154	100.617	155.298	-5.468	0.000	0.000	0.000
200	27.447	119.303	133.173	-2.774	0.000	0.000	0.000
298.15	28.836	130.570	130.570	0.000	0.000	0.000	0.000
300	28.849	130.750	130.570	0.054	0.000	0.000	0.000
400	29.179	139.105	131.708	2.958	0.000	0.000	0.000
500	29.259	145.628	133.863	5.883	0.000	0.000	0.000
600	29.326	150.967	136.281	8.812	0.000	0.000	0.000
700	29.439	155.494	138.712	11.749	0.000	0.000	0.000
800	29.623	159.440	141.059	14.703	0.000	0.000	0.000
900	29.882	162.942	143.302	17.677	0.000	0.000	0.000
1000	30.204	166.105	145.427	20.682	0.000	0.000	0.000
1100	30.581	169.000	147.440	23.719	0.000	0.000	0.000
1200	30.991	171.682	149.348	26.799	0.000	0.000	0.000
1300	31.422	174.180	151.164	29.920	0.000	0.000	0.000
1400	31.861	176.523	152.892	33.083	0.000	0.000	0.000
1500	32.296	178.736	154.540	36.292	0.000	0.000	0.000
1600	32.723	180.832	156.122	39.543	0.000	0.000	0.000
1700	33.137	182.832	157.632	42.836	0.000	0.000	0.000
1800	33.539	184.736	159.088	46.170	0.000	0.000	0.000
1900	33.916	186.560	160.486	49.543	0.000	0.000	0.000
2000	34.280	188.309	161.833	52.953	0.000	0.000	0.000
2100	34.623	189.991	163.134	56.396	0.000	0.000	0.000
2200	34.953	191.606	164.389	59.877	0.000	0.000	0.000
2300	35.263	193.167	165.607	63.388	0.000	0.000	0.000
2400	35.560	194.677	166.787	66.927	0.000	0.000	0.000
2500	35.840	196.133	167.933	70.496	0.000	0.000	0.000
2600	36.112	197.543	169.046	74.094	0.000	0.000	0.000
2700	36.367	198.912	170.126	77.718	0.000	0.000	0.000
2800	36.618	200.238	171.180	81.370	0.000	0.000	0.000
2900	36.857	201.527	172.201	85.044	0.000	0.000	0.000
3000	37.087	202.782	173.201	88.738	0.000	0.000	0.000
3100	37.309	203.999	174.176	92.458	0.000	0.000	0.000
3200	37.526	205.188	175.126	96.203	0.000	0.000	0.000
3300	37.740	206.347	176.054	99.964	0.000	0.000	0.000
3400	37.945	207.476	176.962	103.751	0.000	0.000	0.000
3500	38.150	208.581	177.849	107.554	0.000	0.000	0.000
3600	38.346	209.656	178.720	111.378	0.000	0.000	0.000
3700	38.543	210.710	179.569	115.223	0.000	0.000	0.000
3800	38.735	211.740	180.402	119.089	0.000	0.000	0.000
3900	38.928	212.748	181.217	122.972	0.000	0.000	0.000
4000	39.116	213.740	182.021	126.876	0.000	0.000	0.000
4100	39.300	214.706	182.803	130.796	0.000	0.000	0.000
4200	39.484	215.656	183.577	134.733	0.000	0.000	0.000
4300	39.664	216.585	184.330	138.691	0.000	0.000	0.000
4400	39.844	217.501	185.075	142.666	0.000	0.000	0.000
4500	40.016	218.396	185.807	146.662	0.000	0.000	0.000
4600	40.187	219.279	186.523	150.670	0.000	0.000	0.000
4700	40.355	220.145	187.230	154.699	0.000	0.000	0.000
4800	40.518	220.995	187.924	158.741	0.000	0.000	0.000
4900	40.677	221.831	188.611	162.799	0.000	0.000	0.000
5000	40.827	222.656	189.280	166.875	0.000	0.000	0.000
5100	40.978	223.467	189.945	170.967	0.000	0.000	0.000
5200	41.116	224.262	190.598	175.071	0.000	0.000	0.000
5300	41.250	225.049	191.238	179.188	0.000	0.000	0.000
5400	41.380	225.819	191.874	183.322	0.000	0.000	0.000
5500	41.497	226.580	192.497	187.464	0.000	0.000	0.000
5600	41.610	227.329	193.113	191.619	0.000	0.000	0.000
5700	41.714	228.066	193.719	195.786	0.000	0.000	0.000
5800	41.807	228.794	194.318	199.962	0.000	0.000	0.000
5900	41.890	229.509	194.907	204.146	0.000	0.000	0.000
6000	41.966	230.212	195.489	208.342	0.000	0.000	0.000

TABLE C-7
WATER (H₂O); (IDEAL GAS); M = 18.0152

T(K)	J/(gmol·K)			kJ/gmol			
	$C_p°$	$S°$	$-(G°-H°_{298})/T$	$H°-H°_{298}$	$\Delta H_f°$	$\Delta G_f°$	$Log_{10}K$
0	0.000	0.000	INFINITE	−9.904	−238.911	−238.911	INFINITE
100	33.300	152.281	218.426	−6.615	−240.070	−236.576	123.574
200	33.351	175.377	191.786	−3.280	−240.890	−232.764	60.791
298.15	33.589	188.724	188.724	0.000	−241.814	−228.589	40.047
300	33.598	188.933	188.724	0.063	−241.835	−228.505	39.786
373.15	34.045	196.305	189.514	2.536	-- BOILING	POINT AT P=1	ATM --
400	34.263	198.677	190.050	3.452	−242.831	−223.911	29.239
500	35.225	206.426	192.577	6.925	−243.814	−219.066	22.885
600	36.325	212.945	195.439	10.502	−244.747	−214.028	18.633
700	37.497	218.631	198.355	14.192	−245.618	−208.840	15.584
800	38.723	223.714	201.213	18.004	−246.433	−203.531	13.289
900	39.986	228.350	203.974	21.937	−247.174	−198.121	11.499
1000	41.267	232.630	206.627	25.999	−247.848	−192.636	10.062
1100	42.535	236.622	209.175	30.192	−248.450	−187.083	8.884
1200	43.769	240.375	211.623	34.505	−248.986	−181.481	7.900
1300	44.945	243.927	213.970	38.940	−249.463	−175.837	7.065
1400	46.053	247.300	216.233	43.493	−249.881	−170.155	6.349
1500	47.091	250.513	218.409	48.149	−250.253	−164.448	5.727
1600	48.049	253.580	220.514	52.907	−250.580	−158.720	5.182
1700	48.936	256.521	222.547	57.760	−250.868	−152.967	4.700
1800	49.748	259.341	224.513	62.693	−251.128	−147.201	4.272
1900	50.497	262.052	226.417	67.705	−251.358	−141.419	3.888
2000	51.179	264.659	228.262	72.789	−251.563	−135.629	3.542
2100	51.823	267.174	230.057	77.940	−251.751	−129.825	3.229
2200	52.409	269.596	231.802	83.153	−251.923	−124.018	2.944
2300	52.949	271.939	233.496	88.420	−252.082	−118.198	2.684
2400	53.442	274.203	235.145	93.743	−252.228	−112.374	2.446
2500	53.902	276.395	236.752	99.111	−252.366	−106.541	2.226
2600	54.329	278.516	238.316	104.521	−252.500	−100.705	2.023
2700	54.723	280.575	239.844	109.972	−252.630	−94.864	1.835
2800	55.091	282.571	241.333	115.466	−252.760	−89.023	1.661
2900	55.430	284.512	242.789	120.989	−252.885	−83.174	1.498
3000	55.748	286.395	244.212	126.549	−253.011	−77.316	1.346
3100	56.045	288.227	245.601	132.139	−253.136	−71.459	1.204
3200	56.321	290.010	246.961	137.758	−253.270	−65.597	1.071
3300	56.584	291.750	248.291	143.402	−253.400	−59.727	0.945
3400	56.827	293.441	249.597	149.072	−253.542	−53.861	0.827
3500	57.057	295.093	250.873	154.766	−253.680	−47.982	0.716
3600	57.275	296.704	252.124	160.486	−253.831	−42.104	0.611
3700	57.480	298.273	253.350	166.222	−253.986	−36.221	0.511
3800	57.676	299.809	254.550	171.979	−254.149	−30.334	0.417
3900	57.861	301.311	255.730	177.757	−254.312	−24.443	0.327
4000	58.032	302.779	256.889	183.552	−254.488	−18.543	0.242
4100	58.199	304.214	258.027	189.364	−254.672	−12.644	0.161
4200	58.358	305.616	259.144	195.192	−254.864	−6.736	0.084
4300	58.505	306.993	260.241	201.033	−255.065	−0.828	0.010
4400	58.651	308.340	261.316	206.895	−255.274	5.088	−0.060
4500	58.785	309.658	262.379	212.765	−255.496	11.004	−0.128
4600	58.919	310.951	263.420	218.652	−255.726	16.933	−0.192
4700	59.045	312.218	264.446	224.547	−255.969	22.861	−0.254
4800	59.166	313.465	265.454	230.459	−256.216	28.798	−0.313
4900	59.275	314.687	266.445	236.379	−256.475	34.740	−0.370
5000	59.388	315.884	267.420	242.312	−256.751	40.685	−0.425
5100	59.509	317.064	268.383	248.258	−257.031	46.635	−0.478
5200	59.630	318.218	269.332	254.216	−257.324	52.593	−0.528
5300	59.748	319.356	270.265	260.186	−257.626	58.555	−0.577
5400	59.865	320.473	271.182	266.165	−257.935	64.521	−0.624
5500	59.982	321.574	272.090	272.157	−258.253	70.496	−0.670
5600	60.099	322.653	272.981	278.161	−258.580	76.479	−0.713
5700	60.216	323.720	273.864	284.177	−258.914	82.462	−0.756
5800	60.337	324.766	274.734	290.206	−259.257	88.454	−0.797
5900	60.455	325.800	275.588	296.244	−259.605	94.458	−0.836
6000	60.572	326.816	276.433	302.294	−259.964	100.458	−0.875

TABLE C-8
NITROGEN, DIATOMIC (N₂); (REFERENCE STATE—IDEAL GAS); M = 28.0134

	$C_p°$	$S°$	$-(G°-H°_{298})/T$	$H°-H°_{298}$	$\Delta H_f°$	$\Delta G_f°$	$Log_{10}K$
T(K)	J/(gmol·K)			kJ/gmol			
0	0.000	0.000	INFINITE	−8.669	0.000	0.000	0.000
100	29.104	159.703	217.380	−5.770	0.000	0.000	0.000
200	29.108	179.874	194.163	−2.858	0.000	0.000	0.000
298.15	29.125	191.502	191.502	0.000	0.000	0.000	0.000
300	29.125	191.682	191.502	0.054	0.000	0.000	0.000
400	29.250	200.071	192.644	2.971	0.000	0.000	0.000
500	29.581	206.631	194.807	5.912	0.000	0.000	0.000
600	30.108	212.066	197.242	8.895	0.000	0.000	0.000
700	30.752	216.756	199.702	11.937	0.000	0.000	0.000
800	31.434	220.907	202.100	15.046	0.000	0.000	0.000
900	32.091	224.647	204.401	18.221	0.000	0.000	0.000
1000	32.698	228.061	206.598	21.464	0.000	0.000	0.000
1100	33.242	231.204	208.694	24.761	0.000	0.000	0.000
1200	33.723	234.116	210.694	28.108	0.000	0.000	0.000
1300	34.146	236.835	212.602	31.501	0.000	0.000	0.000
1400	34.518	239.379	214.422	34.936	0.000	0.000	0.000
1500	34.844	241.772	216.166	38.405	0.000	0.000	0.000
1600	35.129	244.028	217.840	41.903	0.000	0.000	0.000
1700	35.380	246.166	219.442	45.430	0.000	0.000	0.000
1800	35.597	248.195	220.982	48.978	0.000	0.000	0.000
1900	35.794	250.124	222.467	52.547	0.000	0.000	0.000
2000	35.970	251.965	223.898	56.137	0.000	0.000	0.000
2100	36.125	253.722	225.275	59.743	0.000	0.000	0.000
2200	36.267	255.408	226.605	63.362	0.000	0.000	0.000
2300	36.397	257.023	227.894	66.994	0.000	0.000	0.000
2400	36.510	258.575	229.141	70.638	0.000	0.000	0.000
2500	36.614	260.065	230.350	74.295	0.000	0.000	0.000
2600	36.715	261.504	231.517	77.965	0.000	0.000	0.000
2700	36.802	262.893	232.656	81.638	0.000	0.000	0.000
2800	36.882	264.232	233.760	85.324	0.000	0.000	0.000
2900	36.957	265.529	234.831	89.015	0.000	0.000	0.000
3000	37.028	266.780	235.877	92.713	0.000	0.000	0.000
3100	37.095	267.998	236.894	96.420	0.000	0.000	0.000
3200	37.158	269.178	237.886	100.136	0.000	0.000	0.000
3300	37.217	270.320	238.848	103.851	0.000	0.000	0.000
3400	37.271	271.433	239.793	107.575	0.000	0.000	0.000
3500	37.321	272.512	240.710	111.307	0.000	0.000	0.000
3600	37.371	273.567	241.609	115.043	0.000	0.000	0.000
3700	37.422	274.592	242.488	118.780	0.000	0.000	0.000
3800	37.464	275.588	243.346	122.524	0.000	0.000	0.000
3900	37.510	276.562	244.182	126.273	0.000	0.000	0.000
4000	37.551	277.512	245.007	130.026	0.000	0.000	0.000
4100	37.589	278.441	245.810	133.783	0.000	0.000	0.000
4200	37.627	279.345	246.597	137.545	0.000	0.000	0.000
4300	37.664	280.232	247.371	141.310	0.000	0.000	0.000
4400	37.702	281.098	248.128	145.076	0.000	0.000	0.000
4500	37.740	281.947	248.869	148.850	0.000	0.000	0.000
4600	37.773	282.776	249.597	152.624	0.000	0.000	0.000
4700	37.807	283.587	250.312	156.406	0.000	0.000	0.000
4800	37.844	284.386	251.011	160.189	0.000	0.000	0.000
4900	37.878	285.165	251.701	163.975	0.000	0.000	0.000
5000	37.911	285.930	252.379	167.762	0.000	0.000	0.000
5100	37.949	286.683	253.044	171.557	0.000	0.000	0.000
5200	37.982	287.420	253.697	175.351	0.000	0.000	0.000
5300	38.012	288.144	254.341	179.151	0.000	0.000	0.000
5400	38.045	288.855	254.973	182.954	0.000	0.000	0.000
5500	38.079	289.554	255.596	186.761	0.000	0.000	0.000
5600	38.116	290.240	256.207	190.569	0.000	0.000	0.000
5700	38.154	290.914	256.810	194.384	0.000	0.000	0.000
5800	38.192	291.579	257.404	198.200	0.000	0.000	0.000
5900	38.233	292.231	257.990	202.024	0.000	0.000	0.000

TABLE C-9
NITRIC OXIDE (NO); (IDEAL GAS); M = 30.008

T(K)	$C_p°$	$S°$	$-(G°-H°_{298})/T$	$H°-H°_{298}$	$\Delta H_f°$	$\Delta G_f°$	$\log_{10}K$
		J/(gmol·K)			kJ/gmol		
0	0.000	0.000	INFINITE	-9.192	89.772	89.772	INFINITE
100	32.305	176.925	237.655	-6.071	89.969	88.935	-46.453
200	30.422	198.644	213.397	-2.950	90.199	87.797	-22.929
298.15	29.844	210.652	210.652	0.000	90.291	86.596	-15.171
300	29.840	210.840	210.656	0.054	90.291	86.575	-15.073
400	29.945	219.426	211.823	3.042	90.333	85.328	-11.142
500	30.489	226.158	214.041	6.058	90.349	84.077	-8.783
600	31.238	231.781	216.543	9.146	90.366	82.822	-7.210
700	32.029	236.655	219.074	12.309	90.379	81.563	-6.086
800	32.769	240.982	221.547	15.548	90.395	80.299	-5.243
900	33.422	244.881	223.928	18.857	90.416	79.036	-4.587
1000	33.987	248.433	226.204	22.230	90.437	77.772	-4.062
1100	34.468	251.697	228.375	25.652	90.458	76.504	-3.633
1200	34.878	254.714	230.446	29.121	90.475	75.233	-3.275
1300	35.225	257.517	232.421	32.627	90.492	73.965	-2.972
1400	35.526	260.140	234.308	36.166	90.504	72.689	-2.712
1500	35.782	262.600	236.111	39.731	90.512	71.417	-2.487
1600	36.003	264.918	237.839	43.321	90.521	70.145	-2.290
1700	36.196	267.107	239.501	46.932	90.521	68.873	-2.116
1800	36.367	269.178	241.090	50.559	90.512	67.597	-1.962
1900	36.514	271.148	242.622	54.204	90.500	66.329	-1.823
2000	36.648	273.027	244.095	57.861	90.483	65.053	-1.699
2100	36.769	274.818	245.517	61.530	90.454	63.781	-1.586
2200	36.874	276.529	246.885	65.216	90.420	62.513	-1.484
2300	36.974	278.169	248.212	68.906	90.379	61.241	-1.391
2400	37.062	279.746	249.492	72.609	90.328	59.982	-1.305
2500	37.141	281.261	250.735	76.320	90.266	58.714	-1.227
2600	37.217	282.717	251.935	80.036	90.199	57.455	-1.154
2700	37.288	284.127	253.103	83.764	90.123	56.199	-1.087
2800	37.351	285.483	254.232	87.492	90.040	54.944	-1.025
2900	37.409	286.792	255.333	91.232	89.948	53.697	-0.967
3000	37.468	288.064	256.404	94.977	89.847	52.446	-0.913
3100	37.522	289.294	257.446	98.726	89.743	51.200	-0.863
3200	37.572	290.483	258.458	102.479	89.625	49.957	-0.815
3300	37.618	291.642	259.450	106.240	89.504	48.718	-0.771
3400	37.664	292.767	260.412	110.002	89.374	47.484	-0.729
3500	37.706	293.859	261.349	113.771	89.241	46.250	-0.690
3600	37.748	294.922	262.270	117.545	89.094	45.028	-0.653
3700	37.790	295.955	263.165	121.323	88.948	43.806	-0.618
3800	37.828	296.964	264.044	125.102	88.793	42.589	-0.585
3900	37.865	297.947	264.897	128.888	88.638	41.376	-0.554
4000	37.899	298.905	265.738	132.675	88.471	40.158	-0.524
4100	37.932	299.842	266.558	136.465	88.299	38.957	-0.496
4200	37.966	300.754	267.362	140.260	88.128	37.756	-0.470
4300	37.999	301.650	268.148	144.059	87.952	36.564	-0.444
4400	38.033	302.524	268.918	147.863	87.768	35.363	-0.420
4500	38.062	303.378	269.676	151.666	87.579	34.179	-0.397
4600	38.095	304.214	270.416	155.473	87.391	33.003	-0.375
4700	38.125	305.035	271.144	159.285	87.199	31.819	-0.354
4800	38.154	305.838	271.860	163.101	87.002	30.644	-0.333
4900	38.179	306.624	272.562	166.916	86.805	29.455	-0.314
5000	38.208	307.398	273.249	170.736	86.605	28.296	-0.296
5100	38.238	308.156	273.926	174.556	86.400	27.129	-0.278
5200	38.263	308.896	274.592	178.381	86.195	25.970	-0.261
5300	38.292	309.624	275.249	182.209	85.985	24.819	-0.245
5400	38.317	310.340	275.889	186.042	85.776	23.656	-0.229
5500	38.342	311.047	276.525	189.874	85.567	22.522	-0.214
5600	38.367	311.737	277.144	193.711	85.354	21.368	-0.199
5700	38.392	312.415	277.759	197.548	85.136	20.230	-0.185
5800	38.417	313.085	278.362	201.388	84.923	19.104	-0.172
5900	38.443	313.741	278.956	205.229	84.705	17.958	-0.159
6000	38.468	314.386	279.541	209.074	84.483	16.836	-0.147

TABLE C-10
NITROGEN DIOXIDE (NO$_2$); (IDEAL GAS); M = 46.008

T(K)	J/(gmol·K)			kJ/gmol			
	$C_p°$	$S°$	$-(G°-H°_{298})/T$	$H°-H°_{298}$	$\Delta H_f°$	$\Delta G_f°$	$Log_{10}K$
0	0.000	0.000	INFINITE	-10.188	35.924	35.924	INFINITE
100	33.275	202.451	271.060	-6.862	34.836	39.936	-20.859
200	34.384	225.744	243.216	-3.494	33.886	45.409	-11.859
298.15	36.974	239.923	239.923	0.000	33.095	51.241	-8.977
300	37.028	240.153	239.927	0.067	33.083	51.354	-8.941
400	40.171	251.232	241.417	3.929	32.510	57.534	-7.513
500	43.208	260.529	244.333	8.100	32.150	63.839	-6.669
600	45.836	268.646	247.722	12.556	31.957	70.199	-6.111
700	47.986	275.880	251.237	17.251	31.874	76.576	-5.714
800	49.710	282.403	254.730	22.138	31.870	82.960	-5.417
900	51.078	288.340	258.140	27.179	31.920	89.349	-5.185
1000	52.166	293.780	261.437	32.342	32.003	95.726	-5.000
1100	53.041	298.796	264.609	37.606	32.108	102.090	-4.848
1200	53.752	303.440	267.655	42.949	32.225	108.453	-4.721
1300	54.329	307.767	270.575	48.350	32.347	114.801	-4.612
1400	54.806	311.813	273.378	53.810	32.472	121.131	-4.519
1500	55.200	315.607	276.069	59.312	32.593	127.461	-4.438
1600	55.534	319.181	278.650	64.848	32.711	133.783	-4.367
1700	55.815	322.557	281.136	70.417	32.828	140.097	-4.304
1800	56.057	325.754	283.525	76.011	32.924	146.407	-4.248
1900	56.262	328.791	285.830	81.626	33.012	152.708	-4.198
2000	56.442	331.682	288.052	87.262	33.087	159.000	-4.152
2100	56.597	334.440	290.194	92.914	33.150	165.297	-4.111
2200	56.735	337.076	292.265	98.579	33.187	171.586	-4.074
2300	56.852	339.599	294.269	104.261	33.217	177.879	-4.040
2400	56.961	342.021	296.210	109.951	33.229	184.175	-4.008
2500	57.053	344.347	298.089	115.650	33.217	190.456	-3.979
2600	57.137	346.586	299.909	121.361	33.192	196.753	-3.953
2700	57.212	348.745	301.679	127.081	33.146	203.045	-3.928
2800	57.279	350.828	303.399	132.804	33.083	209.330	-3.905
2900	57.342	352.837	305.068	138.536	33.003	215.643	-3.884
3000	57.396	354.782	306.691	144.273	32.899	221.940	-3.864
3100	57.446	356.665	308.273	150.013	32.786	228.241	-3.846
3200	57.492	358.489	309.817	155.762	32.648	234.547	-3.828
3300	57.534	360.259	311.319	161.511	32.497	240.852	-3.812
3400	57.572	361.979	312.783	167.268	32.326	247.170	-3.797
3500	57.605	363.648	314.214	173.025	32.146	253.479	-3.783
3600	57.639	365.272	315.607	178.787	31.945	259.814	-3.770
3700	57.668	366.849	316.971	184.552	31.731	266.157	-3.757
3800	57.697	368.389	318.306	190.322	31.501	272.496	-3.746
3900	57.722	369.887	319.607	196.092	31.267	278.834	-3.734
4000	57.743	371.351	320.884	201.865	31.012	285.173	-3.724
4100	57.764	372.778	322.130	207.639	30.740	291.533	-3.714
4200	57.785	374.167	323.356	213.417	30.464	297.905	-3.705
4300	57.806	375.527	324.553	219.200	30.179	304.290	-3.696
4400	57.823	376.857	325.724	224.978	29.874	310.649	-3.688
4500	57.840	378.158	326.875	230.764	29.564	317.030	-3.680
4600	57.856	379.430	328.005	236.547	29.246	323.440	-3.673
4700	57.869	380.673	329.113	242.333	28.916	329.837	-3.666
4800	57.881	381.890	330.201	248.120	28.577	336.251	-3.659
4900	57.894	383.083	331.268	253.910	28.234	342.644	-3.652
5000	57.907	384.254	332.314	259.701	27.878	349.063	-3.646
5100	57.919	385.401	333.343	265.492	27.518	355.498	-3.641
5200	57.932	386.526	334.356	271.282	27.150	361.920	-3.635
5300	57.940	387.631	335.352	277.077	26.778	368.376	-3.630
5400	57.948	388.715	336.331	282.872	26.397	374.799	-3.625
5500	57.957	389.777	337.293	288.667	26.012	381.280	-3.621
5600	57.965	390.819	338.239	294.466	25.614	387.744	-3.617
5700	57.974	391.848	339.168	300.261	25.217	394.196	-3.612
5800	57.982	392.857	340.088	306.060	24.815	400.689	-3.608
5900	57.990	393.848	340.988	311.859	24.405	407.158	-3.605
6000	57.999	394.823	341.879	317.658	23.991	413.655	-3.601

TABLE C-11
OXYGEN, DIATOMIC (O_2); (REFERENCE STATE—IDEAL GAS); M = 31.9988

	J/(gmol·K)			kJ/gmol			
T(K)	C_p°	S°	$-(G^\circ-H_{298}^\circ)/T$	$H^\circ-H_{298}^\circ$	ΔH_f°	ΔG_f°	$Log_{10}K$
0	0.000	0.000	INFINITE	−8.682	0.000	0.000	0.000
100	29.104	173.197	230.982	−5.778	0.000	0.000	0.000
200	29.125	193.376	207.715	−2.866	0.000	0.000	0.000
298.15	29.376	205.037	205.037	0.000	0.000	0.000	0.000
300	29.384	205.221	205.037	0.054	0.000	0.000	0.000
400	30.108	213.761	206.200	3.025	0.000	0.000	0.000
500	31.091	220.585	208.413	6.084	0.000	0.000	0.000
600	32.091	226.342	210.936	9.242	0.000	0.000	0.000
700	32.982	231.358	213.501	12.498	0.000	0.000	0.000
800	33.731	235.810	216.016	15.836	0.000	0.000	0.000
900	34.355	239.823	218.442	19.242	0.000	0.000	0.000
1000	34.869	243.467	220.765	22.702	0.000	0.000	0.000
1100	35.300	246.814	222.982	26.213	0.000	0.000	0.000
1200	35.669	249.902	225.099	29.761	0.000	0.000	0.000
1300	35.987	252.768	227.120	33.342	0.000	0.000	0.000
1400	36.275	255.446	229.049	36.957	0.000	0.000	0.000
1500	36.543	257.956	230.894	40.597	0.000	0.000	0.000
1600	36.798	260.324	232.660	44.267	0.000	0.000	0.000
1700	37.041	262.563	234.354	47.957	0.000	0.000	0.000
1800	37.275	264.688	235.978	51.672	0.000	0.000	0.000
1900	37.510	266.709	237.542	55.413	0.000	0.000	0.000
2000	37.740	268.638	239.049	59.174	0.000	0.000	0.000
2100	37.970	270.487	240.505	62.961	0.000	0.000	0.000
2200	38.196	272.257	241.906	66.768	0.000	0.000	0.000
2300	38.417	273.960	243.262	70.601	0.000	0.000	0.000
2400	38.639	275.600	244.576	74.454	0.000	0.000	0.000
2500	38.857	277.182	245.852	78.329	0.000	0.000	0.000
2600	39.066	278.709	247.086	82.224	0.000	0.000	0.000
2700	39.275	280.186	248.283	86.140	0.000	0.000	0.000
2800	39.476	281.621	249.450	90.077	0.000	0.000	0.000
2900	39.673	283.010	250.584	94.035	0.000	0.000	0.000
3000	39.865	284.357	251.684	98.014	0.000	0.000	0.000
3100	40.049	285.667	252.760	102.010	0.000	0.000	0.000
3200	40.225	286.943	253.810	106.023	0.000	0.000	0.000
3300	40.397	288.181	254.831	110.052	0.000	0.000	0.000
3400	40.560	289.391	255.831	114.102	0.000	0.000	0.000
3500	40.715	290.566	256.806	118.165	0.000	0.000	0.000
3600	40.869	291.717	257.760	122.244	0.000	0.000	0.000
3700	41.012	292.838	258.693	126.340	0.000	0.000	0.000
3800	41.154	293.934	259.605	130.449	0.000	0.000	0.000
3900	41.288	295.005	260.500	134.570	0.000	0.000	0.000
4000	41.422	296.051	261.374	138.704	0.000	0.000	0.000
4100	41.547	297.077	262.232	142.854	0.000	0.000	0.000
4200	41.673	298.081	263.077	147.013	0.000	0.000	0.000
4300	41.798	299.060	263.902	151.189	0.000	0.000	0.000
4400	41.919	300.022	264.709	155.373	0.000	0.000	0.000
4500	42.041	300.968	265.508	159.569	0.000	0.000	0.000
4600	42.162	301.892	266.286	163.783	0.000	0.000	0.000
4700	42.288	302.800	267.056	168.004	0.000	0.000	0.000
4800	42.413	303.691	267.809	172.239	0.000	0.000	0.000
4900	42.543	304.566	268.550	176.485	0.000	0.000	0.000
5000	42.677	305.428	269.278	180.749	0.000	0.000	0.000
5100	42.811	306.273	269.998	185.021	0.000	0.000	0.000
5200	42.957	307.110	270.701	189.309	0.000	0.000	0.000
5300	43.104	307.926	271.395	193.615	0.000	0.000	0.000
5400	43.263	308.733	272.081	197.932	0.000	0.000	0.000
5500	43.426	309.528	272.755	202.267	0.000	0.000	0.000
5600	43.597	310.315	273.416	206.618	0.000	0.000	0.000
5700	43.781	311.089	274.073	210.987	0.000	0.000	0.000
5800	43.974	311.850	274.717	215.376	0.000	0.000	0.000
5900	44.175	312.603	275.353	219.781	0.000	0.000	0.000
6000	44.388	313.348	275.981	224.208	0.000	0.000	0.000

TABLE C-12
SULFUR DIOXIDE (SO$_2$); (IDEAL GAS); M = 64.066

T(K)	C$_p$°	S°	−(G°−H$^{\circ}_{298}$)/T	H°−H$^{\circ}_{298}$	ΔH$_f$°	ΔG$_f$°	Log$_{10}$K
		J/(gmol·K)			kJ/gmol		
0	0.000	0.000	INFINITE	−10.552	−294.307	−294.307	INFINITE
100	33.526	208.915	281.089	−7.217	−294.641	−296.922	155.088
200	36.372	232.923	251.605	−3.736	−295.650	−298.842	78.046
298.15	39.874	248.103	248.103	0.000	−296.842	−300.160	52.585
300	39.945	248.350	248.103	0.075	−296.863	−300.185	52.264
400	43.493	260.337	249.714	4.251	−300.261	−301.026	39.308
500	46.576	270.383	252.868	8.757	−302.738	−300.926	31.436
600	49.049	279.102	256.529	13.544	−304.696	−300.369	26.148
700	50.961	286.813	260.316	18.548	−306.294	−299.415	22.342
800	52.434	293.717	264.069	23.719	−362.305	−303.650	19.825
900	53.580	299.964	267.713	29.024	−362.238	−296.319	17.197
1000	54.484	305.658	271.228	34.430	−362.138	−288.993	15.095
1100	55.204	310.884	274.600	39.915	−362.016	−281.692	13.376
1200	55.794	315.716	277.826	45.463	−361.874	−274.395	11.943
1300	56.279	320.202	280.918	51.070	−361.715	−267.107	10.732
1400	56.689	324.386	283.872	56.718	−361.552	−259.835	9.694
1500	57.036	328.310	286.704	62.404	−361.385	−252.584	8.795
1600	57.338	332.000	289.424	68.124	−361.209	−245.329	8.009
1700	57.601	335.486	292.031	73.873	−361.029	−238.091	7.315
1800	57.831	338.783	294.537	79.642	−360.862	−230.865	6.699
1900	58.040	341.916	296.951	85.437	−360.694	−223.643	6.148
2000	58.229	344.900	299.273	91.249	−360.535	−216.443	5.653
2100	58.400	347.745	301.516	97.081	−360.380	−209.242	5.204
2200	58.555	350.464	303.679	102.931	−360.238	−202.045	4.797
2300	58.702	353.071	305.767	108.792	−360.104	−194.861	4.425
2400	58.840	355.573	307.792	114.671	−359.979	−187.673	4.084
2500	58.965	357.975	309.750	120.562	−359.866	−180.502	3.771
2600	59.086	360.293	311.649	126.461	−359.765	−173.326	3.482
2700	59.199	362.523	313.495	132.378	−359.673	−166.159	3.214
2800	59.308	364.677	315.285	138.302	−359.598	−159.000	2.966
2900	59.413	366.761	317.022	144.239	−359.531	−151.829	2.735
3000	59.513	368.778	318.716	150.185	−359.481	−144.662	2.519
3100	59.609	370.732	320.361	156.143	−359.431	−137.503	2.317
3200	59.706	372.623	321.967	162.109	−359.401	−130.348	2.128
3300	59.794	374.464	323.528	168.084	−359.376	−123.202	1.950
3400	59.881	376.250	325.051	174.067	−359.368	−116.035	1.783
3500	59.969	377.987	326.540	180.058	−359.368	−108.893	1.625
3600	60.053	379.677	327.992	186.058	−359.372	−101.726	1.476
3700	60.137	381.321	329.411	192.071	−359.389	−94.567	1.335
3800	60.216	382.928	330.800	198.087	−359.410	−87.404	1.201
3900	60.296	384.493	332.155	204.112	−359.439	−80.262	1.075
4000	60.375	386.020	333.486	210.146	−359.477	−73.090	0.954
4100	60.450	387.514	334.783	216.187	−359.519	−65.931	0.840
4200	60.530	388.970	336.055	222.237	−359.565	−58.773	0.731
4300	60.605	390.396	337.302	228.292	−359.619	−51.601	0.627
4400	60.676	391.790	338.527	234.358	−359.678	−44.447	0.528
4500	60.752	393.154	339.724	240.429	−359.740	−37.271	0.433
4600	60.823	394.488	340.900	246.509	−359.803	−30.100	0.342
4700	60.894	395.798	342.055	252.592	−359.874	−22.937	0.255
4800	60.969	397.083	343.188	258.688	−359.941	−15.769	0.172
4900	61.036	398.338	344.301	264.789	−360.012	−8.602	0.092
5000	61.107	399.572	345.393	270.893	−360.092	−1.443	0.015
5100	61.178	400.785	346.469	277.010	−360.167	5.753	−0.059
5200	61.250	401.974	347.523	283.131	−360.242	12.916	−0.130
5300	61.317	403.141	348.561	289.257	−360.318	20.108	−0.198
5400	61.388	404.287	349.586	295.395	−360.393	27.271	−0.264
5500	61.455	405.413	350.590	301.537	−360.464	34.464	−0.327
5600	61.522	406.522	351.577	307.683	−360.548	41.639	−0.388
5700	61.588	407.609	352.552	313.842	−360.619	48.806	−0.447
5800	61.655	408.681	353.510	320.001	−360.694	56.011	−0.504
5900	61.727	409.739	354.456	326.172	−360.761	63.174	−0.559
6000	61.793	410.777	355.385	332.348	−360.832	70.387	−0.613

SYMBOLS AND TERMINOLOGY

C_p^o is the ideal state heat capacity.

S^o represents the absolute ideal state entropy computed from

$$S^o = S_0^o + \int_0^T \frac{C_p^o}{T} \, dT$$

$H^o - H_{298}^o$ denotes the sensible ideal state enthalpy relative to 298 K calculated from the conventional relationship

$$H^o - H_{298}^o = \int_{298}^T C_p^o \, dT$$

ΔH_f^o is the standard heat of formation of the particular species based on the reference state elements at the temperature given. Thus, for a reference element such as gaseous O_2, N_2, H_2, or solid graphitic carbon, ΔH_f^o is zero.

$-(G^o - H_{298}^o)/T$, known as the free-energy function, is easily computed from other data in the table [i.e., $S^o - (H^o - H_{298}^o)/T$].

ΔG_f^o denotes the standard Gibbs free energy of formation based on the reference state elements. Thus, as with ΔH_f^o, ΔG_f^o for a reference element is zero.

$\log_{10} K$ stands for the logarithm (to the base 10) of the equilibrium constant for the reaction that forms the compound in question from the reference elements. Values tabulated for K are derived directly from ΔG_f^o through the defining relationship

$$K = e^{-\Delta G_f^o/RT}$$

Thus, as derived from the equilibrium theory, K is the ratio of activities for the formation reaction based on the same standard states as those employed to calculate ΔG_f^o.

The *standard state* for a gas is the ideal gas at one atmosphere pressure. For a liquid or solid, it is the pure compound. Thus, for a gas, the activity is equal to the partial fugacity expressed in atmospheres. For a liquid or solid, the activity is equal to the partial fugacity divided by the fugacity of the pure liquid or solid at the temperature in question.

Index

ECONOMIC NOMENCLATURE

a	size or capacity exponent (see Equation 5-1; dimensionless); also, as a superscript on price data and factors, denotes alloy or special material of construction
A_A	allowances credited to an operation through special tax laws ($/yr)
A_{BD}	depreciation ($/yr)
A_C	constant annual expenses; those that are independent of the optimization variable ($/yr)
A_{CF}	net annual cash flow ($/yr)
A_{CI}	annual cash income ($/yr)
A_{cw}	annual costs for cooling water ($/yr)
A_{DCF}	net annual discounted cash flow ($/yr)
A_{DME}	annual direct manufacturing expense ($/yr)
A_e	annual costs for electricity ($/yr)
A_{FC}	annual expenses due to fixed capital ($/yr)
A_{GE}	annual total general expense ($/yr)
A_I	annual cash investment ($/yr)
A_{IME}	annual indirect manufacturing expense ($/yr)
A_{IT}	annual amount of income tax ($/yr)
A_L	annual costs for labor ($/yr)
A_{ME}	total annual manufacturing expense ($/yr)
A_{NCI}	annual net (aftertax) cash income ($/yr)
A_{NNP}	net annual profit after tax ($/yr)
A_{NP}	net annual profit ($/yr)
A_{OE}	annual operating expense ($/yr)
A_R	annual income recovered through investment of additional capital ($/yr)
A_{RM}	annual raw materials costs ($/yr)
A_s	annual costs for steam ($/yr)
A_S	annual revenue from sales ($/yr)
A_{Su}	annual costs for plant supervision ($/yr)
A_{TE}	total annual expense ($/yr)
A_u	annual cost of a utility ($/yr)
A_V	annual variable expense ($/yr)
(BEP)	break-even period (yr)
C	capital ($)
C_{AB}	capital cost of auxiliary buildings ($)
C_{BM}	bare module capital cost ($; see Table 5-2)
C_C	contingency expense ($)
C_{CFC}	constant fixed capital in Equation 7-2 ($)

C_E	contractor engineering expenses (\$)
C_F	contractor's fee (\$)
C_{FC}	fixed capital cost (\$)
C_{FIT}	freight, insurance, and taxes to procure and install equipment (\$)
C_{GR}	total grass-roots capital (\$)
C_L	labor cost associated with installation of process equipment (\$)
C_l	cost of land and other nondepreciable items (\$)
C_M	cost of materials required to install an item of process equipment (\$)
C_O	construction overhead expense (\$)
C_{OS}	capital cost of offsite facilities (\$)
$C_{P.v.r}$	purchase cost of process equipment having size or capacity v and in the year r (\$) (A superscript, when employed, indicates material of construction. Abbreviations are included in the headings of Table 4-28; cs = carbon steel, as = high alloy steel, ss = stainless steel, etc.)
C_S	selling price per unit of product (\$/unit)
C_{SD}	site development expense (\$)
C_t^a	capital cost of piping materials (\$)
C_{TBM}^a	total bare module cost $= \Sigma\, C_{BM,i}^a$ (\$) (The superscript refers to construction material. When it is absent, costs are for base material, usually carbon steel, construction; see Table 5-5.)
C_{TC}	total capital (\$)
C_{TM}	total capital of a process module (\$)
C_{VFC}	variable fixed capital in Equation 7-2 (\$)
C_{WC}	working capital (\$)
ΔC_{BM}^a	increase in bare module capital above that of carbon steel due to nonbase materials of construction (\$)
(DBEP)	discounted breakeven period (yr)
(DCFRR)	discounted cash flow rate of return (percent)
f_a	annuity factor employed to convert capital cost to annual expense (yr^{-1}; see equation 7-3)
f_d	discount factor based on annual compound interest (dimensionless; see Equation 8-7)
f_d'	discount factor based on continuously compounded interest (dimensionless; see Equation 8-8)
f_o	operating factor (dimensionless); the fraction of elapsed time that a plant is in equivalent full-scale production
f_t	ratio of costs for piping materials relative to those of purchased equipment (dimensionless)
F_{BM}	installation factor; the purchase cost of equipment multiplied by this factor yields bare module cost